X-Ray Binaries and Recycled Pulsars

NATO ASI Series

Advanced Science Institutes Series

A Series presenting the results of activities sponsored by the NATO Science Committee, which aims at the dissemination of advanced scientific and technological knowledge, with a view to strengthening links between scientific communities.

The Series is published by an international board of publishers in conjunction with the NATO Scientific Affairs Division

A Life Sciences	Plenum Publishing Corporation
B Physics	London and New York
C Mathematical	Kluwer Academic Publishers
and Physical Sciences	Dordrecht, Boston and London
D Behavioural and Social Sciences	
E Applied Sciences	
F Computer and Systems Sciences	Springer-Verlag
G Ecological Sciences	Berlin, Heidelberg, New York, London,
H Cell Biology	Paris and Tokyo
I Global Environmental Change	

NATO-PCO-DATA BASE

The electronic index to the NATO ASI Series provides full bibliographical references (with keywords and/or abstracts) to more than 30000 contributions from international scientists published in all sections of the NATO ASI Series.
Access to the NATO-PCO-DATA BASE is possible in two ways:

– via online FILE 128 (NATO-PCO-DATA BASE) hosted by ESRIN, Via Galileo Galilei, I-00044 Frascati, Italy.

– via CD-ROM "NATO-PCO-DATA BASE" with user-friendly retrieval software in English, French and German (© WTV GmbH and DATAWARE Technologies Inc. 1989).

The CD-ROM can be ordered through any member of the Board of Publishers or through NATO-PCO, Overijse, Belgium.

Series C: Mathematical and Physical Sciences - Vol. 377

X-Ray Binaries
and Recycled Pulsars

edited by

E. P. J. van den Heuvel

Astronomical Institute 'Anton Pannekoek',
University of Amsterdam, Amsterdam, The Netherlands and
Center for High Energy Astrophysics (CHEAF),
Amsterdam, The Netherlands and
Institute for Theoretical Physics,
University of California,
Santa Barbara, U.S.A.

and

S. A. Rappaport

Center for Space Research, MIT,
Cambridge, MA, U.S.A. and
Institute for Theoretical Physics,
University of California,
Santa Barbara, U.S.A.

Springer-Science+Business Media, B.V.

Proceedings of the NATO Advanced Research Workshop on
X-Ray Binaries and the Formation of Binary and
Millisecond Radio Pulsars
Santa Barbara, CA, U.S.A.
January 21–25, 1991

Library of Congress Cataloging-in-Publication Data

```
NATO Advanced Research Workshop on "X-ray Binaries and the Formation
  of Binary and Millisecond Radio Pulsars" (1991 : Santa Barbara,
  Calif.)
    X-ray binaries and recycled pulsars : proceedings of the NATO
  Advanced Research Workshop on "X-ray Binaries and the Formation of
  Binary and Millisecond Radio Pulsars," January 21-25, 1991 / edited
  by E.P.J. van den Heuvel and S.A. Rappaport.
       p.   cm.
  Includes bibliographical references and index.
  ISBN 978-0-7923-1940-5     ISBN 978-94-011-2704-2 (eBook)
  DOI 10.1007978-94-011-2704-2
    1. Double stars--Congresses.  2. Pulsars--Congresses.  3. Neutron
  stars--Congresses.  4. X-ray astronomy--Congresses.   I. Heuvel,
  Edward Peter Jacobus van den, 1940-   . II. Rappaport, S. A., 1942-
  . III. Title.
  QB821.N363   1991
  523.8'874--dc20                                    92-26378
                                                       CIP
```

ISBN 978-0-7923-1940-5

Contents

List of participants and observers

Ali Alpar, M.,   Physics Department, Middle East Technical University, Ankara, Turkey

Anderson, S.,   California Institute of Technology, Pasadena CA, USA

Araya, R. A.,   The Johns Hopkins University, Deptartment of Physics and Astronomy, Baltimore MD, USA

Arons, J.,   Department of Astronomy, UC Berkeley, Berkeley CA, USA

Backer, D.,   Department of Astronomy, UC Berkeley, Berkeley CA, USA

Bailyn, C. D.,   Center for Astrophysics, Cambridge MA, USA

Bandiera, R.,   Osservatorio Astrofisico di Arcetri, Firenze, Italy

Banit, M.,   Astrophysics Department, Columbia University, New York NY, USA

Bhattacharya, D.,   Astronomical Institute, University of Amsterdam, The Netherlands

Blandford, R.,   Theoretical Astrophysics, CalTech, Pasadena CA, USA

Blondin, J. M.,   University of Virginia, Department of Astronomy, Charlottesville VA, USA

Cernohorsky, J.,   Center for High Energy Astrophysics, Amsterdam, The Netherlands

Chanmugam, G.,   Louisiana State University, Department of Physics and Astronomy, Baton Rouge LA, USA

Cordes, J.,   Cornell University Centre for Radiophysics & Space Research, Ithaca NY, USA

Dewey, R.,   Jet Propulsion Laboratory, CalTech, Pasadena CA, USA

Eardley, D.,   University of California, Institute for Theoretical Physics, Santa Barbara CA, USA

Edberg, T.,   University of California, Center for Particle Astrophysics, Berkeley CA, USA

Eggleton, P. P.,   Institute of Astronomy, Cambridge, United Kingdom

Eichler, D.,   Ben Gurion University, Department of Physics, Beer Sheva, Israel

Einhorn, M.,   University of California, Institute for Theoretical Physics, Santa Barbara CA, USA

Evans, C. R.,   University of North Carolina, Department of Physics and Astronomy, Chapel Hill NC, USA

Foster, R.,   UC Berkeley, Astronomy Department, Berkeley CA, USA

Frail, D. A.,   National Radio Astronomy Observatory, Soccoro, New Mexico, USA

Fruchter, A.,   Carnegie Institute of Washington, Department of Terrestrial Magnetism, Washington DC, USA

Ghosh, P.,   UIUC Department of Physics, Loomis Laboratory of Physics, Urbana IL, USA

Gorham, P.,   California Institute of Technology, Pasadena CA, USA

Grindlay, J. E.,   Harvard Observatory, Center for Astrophysics, Cambridge MA, USA

Grunsfeld, J. M.,   California Institute of Technology, Pasadena CA, USA

Harding, A. K.   NASA/Goddard Space Flight Center, Greenbelt MD, USA

Harrison, P.,   National Radio Astronomy Laboratories, Jodrell Bank, Macclesfield, United Kingdom

Hasinger, G.,   Max Planck Institute für Extraterrestrische Physik, Garching bei München, BRD

Heuvel, E. P. J. van den,   Astronomical Institute, University of Amsterdam, The Netherlands

Hut, P.,   Institute for Advanced Study, Princeton NJ, USA

Iping, R. C.,   Department of Natural Science, University of Guam, Mangilao, Guam, USA

Johnston, H.,   California Institute of Technology, Pasadena CA, USA

Klis, M. van der, Astronomical Institute, University of Amsterdam, The Netherlands

Kluzniak, W.,   Columbia University, Department of Physics and Astronomy, New York NY, USA

Kulkarni, S. R., Department of Astronomy, CalTech, Pasadena CA, USA

Kuijpers, J.,   Astronomical Institute, University of Utrecht, The Netherlands

Lamb, F. K.,   University of Illinois, Physics Department, Urbana IL, USA

Langer, J.,   University of California, Institute for Theoretical Physics, Santa Barbara CA, USA

Liang, E.   Lawrence Livermore National Laboratories, Livermore CA, USA

Lyne, A. G.,   Nuffield Radio Astronomy Laboratories, Jodrell Bank, Macclesfield, United Kingdom

McClintock, J. E., Harvard Observatory, Center for Astrophysics, Cambridge MA, USA

Miller, C.,   University of Illinois, Department of Physics, Urbana IL, USA

Navarro, J.,   California Institute of Technology, Pasadena CA, USA

Nelson, R. W.,   Cornell University, Ithaca NY, USA

Nomoto, K.,   University of Tokio, Department of Astronomy, Faculty of Science Bunkyo-ku, Tokio, Japan

Pacini, F.,   Osservatorio Astrofisico di Arcetri Largo Enrico Fermi, Firenze, Italy

Parmar, A. N.,   ESTEC, Noordwijk, The Netherlands

Petterson, J. A., Department of Natural Science, University of Guam, Mangilao, Guam, USA

Phinney, E. S.,   Theoretical Astrophysics CalTech, Pasadena CA, USA

Pines, D.,   University of Illinois, Physics Department, Urbana IL, USA

Popham, R.,   University of Arizona, Astronomy Department, Steward Observatory, Tucson AZ, USA

Prince, T.,   California Institute of Technology, Pasadena CA, USA

Radhakrishnan, V., Raman Research Institute, Bangalore, India

Rappaport, S. A., Center for Space Research, MIT, Cambridge MA, USA

Ray, P.,   California Institute of Technology, Pasadena CA, USA

Robinson, C.,   Nuffield Radio Astronomy Laboratories, Jodrell Bank, Macclesfield, United Kingdom

Romani, R.,   Institute for Advanced Study, Princeton NJ, USA

Ruderman, M.,   Department of Theoretical Physics, Oxford, United Kingdom

Ruffert, M.,   Max Planck Institut für Astrophysik, Garching bei München, BRD

Shaham, J.,   Columbia Astrophysics Laboratory, Department of Astronomy & Physics, New York NY, USA

Sigurdsson, S.,   California Institute of Techology, Pasadena CA, USA

Spencer, E.,   Nuffield Radio Astronomy Laboratories, Jodrell Bank, Macclesfield, United Kingdom

Sutantyo, W.,   Astrophysics Department, Institute of Technology, Bandung, Indonesia

Swank, J.,   Goddard Space Flight Center, X-ray Astronomy, Greenbelt MD, USA

Taam, R. E.,   Northwestern University, Department of Physics and Astronomy, Evanston IL, USA

Tanaka, Y.,   Institute for Space and Aeronautical Science, Kanagawa, Japan

Tavani, M.,   Lawrence Livermore National Laboratories, Livermore CA, USA

Taylor, J. H.,   Princeton University, Physics Department, Princeton NJ, USA

Trimble, V.,   Astronomy Program, University of Maryland, College Park MD, USA

Trümper, J.,   Max Planck Institut für Extraterrestrische Physik, Garching bei München, BRD

Webbink, R. F.,   University of Illinois, Department of Astronomy, Urbana IL, USA

Weert, Ch. G. van, Institute for Theoretical Physics, University of Amsterdam, The Netherlands

White, N. E.,   Laboratory for High Energy Astrophysics, Goddard Space Flight Center, Greenbelt MD, USA

Wilkerson, J.   University of California, Center for Particle Astrophysics, Berkeley CA, USA

Wolszczan, A.,   NAIC, Arecibo Observatory, Arecibo PR, USA

Woosley, S. E.,   Lick Observatory, University of California, Santa Cruz CA, USA

Wijers, R.,   Astronomical Institute, University of Amsterdam, The Netherlands

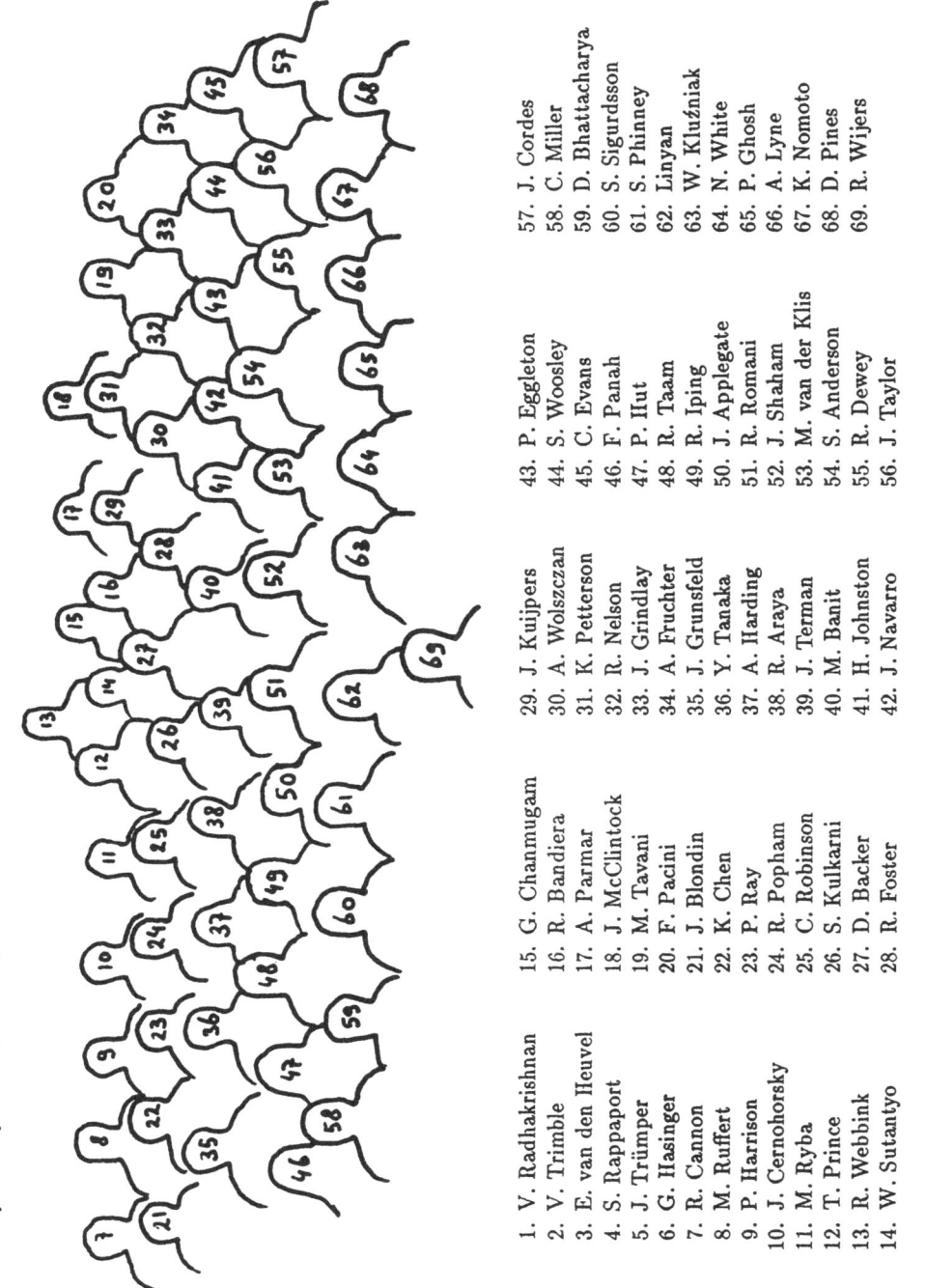

1. V. Radhakrishnan
2. V. Trimble
3. E. van den Heuvel
4. S. Rappaport
5. J. Trümper
6. G. Hasinger
7. R. Cannon
8. M. Ruffert
9. P. Harrison
10. J. Cernohorsky
11. M. Ryba
12. T. Prince
13. R. Webbink
14. W. Sutantyo
15. G. Chanmugam
16. R. Bandiera
17. A. Parmar
18. J. McClintock
19. M. Tavani
20. F. Pacini
21. J. Blondin
22. K. Chen
23. P. Ray
24. R. Popham
25. C. Robinson
26. S. Kulkarni
27. D. Backer
28. R. Foster
29. J. Kuijpers
30. A. Wolszczan
31. K. Petterson
32. R. Nelson
33. J. Grindlay
34. A. Fruchter
35. J. Grunsfeld
36. Y. Tanaka
37. A. Harding
38. R. Araya
39. J. Terman
40. M. Banit
41. H. Johnston
42. J. Navarro
43. P. Eggleton
44. S. Woosley
45. C. Evans
46. F. Panah
47. P. Hut
48. R. Taam
49. R. Iping
50. J. Applegate
51. R. Romani
52. J. Shaham
53. M. van der Klis
54. S. Anderson
55. R. Dewey
56. J. Taylor
57. J. Cordes
58. C. Miller
59. D. Bhattacharya
60. S. Sigurdsson
61. S. Phinney
62. Linyan
63. W. Kluźniak
64. N. White
65. P. Ghosh
66. A. Lyne
67. K. Nomoto
68. D. Pines
69. R. Wijers

By the early eighties it seemed that our knowledge about radio pulsars, their birth events, and their evolution was nearly complete. However, the discovery of several dozen millisecond pulsars and over a dozen binary pulsars in the last few years has shattered part of the "standard scenario" concerning neutron star birth, structure, and evolution, notably the part of the scenario involving magnetic field decay. Also, in the eighties, the part that binaries may play in the formation of neutron stars has come into much sharper focus.

These findings have shown that the binary and millisecond pulsars form a sub-population of neutron stars that underwent an evolution different from the straightforward evolutionary path followed by a pulsar that was born as a single, strongly magnetized neutron star. The high percentage of binaries among the millisecond pulsars (over 40 percent, against only 2–3 per cent in the general pulsar population) and the many characteristics that the binary and millisecond pulsars have in common, such as much weaker magnetic fields and, on average, much faster spin than the general pulsar population, made it clear that together they form one population of objects with a common evolutionary origin.

The finding that in most binary pulsars the companion of the pulsar is itself a "dead" star (neutron star or white dwarf) confirmed the suggestion, first made by Bisnovatyi-Kogan, that neutron stars can be "recycled" in binaries. It also confirms the suggestion of Smarr and Blandford, that the peculiar combination of rapid spin and weak magnetic field, shown by the first-discovered binary pulsar PSR 1913+16, can be explained by such a recycling process. During this process the pulsar was spun-up by accretion of matter with angular momentum from its binary companion, before that star terminated its evolution.

This evolutionary model implies that the progenitors of the binary pulsars must at some time have been binary X-ray sources. A prediction of this "recycling" model was that globular star clusters, which were already known to be particularly rich in X-ray binaries, should also contain many recycled old neutron stars. This prediction was beautifully confirmed by the recent discovery of a couple dozen recycled radio pulsars in globular clusters.

The discovery of Quasi Periodic Oscillations in the X-ray emission of bright low-mass X-ray binaries, strongly suggesting that these systems harbor rapidly spinning weakly magnetized neutron stars (the "Alpar and Shaham model"), has provided a further important evolutionary link between the X-ray binaries and recycled pulsars.

The virtue of these interacting binary systems – X-ray binaries as well as binary pulsars – is that they make very old neutron stars observable, either as an X-ray source or a recycled radio pulsar, while ordinary pulsars spin down rapidly and disappear from view within 10^7 years after their birth. The ordinary pulsar population is, therefore, highly biased towards neutron stars younger than about 10^7 years. Neutron stars in accreting binaries, and recycled pulsars, on the other hand, may be as old as the evolutionary timescales of the binary systems to which they belong, which in low-mass systems may be billions of years. In fact, we know that the recycled pulsars with weak magnetic fields spin down so slowly that they may remain observable for a Hubble time. Therefore, contrary to the ordinary pulsar population, binary and millisecond pulsars offer the unique opportunity to study the evolution of neutron stars and their magnetic fields on a time scale of billions of years.

One of the most striking results of the new discoveries in the field of binary and millisecond pulsars is the realisation that there remains very little support for the original ideas on spontaneous magnetic field decay in isolated neutron stars on a relatively short timescale, of order 10^7 years. This was a "paradigm" that dominated the field of pulsar research for over two decades. In fact, it now appears that, in order to interpret radio pulsar statistics in combination with the entire body of observational knowledge on X-ray binaries and binary and millisecond pulsars, there is no longer a need for spontaneous magnetic field decay. This thought was already emerging in 1990 at the Crete NATO Advanced Study Institute on Neutron Stars (Ventura and Pines 1991), and since then the evidence against spontaneous field decay in isolated neutron stars has further mounted. Only in the case of neutron stars that have gone through an accretion phase in a binary does there appear to be clear evidence for magnetic field decay.

In view of the rapid pace of discovery in this field and the many exciting aspects of neutron-star evolution and neutron-star properties emerging from these findings, the NSF Institute for Theoretical Physics at the University of California Santa Barbara decided in 1990 to bring together a number of experts in this field in order to make, during a 6-month period, an in-depth study of the fundamental problems raised by these new discoveries. The director of the Institute had invited us to direct this 6-month research program, entitled "Neutron Stars in Binary Systems" which was held in Santa Barbara from January through June, 1991.

In order to provide this research program with a "flying start" we decided to organize an intensive one-week Workshop at the beginning of 1991 in Santa Barbara, in which some 35 of the world's experts in the fields of binary X-ray sources, radio pulsars, stellar evolution, plasma astrophysics, and high-energy astrophysics would come together to review the present state of our knowledge and research in this field, from the observational as well as the theoretical point of view, and to formulate key problems for future study.

This book contains the review papers and a selection of the contributed papers presented at the Workshop. Of the exciting new discoveries presented at the Workshop we especially mention:
— The discovery of eleven millisecond pulsars in the globular cluster 47 Tucanae, reported by Lyne.
— The discovery of a number of pulsars with excessively high space velocities, some moving towards the Galactic plane, reported by Harrison.
— The measurement, with high precision, of several relativistic effects in the newly discovered close and eccentric-orbit double neutron star system PSR 1534+12, reported by Wolszczan.
— The very impressive new results from Germany's ROSAT X-ray observatory, reported by Trümper, notably the discovery of many new supernova remnants, and of a new class of ultra-soft binary X-ray sources in the Large Magellanic Cloud.

We wish to thank those without whose support the organisation of this Workshop would not have been possible. In the first place, we acknowledge the NATO Science division for providing financial support, and the Institute for Theoretical Physics at the University of California Santa Barbara, for providing financial, logistical, and organisational support. We especially thank Mrs. Darla Sharp-Fitzpatrick and Mrs. Anne Braddock of the ITP for their invaluable assistance in organisational and administrative matters before, during, and after the Workshop. We thank Mrs. Loeki Lemmens for the preparatory work from the Dutch side during the six months preceding the meeting, and Dr. Sake Hogeveen for his help in the preparation of these proceedings.

E. P. J. van den Heuvel, Amsterdam
S. A. Rappaport, Cambridge, Massachusetts

Reference

Ventura, J., and Pines, D. (eds), 1991, *Neutron Stars: Theory and Observation*, NATO ASI Series C, Vol. 344, Kluwer Academic Publishers, Dordrecht, The Netherlands.

I OBSERVATIONS

CHAPTER 1

X-ray Binaries

THE ORBITAL PERIODS OF LOW-MASS X-RAY BINARIES

A. N. Parmar
Astrophysics Division,
ESTEC/SA,
2200 AG Noordwijk,
The Netherlands.

ABSTRACT. The evidence that LMXRB contain thick azimuthally structured accretion disks is discussed and the ways in which these can give rise to observable orbital modulations presented. The orbital period distribution of LMXRBs is shown to be similar to that of CVs except that there are no LMXRB with orbital periods between 1 and 2 hr corresponding to the SU Uma and AM Her binaries. The evidence that the orbital periods of some LMXRB are evolving on a timescale of $\sim 10^7$ years is presented. Intriguingly of the systems studied in detail, two have *increasing* and two *decreasing* orbital periods.

1. Introduction

Accretion powered X-ray binary systems can be divided into two groups according to the nature of the mass donating star. The first group consists of systems where a compact object (either a neutron star or a black hole), accretes material from a massive early-type star often via a stellar wind. Many of these systems exhibit X-ray pulsations and the optical emission is dominated by that of the companion. The orbital periods of many of these systems have been revealed through the detection of Doppler shifted X-ray pulsations, X-ray eclipses, periodic X-ray flares or periodic optical variability. (See e.g., Stella, White, and Rosner 1986). The second group consists of systems containing a compact object accreting material via Roche lobe overflow from a low-mass companion star. Members of this group include the galactic bulge sources, X-ray burst sources, some X-ray transients and the luminous globular cluster X-ray sources. The optical continua from these systems tend to be blue with high-excitation emission lines superposed, suggesting that the optical light is dominated by reprocessed X-ray radiation.

In contrast to the massive binaries, few low-mass X-ray binary (LMXRB) systems exhibit pulsations or well defined eclipses. In addition, many of these systems have faint optical counterparts or are located in the vicinity of the galactic center making optical observations difficult. Until a few years ago, the orbital periods of only a few (possibly atypical) systems were known. However, with the advent of sensitive CCD detectors and the EXOSAT *Observatory* in the 1980's this changed, and the orbital periods of a substantial number

5

E. P. J. van den Heuvel and S. A. Rappaport (eds.), X-Ray Binaries and Recycled Pulsars, 5–18.
© 1992 *Kluwer Academic Publishers.*

of LMXRB systems have now been established. Studies of the orbital periods of LMXRB provide information on the composition and evolutionary status of these systems. In addition, the properties of the modulations themselves have proved extremely interesting, providing information on the structure of the accretion disks and emission regions present in LMXRB.

2. X-Ray Orbital Modulations from LMXRB

During the late 1970's it became apparent that LMXRB exhibit significantly fewer X-ray eclipses than might be expected if the systems simply consist of a dwarf companion overflowing its Roche lobe and transferring material to a compact object via a thin accretion disk (Joss and Rappaport 1979). Milgrom (1978) suggested that this discrepancy could be resolved if LMXRB contain *thick* accretion disks which obscure the central X-ray sources of systems that are viewed close to the orbital plane. While X-ray eclipses from LMXRB are indeed rare, as discussed below, irregularities in the azimuthal structure of an accretion disk can produce an observable modulation. These modulations may take two forms:

Table 1 – Types of Orbital X-ray Modulation

Source Type	Obscured X-ray Source	Type of Modulation	Typical Energy Dependence	Eclipse (if seen)
Dipper	Point-like	Irregular dips	Strong	Total
ADC	Extended	Smooth variation	Weak	Partial

In the dip sources, the central X-ray source is directly viewed and structure in the disk produces irregular dips, or reductions, in X-ray intensity. If the compact object is hidden from direct view and only X-rays scattered into our line of sight by a corona or wind surrounding the accretion disk (collectively referred to as an accretion disk corona, or ADC) are visible, then obscuration by material in the disk will produce a smooth modulation.

2.1 X-RAY DIP SOURCES

The X-ray source XB 1916-053 was the first LMXRB to reveal its orbital period through the presence of regularly occurring intensity dips (White and Swank 1982; Walter *et al.* 1982). The dips observed from this source are narrow (with a duration of < 10 minutes) and recur every 55 minutes. Occasionally, dips that occur 180° out of phase with the normal dips (called *anomalous* dips) are observed. Figure 1, taken from Smale *et al.* (1988), illustrates the wide variety of dip morphologies observed from XB 1916-053 during three EXOSAT observations. Similar variability is observed from the majority of dip sources. During the first observation the dips are narrow (lasting for < 5 minutes) and their depth varies between 20–100% of the quiescent intensity. During the 1985 May 24 observation, the dips structure is different, with two broad deep dips per cycle. During the first few hours of the 1985 October 13 observation no pronounced dips are visible. About half way through the observation narrow dips, similar to those seen during the first observation appear, followed by the appearance of deep *anomalous* dips towards the end of the observation. The depth of these *anomalous* dips can be contrasted with the shallow *anomalous* dip visible during the first observation at 1983 September 17 18:10 UT.

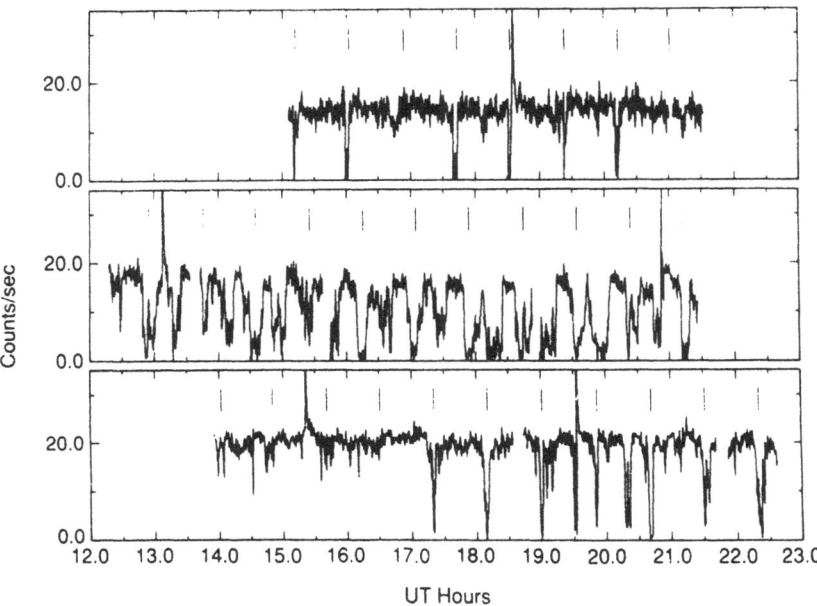

Fig. 1. EXOSAT Medium Energy Instrument (ME) 1–10 keV light curves of XB 1916-053 from observations on 1983 September 17 (upper) 1985 May 24 (middle) and 1985 October 13 (lower), summed in 30 s bins. The X-ray bursts have been truncated. The vertical bars indicate the expected dip occurrence times. (Taken from Smale *et al.* 1988).

Although the dips observed from XB 1916-053 exhibit some variation in occurrence time, the stability of the 55 minute recurrence interval is strongly suggestive of orbital motion. However, the observed variations in dip-depth and duration are not consistent with the dips being caused by eclipses by the companion star. White and Swank (1982) and Walter *et al.* (1982) were the first to suggest that these dips are the result of obscuration of the central X-ray by material located at the point where the gas stream from the companion meets the accretion disk. Variations in the structure at the rim of the accretion disk are then responsible for the observed variations in dip morphology. The presence of *anomalous* dips implies that the accretion disk structure must be complicated with additional vertical structure away from the initial impact point of the gas stream around the disk rim.

XBT 0748-676 is a transient burst source which was discovered serendipitously by EXOSAT. The source exhibits both dips and eclipses, as well as evidence for the presence of an extended X-ray emission region (Parmar *et al.* 1986; Parmar *et al.* 1991). A total of ten EXOSAT and *Ginga* observations of XBT 0748-676 were made between 1985 February and 1989 March, providing a unique data set by which to study the evolution of the system properties. The detection of eclipses allows the phase of the dips to be uniquely related to the orbital geometry. Figure 2 shows the light and hardness ratio curves of just over one orbital cycle of XBT 0748-676. Two 8.3 minute duration eclipses, separated by 3.82 hr, are visible at $\Phi = 0.0$. The light curve shows evidence for variability throughout the orbital cycle. This is most pronounced between $\Phi \sim 0.6\text{–}0.1$ where a series of deep intensity dips are visible. These dips are associated with an increase in hardness ratio, in contrast to the variability observed at other orbital phases. The ratio of the energy dependent to independent variability observed during dips provides an estimate of the abundance of the absorbing material (Table 2).

Table 2 – X-ray Dip Sources

Source	Period (hour)	Abundance Deficiency	Reference
XB 1916-053	0.83	3 ± 2	Smale *et al.* (1988)
XB 1323-619	2.96	15–0.5	Parmar *et al.* (1989)
XBT 0748-676	3.82	7–2	Parmar *et al.* (1986)
XB 1254-690	3.88	2–0.25	Courvoisier *et al.* (1986)
X 1755-338	4.40	> 600	White *et al.* (1984)
XBT 1658-298	7.1	...	Cominsky and Wood (1989)
XB 1624-490	21	1.3–0.5	Jones and Watson (1989)
Her X-1	40.8	...	Tananbaum *et al.* (1972)
Cyg X-2	235	...	Vrtilek *et al.* (1986)

Depending on the assumptions made about mass-radius relation of the companion star to XBT 0748-676, the eclipse duration implies a orbital inclination of 73–83°. This means that the thickened region responsible for the dips must subtend an angle of at least 7° above the plane of the orbit in order to obscure our line of sight. Since dips are sometimes seen for about half the orbital cycle this region must also extend half way around the disk at times. During eclipse 4% of the uneclipsed flux remains. This is most likely X-rays scattered into the line of sight by an ADC.

2.2 ACCRETION DISK CORONA SOURCES

In these systems the central point source is hidden from direct view either by the accretion disk or by the ADC. This means that the value of the X-ray to optical luminosity ratio will be reduced and typical values of $\sim 1\text{–}20$ are measured, compared to $\sim 100\text{–}1000$, seen from the dip sources (Mason 1986). ADC systems can be divided into two groups. Three systems (X 1822-371, XBT 2129+470, and X 0921-630) show both smooth quasi-sinusoidal modulations and partial eclipses. The observation of partial eclipses confirmed the extended nature of the observed X-ray sources in these systems (White *et al.* 1981) and thus provided the first direct evidence for ADC in LMXRB. Modelling of the eclipse profile allows the sizes of the scattering and absorbing regions to be estimated. In a number of

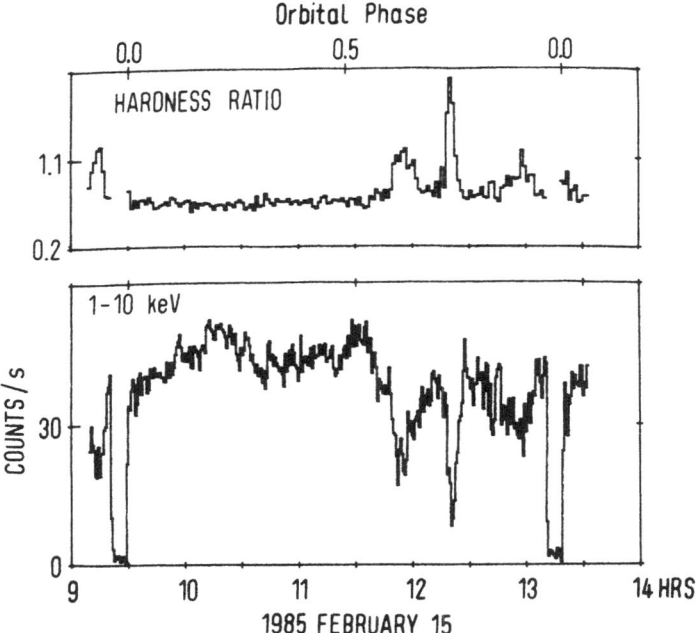

Fig. 2. EXOSAT ME light and hardness ratio curves obtained during a four hour observation of XBT 0748-676 (taken from Parmar *et al.* 1986).

systems such as Cyg X-3, XB 1820-303, X 1728-169 and XB 2127+119 only a sinusoidal-like modulation is seen with no partial eclipses evident (these sources are indicated by an A? in Table 3). While the nature of these modulations remains unconfirmed, it is possible that in these systems the inclinations are lower or the accretion disks are thicker, such that eclipses by the companion star are not observed.

The ADC source X 1822-371 has become a *rosetta stone* in the study of the structure of accretion disks and emission regions in LMXRBs. A 5.57 hr modulation of the optical light curve was discovered by Mason *et al.* (1980). During about one third of this cycle there is a smooth gradual decline before the occurrence of a minimum. The optical spectrum shows high-excitation He II and C III/N IV features typical of LMXRBs as well as broad (~ 1000 km s^{-1}) dish-shaped absorption lines characteristic of accretion disks (Mason *et al.* 1982). The X-ray light curve of X 1822-371 was shown to be modulated with the same period as in the optical by White *et al.* (1981). The upper left hand panel of Figure 3 shows the EXOSAT ME folded light curve of X 1822-371 from an observation on 1985 May 6. The light curve consists of a broad quasi-sinusoidal modulation which has a minimum ~ 0.2 cycles before the occurrence of a gradual partial eclipse. Comparison of the folded optical (Figure 3, right hand panel) and X-ray light curves shows that X-ray eclipse occurs simultaneously with the optical minima. The partial nature of the eclipse results from the fact that the central source is not seen directly, but rather only X-rays scattered in an ADC are observed (White *et al.* 1981; White and Holt 1982). The other feature of the X-ray light curve is the quasi-sinusoidal modulation. This is explained as being the result

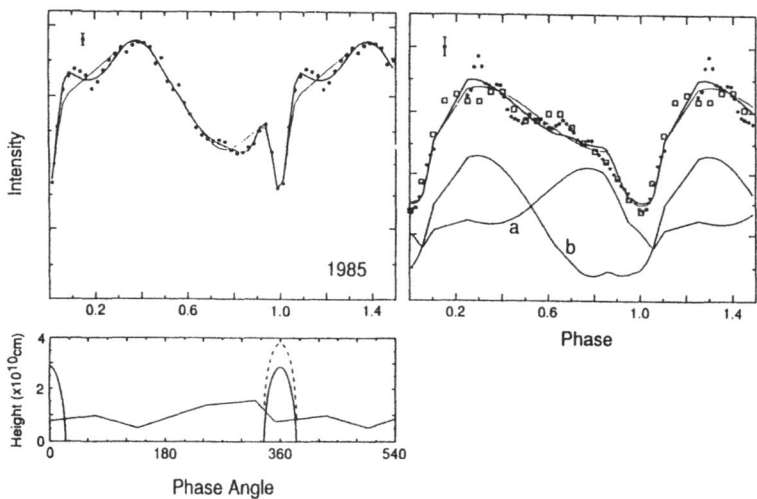

Fig. 3. The observed X-ray (upper left panel) and optical (right panel) folded light curves of X 1822-371 and the results of model fitting of Hellier and Mason (1989). The curves (a) and (b) show the contributions from the inner and outer regions of the accretion disk, respectively. The lower panel shows the fitted disk rim profile and the relative sizes of the ADC (dotted) and the secondary.

of partial obscuration of the ADC by a thick azimuthally structured accretion disk. Since the minimum of this modulation occurs at $\Phi = 0.8$, the thickest part of the accretion disk must be in the line of sight at this phase. This phase corresponds to the expected impact point of the gas stream.

If the companion is assumed to fill its Roche lobe and if $M_x/M_c < 0.8$ (where M_x is the mass of the compact object and M_c the mass of the companion), then combining Kepler's third law with the mass-radius relation of Paczyński (1971) and the observed eclipse depth and width gives a radius for the ADC of $0.3R_\odot$ (White and Holt 1982). The broader dip seen in the optical waveband implies that the optical disk has a size about a factor of two larger than that of the ADC. Modelling of the X-ray light curve allows constraints to be placed on the location and size of the occulting region responsible for the quasi-sinusoidal modulation. This modelling shows that the obscuring material must be located at the rim of the accretion disk with a maximum vertical extent of $0.15R_\odot$ (Mason and Córdova 1982; White and Holt 1982; Hellier and Mason 1989). In some cases the fits to the X-ray and optical light curves are significantly improved if a second, smaller thickened region is included in the model at $\Phi = 0.2$, again suggesting a much more complex structure at the disk rim (Mason and Córdova 1982; White and Holt 1982).

Figure 4 shows a comparison between the folded light curves of the ADC source X 1822-371 and the dip source XBT 0748-676. The eclipses in XBT 0748-676 are sharp (with a typical transition time of ~ 6 s) and almost total, and the dips are highly irregular. The eclipses observed from X 1822-371 are gradual and partial, and instead of irregular dips a smooth modulation is visible over approximately the same range of orbital phases as the dips seen from XBT 0748-676. These differences are consistent with the view that the observed X-ray source is point-like in the case of XBT 0748-676 and extended in the case of X 1822-371, and that the orbital modulation seen in both systems has a similar origin.

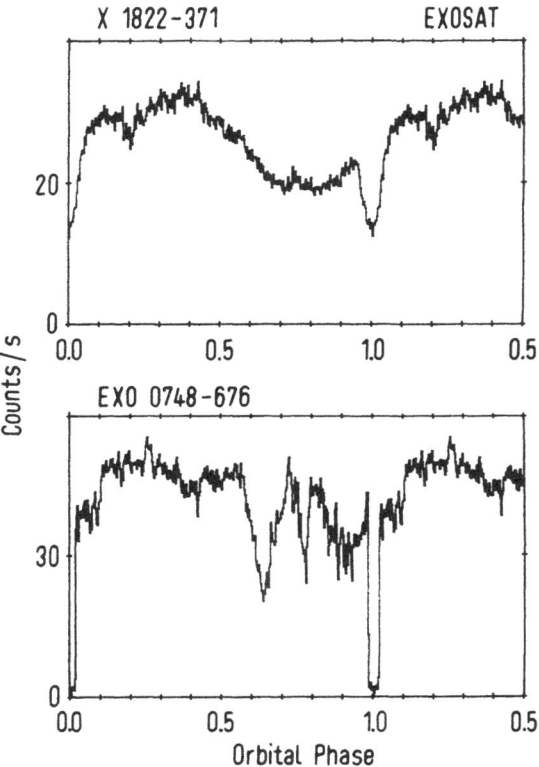

Fig. 4. The folded 1–10 keV light curves of X 1822-371 and XBT 0748-676 (taken from Parmar *et al.* 1986). One and a half orbital cycles are shown.

3. The Orbital Period Distribution of LMXRB

Table 3 lists the 26 LMXRB with confirmed orbital periods. The orbital periods range from 0.19 to 235 hr. LMXRB can be divided into two classes depending on the intrinsic luminosity of the central X-ray source. The low ($\sim 10^{37}$ ergs s^{-1}) luminosity systems (which include the X-ray burst sources and the transient systems), tend to have orbital periods $\lesssim 15$ hr. However, this cut-off may result from increased selection effects against detecting modulations with periods longer than this. The orbital period distribution of the high ($\sim 10^{38}$ ergs s^{-1}) luminosity systems (mostly located in the galactic bulge) is still largely unknown, but in two cases (Sco X-1 and Cyg X-2) orbital periods of 19 hr and 235 hr have been found from optical studies.

Figure 5 compares the distributions of the orbital periods of LMXRB and cataclysmic variables (CV) systems. The orbital periods show a clustering between 3–10 hr, which in the case of the CV's is predominately caused by the presence of classical novae and U Gem dwarf novae. There are no X-ray binaries with orbital periods corresponding to the well-known gap in the CV period distribution between 2–3 hr. In the case of the LMXRB the period gap may extend down to $\lesssim 1$ hr, which is populated by the SU Uma and AM Her type binaries (White 1985; White and Mason 1985). The reason for this difference is not clear, but it seems to be real since there is no known selection effect against detecting LMXRB with periods in this range.

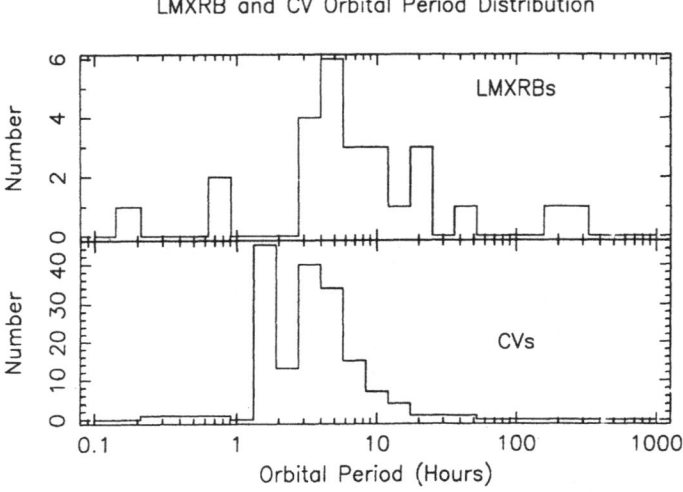

Fig. 5. The orbital period distributions of cataclysmic variables (CV) and LMXRB systems. The CV periods are taken from Ritter (1990).

4. Orbital Period Changes in LMXRB

A number of CVs show evidence for decreasing orbital periods with typical timescales, $-P_{orb}/\dot{P}_{orb}$, of $\gtrsim 10^7$ years (e.g., Patterson 1984). As pointed out by Patterson, this effect appears to be restricted to those systems with companion masses $\gtrsim 0.6\ M_\odot$. Recently a number of precise measurements of the orbital periods of four LMXRB that exhibit stable features in their lightcurves have been made. These reveal an interesting situation in that two of the sources show *decreasing* and two show *increasing* orbital periods (Table 4). In particular, the 4.8 hr orbital period of Cyg X-3 appears to be *increasing* on a very short timescale of 4.5×10^5 years (Kitamoto et al. 1987; Molnar 1988; van der Klis and Bonnet-Bidaud 1989) as does the 5.6 hr orbital period of X 1822-371 with a timescale of 2.9×10^6 years (Hellier et al. 1990). In contrast, the 11 min orbital period of XB 1820-303 (which probably has a white dwarf companion), appears to be *decreasing* on a timescale of

Table 3 – The Orbital Periods of LMXRB

Source	Period (hr)	Optical	X-ray	Discovery	Reference
XB 1820-303	0.19	-	A?	X	1,2,3
X 1626-673	0.70	o	-	O	4
XB 1916-053	0.83	o	D	X	5,6,7
XB 1323-619	2.93	-	D	X	8
XB 1636-536	3.8	o	-	O	9,10
XBT 0748-676	3.86	o	DE	X	11,12
XB 1254-690	3.9	o	D	O/X	13,14
X 1728-169	4.2	o	A?	X	15,16
X 1755-338	4.4	o	D	X	17,18
XB 1735-444	4.6	o	-	O	19
Cyg X-3	4.8	(IR)	A?	X	20,21
XBT 2129+470	5.2	o	AE	O	22,23
X 1822-371	5.6	o	AE	O	24,25,26
XB 1658-298	7.2	-	DE	X	27
XT 0620-003	7.3	o	-	O	28
XB 2127+119 (M15)	8.5	o	A?	O	29,30,31
X 1957+115	9.3	o	-	O	32
Cal 87	10.6	o	-	O	33,34
X 1659-487	14.8	o	-	O	35
Cen X-4	15.1	o	-	O	36,37
Sco X-1	19.2	o	-	O	38,39
X 1624-490	21	-	D	X	40
Cal 83	25.0	o	-	O	41
Her X-1	40.8	o	DE	X	42,43,44
X 0921-630	216	o	AE	O	45,46,47
Cyg X-2	235	o	D	O	48,49

NOTES. — The letters 'B' and 'T' in the source name indicate a burst source or transient nature, respectively. An 'o' indicates that an orbital periodicity is seen in the optical band. The type of X-ray modulation (if any) observed is indicated with an 'A' - ADC, 'E' - eclipses, or 'D' - dips. The letters 'O', and 'X' indicate the discovery waveband.

References: 1 Stella et al. 1987; 2 Sansom et al. 1989; 3 Tan et al. 1991; 4 Middleditch et al. 1981; 5 Walter et al. 1982; 6 Grindlay & Cohn 1987; 7 White & Swank 1982; 8 Parmar et al. 1989; 9 Pedersen et al. 1981; 10 Smale & Mukai 1988; 11 Parmar et al. 1986; 12 Parmar et al. 1991; 13 Courvoisier et al. 1986; 14 Motch et al. 1987; 15 Hertz & Wood 1988; 16 Schaefer 1987; 17 White et al. 1984; 18 Mason et al. 1985; 19 Corbet et al. 1986; 20 Parsignault et al. 1972; 21 Sanford & Hawkins 1972; 22 Thorstensen et al. 1979; 23 Ulmer et al. 1980; 24 White et al. 1981; 25 Mason et al. 1980; 26 Hellier et al. 1990; 27 Cominsky & Wood 1984; 28 Mc Clintock & Remillard 1986; 29 Ilovaisky et al. 1987; 30 Naylor et al. 1986; 31 Hertz 1987; 32 Thorstensen 1987; 33 Callanan et al. 1989; 34 Cowley et al. 1989; 35 Honey et al. 1988; 36 Ilovaisky & Chevalier (1986); 37 Cowley et al. 1988; 38 Gottlieb et al. 1975; 39 Cowley & Crampton 1975; 40 Jones & Watson 1989; 41 Naylor et al. 1989; 42 Tananbaum et al. 1972; 43 Bahcall et al. 1974; 44 Voges et al. 1985; 45 Mason et al. 1987; 46 Branduardi-Raymont et al. 1983; 47 Chevalier & Ilovaisky 1982; 48 Cowley et al. 1979; 49 Vrtilek et al. 1986.

1×10^7 years (Sansom *et al.* 1989; Tan *et al.* 1991) as does the 3.82 hour orbital period of XBT 0748-676 with a timescale of 5×10^6 years (Parmar *et al.* 1991).

There are problems with simply interpreting these variations as being evolutionary. Firstly, other close binary systems show departures from their long-term trends that are possibly caused by the effects of solar-type magnetic cycles (Applegate and Patterson 1987; Warner 1988). Secondly, in the case of XB 1820-303 and Cyg X-3, although the orbital modulations appear stable over a timescale of 20 years, their origin is not well understood. This means that if the modulations are not primarily due to obscuration by the companion but instead result from variable obscuration by the accretion disk, then changes in the disk structure could mimic a changing orbital period. In the case of X 1822-371 and XBT 0748-676 the situation is better since the eclipses (partial in the case of X 1822-371) provide the fiducial markers. However, in the case of XBT 0748-676, the evidence for a changing P_{orb} strongly relies on a single measurement made in 1989 (see Figure 6). Clearly further observations are required to confirm the change in P_{orb}.

Table 4 – Changes in Orbital Periods of LMXRB

Source	P_{orb} (hrs)	$-P_{orb}/\dot{P}_{orb}$ (yrs)	Change	Reference
XB 1820-303	0.18	-1×10^7	Decrease	Tan et al. (1991)
				Sansom et al. (1989)
XBT 0748-676	3.8	-5×10^6	Decrease	Parmar et al. (1991)
Cyg X-3	4.8	$+6 \times 10^5$	Increase	Kitamoto et al. (1987)
				Molnar (1988)
				vdKlis & Bonnet-Bidaud (1989)
X 1822-371	5.6	$+3 \times 10^6$	Increase	Hellier et al. (1990)

Standard theories of LMXRB evolution predict that mass transfer is driven by the loss of orbital angular momentum via gravitational radiation (Faulkner 1971; Rappaport, Joss, and Webbink 1982), possibly supplemented by magnetic breaking in the cases of main-sequence or sub-giant companions with sufficiently large magnetic fields and intrinsic mass loss rates (Verbunt 1984; Rappaport, Verbunt, and Joss 1983). For LMXRB with non-degenerate companions, the range of timescales predicted by these theories is between $10^8 - 10^9$ years, much longer than those observed (Table 4). In the case of XB 1820-303, where the companion is thought to be a white dwarf, binary evolution theories predict an *increasing* orbital period, whereas the opposite is observed.

A possible explanation for the rapid spin-down observed from XBT 0748-676 has been proposed by Parmar *et al.* (1991) who assume that the companion obeys the main-sequence mass-radius relation, that it continues to fill its Roche lobe and that the mean mass transfer rate is $10^{-9} M_\odot$/year (obtained assuming the peak X-ray burst luminosity is Eddington). The observed rate of change of orbital period is then a factor ~ 100 times faster then predicted given simple LMXRB evolution and the above assumptions. Parmar *et al.* (1991)

propose that this rapid change in orbital period results from the expansion of the companion due to the effects of X-ray heating. This change requires an increase in radius of the companion $\Delta R/R \simeq 6 \times 10^{-6}$ which is quite plausible given that $\sim 2\% - 4\%$ of the X-ray luminosity of 10^{37} erg/sec is incident on the companion.

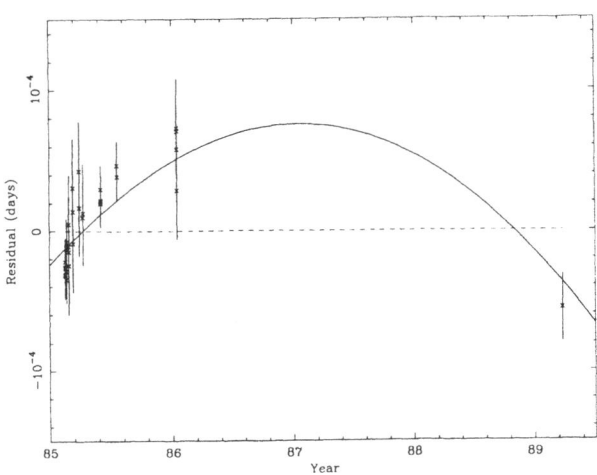

Fig. 6. The residuals with respect to a linear ephemeris of the eclipses of XBT 0748-676 with the best fitting parabolic ephemeris shown (taken from Parmar *et al.* 1991).

Tavani (1991) proposes that the mechanism driving the mass-loss in LMXRB is not predominately gravitational radiation but X-ray irradiation of the companion and outer edge of the accretion disk. The typical lifetime of a radiation-driven LMXRB is expected to be of order $10^6 - 10^7$ years. This relatively short LMXRB lifetime is appealing from the point of view of reconciling the apparent discrepancy between the number of binary millisecond pulsars and their potential LMXRB progenitors (Narayan *et al.* 1990). Tavani's model predicts two classes of radiation driven LMXRB depending on the fraction of mass lost from the system. In systems with low binary mass-loss rates the orbital periods will tend to increase while the opposite is the case if significant amounts of mass are lost from the system. Thus, Cyg X-3 and X 1822-371 may belong to the first class and XBT 0748-676 and XB 1820-303 to the second.

5. References

Applegate, J. H., and Patterson, J. 1987, *Ap. J. (Letters)*, **322**, L99.

Bahcall, J. N., Joss, P., C., and Avni, Y. 1974, *Ap. J.*, **191**, 211.

Branduardi-Raymont, G., Corbet, R. H. D., Mason, K. O., Parmar, A. N., Murdin, P. G., and White, N. E., 1983, *M.N.R.A.S.*, **205**, 403.

Callanan, P. J., Fabian, A. C., Tennant, A. F., Redfern, R. M., Shafer, R. A. 1987, *M.N.R.A.S.*, **224**, 781.

Callanan, P.J., Machin, G., Naylor, T., and Charles, P. A. 1989, *M.N.R.A.S.*, **241**, 37p.

Chevalier, C., and Ilovaisky, S. A. 1982, *Astr. Ap.*, **112**, 68.

Corbet, R. H. D., Thorstensen, J. R., Charles, P. A., Menzies, J. W., Naylor, T., and Smale, A. P. 1986, *M.N.R.A.S.*, **222**, 15p.

Cominsky, L., and Wood, K. 1984, *Ap. J.*, **283**, 765.

Cominsky, L., and Wood, K. 1989, *Ap. J.*, **337**, 485.

Courvoiser, T. J-L., Parmar, A. N., Peacock, A., and Pakull, M. 1986, *Ap. J.*, **309**, 265.

Cowley, A. P., and Crampton, D. 1975, *Ap. J. (Letters)*, **201**, L65.

Cowley, A. P., Crampton, D., and Hutchings, J. B. 1979, *Ap. J.*, **231**, 539.

Cowley, A. P., Hutching, J. B., Schmidtke, P. C., Hartwick, F. D. A., Crampton, D., and Thompson, I. B. 1988, *Astron. J.*, **95**, 4, 1231.

Cowley, A. P., Schmidtke, P. C., Crampton, D., and Hutchings, J. B. 1989, *IAU Circ. No.* 4722.

Faulkner, J. 1971, *Ap. J. (Letters)*, **170**, L99.

Gottlieb, E. W., Wright, E. L., and Liller, W. 1975, *Ap. J. (Letters)*, **195**, L33.

Grindlay, J. E., and Cohn, H. 1987, *IAU Circ. No.* 4393.

Hellier, C., and Mason, K. O. 1989, *M.N.R.A.S.*, **239**, 715.

Hellier, C., Mason, K. O., Smale, A. P., and Kilkenny, D. 1990, *M.N.R.A.S.*, **244**, 39p.

Hertz, P. 1987, *Ap. J.*, **315**, L119.

Hertz, P., and Wood, K. S. 1987, *IAU Circ. No.* 4221.

Honey, W. B., Charles, P. A., Thorstensen, J. R., and Corbet, R. H. D. 1988, *IAU Circ. No.* 4532.

Ilovaisky, S. A., Aurière, M., Chevalier, C., Koch-Miramond, L., Cordini, J. P., and Angebault, L. P. 1987, *Astr. Ap.*, **179**, L1.

Ilovaisky, S. A, and Chevalier, C. 1986, *IAU Circ. No.* 4264.

Jones, M. H., and Watson, M. G., in *Lecture Notes in Physics*, Vol. **266**, *The Physics of Accretion onto Compact Objects*, ed. K. O. Mason, M. G. Watson, and N. E. White (Berlin: Springer-Verlag), p. 439.

Joss, P. C., and Rappaport, S. A. 1979, *Astr. Ap.*, **71**, 217.

Kitamoto, S., Miyamoto, S., Matsui, W., and Inoue, H. 1987, *Pub. Astr. Soc. Japan*, **39**, 259.

Mason, K. O. 1986, in *Lecture Notes in Physics*, Vol. **266**, *The Physics of Accretion onto Compact Objects*, ed. K. O. Mason, M. G. Watson, and N. E. White (Berlin: Springer-Verlag), p. 29.

Mason, K. O., Branduardi-Raymont, G., Córdova, F. A., and Corbet, R. H. D. 1987, *M.N.R.A.S.*, **226**, 423.

Mason, K. O., and Córdova, F. A. 1982, *Ap. J.*, **262**, 253.

Mason K. O., Middleditch, J., Nelson, J. E., White, N. E., Seitzer, P., Touhy, I. R., and Hunt, L. K. 1980, *Ap. J. (Letters)*, **242**, L109.

Mason, K. O., Murdin, P. G., Tuohy, I. R., Seitzer, P., and Branduardi-Raymont, G. 1982, *M.N.R.A.S.*, **200**, 793.

Mason, K. O., Parmar, A. N., and White, N. E. 1985, *M.N.R.A.S.*, **216**, 1033.

McClintock, J. E., and Remillard, R. A. 1986, *Ap. J.*, **308**, 110.

McClintock, J. E., Remillard, R. A. and Margon, B. 1981, *Ap. J.*, **243**, 900.

Middleditch, J., Mason, K. O., Nelson, J. E., and White, N. E. 1981, *Ap. J.*, **244**, 1001.

Milgrom, M. 1978, *Astr. Ap.*, **208**, 191.

Molnar, L. A. 1988, *Ap. J. (Letters)*, **331**, L25.

Motch, C., Pedersen, H., Beuermann, K., Pakull, M. W., and Courvoisier, T. J.-L. 1987, *Ap. J.*, **313**, 792.

Narayan, R., Fruchter, A. S., Kulkarni, S. R., and Romani, R. W. 1990, in *Accretion Powered Compact Binaries*, ed. C. W. Mauche, (Cambridge, Cambridge University Press), p. 451.

Naylor, T., Charles, P. A., Callanan, P. J., and Redfern, R. M. 1986, *IAU Circ. No.* 4263.

Paczyński, B. 1971, *Ann. Rev. Astr. Ap.*, **9**, 183.

Parmar, A. N., Gottwald, M., van der Klis, M., and van Paradijs, J. 1989, *Ap. J.*, **338**, 1024.

Parmar, A. N., Smale, A. P., Verbunt, F., and Corbet, R. H. D. 1991, *Ap. J.*, **366**, 253.

Parmar, A. N., White, N. E., Giommi, P., and Gottwald, M. 1986, *Ap. J.*, **308**, 199.

Parsignault, D. R. *et al.* 1972, *Nature (Phys. Sci.)*, **239**, 123.

Patterson, J., 1984 *Ap. J. Suppl.*, **54**, 443.

Pedersen, H., van Paradijs, J., and Lewin, W. H. G. 1981, *Nature*, **294**, 725.

Rappaport, S. A., Joss, P. C., and Webink, R. F. 1982, *Ap. J.*, **254**, 616.

Rappaport, S., A., Verbunt, F., and Joss, P. C. 1983, *Ap. J.*, **275**,, 713.

Ritter, H. 1990, *Astr. Ap. (Suppl)*, **85**, 1179.

Sansom, A. E., Watson, M. G., Makishima, K., Dotani, T. 1989, *Pub. Astr. Soc. Japan*, **41**, 595.

Sanford, P. W., and Hawkins, F. J. 1972, *Nature (Phys. Sci.)*, **239**, 135.

Schaefer, B. E. 1987, *IAU Circ. No.* 4478.

Smale, A. P., Mason, K. O., White, N. E., and Gottwald, M. 1988, *M.N.R.A.S.*, **232**, 647.

Smale, A. P., and Mukai, K. 1988, *M.N.R.A.S.*, **231**, 663.

Stella, L., Priedhorsky, W., and White, N. E. 1987, *Ap. J. (Letters)*, **312**, L17.

Stella, L., White, N. E., and Rosner, R., 1986, *Ap. J.*, **308**, 669.

Tan, J. *et al.* 1991, *Preprint.*

Tananbaum, H., Gursky, H., Kellogg, E. M., Levinson, R., Schreier, E., and Giacconi, R. 1972, *Ap. J. (Letters)*, **174**, L143.

Tavani, M. 1991, *preprint.*

Thorstensen, J. R. 1987, *Ap. J.*, **312**, 739.

Thorstensen, J. R., Charles, P. A., Bowyer, S., Briel, U. G., Doxsey, R. E., Griffiths, R. E., Schwartz D. 1979, *Ap. J. (Letters)*, **233**, L57.

Ulmer, M. P. *et al.* 1980, *Ap. J. (Letters)*, **235**, L159.

Van der Klis, M., and Bonnet-Bidaud, J. M. 1989, *Astr. Ap.*, **214**, 203.

Verbunt, F. 1984, *M.N.R.A.S.*, **209**, 227.

Voges, W., Kahabka, H., Ögelmann, H., Pietsch, W., and Trümper, J. 1985, *Space Sci. Rev.*, **40**, 339.

Vrtilek, S. D., Kahn, S. M., Grindlay, J. E., Helfand, D. J., and Seward, F. D. 1986, *Ap. J.*, **698**, 710.

Walter, F. M., Bowyer, S., Mason, K. O., Clarke, J. T., Henry, J. P., Halpern, J., and Grindlay, J. E. 1982, *Ap. J. (Letters)*, **253**, L67.

Warner, B. 1988, *Nature,* **336**, 129.

White, N. E. 1985, in *The Evolution of Galactic X-ray Binaries*, ed. J. Trümper, W. H. G. Lewin and W. Brinkmann (NATO ASI: Reidel), p. 227.

White, N. E., Becker, R. H., Boldt, E. A., Holt, S. S., Serlemitsos, P. J., and Swank, J. H. 1981, *Ap. J.*, **247**, 994.

White, N. E., and Holt, S. S. 1982, *Ap. J.*, **257**, 318.

White, N. E., and Mason, K. O. 1985, *Space Sci. Rev.*, **40**, 167.

White, N. E., Parmar, A. N., Sztajno, M., Zimmermann, H. U., Mason, K. O., and Kahn, S. M. 1984, *Ap. J. (Letters)*, **283**, L9.

White, N. E., and Swank, J. H. 1982, *Ap. J. (Letters)*, **253**, L61.

ROSAT – EARLY RESULTS

Joachim E. Trümper
Max Planck Institut für Extraterrestrische Physik
D-8046 Garching bei München
Bundes Republik Deutschland

Introduction and Abstract

The general scientific objectives of ROSAT [1] are to peform (*a*) the first all sky surveys using imaging X-ray and EUV telescopes and (*b*) detailed investigations of interesting sources in a guest investigator programme.

The survey operations commenced in August 1990 and at the time of this meeting about 75 % of the sky has been scanned. In February 1991 we shall start the first half year (AO-1) guest observer programme for which in total 738 proposals have been received. A glimpse of what can be achieved by pointed observations was obtained during the calibration and verification measurements in the early phase of the mission (June/July 1990).

Before I give a brief summary of the results obtained with the X-ray telescope so far, a few remarks on the scientific instruments may be appropriate.

ROSAT

The ROSAT spacecraft is the largest scientific satellite so far built in Western Europe (2.4 tonnes). It carries two main instruments. The large one is a four fold nested X-ray Wolter telescope (XRT) with two position sensitive proportional counters (PSPC) and one high resolution imager (HRI) wich can be brought alternatively into the focus. The smaller one is the threefold nested Wolter-Schwarzschild mirror system (WFC) with two channel plate detectors in the focal plane. Both instruments are oriented parallel and observe simultaneously in adjacent energy bands (XRT: 0.1–2.4 keV; WFC: 0.03–0.1 keV).

All instruments are working well. The main new aspects compared to previous X-ray telescope missions (EINSTEIN, EXOSAT) are as follows:
- ROSAT performs the first all sky survey in X-rays with an imaging telescope having a sensitivity of $\sim 100\times$ HEAO-1,
- it performs the first all sky survey in EUV,
- it provides an increased sensitivity (factor 3–10) because of the large collecting area of the X-ray telescope,
- the spectral resolution of the PSPC ($E/\Delta E \sim 2.5$) allows broad band spectroscopy: four color bands,
- the low intrinsic background of the PSPC ($\sim 1.5 \times 10^{-5}$ cts/cm^{2}s) which together with the extreme uniformity of the detector response allows direct imaging of the diffuse galactic and extragalactic X-ray backgrounds.

19

E. P. J. van den Heuvel and S. A. Rappaport (eds.), X-Ray Binaries and Recycled Pulsars, 19–26.

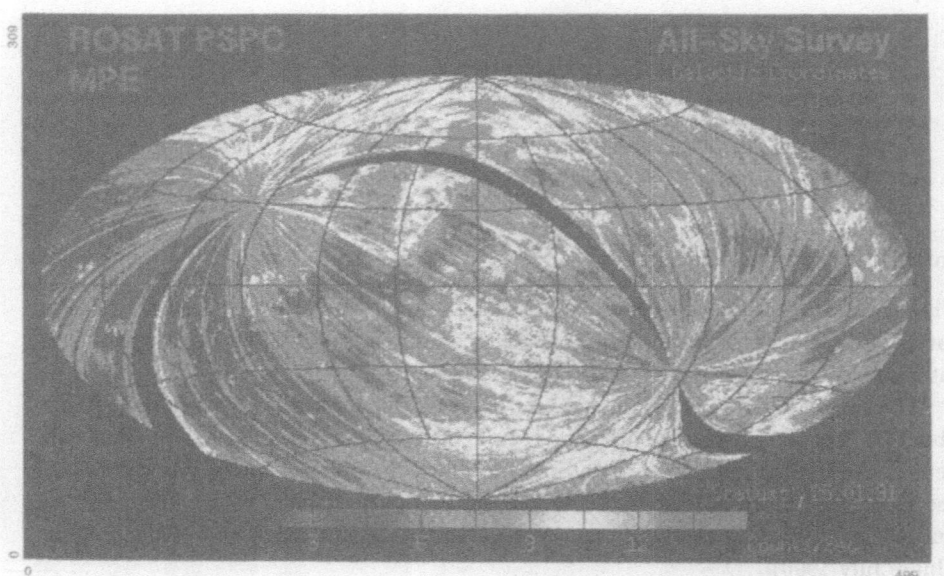

Figure 1 Status of the ROSAT all sky survey in X-rays. Plotted is the PSPC count rate in galactic coordinates with the galactic center in the middle.

ROSAT results

The present status of the sky survey is shown in figure 1, in which the total PSPC count is plotted as a function of galactic coordinates. The map shows some of the large scale features of the X-ray sky such as the Cygnus superbubble, with the adjacent Cygnus Loop, the North Polar Spur, the Vela Supernova Remnant, and the Crab Nebula.

Preliminary estimates of the total number of sources visible in the sky survey are close to 60 000 [2]. Most of them are expected to be AGNs (20–30 000), followed by normal stars. The accuracy of location will be much better than 1 arcmin. Figure 2 shows the f_X/f_o ratio as a function of visual magnitude for a compilation of a large number of sources which have been identified with SIMBAD catalogue sources.

Galactic Sources

The ROSAT survey provides us for the first time with a view of the X-ray sky which is both wide and deep. Figure 3 shows a rather large region (250 square degrees) near the galactic center observed in the XRT survey. Besides the well known bright X-ray binaries it contains Kepler's Supernova remnant, and the Moon which is scanned every fortnight in two consecutive ROSAT orbits. Some diffuse emission regions and

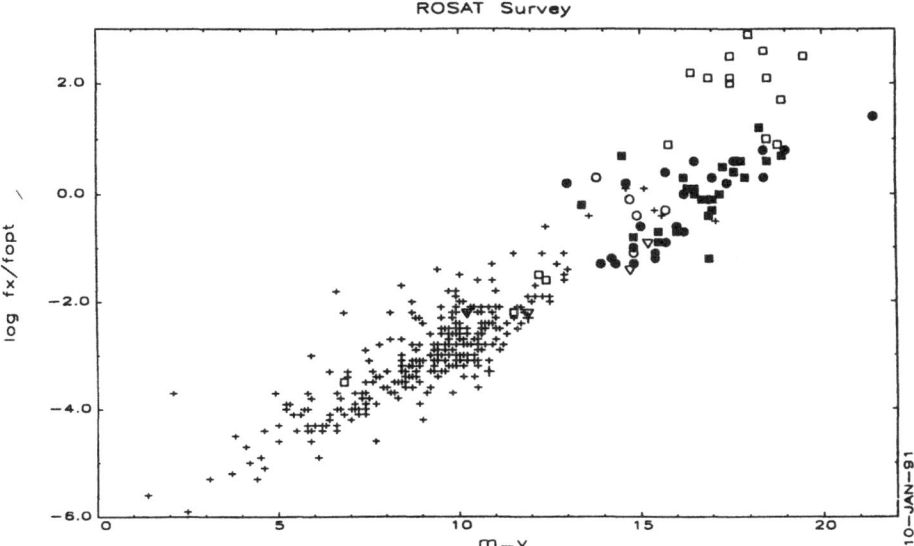

Figure 2 Ratio of X-ray to optical flux (f_X/f_o) for a sample of a large number of sources derived from the ROSAT all sky survey and identified with SIMBAD catalogue sources: + stars, o white dwarfs, ▽ cataclysmic variables, □ neutron star binaries, ■ AGN, • clusters of galaxies.

many new faint X-ray sources are seen as well. Figure 4 shows the Orion region with the bright O stars of the belt and diffuse emission from the well-known star formation region. Preliminary estimates indicate that 15–20 000 normal stars of all spectral types will be visible in the survey.

The most photogenic sources are, of course, the old and extended supernova remnants which are imaged by ROSAT for the first time in a homogeneous way. Figure 5 shows the Cygnus Loop as an example. The first of our newly discovered supernova remnants is shown in figure 6. It has an almost perfect spherical shape with a diameter of 108 arcmin and delivers an X-ray flux of $\sim 2 \times 10^{-10}$ erg/cm^2 (0.1–2.4 keV), ranking it among the 10 X-ray brightest galactic SNRs. A preliminary Sedov analysis [3] indicates that it exploded $\sim 3 \times 10^4$ years ago in a low desity region (0.01 cm^{-3}). It is neither visible in the Condon radio survey, nor on the optical sky plates. Deep observations with the Effelsberg telescope have revealed a shell radio source of low surface brightness. The X-ray/radio brightness ratio of this object is a factor 4 higher than for any other known SNR. It is clear that ROSAT is very powerful in picking up new SNRs. We expect to discover a few dozen of them in the all sky survey.

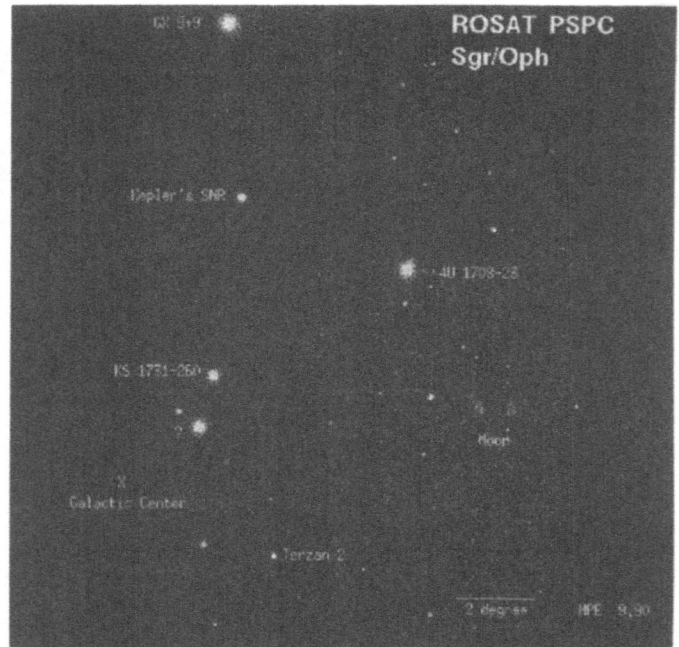

Figure 3 X-ray survey image of the Sagittarius Ophiuchus region.

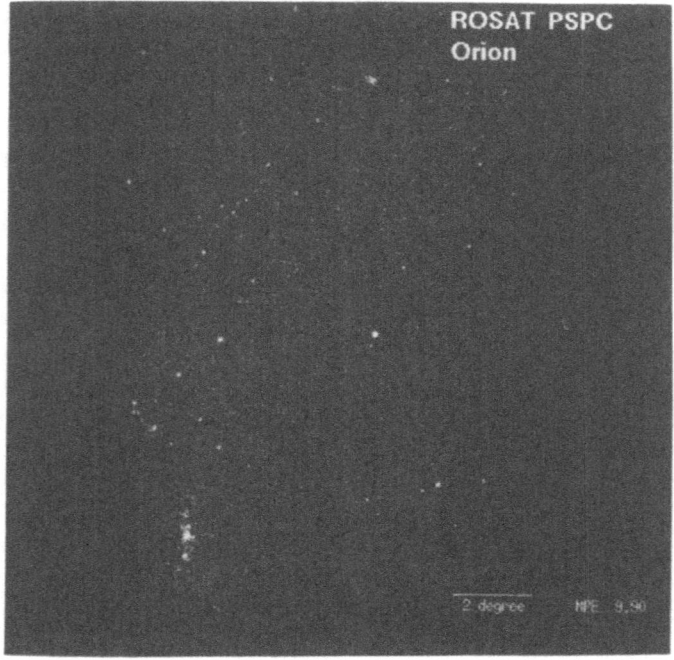

Figure 4 X-ray survey image of the Orion region.

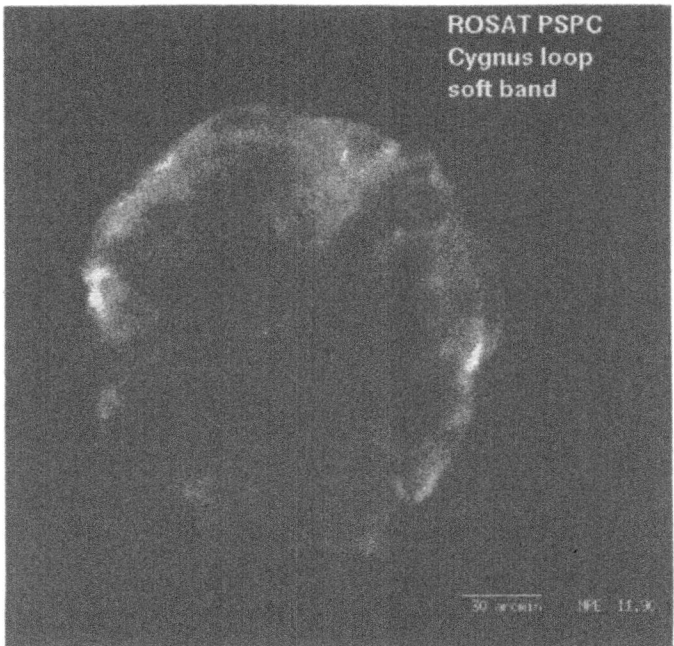

Figure 5 X-ray survey image of the Cygnus Loop (0.1–0.28 keV).

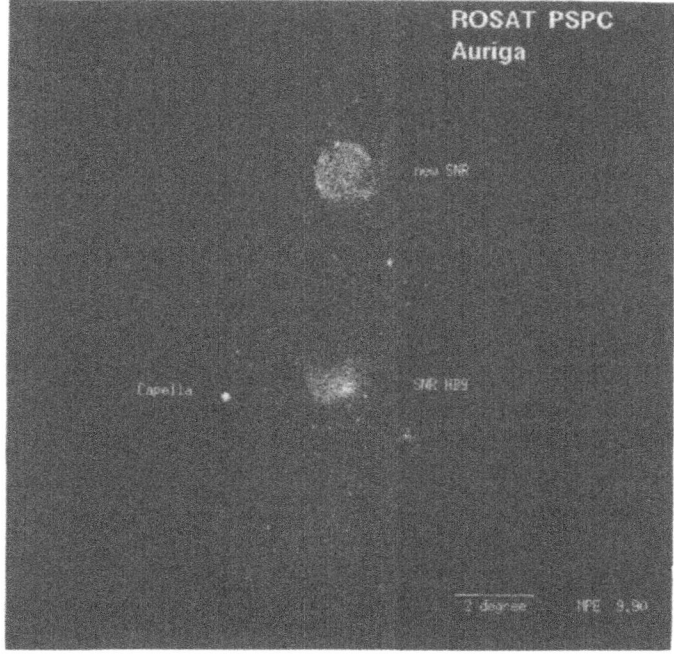

Figure 6 A new Supernova Remnant discovered close to HB9 and Capella in the X-ray survey.

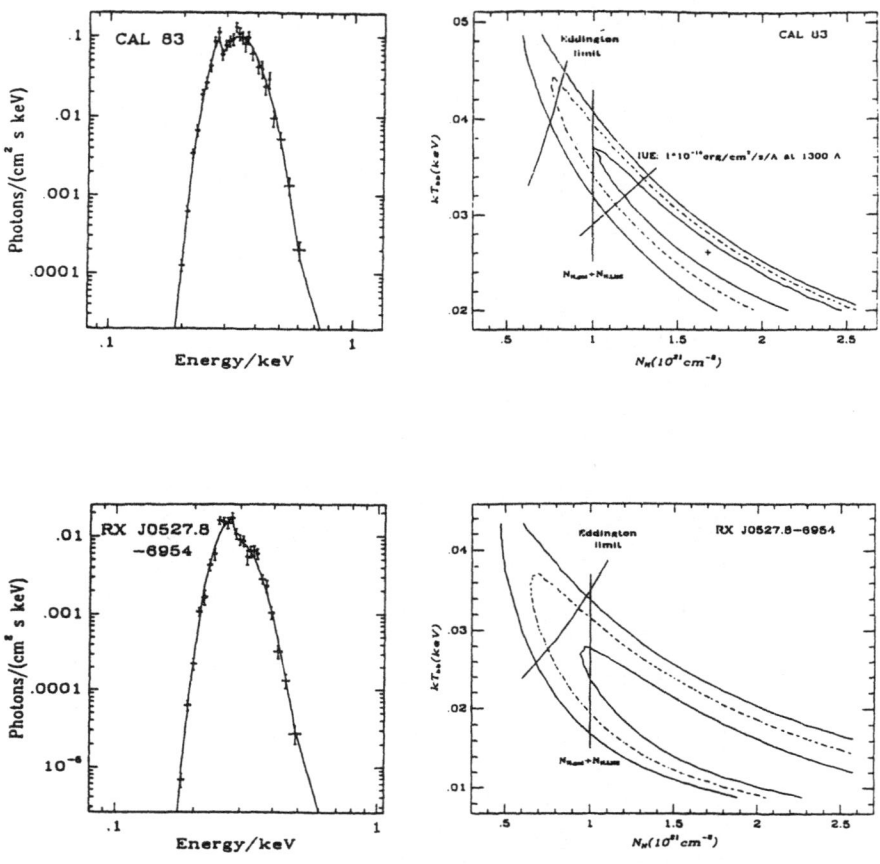

Figure 7 Photon spectrum (top) and error ellipse for the black-body temperature versus N_H (right) of CAL 83. The Eddington and the IUE limit assume a LMC distance. The contours in b correspond to the 68 %, 95 %, and 99 % confidence levels. Photon spectrum (bottom) and error ellipse for the black-body temperature versus N_H (right) of the new supersoft source RX J0527.8−6954. The Eddington limit assumes a LMC distance. The contours in b correspond to the 68 %, 95 %, and 99 % confidence levels.

Large Magellanic Cloud

The LMC was our "first light" target because we were eager to look at SN 1987A. A MPE rocket flight in August 1979 had given an upper limit of the soft X-ray flux [4]. Despite the much larger collecting power and longer observation time with ROSAT again only an upper limit could be obtained ($L_X < 3 \times 10^{34}$ erg/s) which was confirmed by later scanning observations. The null result is consistent with the fact that the X-ray emission from the SN shock wave is still weak because of the low density of the progenitor's stellar wind. At the same time it means that no neutron star activity [5] can be seen so far.

On the other hand a total number of 45 sources was detected in the 8 square degree field [6]; 34 had been known from EINSTEIN and EXOSAT observations. Among the 11 new sources there is an extremely soft bright source which is a transient because it was not seen by the EINSTEIN observatory. The spectra of this source (RX I0527.8−6954) and of CAL 83, which is probably a low mass X-ray binary, are very similar. The spectra of both sources indicate temperatures in the range 20–40 eV and hydrogen column densities consistent with a LMC membership (figure 7). The luminosities of both sources are close to the Eddington limit. They may represent a new class of low mass X-ray binaries which remained largely undetected so far. If true, this could help solve the millisecond birthrate problem [7].

Neutron Stars

Because of the instrument characteristics of the ROSAT PSPC, in particular its high sensitivity, low background and spectral resolution, this instrument is well-suited for a number of neutron star studies:

- A search for thermal emission from neutron stars,
  1. by looking at all radio pulsars,
  2. by searching for point sources in SNR,
  3. by searching for weak supersoft sources.
- A search for X-ray emission from gamma-ray bursters.
- A synoptic investigation of all X-ray binaries with respect to their soft X-ray spectra and time variability.

These investigations will be performed both using survey data and in the course of the guest observer programme. So far no concrete results are available with the exception of the supersoft LMC sources discussed above.

Acknowledgement

The ROSAT project is supported by the Bundesministerium für Forschung und Technologie (BMFT), by NASA, and by SERC. I acknowledge the contributions of many people in DARA, in DLR, in industry and in the participating scientific institutes who helped make ROSAT possible.

I gratefully acknowledge the efficient mission support provided by the ROSAT team of the German Space Operations Center (GSOC). In particular, I would like to thank many collegues at the Max Plack Institut für Extraterrestrische Physik who worked with dedication on the ROSAT hardware and software.

References

[1] J. Trümper, 1983, *Adv. Space Res.* **2**, 241

[2] W. Voges, private communication

[3] B. Aschenbach, to be published in *Astron. Astrophys. Letters*

[4] B. Aschenbach *et al.*, 1987, *Nature* **330**, 232

[5] F. Pacini, 1987, *ESO Conf. on SN 1987A*

[6] J. Trümper *et al.*, to be published in *Nature*

[7] J. Greiner *et al.*, to be published in *Astron. Astrophys. Letters*

BLACK HOLES IN BINARY SYSTEMS

Jeffrey E. McClintock
Harvard-Smithsonian Center for Astrophysics
60 Garden Street
Cambridge, MA 02138, USA

ABSTRACT. The best evidence for black holes comes from studies of three x-ray binaries: Cyg X-1, LMC X-3 and A0620-00. Recent spectroscopic studies of A0620-00 have constrained the ratio of the primary mass to the secondary mass. Two color light curves of A0620-00 are presented for the period 1981-1989. The spectra of Cyg X-1 and two additional sources (Nova Muscae 1991 and 1E1740.7-2942) occasionally contain features near 0.5-1 MeV. These features, which are probably due to e^+e^- annihilation, may prove to be a reliable signature of a black hole. Cyg X-1, LMC X-3 and A0620-00 are excellent black hole candidates. Nevertheless, the evidence for black holes is indirect and therefore not conclusive. The task remains of discovering a relativistic effect that is peculiar to black holes alone.

1. Introduction

There is no decisive evidence for black holes, although it is presumed they exist. On the one hand, general relativity strongly favors their formation: In a neutron star the internal pressure forces, which normally forestall stellar collapse, also act as gravitating matter to *promote* collapse (Misner et al. 1973). For a massive neutron star this effect is sizable and it leads to a mass limit of ~ 3 $M_\odot$, a limit that does not depend on the unknown properties of nuclear forces at high density (Zwicky 1939; Rhoades and Ruffini 1974; Chitre and Hartle 1976). On the other hand, however, general relativity may fail in the strong field of a neutron star or a black hole ($GM/Rc^2 \sim 1$), since the theory has been tested only in weak fields ($GM/Rc^2 \lesssim 10^{-6}$).[1] Nevertheless, our working hypothesis is that black holes exist and are described by general relativity. Whether or not this hypothesis is true can be decided only by observation.

---

[1] The surface potential of the white dwarf 40 Eri B is $GM/Rc^2 \approx 8 \times 10^{-5}$ and the gravitational redshift has been measured to a precision of 5% (Wegner 1989). However, a gravitational redshift experiment is really not a test of general relativity; it is a test of the Einstein equivalence principle, and hence a test of all metric theories of gravity (Will 1984).

E. P. J. van den Heuvel and S. A. Rappaport (eds.), X-Ray Binaries and Recycled Pulsars, 27–36.
© 1992 Kluwer Academic Publishers.

2. The Three Leading Black Hole Candidates

Currently the clearest evidence for black holes comes from dynamical studies of three x-ray binaries: Cyg X-1, LMC X-3 and A0620-00. Three lines of evidence make the case that these binaries contain black holes: (1) A luminous ($\sim 10^{38}$ erg s^{-1}) and rapidly variable x-ray source establishes beyond doubt that each binary contains a compact object at least as dense and massive as a neutron star. (2) The mass function of the optical star and some additional data imply that the mass of the compact object exceeds $\sim$3 M$_\odot$. (3) According to theory, a neutron star more massive than 3 M$_\odot$ is unstable and will collapse to form a black hole.

The nominal characteristics of the three binary systems are summarized in Table 1. (For references see Fig. 1.) The x-ray luminosities are the maximum that have been observed. The visual magnitude of A0620-00, an x-ray nova, is for the quiescent state. For all three systems, the orbital solutions are consistent with circular orbits, and x-ray eclipses are not observed.

The most important and reliable observable is the optical mass function (Mc-Clintock 1991):

$$f(M) \equiv (M_x \sin i)^3 \ / \ (M_x + M_c)^2 = PK^3/2\pi G,$$

M_x and M_c are the masses of the x-ray source and the optical companion star, respectively, and i is the orbital inclination angle. The value of the mass function, $PK^3/2\pi G$ (Table 1, bottom line), depends only on the orbital period, P, and the velocity semiamplitude, K. Thus the mass function is a cubic equation in M_x, with M_c and i as parameters.

For the discussion at hand, the following fact is crucial: M_x cannot be less than the value of the mass function, i.e., $M_x \geq f(M)$. This rock-bottom limit on M_x corresponds to a system with a zero-mass companion ($M_c = 0$) viewed at the maximum inclination angle ($i = 90°$). Thus, based solely on the observed values of the mass function (Table 1), one obtains the following 3σ lower limits on the masses of the compact x-ray sources: M_x (Cyg X-1) > 0.22 M$_\odot$, M_x (LMC X-3) > 1.4 M$_\odot$, and M_x (A0620) > 2.7 M$_\odot$. Therefore, A0620-00 is the only candidate that qualifies as a probable black hole by virtue of having a large mass function. The other candidates require additional supporting evidence (e.g., large values of M_c and/or small values of i).

TABLE 1

PROPERTIES OF THREE BLACK HOLE BINARIES

	Cyg X-1	LMC X-3	A0620-00
L_x(erg s^{-1})	2×10^{37}	3×10^{38}	1×10^{38}
MK type	O9.7I$_{ab}$	B3V	K5V
d(kpc)	2.5(?)	55	1(?)
m_v	9	17	18
e^a	0.00 ± 0.01^b	0.13 ± 0.05	0.01 ± 0.01
K(km s^{-1})	74.7 ± 1.0^b	235 ± 11	443 ± 4
P(days)	5.6	1.7	0.32
f(M/M$_\odot$)	0.25 ± 0.01	2.3 ± 0.3	2.91 ± 0.08

^{a}Orbital eccentricity.

^{b}The values of e and K for Cyg X-1 were derived from the high-excitation line data (Gies and Bolton 1982).

A comparison of the three candidates, based on published models, is shown in Figure 1. In each case the model depends on the mass function and additionally on two less reliable parameters, the orbital inclination angle, i, and the mass of the companion star, M_c. The value of M_c is determined primarily from the MK spectral class of the companion star, a method that is reasonably reliable if a star is normal. However, the companion star in an x-ray binary system has had a violent and uncertain history and cannot be presumed normal. Moreover, in pulsating x-ray binaries, the companion is often found to be undermassive by a factor of two or more (Hutchings et al. 1979). Conceivably, the companions of black holes are severely undermassive (van den Heuvel 1983). This possibility is not too important in the case of A0620-00, since the value of M_x depends weakly on the value of M_c. It may be important, however, for LMC X-3 and Cyg X-1 (Mazeh et al. 1986; Trimble et al. 1973). (For a further discussion of the dynamical data see McClintock 1991.)

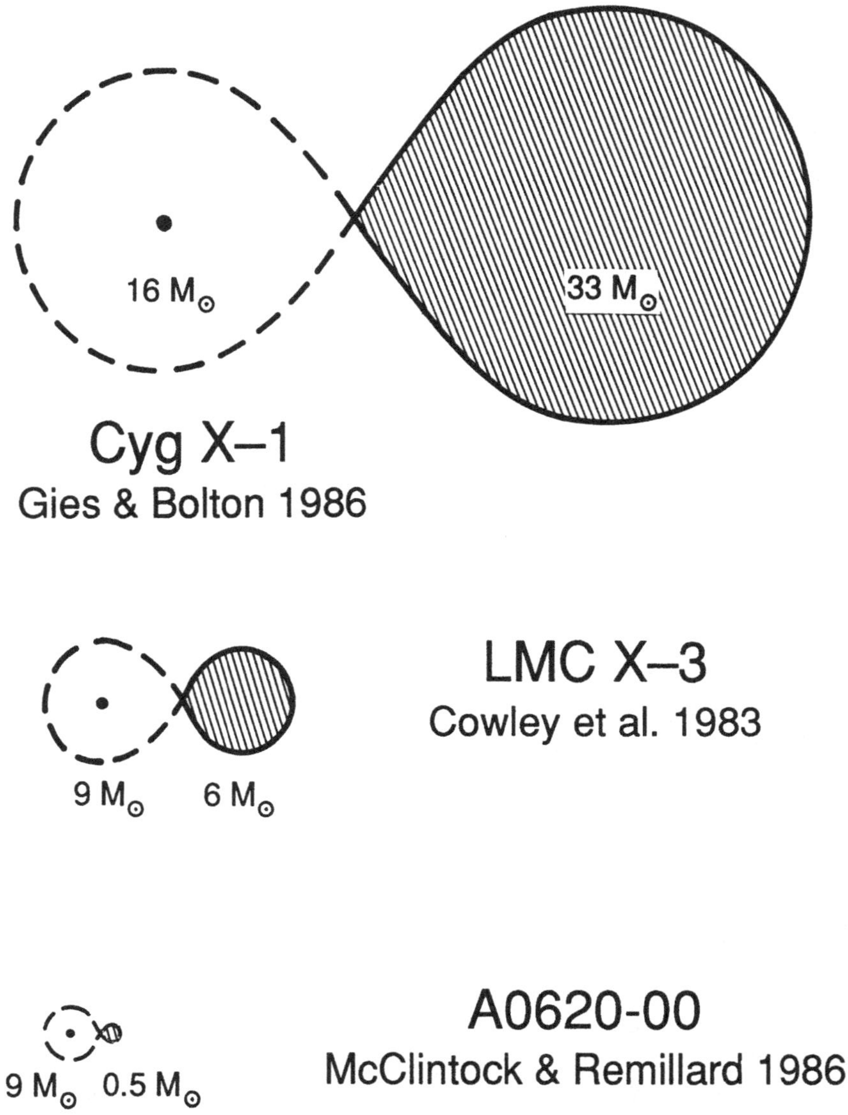

30

Cyg X–1
Gies & Bolton 1986

16 M$_\odot$ 33 M$_\odot$

LMC X–3
Cowley et al. 1983

9 M$_\odot$ 6 M$_\odot$

A0620-00
McClintock & Remillard 1986

9 M$_\odot$ 0.5 M$_\odot$

Figure 1. Schematic sketch, to scale, of plausible models for three dynamical black hole candidates. The optical companions (shaded regions) are shown filling their critical Roche equipotential lobes. The value of the mass function (Table 1) and the masses given in the figure determine the following values of the inclination angle: i(Cyg X-1) = 32°, i(LMC X-3) = 63°, and i(A0620-00) = 45°. In addition to the references in the figure see Gies and Bolton (1982) and McClintock and Remillard (1990).

2.1. A0620-00

The transient x-ray source A0620-00 is probably the most secure black hole candidate because of the large value of its mass function (e.g., Watson 1986; Bailyn 1990). The value of M_x depends weakly on the mass of the K5V secondary and strongly on the orbital inclination angle. For example, if $M_c = 0.5 \ M_\odot$, then $M_x(90°) = 3.7 \ M_\odot$, $M_x(60°) = 5.4 \ M_\odot$, and $M_x(40°) = 11.9 \ M_\odot$. Conversely, the inclination angle is well determined by the mass of the black hole via the value of the mass function.

The x-ray nova A0620-00 (1917, 1975) was first detected on 1975 August 3. By mid-August the source was ~50 times as bright as the Crab Nebula. Thereafter, its x-ray intensity decayed with an e-folding time of ~1 month, and by 1978 its x-ray luminosity had fallen by more than 6 orders of magnitude (Long et al. 1981). Of special interest is an x-ray precursor, which was observed for about one day during the rise to maximum; it had a hard spectrum (kT ~30 keV) compared to the very soft spectrum observed during outburst (kT ~1 keV; Ricketts et al. 1975).

Near the time of x-ray maximum, A0620-00 was identified with a 12th-magnitude blue star, which is now known as V616 Mon. Over the course of a year, the star faded to its present magnitude, V ~18.3. In quiescence, about half the V-band light of V616 Mon comes from the K-dwarf secondary; the other half comes from an accretion disk, as evidenced by a blue continuum and double-peaked Balmer emission superimposed on the spectrum of the secondary (Oke 1977; McClintock and Remillard 1986; Johnston et al. 1989; Haswell and Shafter 1990).

In the following we discuss some recent results of optical studies. First, we consider three techniques that have yielded constraints on the mass ratio, M_x/M_c; the first two make use of the H_α emission line from the accretion disk, and the third technique uses the absorption-line spectrum of the K dwarf. Finally, we describe work in progress that is aimed at constraining the orbital inclination angle, i.

Johnston et al. (1989) modeled the optical emission line profiles (principally H_α) arising from the accretion disk. Using their determination of K_c (468 ± 44 km s^{-1}), they concluded that $M_x/M_c \gtrsim 13.3$, which implies $M_x > 6.6 \ M_\odot$ for a plausible, assumed value of $M_c = 0.5 \ M_\odot$ (limits are 3σ). Their result strengthens the case for a black hole; however, the result should be viewed with reserve for two reasons. First, if one uses a more refined value for K_c (Table 1), then one finds an extraordinarily large value of the mass ratio, $M_x/M_c \gtrsim 40 \ (3\sigma)$, which disagrees with the determination quoted below. Second, the model for the profile of disk emission lines is known to have shortcomings (Horne and Marsh 1986).

Haswell and Shafter (1990) inferred the orbital radial velocity variations of the compact star from observations of the H_α disk emission line. They report $K_x =$

43 ± 8 km s^{-1}, which implies $M_x/M_c = 10.3 \pm 1.9$ (for $K_c = 443$ km s^{-1}). They conclude that the *minimum* masses (for $i = 90°$) are $M_x = 3.50 \pm 0.21$ M$_\odot$ and $M_c = 0.34 \pm 0.08$ M$_\odot$. This result is less model dependent than the result discussed above; nevertheless, the extraction of reliable orbital parameters from observations of the broad, multicomponent H$_\alpha$ line is by no means straightforward, as Haswell and Shafter readily acknowledge.

The rotational broadening of the K dwarf's absorption lines also gives a measure of $M_x/M_c(\equiv q)$, if one makes the reasonable assumption that the secondary is in corotation:

$$v \sin i/K_c \leq 0.462 \ (1 + q)^{2/3}/q$$

The [in] equality holds if the K star [under] fills its Roche lobe. McClintock and Remillard (1990) measured $v \sin i$ and found $M_x/M_c > 9.1$, which corresponds to $M_x > 4.5$ M$_\odot$ for an assumed value of $M_c = 0.5$ M$_\odot$ (limits are 3σ). All three of the above constraints on M_x/M_c support a black hole model for A0620-00 provided $M_c \gtrsim 0.3\text{-}0.4$ M$_\odot$.

As pointed out above, the orbital inclination angle is the parameter with the most leverage on the value of M_x. Consequently, we are attempting to constrain i by modeling the light curves. Several 2-color light curves, which Ronald Remillard and I have collected between 1981 and 1989, are shown in Figure 2. The color bands are B+V and I, which are centered at 5200Å and 9000Å, respectively. All of the intensity values are normalized to those of a nearby field star. Phase 1.0 corresponds to conjunction, with the K star viewed behind the black hole.

Two observations: First, the blue continuum decreased by $\sim 20\%$ between 1981 and 1989, while the I-band flux appears roughly constant. This may indicate a reduction in the size or the temperature of the accretion disk. Second, the light curve has evolved significantly during the last few years: in 1988, the primary and secondary minima were about equal, while in 1989, the secondary minimum became significantly deeper than the primary minimum. In 1990 and 1991 the light curves (not shown) have reverted to their earlier form. The near-normal ellipsoidal variations observed during 1981-1987 are consistent with a fairly straightforward model of the binary system in which the K dwarf is tidally distorted by its massive partner. Such a model depends on only three parameters: i, M_x/M_c, and the fraction of the Roche lobe which is filled by the secondary star. At present we have no satisfactory explanation for the changes in the appearance of the light curves during 1988-89.

2.2. CYGNUS X-1

There has been renewed interest in the high-energy spectra of Cyg X-1 and a few

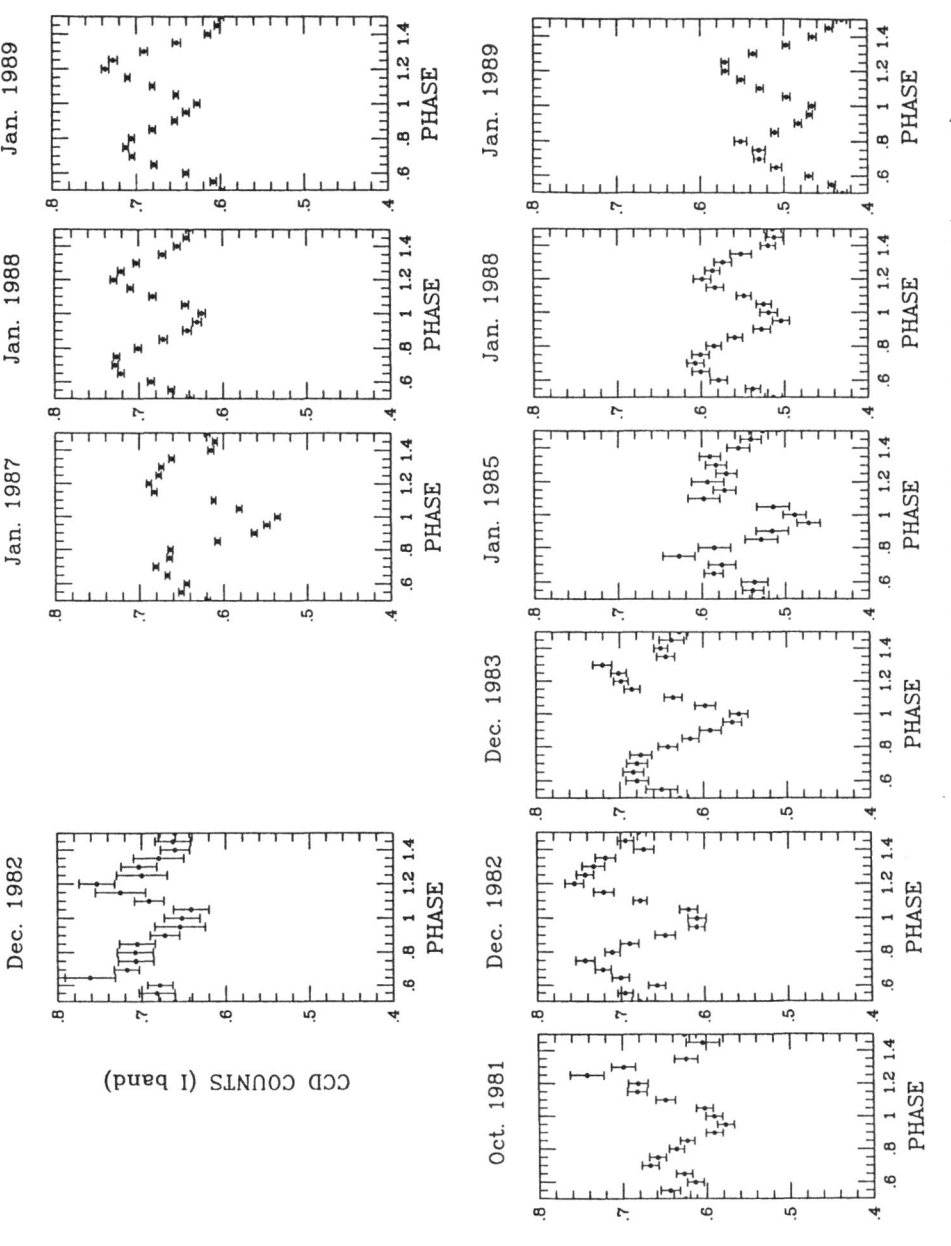

Figure 2. The light curves of A0620-00 folded module the 7.8-hour orbital period (see text).

other x-ray binaries. A reanalysis of HEAO-3 γ-ray data for Cyg X-1 has revealed a complex 3-state spectrum which extends to $\sim$1.5 MeV (Ling et al. 1987). Most remarkable is the spectrum of the low (γ_1) state which contains a strong, broad "bump" centered at $\sim$1 MeV. The bump has been interpreted as a emission from a $\approx$400 keV plasma that is dominated by positron-electron pairs (Liang and Dermer 1988). Despite the good evidence for a hot pair plasma, there is only very weak evidence (1.9σ) that Cyg X-1 emits a narrow 511 keV annihilation line (Ling and Wheaton 1989).

Recently the SIGMA/GRANAT experiment has pinpointed the locations of two additional x-ray sources that probably contain e^+e^- pair plasmas: 1E1740.7-2942 (Mandrou et al. 1990) and Nova Muscae 1991 (GRS1124-68; Sunyaev et al. 1991a,b). On occasion both sources have exhibited spectra that can be described roughly as follows: the spectra fall in a normal fashion below $\sim$200 keV, flatten above $\sim$200 keV, rise to a shoulder at $\sim$500 keV, and then fall again above 500 keV. The broad feature at $E \lesssim 500$ keV is presumably due to the decay of singlet (511 keV line) plus triplet (3-photon continuum) positronium as observed with the modest resolution of a NaI scintillator ($\Delta E/E \sim 10\%$). The spectrum of Nova Muscae 1991 is reported as containing two line features, including one near 500 keV. The source 1E1740.7-2942 (located at $l = 359.1°$, $b = -0.1°$) is widely regarded as the probable counterpart of the narrow-line 511 keV source, that was observed repeatedly from the direction of the Galactic Center during the 1970's and 1980's (Leventhal 1987; Schwarzschild 1991).

It is reasonable to regard these 0.5-1 MeV spectral features, which are apparently due to e^+e^- annihilation, as a plausible black hole signature because Cyg X-1 is a top black hole candidate and because black hole models for the 511 keV Galactic-Center source have been favored strongly for many years (Lingenfelter and Ramaty 1989). It appears that this annihilation feature is present only infrequently: the 1 MeV bump in the spectrum of Cyg X-1 has been observed just once, and a $\sim$0.5 MeV feature was present during only one of six observations of 1E1740.7-2942 (Mandrou et al. 1991).

2.3. LMC X-3

The x-ray intensity of LMC X-3 is modulated by a factor of $\sim$3-4 with a cycle time of $\sim$99 days (or possibly 198 days; Cowley et al. 1991). Four consecutive maxima are evident in $\sim$500 days of HEAO-1 archival data, although one predicted maximum ($\sim$JD2443420) apparently did not occur (R. Remillard, private communication). Additional cycles have been observed with Ginga.

Cycles of $\sim$30-300 days are observed in several other x-ray binaries (Priedhorsky and Holt 1987). The best and most relevant examples are LMC X-4 (30.4 d), Her X-

1 (35 d), SS433 (164 d) and Cyg X-1 (294 d). It is thought that all of these systems contain a precessing element — the outer rim of an accretion disk, a misaligned companion star, or a neutron star. An important clue to the operative mechanism in LMC X-3 should be the positive correlation that is observed between the source intensity and the spectral hardness (1-13 keV).

3. Conclusion

Cygnus X-1, A0620-00 and LMC X-3 are excellent black hole candidates. Nevertheless, the evidence for black holes remains inconclusive because it is indirect. We have not yet discerned a strong-field phenomonen that is unique to black holes. Possibly some new observation of known candidates will yield decisive evidence, or possibly new candidates in new settings will offer the chance for a breakthrough. The annihilation-radiation sources (discussed above) appear promising.

Acknowledgements

I am grateful to George Blumenthal, Saul Rappaport, Gary Wegner, and especially Ronald Remillard for helpful discussions.

References

Bailyn, C.D. (1990), Nature 347, 426.

Chitre, D.M., and Hartle, J.B. (1976), Ap.J. 207, 592.

Cowley, A.P., Schmidtke, P.C., Ebisawa, K., Makino, F., Crampton, D., Hutchings, J.B., Remillard, R.A., Kitamoto, S., and Treves, A. (1991), Ap.J. (submitted).

Cowley, A.P., Crampton, D., Hutchings, J.B., Remillard, R., and Penfold, J.E. (1983), Ap.J. 272, 118.

Gies, D.R., and Bolton, C.T. (1982), Ap.J. 260, 240.

Gies, D.R., and Bolton, C.T. (1986), Ap.J. 304, 371.

Haswell, C.A., and Shafter, A.W. (1990), Ap.J.(Letters) 359, L47.

Horne, K., and Marsh, T.R. (1986), MNRAS 218, 716.

Hutchings, J.B., Cowley, A.P., Crampton, D., van Paradijs, J., and White, N.E. (1979), Ap.J. 229, 1079.

Johnston, H.M., Kulkarni, S.R., and Oke, J.B. (1989), Ap.J. 345, 492.

Leventhal, M. (1987), in M.P. Ulmer (ed.), 13th Texas Symposium on Relativistic Astrophysics, World Scientific, Singapore, p. 382.

Liang, E.P., and Dermer, C.D. (1988), Ap.J. (Letters) 325, L39.

Ling, J.C., Mahoney, W.A., Wheaton, Wm.A., and Jacobson, A.S. (1987), Ap.J. (Letters) 321, L117.

Ling, J.C., and Wheaton, Wm.A. (1989), Ap.J. (Letters) 343, L57.

Lingenfelter, R.E., and Ramaty, R. (1989), Ap.J. 343, 686.

Long, K.S., Helfand, D.J., and Grabelsky, D.A. (1981), Ap.J. 248, 925.

Mandrou, P., Roques, J.P., Sunyaev, R., Churazov, E., Paul, J., and Cordier, B. (1990), IAU Circ. No. 5140.

Mazeh, T., van Paradijs, J., van den Heuvel, E.P.J., and Savonije, G.J. (1986), Astron. Astrophys. 157, 113.

McClintock, J.E. (1991), to appear in the Proceedings of the Texas-ESO-CERN Symposium on Relativistic Astrophysics, New York Academy of Science Press, New York.

McClintock, J.E., and Remillard, R.A. (1990), B.A.A.S. 21 (No.4), 1206.

McClintock, J.E., and Remillard, R.A. (1986), Ap.J. 308, 110.

Misner, C.W., Thorne, K.S., and Wheeler, J.A. (1973), Gravitation, Freeman Publishers, San Francisco, p. 605.

Oke, J.B. (1977), Ap.J. 217, 181.

Priedhorsky, W.C., and Holt, S.S. (1987), Space Sci. Rev. 45, 291.

Rhoades, C.E., and Ruffini, R. (1974), Phys. Rev. Lett. 32, 324.

Ricketts, M.J., Pounds, K.A., and Turner, M.J.L. (1975), Nature 257, 657.

Sunyaev, R., Jourdain, E., and Laurent, P. (1991a), IAU Circ. No. 5176.

Sunyaev, R., Jourdain, E., and Goldwurm, A. (1991b), IAU Circ. No. 5201.

Schwarzschild, B. (1991), Physics Today, 44 (March), 17.

Trimble, V., Rose, W.K., and Weber, J. (1973), MNRAS 162, 1P.

van den Heuvel, E.P.J. (1983), in W.H.G. Lewin and E.P.J. van den Heuvel (eds.) Accretion-Driven Stellar X-ray Sources, Cambridge U. Press, Cambridge, p. 303.

Watson, M.G. (1986) Nature 321, 16.

Wegner, G. (1989), in G. Wegner (ed.), Lecture Notes in Physics, 328: White Dwarfs, Springer-Verlag, New York, p. 401.

Will, C.M. (1984), Phys. Rep. 113, 345.

Zwicky, F. (1939), Phys. Rev. 55, 726.

Recent Results from *Ginga* on X-Ray Binaries: Selected Topics

Y. Tanaka
Institute of Space and Astronautical Science
3-1-1 Yoshinodai, Sagamihara, Kanagawa-ken 229
Japan

Abstract Some selected topics on X-ray binaries are discussed based on the recent *Ginga* results. These include (1) the magnetic fields of neutron stars and the question of whether they decay with time, and (2) the black hole candidates and in particular the origin of these in low-mass binary systems.

1. Introduction

The X-ray astronomy observatory *Ginga* [1], launched in February 1987, is nearly completing its fourth year in orbit and continues to function normally (as of January 1991). The main instrument of *Ginga* is an array of large area proportional counters (LAC) with a total of 4000 cm^2, together with an all sky monitor and a gamma-ray burst detector. The pronounced capability of *Ginga* is to obtain high photon statistics over a wide energy range from 2 to 50 keV which covers the most important part of the spectra of the galactic X-ray binaries.   Observations of many objects for a variety of objectives have been conducted. *Ginga* has also been serving for many international investigators; U.S. as well as European scientists in addition to our U.K. collaborators with whom the LAC was jointly built.   The orbital life of *Ginga* is expected to end late 1991 or at most early 1992. Unfortunately, further development of the important science explored with the *Ginga* capability will have to wait until the second half of the 1990s for the launch of *XTE*.

In this paper, I shall discuss a few selected topics based on the recent *Ginga* results.

2. Magnetic Field of Neutron Stars

2.1. Cyclotron resonance features

Murakami *et al*. [2] discovered two "absorption" lines separated by a factor of 2 in energy in the spectrum of a gamma-ray burst. They identified these to be the fundamental and the second harmonic lines of the cyclotron resonance (transition between the quantized Landau levels). Subsequently, the observed structure of these lines was explained satis-

E. P. J. van den Heuvel and S. A. Rappaport (eds.), X-Ray Binaries and Recycled Pulsars, 37–48.

factorily by the radiative transfer theory involving multiple resonant scattering of the "Landau photons" [3][4][5]. This case has made it convincing that the site of gamma-ray bursts is highly ($> 10^{12}$ Gauss) magnetized neutron stars.

As regards the binary X-ray pulsars, a significant spectral feature interpreted as due to the cyclotron resonance was first discovered from Her X-1 by Trümper *et al* [6]. However, whether the structure is due to absorption or emission has been left unsettled. As a matter of fact, the binary X-ray pulsars commonly exhibit a characteristic shape in their energy spectra; a hard power-law like spectrum which is abruptly steepend above a certain energy typically in the range 10 - 20 keV. Previously, binary X-ray pulsar spectra used to be approximated by the following empirical formula [7]:

$$E^{-\gamma} \qquad\qquad\qquad \text{if } E < E_c, \text{ or}$$
$$E^{-\gamma} \cdot exp[-(E-E_c)/E_f] \qquad \text{if } E > E_c,$$

while the physical meaning of the parameters, E_c and E_f was unclear.

The first attempt to interpret the high-energy cut off of the X-ray pulsar spectra in terms of the cyclotron resonance was made with the high-quality spectra obtained from *Tenma* [8]. Recently, thanks to the high statistical accuracy obtainable from the *Ginga* LAC, clear spectral features explicable in terms of the cyclotron resonant scattering have been revealed from as many as ten binary X-ray pulsars. It has been shown that the observed spectra of X-ray pulsars are in general fitted satisfactorily with the following formula [9]:

$$E^{-\gamma} \cdot exp[-H(E)],$$
where $\quad H(E) = A_1(W_1 E/E_1)^2/[(E-E_1)^2+W_1^2]+A_2(W_2 E/E_2)^2/[(E-E_2)^2+W_2^2].$

The second formula represents a classical cyclotron resonant absorption coefficient including the second harmonic, if $E_2 = 2\times E_1$. For the fitting, E_1 and E_2 are dealt with as free parameters. Table 1 lists the results of the fitting in which E_n (E_1 only or both E_1 and E_2) were uniquely determined [9]. Whenever both line energies are determined, E_2 is indeed found to be approximately $2\times E_1$. It is also interesting to note that, even in the cases where only E_1 was determined uniquely (E_2 is presumably out of the *Ginga* range), a better fit was obtained by including the second term of $H(E)$ in the fitting. Examples of these fits are shown in Fig. 1.

Thus, we are convinced that the high-energy cut off characteristic of the binary X-ray pulsar spectrum is formed by a modification due to the cyclotron resonant scattering in a highly magnetized plasma in these systems. Then, the magnetic field strength, B, is obtained from the values of E_1; $B = (E_1/11.6 \text{ keV})\times 10^{12}$ Gauss. It is of great interest to find from the E_1 values in Table 1 that field strengths obtained in this way are confined to the range $(1 - 4)\times 10^{12}$ Gauss.

For those cases in which E_1 is not uniquely determined, we can still estimate the magnetic field strength, though crude, in the following way. As noticed in Fig. 2, E_1 is roughly proportional to E_c [9], hence E_1 can be estimated from E_c within an accuracy of a factor of 2. Since the E_c values of all known binary X-ray pulsars lie below 30 keV,

Table 1. Spectral parameters of binary X-ray pulsars

Source Name	Companion	Orbital Period(d)	Pulse Period(s)	Spectral Parameters E_c(keV)	E_n(keV)
Her X-1	A9-B (HZ Her)	1.7	1.24	17-21	35
4U0115+63	Be-transient	24.3	3.6	7-9	12, 23, 36
X0331+53	Be(BQ Cam)-transient	34.3	4.38	14-17	28.0, $\sim$53
Cep X-4	Be?-transient	?	66.3	15-17	32
4U1907+09	OB or Be	8.4	438	14-16	20
4U1538-52	BOI (QV Nor)	3.74	530	14-16	20
Vela X-1	B0.5Ib (GP Vel)	8.96	283	15-20	27, 53
GX301-2	B1.5Ia (BP Cru)	41.5	690	19-21	40
X2259+58	single?	?	6.9		$\sim$7?, $\sim$13?

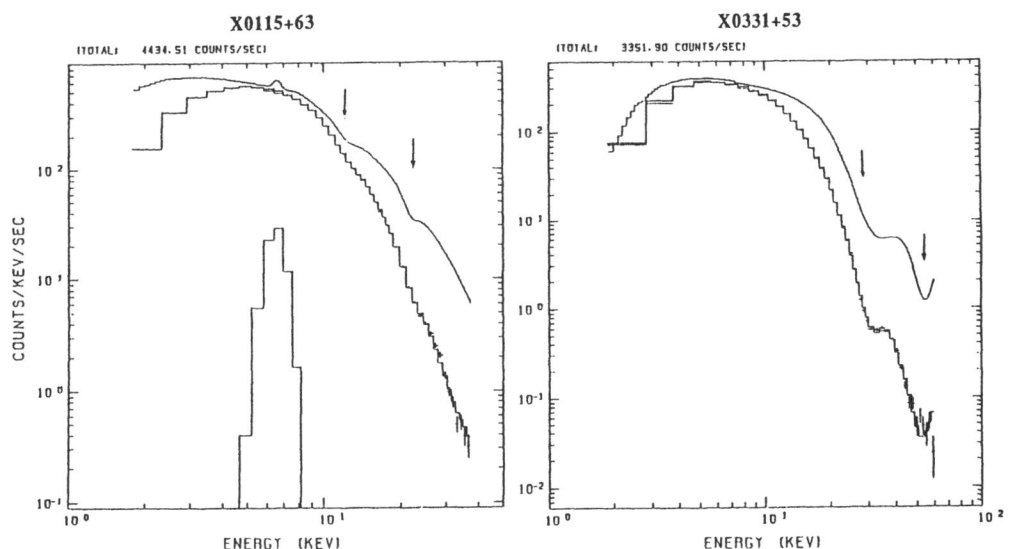

Fig. 1. Examples of the model fitting to the observed binary X-ray
pulsar spectra. Solid curves are the best-fit model spectra,
and the histograms after folding with the counter response.

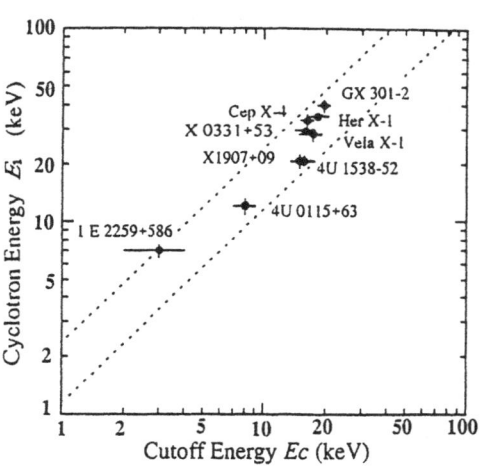

Fig. 2.

Relation between E_c and E_l [9].

few of them would have a magnetic field strength higher than 6×10^{12} Gauss. On the other hand, it is unlikely that X-ray pulsars with magnetic fields smaller than 10^{12} Gauss are numerous, for those with E_l in the range 1-10 keV would hardly escape detection.

2.2. Does magnetic field decay?

It is important to mention that at least two binary X-ray pulsars, X1626-67 and Her X-1, and possibly GX1+4 have low-mass companions. This fact and the concentration of the field strength within $(1 - 4) \times 10^{12}$ Gauss would imply that the magnetic field of neutron stars does not decay substantially within 10^7 years, or even within 10^8 years.

As regards the gamma-ray burst sources, at least one source is shown to have magnetic field of $\sim 2 \times 10^{12}$ Gauss [2]. Could this highly magnetized neutron star be exceptionally young among other gamma-ray burst sources, much younger than 10^8 years? It is very unlikely. The total number of gamma-ray burst sources in our galaxy is estimated from the observed *log N-log P* relation to be as many as 10^9. Suppose gamma-ray burst sources are single neutron stars (as no gamma-ray burst source has been identified with an X-ray source) created by supernovae. The total number of supernovae over the past 10^{10} years would not be greater than 10^9. Hence, most of the neutron stars created by supernovae must be gamma-ray burst sources, and the chance of picking up a neutron star which is younger than 10^8 years is negligible. This supports the hypothesis that most gamma-ray burst sources are highly magnetized neutron stars, and consequently that the magnetic field does not decay in a period of the order of 10^8 years. They must be slow rotators, otherwise gamma-ray burst sources would be radio pulsars.

Thus, observationally, we have not as yet any evidence that the magnetic field of neutron stars, whether accreting or not, decays within 10^8 years. Therefore, a possibility is still reserved that the neutron stars of millisecond radio pulsars and the low-mass X-ray binaries (LMXB) are generically different than those of ordinary radio pulsars, X-ray pulsars, and the gamma-ray burst sources.

3. Search for Millisecond Pulsations from LMXB

In order that the "recycled pulsar" scenario be valid, neutron stars in LMXBs must be millisecond rotators. Extensive attempts have been made to search for millisecond X-ray pulsations from LMXBs with the *Ginga* LAC. As listed in Table 2, results are so far negative at a level even below 1% in amplitude [10]. However, these results cannot exclude fast rotation of neutron stars in these systems. After the fact, it is not difficult to explain the absence of millisecond modulations in X-rays in terms of several possible effects.

An alternative means to test for millisecond rotation of neutron stars in LMXBs could be to search for radio pulsations from transient LMXBs when they are X-ray dormant (absence of mass accretion). Several candidate sources are available for such an attempt. Since the spin rates of neutron stars in LMXBs should provide a very important clue to the problem of the origin of millisecond radio pulsars, such observations are highly warranted.

Table 2. Limits on pulse fraction in low-mass X-ray binaries

Source	Modulation depth at 50 Hz	upper limit* at 400Hz	Min. orbital period surveyed (hours)
Sco X-1	0.0019	0.0026	(a)
GX340+0	0.0069	0.014	3
GX5-1	0.0039	0.0077	3
	0.0031	0.0042	6
GX9+1	0.0057	0.011	3
GX17+2	0.0055	0.011	3
4U1820-30	0.006		(b)
Cyg X-3	0.012	0.040	(c)
Cyg X-2	0.0051	0.010	(d)

* 95% confidence. (a) known period: 19.2 hr. (b) known period: 685 s.
(c) known period: 4.8 hr. (d) known period: 9.8 d.

4. Transient Sources

4.1. Transient X-ray pulsars

We have discovered seven transient X-ray pulsars with *Ginga* thus far [11]. These new X-ray pulsars have been detected mostly from several scanning observations with the LAC along the glactic plane, and are of relatively low flux values. The distribution of the binary X-ray pulsars, including these new ones, is shown in Fig. 3. As noticeable in the figure, nearly a half of them are transient sources (marked *), of which several are known to be recurrent. In fact, most of the recurrent X-ray pulsars are known to be Be-binaries with highly eccentric orbits [11]. Their recurrence is explained as due to temporary accretion which occurs when the neutron star passes near the periastron or crosses the equatorial plane of the Be companion where the stellar wind density is the largest.

We suspect, from the rate of new detections, that there may still be a considerable number of transient X-ray pulsars yet to be found in our galaxy.

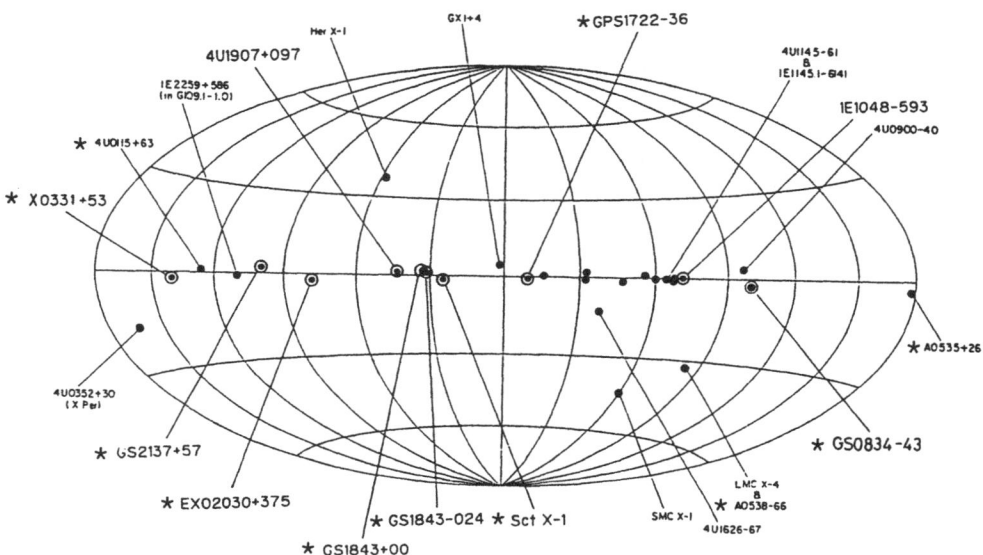

Fig. 3. Galactic distribution of binary X-ray pulsars. Transient pulsars are marked with asterisks (*).

4.2. Transient black hole candidates

In the past four years, three extremely bright transient sources have been detected by the *Ginga* All Sky Monitor; GS2000+25 (April, 1988)[12], GS2023+33 (May, 1989) [13], and GS1124-68 (January, 1991) [14] which was also discovered simultaneously by *GRANAT* [15]. These three sources are all low-mass binaries of which the compact objects are suspected to be black holes, as discussed below.

The X-ray light curves of these three transients are shown in Fig. 4, together with that of a black hole candidate A0620-00 [16]. These light curves are strikingly similar to each other. Excluding a secondary

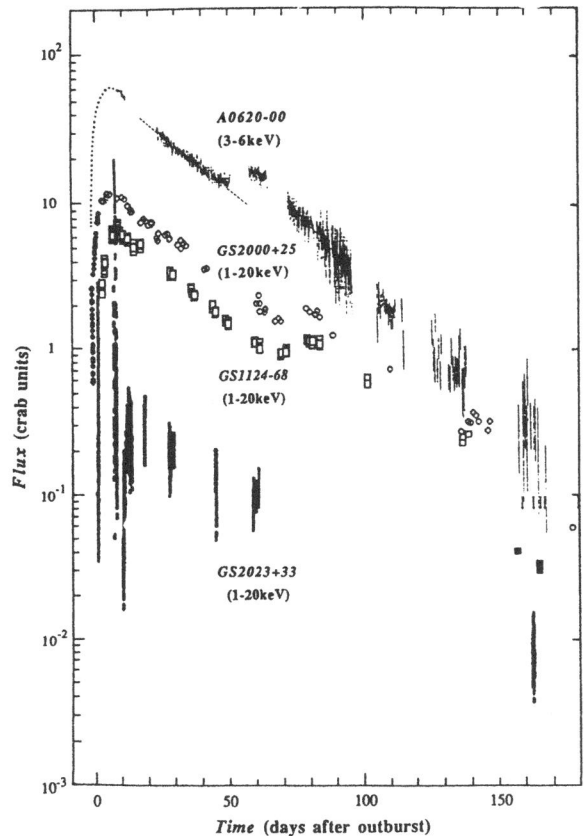

Fig. 4.

X-ray light curves of three bright transiens, together with that of A0620-00 [16].

enhancement seen in A0620-00 and GS2000+25, the decay curves are all nearly exponential with essentially the same e-folding time of 30 - 40 days. Although what governs the decay time is not fully understood yet, this fact may probably be indicating close similarities among these systems. The optical counterparts of these transients have been identified with very faint stars which underwent a dramatic brightening during the X-ray outbursts by several V-magnitudes [17][18][19]. Hence, the companions are obviously low-mass (probably late K) stars. Associated radio outbursts were also reported [20][21][22]. The nature of the transient outbursts of low-mass companion systems is considered to be quite different from that of massive companion systems (e.g. Be transients), which is to be investigated as an important subject by itself.

There are several galactic binary sources which have so far been considered as black hole candidates, while genuine black hole signatures are not confirmed as yet. The strongest support for considering a source to be a black hole has been a lower limit on mass of the compact object which exceeds 3Mo, the absolute theoretical upper limit for a stable neutron star. According to McClintock [23], among four black hole candidates (Cyg X-1, LMC X-1, LMC X-3, and A0620-00) for which lower limits on the mass of the compact objects are available, those for Cyg X-1 (>3.4Mo) and A0620-00 (>4.5Mo) are most reliable. Qualification of the other black hole candidates are more or less based on their characteristic X-ray properties which either Cyg X-1 or A0620-00 exhibits and are distinct from those of the X-ray binaries known to include neutron stars [24]. As listed in Table 3, Cyg X-1 exhibits an energy spectrum approximated by a single power law, and rapid (down to millisecond time-scales) chaotic fluctuations (flickering). A0620-00 shows a distinct energy spectrum which is composed of an ultrasoft component (much softer than those of neutron star LMXB) and a hard power-law tail. As a matter of fact, GS2023+33 closely resembles Cyg X-1, and both GS2000+25 and GS1124-68 resemble A0620-00 with respect to their characteristic X-ray properties.

Table 3. X-ray properties of the black hole candidates

Spectrum	Time Variation	Sources
Hard-state		
Approximately a single power-law	Flickering	Cyg X-1* GX339-4* GS2023+33 GS1826-24
Ultrasoft-state		
Ultrasoft + hard tail	Little flickering	Cyg X-1* GX339-4* LMC X-3 LMC X-1 A0620-00 GS2000+25 GS1124-68

* Occasionally switches between the hard- and ultrasoft-states.

The X-ray properties of the above three bright transients have been investigated extensively in comparison with the previously known black hole candidates with the *Ginga* LAC. Concerning the A0620-00 type energy spectrum, the shape of the ultrasoft component is clearly characteristic of a thermal origin, and even of blackbody nature. In general, the ultrasoft component can be well fitted with the multicolor blackbody disk model developed by Mitsuda *et al.* [26] which qualitatively represents the emission spectrum expected for an optically-thick accretion disk. This model assumes that the gravitational energy released is locally dissipated into blackbody (or quasi-blackbody) radiation and that the accretion flow is continuous throughout the disk. Then, the temperature $T(r)$ will depend on the distance r from the center as $r^{-3/4}$, and the energy spectrum is given by the surface integral of the blackbody spectrum $B[E, T(r)]$. This model includes two parameters; $r_{in}^2 cos\theta$ with r_{in} and θ being respectively the innermost radius and the inclination angle of the disk, and T_{in} the temperature at r_{in}.

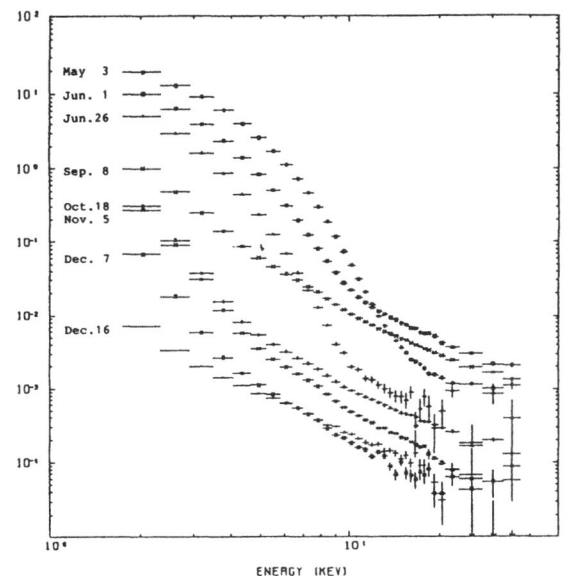

Fig. 5.

Evolution of the energy spectrum of GS2000+25.

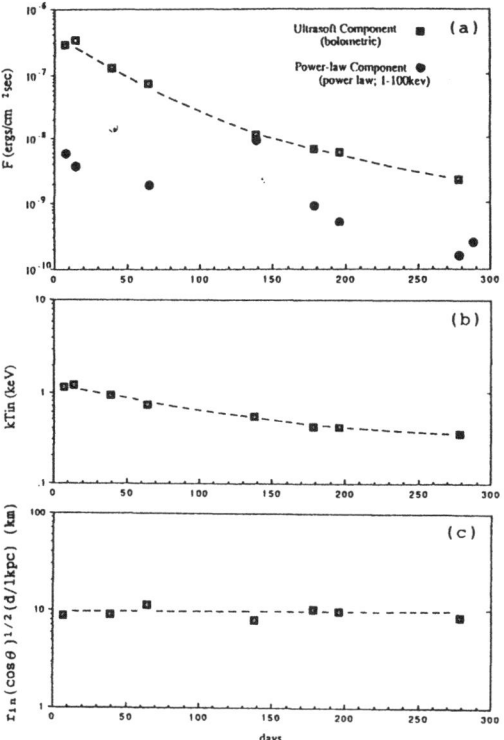

Fig. 6. Time histories of spectral parameters of GS2000+25.

Figure 5 shows the evolution of the energy spectrum of GS2000+25 with itme. Evidently, these spectra are composed of (1) an ultrasoft component which softens gradually as the intensity decreases and (2) a hard power-law component. The time history of the parameter values determined from fitting with the multicolor blackbody disk model is shown in Fig. 6. The bolometric flux of the ultrasoft component monotonously decreases with time, representing a gradual decrease of $\dot{M}$. On the other hand, the flux of the power-law component changes independently of the ultrasoft flux. The origin of the power-law component and its variation still remain uncertain [24].

An important finding concerning the ultrasoft component is that the value of $r_{in}(\cos\theta)^{1/2}$ is found to remain remarkably constant while the bolometric ultrasoft flux ($\propto \dot{M}$) decreased through two orders of magnitude. This invariance of $r_{in}(\cos\theta)^{1/2}$ against large changes of $\dot{M}$ is found to hold for LMC X-3 [27] and, though preliminary, for the most recent ultrasoft transient GS1124-68. Although the face value of this parameter should not be taken seriously, the invariance of it over a wide range of M must imply something very important. We believe that the parameter r_{in} is representing (proportional to) the radius of the innermost stable orbit, $3r_g$ ($r_g = 2GM/c^2$), since this is the only length which could possibly stay constant. The best-fit values of $r_{in}(\cos\theta)^{1/2}$ are listed in Table 4 for those sources with an ultrasoft component + a hard tail for which the distance estimates are available.

Table 4. Values of $r_{in}(\cos\theta)^{1/2}$

Black Hole Candidates			Neutron Star LMXBs		
Source	Distance (kpc)	$r_{in}(\cos\theta)^{1/2}$ (km)	Source	Distance (kpc)	$r_{in}(\cos\theta)^{1/2}$ (km)
LMC X-1	50	~40 [27]	4U1608-52	4.1*	6.2[26]
LMC X-3	50	24 [27]	4U1636-53	6.9*	8.0
A0620-00	~1	25-30(d/1kpc)+	4U1820-30	8.5	6.1
GS2000+25	2-3[25]	25(d/2.5kpc)	Sco X-1	~0.7	4.1(d/0.7kpc) [26]
GS1124-68	~3	30(d/3kpc)	LMC X-2	50	10±1

+ Estimated from the spectrum
   shown in Ref.[29]

* Estimated from peak luminosity of
   X-ray bursts, for $M_{NS}=1.4M_0$ [28].

How about neutron star LMXBs? The only fundamental difference from an accreting black hole is the presence of a solid surface. Since matter while accreting will not know if the compact object is a neutron star or a black hole, an ultrasoft component from the disk is expected to be present also in the neutron star LMXBs of which the magnetic field is considered to be weak. Indeed, it has been shown that the energy spectra of typical neutron star LMXBs are composed of an ultrasoft component and a blackbody component [26], the latter being expected for the emission from the neutron star surface. Table 4 also contains the values of $r_{in}(\cos\theta)^{1/2}$ determined for several neutron star LMXBs for which the distance estimates are available.

Remarkably, one notices in Table 4 a large systematic difference by more than a factor of two in the value of $r_{in}(\cos\theta)^{1/2}$ between the black hole candidate sources and the neutron star sources. Based on the above interpretation that r_{in} represents $3r_g$, this result indicates that the mass of the compact objects of these black hole candidates are twice or more that of a neutron star; 3Mo or more for the neutron star mass of 1.4Mo. In fact, the mass of the compact object of A0620-00 estimated this way is consistent with the mass lower limit obtained optically [23]. The discrepancy between the $r_{in}(\cos\theta)^{1/2}$ value and the actual $3r_g$ is no surprise considering the simplified assumptions made in the model, and not difficult to reconcile. For instance, T_{in} is in reality the colour temperature rather than the effective temperature, from which an underestimation of r_{in} by as large as a factor of two is expected, in addition to the effect from inclination (θ) of the disk.

Thus, we believe that this parameter provides a useful mass indicator of the compact object. Ebisawa et al. [28] reached the same conclusion with a more sophisticated version of the multicolor blackbody disk model. For further confirmation of the validity of this new method for mass estimation, determination of distances of more black hole candidates is extremely valuable.

4.3. How many black hole binaries in our galaxy?

It is interesting to try to estimate how many black hole binaries exist in our galaxy. At least eight very bright transients which may be qualified as black hole candidates have so far been recorded. Seven of them (A0620-00, H1705-25, H1743-32, X1630-47, GS1354-64 [29], GS2000+25, and GS1124-68) showed an energy spectrum composed of an ultrasoft component and a power-law tail, and one (GS2023+33) exhibited a single power-law spectrum and pronounced flickering (Cyg X-1 type). (See [30] for a previous review on X-ray transients.) The occurrence rate of such bright transients has been roughly 0.5 - 1 per year. Except for X1630-47 (600-day recurrent), the reccurrence periods of them are unknown. GS1354-64 was detected a few times [29]. However, no X-ray outburst was previously recorded from the others in the last 20 years. Considering the coverage by the Vela satellites, Ariel V All Sky Monitor, and wide field of view monitoring from Hakucho, Tenma, and Ginga, recurrence periods shorter than 10 years seem unlikely. On the other hand, at least two of them, A0620-00 and GS2023+33, were spotted previously as optical novae; A0620-00 (V616 Mon) in 1917 and GS2023+33 (V404 Cyg) in 1938, although intermediate outbursts may have been missed. From these facts, it would be fair to assume an average recurrence period of 10 - 50 yrs. It is rather difficult to evaluate the sky coverage; probability that the field of view of a detector swept the source during an X-ray outburst. We here assume the probability to be between 20 and 50 %. Assuming a peak luminosity of the order a few times 10^{38} ergs/s, most of these eight transients happen to be within about 3 kpc from us. Therefore, the effective "search" volume is of the order of 10% of the entire galactic disk.

All these factors combined, the total number of black hole candidate binaries in our galaxy is estimated to be in the range 100 - 3000. Although this estimation is crude, we may still conclude that the black hole candidate binaries are at least as numerous as the neutron star binaries. Furthermore, it is remarkable that at least six out of eight were found to be associated with optical novae, which strongly suggests that the black hole candidate binaries are predominantly of low-mass systems. In addition, another black hole candidate GX339-4 is also believed to have a low-mass companion.

If, eventually, the compact objects in these systems indeed turn out to be black holes, the origin of the black hole low-mass binaries is a very interesting problem. Suppose, as it seems plausible, the process of formation of the black hole low-mass binaries is essentially the same as that of the neutron star low-mass binaries. In this case, accretion induced collapse would not work, for it would be impossible to produce a black hole of >3 M_0 (at least as A0620-00 is the case) by this process. Obviously, it is extremely important to determine the mass of the remaining black hole candidates and their distances. In particular, those with low-mass companions are advantageous for obtaining a secure lower mass limit, if the mass function is determined [23].

REFERENCES

[1] F.Makino and the Astro-C Team (1987), Astrophys. Letters Commun. 25, 223.
[2] T.Murakami *et al.* (1988), Nature 335, 234.
[3] A.Yoshida (1989), Ph.D. thesis.
[4] J.Nishimura (1990), private communication.
[5] D.Q.Lamb, J.C.L.Wang, & I.M.Wasserman (1990), Astrophys. J. 363, 670.
[6] J.Trümper *et al.* (1978), Astrophys. J. Letters, 219, L105.
[7] N.E.White, J.H.Swank, and S.S.Holt (1983), Astrophys. J. 270, 711.
[8] Y.Tanaka (1985), Lecture Note in Physics 255: Radiation Hydro-dynamics in Stars and Compact Objects, eds. D.Mihalas and K.-H.A.Winkler (Springer Verlag)
[9] K.Makishima (1990), to appear in Proc. U.S.-Japan Joint Seminar on the Structure and Evolution of Neutron Stars, 1990.
[10] K.S.Wood *et al.* (1990), preprint, submitted to Astrophys. J. in July 1990.
[11] F.Nagase (1989), Proc. 23rd. ESLAB Symp. (ESA SP-296, Nov. 1989), p45.
[12] H.Tsunemi *et al.* (1989), Astrophys. J. Letters 337, L81.
[13] F.Makino and the Ginga Team (1989), IAU Circ. 4782.
[14] F.Makino and the Ginga Team (1991), IAU Cir. 5161.
[15] N.Lund and S.Brandt (WATCH/Granat Team) (1991), IAU Circ. 5161.
[16] M.Elvis *et al.*(1975), Nature 257, 656; L.J.Kaluzienski *et al.*(1977) Astrophys. J. 212, 203.
[17] S.Okamura and T.Noguchi (1988), IAU Cir. 4606.
[18] R.M.Wagner, S.G.Starfield, and A.Cassatella (1989), IAU Circ. 4783.
[19] M.Della Valle and B.Jarvis (1991), IAU Circ. 5165
[20] R.M.Hjellming, T.A.Calovini, and F.A.Cordova (1988), IAU Circ. 4607.
[21] R.M.Hjellming, X.-h.Han, and F.A.Cordova (1989), IAU Circ. 4790.
[22] M.J.Kesteven and A.J.Turtle (1991), IAU Circ. 5181.
[23] J.E.McClintock (1990), preprint, to appear in Proc. Texas-ESO-CERN Symp. on Relativistic Astrophysics, 1990. (N.Y. Acad. of Sci.)
[24] Y.Tanaka (1989), Proc. 23rd. ESLAB Symp. (ESA SP-296, Nov. 1989), p3.
[25] C.Chevalier and S.A.Ilovaisky (1990), Astron. Astrophsy. 238, 163.
[26] K.Mitsuda *et al.* (1984), Publ. Astron. Soc. Japan 36, 741.
[27] K.Ebisawa (1991), Ph.D. Thesis.
[28] K.Ebisawa, K.Mitsuda, and T.Hanawa, (1991), Astrophys. J. 367, 213.
[29] S.Kitamoto et al. (1990), Astrophys. J. 361, 590.
[30] N.E.White, J.L.Kaluzienski, and J.H.Swank (1984), High Energy Transients in Astrophysics, ed. S.E.Woosley (New York: AIP), p31.

Z AND ATOLL SOURCES AS A CLUE TO THE EVOLUTION OF LOW-MASS X-RAY BINARIES

M. van der Klis

Astronomical Institute "Anton Pannekoek", University of Amsterdam, and Center for High-Energy Astrophysics

Kruislaan 403, 1098 SJ Amsterdam, The Netherlands

ABSTRACT. Observations of the rapid X-ray variability (including QPO), X-ray spectral variations, optical, UV and radio emission and X-ray bursts in bright LMXB indicate the existence of two types of sources, the Z and the atoll sources. Two types of QPO are seen in the Z sources and absent in the atoll sources. Models for these QPO suggest that Z sources have both higher neutron-star magnetic-field strengths and higher accretion rates than atoll sources. Z sources tend to have orbital periods >10 hrs and atoll sources < 10 hrs. This information is discussed within the framework of the predictions of binary evolution theory. An evolutionary origin of the observed correlations seems possible; this requires that the neutron-star magnetic-field strength couples to the details of the binary evolution.

1. Introduction: the phenomenology

The discovery with EXOSAT of two types of quasi-periodic oscillations (QPO) in the bright LMXB, and the realization that these QPO correlate with variations in the X-ray spectral state of the sources (see van der Klis 1989 for a historical overview), have led to the insight that practically all of the brightest LMXB (Table 1) fall into one of two groups, the Z sources and the atoll sources (Hasinger and van der Klis 1989, hereafter HK89).

The **Z sources** show three X-ray spectral states, corresponding to three branches in an X-ray colour-colour diagram which are arranged in a rough Z pattern. The upper branch in the Z is called the horizontal branch (HB), the middle branch the normal branch (NB) and the lower branch the flaring branch (FB). Z-patterns can vary in their detailed shape. On a time scale of hours to a day a Z source moves stochastically along this pattern, never jumping from one branch to the other without passing by the branch junction. This motion probably reflects variations in the accretion rate. Optical and UV emission monotonically increase in the sense HB–NB–FB (Vrtilek *et al.* 1989, 1990) and $\dot{M}$ probably does as well (Hasinger *et al.* 1990, hereafter H90). There is no simple relation between X-ray intensity and $\dot{M}$.

On the HB, the power spectrum of the X-ray variability shows a QPO peak

49

E. P. J. van den Heuvel and S. A. Rappaport (eds.), X-Ray Binaries and Recycled Pulsars, 49–59.
© 1992 *Kluwer Academic Publishers.*

that shifts in frequency from ~15 Hz when the source is at the left end of the HB, to ~60 Hz at the HB–NB junction. The peak sometimes persists (without increasing further in frequency) when the source moves into the NB. A broad noise component called low-frequency noise (LFN) always accompanies this QPO peak. Halfway down the NB another QPO peak appears near 6 Hz (sometimes while the HB QPO peak described above is still present; H90). This peak remains at 6 Hz down to the NB–FB junction, where it gradually moves up to ~10 Hz at the low end, and further increases to ~20 Hz further up the FB. When, on the FB, the source moves away from the NB–FB junction, the peak becomes wider and finally dissolves into the noise. No separate broad noise component such as the LFN accompanies the NB–FB QPO.

Table 1. Bright low-mass X-ray binaries[a]

Source	I_x[b] (μJy)	P_{orb}[c] (hr)	Type[d]	Phenomena[e]
Sco X-1	12400	19.2	Z	QPO
GX 5-1	1200	—	Z	QPO
GX 349+2	780	—	Z	QPO
GX 17+2	680	19.8?	Z	QPO,(bu)
GX 340+0	490	—	Z	QPO
Cyg X-2	430	235.	Z	QPO,(bu)
GX 9+1	650	—	A	—
GX 3+1	430	—	A	(Bu)
GX 13+1	340	—	A	—
GX 9+9	290	4.2	A	—
4U 1820-30	260	0.2	A	(Bu)
4U 1705-44	260	—	A	Bu
4U 1636-53	220	3.8	A	Bu
4U 1735-44	160	4.6	A	Bu
4U 1728-33	170	—	A	Bu
Cyg X-3	220	4.8	P	—
Ser X-1	200	—	—	Bu
GCX-1	170	—	—	—
GX339-4	160	14.8	BH	QPO

[a] All variable objects in 3A catalogue (McHardy *et al.* 1981, Warwick *et al.* 1981) with an average flux $\geq 100\mu$Jy not identified with an early type star.

[b] Mean intensity converted from Ariel V ASM counts into μJy (2–11 keV) according to 1 ASM c/s = 2.6 μJy (Bradt and McClintock 1983).

[c] See Parmar and White (1988).

[d] Z: Z source, A: atoll source, P: peculiar, BH: black hole candidate

[e] Bu: regular X-ray bursts; (Bu): has shown (an) episode(s) of regular X-ray bursts; (bu): occasional X-ray bursts reported.

Atoll sources probably show only one, curved, branch in the X-ray colour-colour diagram. Particularly in the left (soft) end of this branch, they move much slower through the diagram than Z sources. Therefore, typical X-ray observations of a day or less only show a branch when the source is in the right-hand (hard) part of the branch ("banana state"); when it is in the left-hand part of the branch, such an observation leads to an "island" with no discernable motion in the colour-colour diagram. No "branch jumping" is observed. The island state often (but not always) shows lower X-ray intensities than the banana, and also optical data (Corbet *et al.* 1989) and properties of the X-ray bursts (van der Klis *et al.* 1990) suggest that accretion rate increases from the island to the banana state. As in the Z sources, X-ray intensity is not a good indication for $\dot{M}$.

Just like in the Z sources, the power spectral shape in atoll sources correlates with the position in the colour-colour diagram. It can be described in terms of just two broad noise components: a power-law shaped component dominating the power spectrum at the lowest frequencies called very-low frequency noise (VLFN) and a shoulder- or hump-like component with the shape of a flat or even slightly rising power-law with a high-frequency cut-off near 100 Hz that is called high-frequency noise (HFN). The strength of both components depends monotonically on accretion rate, VLFN increasing and HFN decreasing with $\dot{M}$. Similar components, with a similar dependence on $\dot{M}$, are also observed in the Z sources in addition to QPO and LFN. Both LFN in Z sources and HFN in atoll sources sometimes produce a local maximum in the power spectrum; they are then called "peaked" noise.

It is important to stress that it is not always possible to make the distinction between Z and atoll sources nor indeed between an island state and a banana state on the basis of the colour-colour diagrams alone. In fact, banana branches have sometimes been identified with HB–NB patterns. The additional information about the rapid variability is crucial.

2. Interpretation of the observed phenomena

HFN (HK89) includes the fastest variability detected in X-ray binaries and can be very strong, with amplitudes up to 20% of the total flux in the island state (all the other components discussed here have amplitudes of typically a few percent). Similarly-shaped components have been observed in black hole candidates and X-ray pulsars (Belloni and Hasinger 1990), and HFN may be a general feature of disk accretion. As the cut-off frequency does not vary a great deal between sources with very different B fields, this frequency is probably *not* related to the inner disk radius. The strength of the HFN may be related to accretion rate; this would explain its relative weakness in Z sources. Such a relation could come about by an inverse dependence on disk thickness or on column depth in an accretion disk corona that by scattering the emerging radiation smears the variability. VLFN is usually attributed to external variations in accretion rate.

The increase of QPO frequency with $\dot{M}$ in the HB suggests that these QPO are related to the orbital motion of matter in the accretion disk just outside a magnetosphere (Alpar and Shaham 1985). In the "modulated-accretion beat-frequency model" (Lamb *et al.* 1985), magnetically gated accretion from the inhomogenous flow in the disk explains various observed properties of the QPO, including the associated LFN. This model requires the presence of a neutron-star magnetic field strong enough to disrupt the accretion flow and form a small magnetosphere. The fact that the expected pulsations due to the neutron star rotation are not observed is attributed at least partly to scattering in matter surrounding the star, which smears pulsations (due to beaming) much more effectively than oscillations (due to intensity variations) (Brainerd and Lamb 1987, Kylafis and Klimis 1987). Other effects (polar cap size; van der Klis *et al.* 1985, graviational bending of photon paths; Wood *et al.* 1988) may also be important. The model predicts that $\dot{M}$ increases with QPO frequency in the HB, which was later confirmed through optical and UV observations (below).

The luminosity of the Z sources as a group is of the order of the Eddington limit. Therefore, the effects of radiation pressure are expected to begin to show up somewhere along the Z track. It is believed that the NB QPO are in fact just that (Fortner *et al.* 1989). According to a model supported by numerical simulations, oscillations near 6 Hz can be set up in a near-Eddington accretion flow onto a neutron star by the force exerted by the accretion-induced radiation on the flow itself. The FB QPO, which merge smoothly in frequency with the NB QPO, might be explained as photohydrodynamic modes in a super-Eddington flow (Lamb 1989).

The X-ray spectral variations underlying the patterns in the colour-colour diagrams are usually interpreted as due to changes in the accretion flow that occur as a function of $\dot{M}$. In Z sources, the HB–NB transition might correspond to the inner disk becoming geometrically thick due to radiation pressure and "engulfing" the magnetosphere (HK89) and the transition to an optically thick radial-flow accretion regime (Lamb 1989). The NB–FB transition might indicate the onset of a super-Eddington flow (Hasinger 1988, see also van der Klis *et al.* 1987). The fact that such transitions do not occur in atoll sources is discussed in the next Section.

As mentioned above, clear correlations of the optical and UV properties of optically observable Z sources with source state have been seen. In general, the optical and UV flux increase in the order HB–NB–FB. H90 show that this provides evidence for an increase in the accretion rate in this sense. The accretion disk, whose optical and UV emission due to reprocessing of X-rays dominates that of the system, has a "better view" of the X-ray source than we have on Earth (our line of sight being obstructed in a source-state dependent way), so that optical and UV flux are better indicators for the X-ray luminosity, and therefore for the accretion rate, than is the X-ray intensity. Simultaneous X-ray-optical observations by Corbet *et al.* (1989) of the atoll source 4U 1735-44 on the banana branch also show a correlation of optical flux with accretion rate.

When interpreted in terms of models for X-ray bursts that include steady thermonuclear burning in the bursting layers, observations of the relation between X-ray burst properties and source state in the atoll sources 4U 1636-53, 4U 1705-44, 4U 1820-30 and 4U 1735-44 support the idea that atoll source state is strictly determined by accretion rate (van der Klis *et al.* 1990, 1991).

3. The difference between Z and atoll sources

As described in the previous sections, a large array of phenomena now supports the view that two groups exist among the bright LMXB, in each of which well-defined, but different, changes take place as a function of $\dot{M}$. The predictive value of the classification, is considerable: there have been many cases of power spectral features and colour-colour diagram branches that were expected in a source on the basis of its classification, and subsequently found. In this Section, I discuss the

Z/atoll classification in terms of the possible physical differences of the compact objects in these systems.

A possible interpretation of the classification (HK89) can be constructed using the models for HB and NB QPO described in Section 2 as the main clue. According to these models, HB QPO require a neutron-star magnetic field strong enough to form a magnetosphere, whereas NB QPO require near-Eddington accretion rates. Suppose that the Z sources have *both* higher neutron-star magnetic-field strengths than atoll sources, *and* higher accretion rates. The higher neutron-star magnetic-field strength explains why HB QPO/LFN occur in Z sources only. The higher accretion rate can explain why NB/FB QPO have also only been observed in Z sources, and why as a group Z sources are brighter (Table 1).

It is hard to see how other differences such as orbital inclination (a gradual range in properties would be expected among bright LMXB rather than a division into two groups), neutron star rotation rate (HB QPO are expected in magnetospheric accretion at any rotation rate), or *just* neutron-star magnetic-field strength (why are Z sources brighter, and why do only Z sources show NB/FB QPO), could consistently explain all the observed differences.

A difference in *just* accretion rate, with the Z sources having the higher $\dot{M}$ values is also unlikely. First, there is no reason why atoll sources would show no HB QPO if their field strengths are similar to the Z sources. Second, one would expect the existence of sources that switch between Z and atoll behaviour as their accretion rate changes, and such sources have not been observed. Third, one of the reasons why pulsation amplitudes are so low in Z sources is believed to be the scattering in material in the line of sight. The lower accretion rates in atoll sources would be expected to lead to less dense scattering material and consequently, if they have similar fields as the Z sources, to higher pulse amplitudes. No pulsations have been seen from either Z or atoll sources, however, with upper limits down to a fraction of a percent for members of both classes (Wood et al. 1990). Finally, at their lowest accretion rates, i.e., at the left end of the HB, Z sources can have an X-ray intensity that is less than 40% of that at the HB–NB junction; their power spectral properties there show no indication whatsoever to begin resembling those of an atoll source. On the contrary, very strong HB QPO are seen (Tan et al. 1991).

The fact that Z sources show three states with two distinct transitions between them, whereas atoll sources show no such transitions but only a gradual range of properties, can in this scenario be explained by the fact that the two important physical changes that can occur in the accretion stream of a Z source (the "engulfing" of the magnetosphere by the inner disk and the onset of a considerable radial accretion flow when with increasing accretion rate the disk swells up, and the onset of mass ejection as the source reaches the Eddington limit) do not happen in an atoll source, (where the neutron star is always "engulfed" and accretion remains sub-Eddington at all times) (HK89, Lamb 1989).

As a class, the Z sources are clearly brighter radio sources than the atoll sources (Penninx 1989). All of the known Z sources have now been detected as radio sources at 4.8 GHz (Penninx 1989, Zwarthoed et al. 1991, Cooke and Ponman 1991). Their maximum observed fluxes are between 1 and 16 mJy. By contrast, only one atoll source has been seen at this frequency (GX 13+1, at 2.2 mJy, Grindlay and Seaquist 1986) whereas upper limits of about 0.4 mJy have been obtained for more than ten of them (Penninx 1989; 4U 1820-30 has been detected at 1.5 GHz). In the NB, all Z sources may have roughly the same radio luminosity (Penninx 1989). This stronger radio emission of the Z sources might be related to their stronger magnetic fields, allowing a magnetospheric acceleration mechanism to produce the relativistic particles required for this emission.

The scarcity of bursts in Z sources and the fact that they are common in atoll sources, finally, also finds a natural explanation in burst models predicting an absence of bursts above certain critical values of $\dot{M}$ (Joss 1978) and B (Joss and Li 1980).

4. Comparison with previous classifications

Earlier attempts at classifying the X-ray properties of LMXB that did not consider the rapid variability (Parsignault and Grindlay 1978, Ponman 1982), including those based on detailed X-ray spectral fits (White and Mason 1985, White et al. 1988) or hardness-intensity diagrams and colour-colour diagrams of EXOSAT data (Schulz et al. 1989), in general led to a division of the LMXB in a high X-ray luminosity (hereafter "hi") and a low X-ray luminosity ("lo") group different from the Z/atoll classification. With few exceptions, the "hi" group included the bright galactic bulge sources. The "lo" group included X-ray burst sources, sources showing periodic X-ray dips and sources which are located in globular clusters (these groups partly overlap). There is a good correspondence between these earlier classifications, with in general the GX sources ending up in the "hi" group (van der Klis 1991).

By contrast, in the Z/atoll classification (Table 1) only four of the GX sources are grouped together with Sco X-1 and Cyg X-2 in what might be identified as the "hi" group (the Z sources), and four other GX sources are grouped together with classical X-ray burst sources such as 4U 1636-53 in the "lo" group (atoll sources). As will be clear from the previous Sections, I feel that the earlier classifications, which could not take advantage of our present knowledge of the rapid variability, are unsatisfactory, as they group together sources with very different QPO and noise characteristics, colour-colour diagrams and radio properties. The "GX" atoll sources do stand out as the brightest, and usually non-bursting atoll sources; they are nearly always in the banana state (although GX 3+1 has shown a bursting, low-intensity state [Makishima et al. 1983] that may have been an island state) and it seems reasonable to suppose that these sources are the highest accretion rate atoll sources. They remain completely different from Z sources by the absence of both NB/FB QPO ($\dot{M}_{Edd}$ is not attained) and HB QPO/LFN (magnetosphere is not formed).

5. The link with low-mass X-ray binary evolution

If we accept the interpretation of the differences between the Z and the atoll sources in terms of a difference in *two* basic source characteristics, accretion rate and neutron-star magnetic-field strength, then it remains to be explained why among the brightest LMXB these two characteristics correlate. The answer to this may be related with LMXB evolution, and an observational clue may consequently be the orbital period distribution of these sources (Table 1). Of the six Z sources two (maybe three) orbital periods are known and they are all between 19 hrs and

10 days. Among the twelve atoll sources, five orbital periods are known, and all are shorter than 10 hrs.

As is well known, Webbink *et al.* (1983, hereafter WRS) proposed that for Eddington accretion rates to be sustained in LMXB, the mass transfer should likely be driven by the evolutionary expansion of a low-mass star after core hydrogen exhaustion (hereafter "evolved companion"), rather than by gravitational radiation and magnetic braking in a system containing a main-sequence or degenerate dwarf companion. They specifically mentioned that the bright galactic bulge sources (the "GX" sources or the ones called "hi" in Section 4) are the ones to be identified with the systems containing evolved companions. A problem with this scenario is that predicted mass transfer rates for binaries with orbital periods less than a few tens of days are much too low (1 to 10% of the Eddington value) to explain sources such as Sco X-1 and Cyg X-2.

However, WRS only considered cases where the donor star was less massive than the neutron star at the onset of mass transfer. It was shown by Pylyser and Savonije (1988a, b; hereafter P&S) that if the donor star is more massive than the neutron star at the onset of mass transfer, a qualitatively different type of evolution involving what they call rapid mass transfer (RMT) ensues due to the fact that in that situation the mass transfer causes the binary orbit, and the Roche lobe of the donor star, to shrink. Although initiated by the evolution of the donor star, the RMT is not driven by it, but rather by the adjustment of the donor star radius to the mass loss. Mass transfer proceeds on the thermal time scale of the donor star until several 0.1 $M_\odot$ have been transferred, well beyond the point where the donor reaches the same mass as the neutron star. P&S mention that such a RMT model may apply to Sco X-1. In some cases a system containing an evolved companion may in fact evolve to a shorter period. P&S call these systems "converging".

According to this scenario, there are several ways to produce LMXB with **low** accretion rates with both long and short orbital periods. Short-period low-accretion-rate systems can result from the "classical" scenario involving a main sequence or degenerate dwarf companion with angular momentum loss due to gravitational radiation and magnetic braking, but also from converging evolved-companion systems, that may or may not have undergone an RMT phase dependent on the companion mass at the onset of mass transfer. Long-period low-accretion-rate systems can be either WRS-type systems with orbital periods less than a few tens of days where the evolved companion was less massive than the neutron star at the onset of mass transfer, or systems in a post-RMT phase.

However, within the same scenario there are only two types of LMXB that have **Eddington** accretion rates, and both have orbital periods longer than would be consistent with a main-sequence companion (~10 hrs). One type consists of the WRS systems with orbital periods longer than several tens of days, the other of P&S systems in the RMT phase. Note that, even if the latter type of system eventually converges, the orbital period is still long during the RMT phase.

The six Z sources, with their Eddington accretion rates, must be found among these last two groups. The two Z sources with known orbital periods, Sco X-1 and Cyg X-2, have periods that are only consistent with the RMT scenario. On the basis of this one might speculate that all Z sources are RMT phase systems. The WRS systems with very long (more than several tens of days) orbital periods are perhaps smothered by persistent excess mass transfer (perhaps they could be identified with the ROSAT soft sources, Hasinger, this volume) whereas the instability of the mass transfer in the RMT phase allows the Z sources to be observed around the

Eddington limit. It remains to be seen whether the population statistics of LMXB are consistent with this speculation.

It is interesting to note, that the evolutionary scenario sketched above predicts further subdivisions among the low accretion-rate sources into systems with main-sequence and evolved companions, where the latter may have either long or short periods (converging or diverging systems), and may or may not have undergone a RMT phase. This might have something to do with the difference between the "GX" and the other atoll sources. Note also, that long-period, low-$\dot{M}$ systems are possible; the fact that no atoll sources with long periods have been identified may be due to selection effects.

6. The relation between accretion rate and neutron-star magnetic-field strength

It was proposed shortly after the discovery of the QPO in GX 5-1, and their interpretation as a magnetospheric phenomenon, that the presence of a neutron-star magnetic field strong enough to form a magnetosphere (10^{9-10} G) might be related to the high accretion rates possible from an evolved companion (van der Klis *et al.* 1985) and that consequently QPO and evolved companions should correlate. As argued in the previous Sections, for the Z sources this link between evolved companions, high accretion rates and higher neutron-star magnetic-field strengths appears to hold.

However, the mechanism originally proposed for the evolutionary link between these properties, accretion-induced collapse of a white dwarf in high-accretion-rate systems that produces young neutron stars with less decayed fields in the evolved-companion systems, has meanwhile become much less attractive, as the evidence for spontaneous gradual field decay seems to have evaporated (see van den Heuvel, this volume and Bhattacharya, this volume). As noted by HK89, the observational evidence suggests that we may have to look for another way to link companion type, accretion rate and neutron-star magnetic-field strength.

The mechanism for spin-rate-change-induced field decay proposed by Srinivasan *et al.* (1990) might provide such a way. In this mechanism, the residual (core) field strength is proportional to the lowest frequency to which the neutron star has been spun down in its life (it is not important whether it has been spun back up later). If a neutron star would be spun down by propellor braking to less low spin rates during the onset of mass transfer in a system driven by the P&S RMT mechanism (because the mass transfer grows to large values on a short time scale) than in other systems, then this might provide the required link between neutron-star magnetic-field strength and orbital period. It remains to be seen how GX 1+4, an X-ray pulsar with a strong field and probably a very long orbital period fits in with this scenario.

7. Other sources

As can be seen in Table 1, the Z/atoll classification encompasses all the brightest persistent LMXB with the exception of Ser X-1 (which is probably an atoll source, Hasinger and van der Klis, priv. comm.), Cyg X-3 (see below), GCX-1 (source near galactic center) and GX 339-4 (a black-hole candidate). Among these ~20 brightest LMXB, no X-ray dip sources occur (Cyg X-3 may be an accretion-disk corona source, which is not the same [White and Mason 1985]; in general ADC sources need to be treated separately in any classification as the fact that we do not see the X-ray source directly strongly modifies the observed characteristics). The absence of dippers among the brightest sources suggests that the structure in the accretion stream that is causing the dips (outer disk rim [White and Mason 1985], deflected stream [Frank, King and Lasota, 1987]) is shadowed in these sources, possibly by the ADC, or that it is absent. There is only one known globular cluster source among the brightest LMXB (4U 1820-30); it is an atoll source (HK89).

On the basis of the interpretation for the differences between Z and atoll sources given in Section 3, one might expect most burst sources to belong to the class of the atoll sources (if burst models predicting an absence of bursts in high accretion rate neutron stars [Joss 1978] are correct – note, however, that GX 17+2 has shown X-ray bursts [Tawara *et al.* 1984]).

Indeed, the burst source AC 211 in the globular cluster M15 (Van Paradijs *et al.* 1990) and the burst source 4U 1702-42 (Oosterbroek *et al.* 1990) were recently classified as atoll sources in the island and the banana state, respectively, on the basis of their X-ray colour-colour diagram and rapid X-ray variability. The M15 source has an orbital period of 8.5 hr, consistent with a main sequence companion. I note, that the classification of AC 211 is mainly based on the presence of a strong noise component, which may be risky (below).

Even if binary evolution causes a correlation between accretion rate and neutron-star magnetic-field strength, with Z sources representing the high, and atoll sources the low niche in this correlation, one might look for peculiar sources that, through some tortuous evolutionary path, came to occupy the two remaining niches (high $\dot{M}$ low B, and *vice versa*). A high $\dot{M}$, low B source would be predicted to show NB/FB QPO, but no HB QPO; for a low $\dot{M}$, high B source the opposite would be expected. GX 339-4 is a black hole candidate that shows 6 Hz QPO (Miyamoto *et al.* 1991) and could be an example of a low B high $\dot{M}$ system. Apart from this source, the persistently bright sources are mostly covered, so that other candidate high $\dot{M}$, low B sources should be transient or strongly variable. Cir X-1, which has shown NB/FB-like QPO and hardness-intensity diagram properties (Tennant 1988), may fill this ticket (van der Klis 1991). No examples spring to mind of high B, low $\dot{M}$ LMXB – the way to find one would be to follow a soft transient that displays Z behaviour when it is bright, down its decay. It is important to note that consequently we have no information about the power spectrum of such a source. It is conceivable that its HFN, just like in a wide range of classes of faint sources, becomes strong (several 10%). So, just the fact that a faint source shows strong HFN is not enough to classify it as an atoll source. I note that a "low-$\dot{M}$ Z source" would be a prime candidate for the discovery of millisecond X-ray pulsations ($B\sim 10^{9-10}$, $\dot{M} \ll L_{Edd}$). However, it is not clear that such sources exist. Possibly, dwarf companions are required for transient behaviour.

From empirical considerations it is far from clear yet how well the binary period is correlated to the Z/atoll distinction. Considerations of binary evolution (Section 5) appear to predict that the near-Eddington Z sources should have long orbital periods, whereas less luminous sources might have either long or short periods. It is not clear how the atoll sources fit in with the various predicted classes of low-luminosity LMXB, although the observed periods show that a fair number of them have dwarf companions. To clarify these matters it is important to measure more orbital periods among known Z and atoll sources. The two known orbital periods of Z sources have both been measured in the radial velocity of the optical emission lines. Until very recently there was no hope to do this in the other Z sources, as they had not been optically identified. Therefore, the recent identification of a third Z source, GX 349+2 (Penninx and Augusteijn 1991), the first Z source identified in 23 years, is important. Inversely, it is of interest to study the characteristics of LMXB with known orbital periods within the framework of the Z/atoll phenomenology. In particular the unclassified long-period LMXB deserve further attention. The 9.8-day period source X 0921-63 (an ADC source) could be a Z source edge-on. Although it is too weak in X-rays to classify it, it might turn out to be a radio source similar to other Z sources. The 21-hr X-ray dip source 4U 1624-49 (Jones and Watson 1989) is an example of a low-luminosity long-period system.

Acknowledgements. This work was supported in part by the Netherlands Organization for Scientific Research (NWO) under grant PGS 78-277.

8. References

Alpar, M.A., Shaham, J., 1985, *Nature* **316**, 239.

Belloni, T., Hasinger, G., 1990, *Astron. Astrophys.* **230**, 103.

Bradt, H.V.D., McClintock, J.E., 1983, *Annu. Rev. Astron. Astrophys.* **21**, 13.

Brainerd, J., Lamb, F.K., 1987, *Astrophys. J. (Letters)* **317**, L33.

Cooke, B.A., Ponman, T.J., 1991, *Astron. Astrophys.* , in press.

Corbet, R.H.D., Smale, A.P., Charles, P.A., Lewin, W.H.G., Menzies, J.W., Naylor, T., Penninx, W., Sztajno, M., Thorstensen, J.R., Trümper, J., Van Paradijs, J., 1989, *Monthly Notices Roy. Astron. Soc.* **239**, 533.

Fortner, B., Lamb, F.K., Miller, G.S., 1989, *Nature* **342**, 775.

Frank, J., King, A.R., Lasota, J.-P., 1987, *Astron. Astrophys.* **178**, 137.

Grindlay, J.E., Seaquist, E.R., 1986, *Astrophys. J.* **310**, 172.

Hasinger, G., 1988, *Advances in Space Research* **8(2)**, 377.

Hasinger, G., van der Klis, M., 1989, *Astron. Astrophys.* **225**, 79. [HK89]

Hasinger, G., van der Klis, M., Ebisawa, K., Dotani, T., Mitsuda, K., 1990, *Astron. Astrophys.* **235**, 131. [H90]

Jones, M.H., Watson, M.G., 1989, Proc. 23rd ESLAB Symp. *Two Topics in X-Ray Astronomy*, *ESA SP-* **296**, 439.

Joss, P.C., 1978, *Astrophys. J.* **225**, L123.

Joss, P.C., Li, F.K., 1980, *Astrophys. J.* **238**, 287.

Kylafis, N.D., Klimis, G.S., 1987, *Astrophys. J.* **323**, 678.

Lamb, F.K., Shibazaki, N., Alpar, M.A., Shaham, J., 1985, *Nature* **317**, 681.

Lamb, F.K., 1989, Proc. 23rd ESLAB Symp. *Two Topics in X-Ray Astronomy*, *ESA SP-* **296**, 215.

Makishima, K., et al., 1983, *Astrophys. J.* **267**, 310.

McHardy, I.M., Lawrence, A., Pye, J.P., Ponds, K.A., 1981, *Monthly Notices Roy. Astron. Soc.* **197**, 893.

Miyamoto, S., Kimura, K., Kitamoto, S., Dotani, T., Ebisawa, K., 1991, *Astrophys. J.* , in press.

Oosterbroek, T., Penninx, W., van der Klis, M., Van Paradijs, J., Lewin, W.H.G., 1990, *Astron. Astrophys.* , submitted.

Parmar, A.N., White, N.E., 1988, *MEMORIE della Società Astr. Italiana* **59**, 147.

Parsignault, D.R., Grindlay, J.E., 1978, *Astrophys. J.* **225**, 970.

Penninx, W., 1989, Proc. 23rd ESLAB Symp. *Two Topics in X-ray Astronomy, ESA SP-* **296**, 185.

Penninx, W., Augusteijn, T., 1991, *Astron. Astrophys.* , submitted.

Ponman, T., 1982, *Monthly Notices Roy. Astron. Soc.* **201**, 769.

Pylyser, E., Savonije, G.J., 1988a, *Astron. Astrophys.* **191**, 57. [P&S]

Pylyser, E.H.P., Savonije, G.J., 1988b, *Astron. Astrophys.* **208**, 52. [P&S]

Schulz, N.S., Hasinger, G., Trümper, J., 1989, *Astron. Astrophys.* **225**, 48.

Srinivasan, G., Bhattacharya, D., Muslimov, A.G., Tsygan, A.I., 1990, *Current Science* **59**, 31.

Tan, J., Lewin, W.H.G., Penninx, W., Van Paradijs, J., van der Klis, M., Mitsuda, K., 1991, *Astrophys. J.* , submitted.

Tawara, Y., Hirano, T., Kii, T., Matsuoka, M., Murakami, T., 1984, *Proc. Astron. Soc. Japan* **36**, 861.

Tennant, A.F., 1988, *Monthly Notices Roy. Astron. Soc.* **230**, 403.

Van Paradijs, J., Dotani, T., Tanaka, Y., Tsuru, T., 1990, *Proc. Astron. Soc. Japan* , in press.

Van der Klis, M., Jansen, F., Van Paradijs, J., Lewin, W.H.G., van den Heuvel, E.P.J., Trümper, J.E., Sztajno, M., 1985, *Nature* **316**, 225.

Van der Klis, M., Stella, L., White, N., Jansen, F., Parmar, A.N., 1987, *Astrophys. J.* **316**, 411.

Van der Klis, M., 1989, *Annu. Rev. Astron. Astrophys.* **27**, 517.

Van der Klis, M., Hasinger, G., Damen, E., Penninx, W., Van Paradijs, J., Lewin, W.H.G., 1990, *Astrophys. J. (Letters)* **360**, L19.

Van der Klis, M., *et al.*, 1991, in prep.

Van der Klis, M., 1991, Proc. NATO ASI *Neutron Stars: An Interdisciplinary Field*, Agia Pelagia, Crete, Greece, September 1990; in press.

Vrtilek, S.D., Penninx, W., Raymond, J.C., Verbunt, F., 1989, Proc. 23rd ESLAB Symp. *Two Topics in X-Ray Astronomy, ESA SP-* **296**, 671.

Vrtilek, S.D., Raymond, J.C., Garcia, M.R., Verbunt, F., Hasinger, G., Kürster, M., 1990, *Astron. Astrophys.* **235**, 162.

Warwick, R.S., Marshall, N., Fraser, G.W., Watson, M.G., Lawrence, A., *et al.*, 1981, *Monthly Notices Roy. Astron. Soc.* **197**, 865.

Webbink, R.F., Rappaport, S., Savonije, G.J., 1983, *Astrophys. J.* **270**, 678. [WRS]

White, N.E., Mason, K.O., 1985, *Space Sci. Rev.* **40**, 167.

White, N.E., Stella, L., Parmar, A.N., 1988, *Astrophys. J.* **324**, 363.

Wood, K.S., Ftaclas, C., Kearny, M., 1988, *Astrophys. J. (Letters)* **324**, L63.

Wood, K.S., Norris, J.P., Hertz, P., Vaughan, B.A., Michelson, P.F., Mitsuda, K., Lewin, W.H.G., Van Paradijs, J., Penninx., W., van der Klis, M., 1990, *Astrophys. J.* , in press.

Zwarthoed, R., *et al.*, 1991, in prep.

X-RAY DIAGNOSTICS OF ACCRETION DISKS

G. HASINGER
Max-Plank-Institut für Extraterrestrische Physik
8064 Garching
BRD

1 Introduction and Abstract

Within this article the very wide field of accretion phenomena observed in X-rays is restricted to bright low-mass X-ray binaries (LMXB), where recent observational evidence leads to a global picture that for the first time sheds light on the innermost regions of an accretion disk. LMXB are actually very similar to cataclysmic variables, one just has to replace the central white dwarf by a neutron star. Given a small enough neutron star magnetosphere the disk then reaches to much smaller radii and therefore much higher temperatures. Consequently, radiation pressure can become important in the inner part of accretion disks around neutron stars.

Table 1 from the recent review by van der Klis (1989b) shows the known LMXB sorted according to apparent X-ray brightness and marked as "Z-sources" and "Atoll-Sources" according to the classification scheme by Hasinger and van der Klis (1989). The class of LMXB we are concerned here – the Z-sources – belong to the brightest X-ray objects in the sky. They are characterized by Z-shaped X-ray colour-colour diagrams and comparatively narrow peaks of quasi-periodic oscillations (QPO) at frequencies between 5 and 50 Hz in their power spectra.

In the recent years it turned out that in the handful of Z-sources nature has provided us with a magnificent laboratory to study the innermost regions of accretion disks – not the least at X-ray wavelengths.

E. P. J. van den Heuvel and S. A. Rappaport (eds.), X-Ray Binaries and Recycled Pulsars, 61–74.
© 1992 *Kluwer Academic Publishers.*

Table 1. The brightest low-mass X-ray binaries (adapted from van der Klis, 1989b)

Source Name	I_x [μJy]	P_{orb} [hr]	Type A/Z	Phenomenology
Sco X-1 (1617-155)	12400	19.2	Z	QPO
GX 5-1 (1758-259)	1200	—	Z	QPO
GX 349+2 (1702-363)	780	—	Z	QPO
GX 17+2 (1813-140)	680	19.8?	Z	QPO, (bu)
GX 9+1 (1758-205)	650	—	A	—
GX 340+0 (1642-455)	490	—	Z	QPO
GX 3+1 (1744-265)	430	—	A	(QPO),(Bu)
Cyg X-2 (2142+380)	430	235.	Z	QPO,(Bu)
GX 13+1 (1811-171)	340	—	A	—
GX 9+9 (1728-169)	290	4.2	A	Mo
4U 1820-30 (NGC 6624)	260	0.2	A	(QPO),(Bu),Mo
4U 1705-44	260	—	A	Bu
4U 1636-53	220	3.8	A	Bu
Ser X-1 (1837+049)	200	—	-	Bu
GCX-1 (1742-294)	170	—	A	Bu?
4U 1728-33	170	—	A	Bu
GX 339-4 (1659-487)	160	14.8?	-	QPO,BH?
4U 1735-44	160	4.6	A	Bu

2 X-ray Spectrophotometry

The technique we utilize here is rather simple and similar to the UBV spectroscopy in optical astronomy. The raw X-ray spectrum is divided into three rather broad energy bands - typically 1-3 keV, 3-6 keV and 6-15 keV. From the ratio between the middle and the soft X-ray band a so-called *soft colour* is derived, the ratio between the hard and the middle X-ray band gives the *hard colour*.

Figure 1 shows X-ray data from Cyg X-2 measured over four days with the Japanese satellite *Ginga* during a multifrequency campaign in 1988 (from Hasinger et al., 1990, hereafter HA90). The lowest panel shows the X-ray intensity as a function of time, the middle one gives the hard colour. The uppermost panel shows the times of simultaneous observations in other wavebands (see below). Both X-ray intensity and X-ray colour vary quite erratically on a variety of

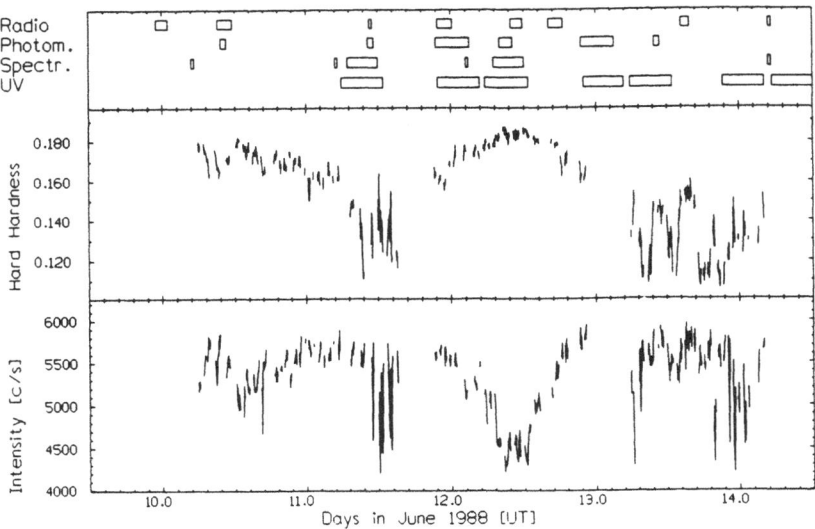

Fig. 1. X-ray lightcurve (lower panel) and hard colour (middle panel) of Cyg X-2 as a function of time for the 4-day Ginga observation in June 1988. The upper panel shows the time interval of observations in other wavebands. Taken from Hasinger et al., 1990

different timescales. However, the source has been observed in different modes of activity. Phases of slowly varying intensity and hardness are interrupted by periods of rapidly varying color or intensity.

3 The "Z-Diagram"

The chaotic nature of the temporal variability changes into very well structured and characteristic patterns when one plots the same data into a colour-intensity diagram (similar to the Hertzsprung-Russel diagram of optical astronomy) or into a colour-colour diagram. These two diagrams are given in Figs. 1 and 2. Now three well separated "spectral branches" are visible which form an elongated "Z-shape" in the colour-colour diagram which gave these sources their name.

The three branches are called "horizontal branch" (HB; upper left) where the hard colour stays almost constant while the intensity varies, "normal branch" (NB; middle) where both soft and hard colour change in a correlated fashion while the X-ray intensity stays almost constant, and "flaring branch" (FB; bottom) which again is characterized by a correlated variation of soft and hard colour, however along a different track. (The dramatic intensity decrease along the flaring branch is peculiar to Cyg X-2, in most other Z-sources a similarly dramatic intensity increase has been observed, thus the term "flaring".)

64

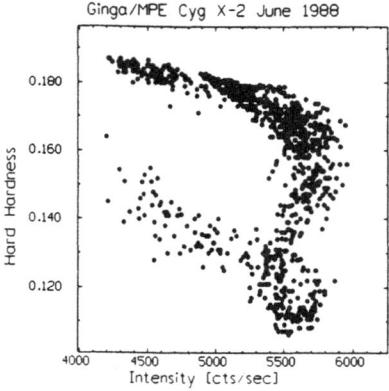

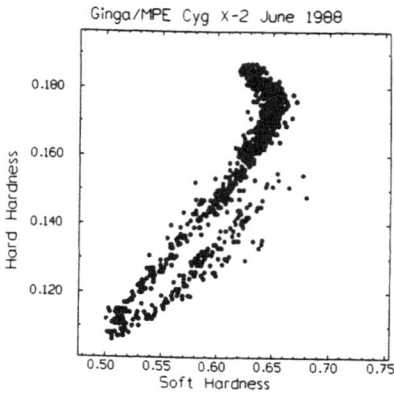

Fig. 2. Colour-intensity diagram from the data in Fig. 1 (HA90)

Fig. 3. Colour-colour diagram for the same data (HA90).

The same technique has been applied to a large body of data from the European X-ray satellite EXOSAT (Hasinger and van der Klis, 1989, hereafter HK89) and a total of 6 bright LMXB have been identified as Z-sources. Figure 4 shows their colour-colour diagrams which display a similar complete Z-pattern or at least fragments of it.

The existence of these characteristic, well confined tracks of spectral and intensity evolution calls for a simple explanation. One interpretation assumes that the observed changes are governed by the variation of a single parameter, which most likely is the mass-accretion rate $\dot{M}$ fed through the disk onto the central engine (Priedhorsky et al., 1986). If this is the case, then there are only two choices how $\dot{M}$ can vary along the "Z": either in the "positive" sense (from the upper left to the lower right) or in the opposite direction. In order to solve this question simultaneous observations in other wavebands had to be performed.

4 Observations in Other Wavebands

Simultaneously with the X-ray measurements two multifrequency campaigns with observations of Cyg X-2 in the UV, optical, and radio range have been performed (see figure 1). The IUE observations (Vrtilek et al., 1990), the optical spectroscopy from La Palma (van Paradijs et al., 1990) and the VLA radio measurements (Hjellming et al., 1990) will be published together with the Ginga X-ray data (HA90) soon. Luckily enough the source cooperated: it displayed all three spectral states during the simultaneous observations and

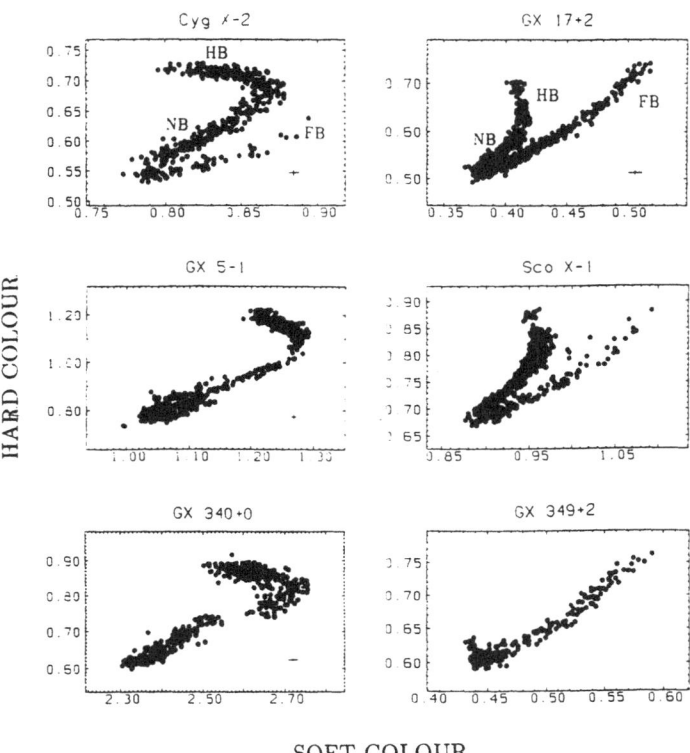

Fig. 4. Colour-colour diagram of EXOSAT data from 6 Z-sources (HK89)

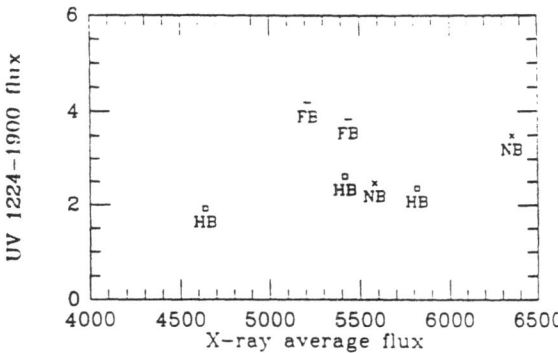

Fig. 5. UV continuum flux as (IUE) versus X-ray flux (Ginga). The X-ray spectral state is indicated for each measurement. The UV flux increases continuously from horizontal to normal branch. Taken from Vrtilek et al., 1990

in all wavebands clear correlations with the source position in the Z-diagram were found.

The UV continuum and line flux increase steadily by a factor of about 2 as the source moves from the horizontal branch to the flaring branch (Vrtilek et al., 1990, see fig. 5). The same is true for the equivalent width of the He II $\lambda 4686$ and $\lambda 4640$ Bowen emission lines as well as the Hβ and Hγ absorption lines (van Paradijs et al., 1990). A similar behaviour was found already in the seventies for Sco X-1, where the optical flux increases by about one magnitude from the top to the bottom of the normal branch normal and further up the flaring branch (Canizares et al., 1975). Since the optical and UV flux of the X-ray irradiated disk is dominated by reprocessed radiation, one can conclude that the X-ray (or EUV) illumination of the disk increases from HB to FB – and thus the mass accretion rate has to increase in the same way (Vrtilek et al., 1990, HA90)

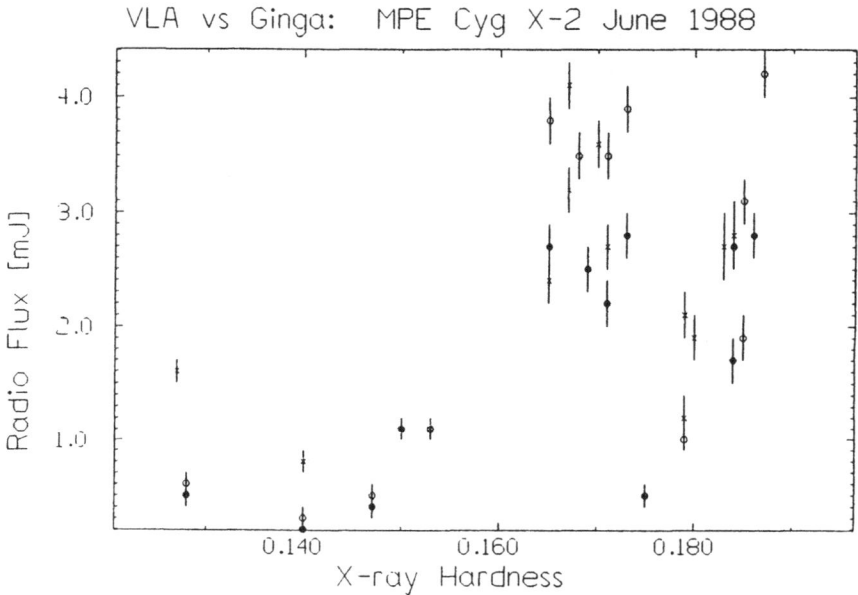

Fig. 6. Radio flux density (VLA) versus X-ray hardness (Ginga). Hardness is the same 'hard hardness' as in fig. 1-3, thus the source is on the horizontal branch for a hardness larger than 0.16. The radio flux density is high and strongly variable on the horizontal branch and low and quiescent on the normal and flaring branch.

The radio flux behaves in a completely different manner: as Figure 6 shows, the radio flux is loud and highly variable on the horizontal branch, and quiet and quiescent on the normal and flaring branch (Hjellming et al., 1990). This is similar as observed before in the radio/X-ray correlation in GX 17+2 (Penninx et al., 1988).

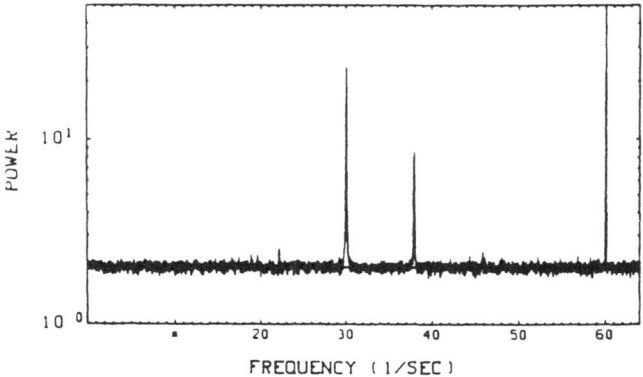

Fig. 7. Power spectrum of the Crab Nebula and its pulsar as observed with EXOSAT. The fundamental pulsar frequency (30 Hz) and its first harmonic (60 Hz) are visible as sharp spikes. A third peak is seen at a frequency of 38Hz, it is the aliased signal of the second harmonic (90 Hz).

5 Quasi-Periodic Oscillations and Noise

The last pieces to solve the puzzle of the Z-sources comes from a systematic analysis of their X-ray flickering in the sub-second range. As diagnostic tool the segmented Fourier analysis technique (see van der Klis, 1989a) is utilized: the observation is divided into small continuous stretches of data (typically 8-32 seconds), the power spectra of which are averaged to enhance the statistical significance of the signal. If this method is applied to the 33ms- pulsar in the Crab Nebula a series of sharp spikes at the fundamental pulsar frequency and its harmonics is observed (see figure 6) on top of a constant (white noise) signal which is due to the counting statistics of the data.

In a search for possible millisecond pulsars in low-mass X-ray binaries van der Klis et al. (1985) analysed data of the bright galactic bulge source GX 5-1. However, instead of several sharp spikes they found a single broad peak in the power spectra, signalling that not a strictly periodic, but a *quasi periodic oscillation* (QPO) is present in the data. Additionally a component called 'red noise' or 'low-frequency noise' (LFN), i.e. excess power which rises continuously towards the lowest frequencies, is visible in the power spectra (see figure 7).

Moreover, as the X-ray source intensity varies, the centroid frequency of the QPO peak is not stable but varies as a function of time, roughly in the frequency range 20-45 Hz – the two quantities go hand in hand.

It turns out, that the presence of QPO peaks in the power spectra is a class property of all Z-sources (see HK89). However, the rapid, intensity-dependent QPO are mainly visible in the horizontal branch. As soon as the source 'turns the corner' into the normal branch the power spectra change dramatically. Figure 8 shows a comparison of two power spectra of Cyg X-2, taken in the horizontal and normal branch, respectively. Compared to figures 6 and 7 the

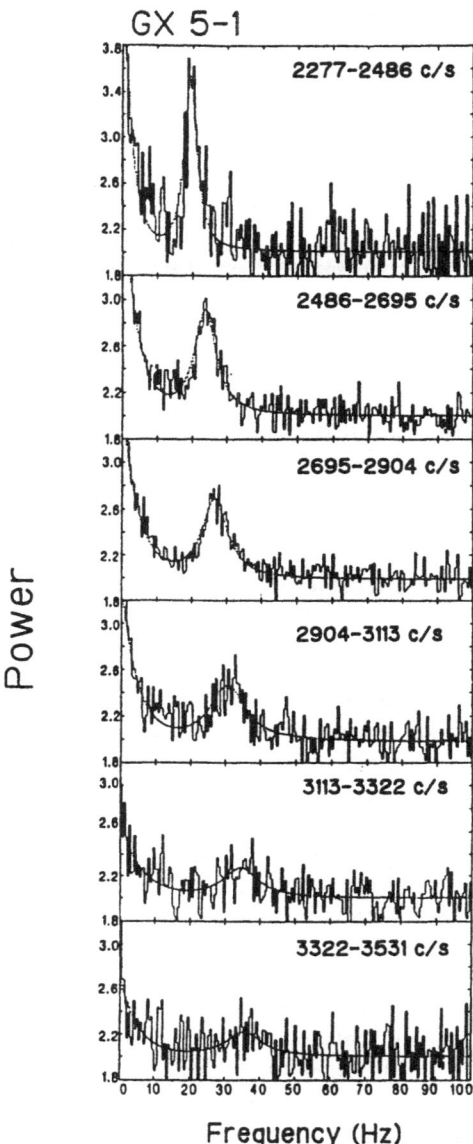

Fig. 8. Six different power spectra of the bright galactic bulge source GX 5-1 in the same representation as figure 7. The X- ray source intensity rises from approximately 2200 cts/s to 3800 cts/sec in the EXOSAT detector from top to bottom (from van der Klis et al., 1985).

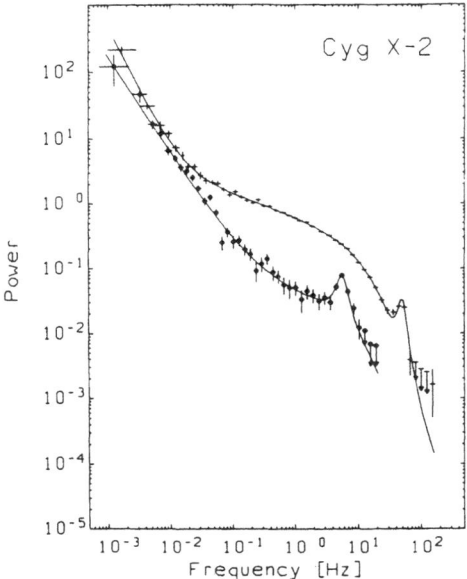

Fig. 9. Two power spectra of the Z-source Cyg X-2 taken in the horizontal (upper) and normal (lower) branch. The power spectra are displayed on a logarithmic scale, with the white noise level due to counting statistics removed. Note that the power spectra are not shifted with respect to each other (from Hasinger, 1987a).

power spectra here are displayed on a logarithmic scale and the white noise due to counting statistics has been subtracted. One sees, that the strong and characteristically shaped low-frequency noise component, dominating in the horizontal branch at frequencies 0.05 - 50 Hz, is almost completely gone in the normal branch. The same is true for the fast (50 Hz) QPO peak. Roughly in the middle of the normal branch another, slower QPO peak occurs, its centroid remains rather stable at frequencies 5-7 Hz throughout the lower part of the normal branch.

These dramatic changes of the different power spectral components are further highlighted in figure 9 where all Cyg X-2 power spectra from the EXOSAT and Ginga X-ray observations have been systematically analysed to trace the source behaviour throughout the whole Z-diagram. Particularly interesting seems the variation of the LFN, which first increases on the horizontal branch and the abruptly dies out close to the HB/NB-corner. This transition has become one of the most important ingredients in the interpretation discussed below. The changes in the HBO (horizontal-branch oscillation) and NBO (normal-branch oscillation) power are anticorrelated, while the former continuously decreases from HB to NB, the latter increases abruptly along the normal branch and is most prominent in the middle of the NB. There are actually NB power spectra which show both HBO and NBO peaks simultaneously (HA90).

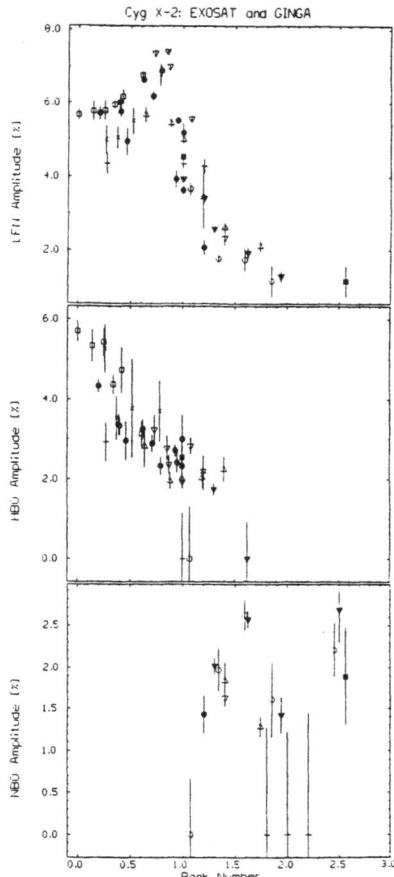

Fig. 10. Variation of the main power spectral components of Cyg X-2. The figures display the rms. variability in the red noise (upper), in the horizontal-branch QPO (middle) and in the normal- branch QPO (lower panel) as a function of position along the Z- diagram. The abscissa (called 'rank number') gives an arbitrary coordinate that increases monotonically from the left of the HB (0.0) over the HB/NB-transition (1.0) and the NB/FB-transition (2.0) to the upper right of the flaring branch (3.0), following the change in mass-accretion rate. Data from the EXOSAT and Ginga X-ray satellites have been merged (Hasinger, et al. 1990, in prep.

The same analysis applied to other Z-sources reveals almost carbon-copy results. Figure 10 shows a selection of power spectra from the same set of sources the colour-colour diagrams of which are given in figure 4 (HK89). Where available, power spectra have been accumulated separately for each branch. The sense and magnitude of variation in the different power spectral components is very similar to Cyg X-2.

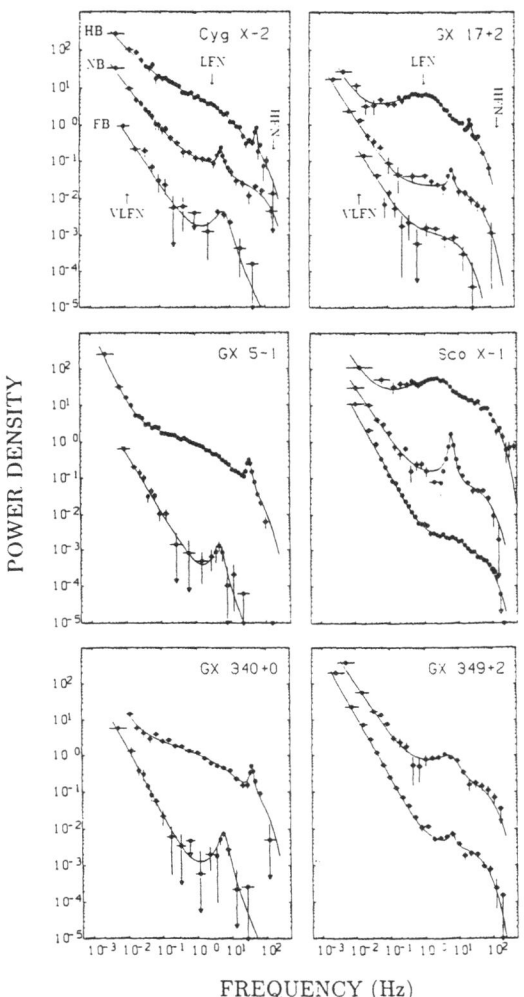

Fig. 11. Power spectra of six Z-sources in the different spectral branches, corresponding to figure 4. Power spectra are given on logarithmic scales with white noise background subtracted. For clarity power spectra from different branches (from top to bottom: HB, NB, FB) are shifted by two decades with respect to each other (from HK89).

6 The Big Picture: a Jumping Accretion Disk ?

Figure 11 summarises the complex variation of some of the observable quantities as a function of mass accretion rate along the Z-diagram: $\dot{M}$ increases monotonically from HB to FB by about a factor of 2 (panel 1). The UV and blue optical light and the emission lines follow hand in hand (panel 2). This is also true for the X-ray flux on the horizontal branch, however, at the HB/NB transition it starts to deviate from this one-to-one correspondence (panel 3). The latter three quantities all show marked transitions close to the corner between horizontal and flaring branch: horizontal-branch QPO and normal branch QPO change importance in the upper normal branch, where they can coexist simultaneously (panel 4). The LFN, which is strong on the HB, dies out at the same transition (panel 5; see also Figure 9), and finally, strong nonthermal radio flares occur only on the horizontal branch.

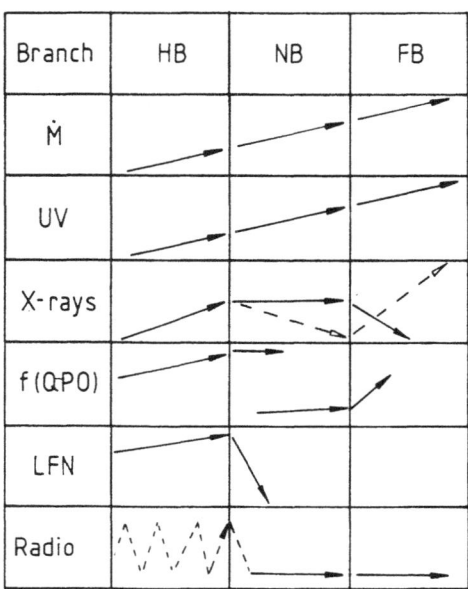

Fig. 12. Schematic display of the variation of important observables and the mass-accretion rate as a function of source position along the Z-profile (see text) (from HA90).

We have now collected the ingredients neccessary to interpret the complex spectral, power spectral and multifrequency behaviour of Z-sources. Most important is the knowledge that the observed variations are most likely driven by a simple change of mass accretion rate. The second most important feature - at least in my judgement - is the dramatic phase transition when the source moves from the horizontal to the normal branch. X-ray spectrum, and intensity, noise and QPO as well as the radio flux all change dramatically and almost simultaneously there. These phenomena suggest a scenario (Hasinger 1988, HA90), which is sketched in Figure 12, and which is similar to the 'unified QPO model' by Lamb (1989, 1990) in some important features.

At low accretion rates (HB) the source starts out in a configuration with a geometrically thin inner disk which is cutting into a small magnetosphere as shown in figure 12a. This is the geometry required for the 'accretion modulated beat frequency model' for the horizontal-branch QPO (Alpar and Shaham, 1985; Lamb et al., 1985). Turbulence at the magnetospheric boundary

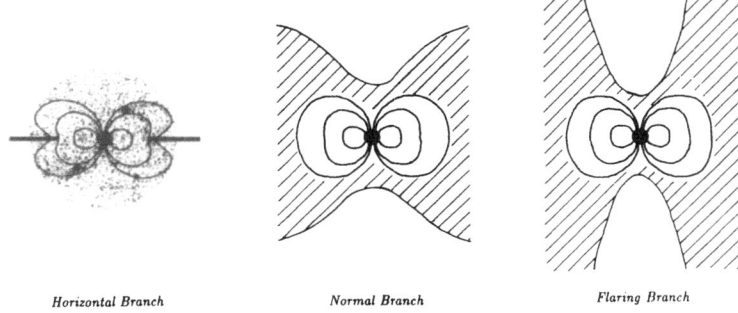

Horizontal Branch   Normal Branch   Flaring Branch

Fig. 13. Schematic display of the accretion geometry in the adopted scenario for the (a) horizontal, (b) normal, and (c) flaring branch (from HA90).

produces clumps, causing a non-stationary accretion which is responsible for the LFN. Interaction between the clumps in Keplerian motion and between the rotating neutron star magnetic field cause quasi-periodic intensity enhancements leading to HBO. The turbulent magnetospheric boundary may also be the site of electron acceleration which we observe through the HB radio flares.

When a critical value of mass accretion rate is exceeded, but still well below the neutron star Eddington limit, the inner disk suddenly flares up to form a thick torus (see Fig. 12b), engulfing the whole magnetosphere (Hasinger, 1988; HK 1989). All the dramatic changes observed at this transition can be explained by this postulated instability. The additional cool disk material in the line of sight causes the softening of the observed spectrum by Compton degradation (Hasinger, 1987b; Lamb, 1990). The interface between the disk torus and the magnetosphere is probably much less turbulent than in the horizontal branch and matter can accrete smoothly onto the neutron star. Therefore LFN and HB QPO are reduced in strength (although the existence of weak HB QPO in the upper normal branch indicates that some inhomogeneities are still present). The flux of energetic electrons producing the HB radio flares may be quenched in two ways: first, the reduced turbulence at the magnetospheric boundary may be less efficient in accelerating particles, and second, the thick inner disk torus will trap the electrons.

Down the normal branch the mass accretion rate is increasing even further and approaches the Eddington limit. At this moment the effective gravity is reduced by the strong radiation pressure, enough to slow down the dynamic timescales so that sound waves (Hasinger 1987a, Alpar et al., 1990) or other resonances in the flow (Lamb 1989, 1990) will become important and possibly form the 6 Hz normal-branch QPO.

Finally, on the flaring branch the mass accretion rate has increased above the Eddington limit and matter will be ejected, probably through the narrow tunnel formed by the thick inner disk torus (see Fig. 12c). The funnel might also effectively beam the X-ray radiation, so that the line-of sight inclination becomes important for a distant observer. This may explain the different intensity behaviour on the flaring branch between e.g. Sco X-1 and Cyg X-2 (Hasinger et al., 1989)

References

Alpar, M.A., Shaham, J.: 1985, *Nature* **316**, 239

Alpar, M.A., Hasinger, G., Shaham, J., Yancopoulos, S., 1990 (in prep.)

Canizares C.R. et al.: 1975 *Ap. J.* **197**, 457

Hasinger, G.: 1987 a, *Astron. Astrophys.* **186**, 153.

Hasinger, G.: 1987 b, *IAU Symp.* **125**, 333.

Hasinger, G.: 1988 Proc. Symp. *"Physics of Neutron Stars and Black Holes"*, Tokyo: Universal Academy Press, p. 97

Hasinger, G., Priedhorsky, W.C., Middleditch, J.: 1989, *Ap. J.* **337**, 843.

Hasinger, G., van der Klis, M.: 1989, *Astron. Astrophys.* **225**, 79

Hasinger, G., van der Klis, M., Ebisawa, K., Dotani, T., Mitsuda, K.: 1990 *Astron. Astrophys.* (in press)

Hjellming, R.M., Han, X.H., Cordova, F.A., Hasinger, G.: 1990 *Astron. Astrophys.* (in press)

Lamb, F.K., Shibazaki, N., Alpar, M.A., Shaham, J.: 1985 *Nature* **317**, 681

Lamb, F.K.: 1989, Proc. 23rd ESLAB Symposium, ESA SP-296, Part 1., p. 215

Lamb, F.K.: 1990, *Ap. J.* (in press)

Penninx, W., Lewin, W.H.G., Zijlstra, A.A., Mitsuda, K., van Paradijs, J., van der Klis, M.: 1988 *Nature* **336**, 146

Priedhorsky, W., Hasinger, G., Lewin, W.H.G., Middleditch, J., Parmar, A., Stella, L., White, N.: 1986, *Ap. J. Lett.* **306**, L91.

Shakura, N.I., Sunyaev, R.A.: 1973 *Astron. Astrophys.* **24**, 337

Van der Klis, M., Jansen, F., van Paradijs, J., Lewin, W.H.G., van den Heuvel, E.P.J., Trümper, J.E., and Sztajno, M.: 1985 *Nature* **316**, 225.

Van der Klis, M.: 1989a, Proc. NATO ASI *"Timing Neutron Stars"*, Çeşme, Turkey. Dordrecht: Reidel p.27

Van der Klis, M.: 1989b, *Ann. Rev. Astron. Astrophys.* **27**, 517

Van Paradijs, J., Allington-Smith, J., Callanan, P., Charles, P.A., Hassal, B.J.M., Machin, G., Mason, K.O., Naylor, T., Smale, A.P.: 1990 *Astron. Astrophys.* (in press)

Vrtilek, S.D., Raymond, J.C., Garcia, M.R., Verbunt, F., Hasinger, G., Kürster, M.: 1990 *Astron. Astrophys.* (in press)

SIGHT: A SCINTILLATION IMAGING GAS-FILLED HARD X-RAY TELESCOPE

T.K. Edberg[†], B. Sadoulet, K. Hurley, R.P. Lin, A. Parsons, S. Weiss, J. Wilkerson
*Center for Particle Astrophysics, Space Sciences Laboratory, and
University of California, Berkeley*

G. Smith
Lawrence Berkeley Laboratory

T.A. Prince, S.M. Schindler, W.R. Cook
California Institute of Technology

ABSTRACT. We are building a Scintillation Imaging Gas-filled Hard X-ray Telescope (SIGHT), a hard X-ray telescope optimized for 30 keV - 300 keV. The central detector of this telescope is a high pressure xenon scintillation time projection chamber, using waveshifting fibers in a novel scheme to read out the light. This detector should have good energy resolution (~ 5 times better than NaI, ~ 3 times worse than Ge), excellent imaging capabilities (~ 200 μm position resolution in the detector leading to ~ 1'20" mapping resolution on the sky), and outstanding sensitivity due to the large effective area and low detector background (the sensitivity is competitive with many satellites even though the SIGHT integration time is assumed 100 times shorter). With these detector capabilities, SIGHT is well-suited for observations of X-ray binary systems, as well as nuclear lines from supernovae, active galactic nuclei, and the diffuse X-ray flux. SIGHT is funded by NASA for a proposed first flight in 1993. If proven successful, this technology can evolve in several directions for future experiments: extrapolation to larger detectors for deep pointed observations, hemispherical coded aperture masks for continual coverage of half the sky, and Compton telescopes utilizing imaging of the recoil electron.

† *Center for Particle Astrophysics, 301 LeConte Hall, University of California at Berkeley, Berkeley, CA 94720. (Bitnet: edberg@lbl.)*

E. P. J. van den Heuvel and S. A. Rappaport (eds.), X-Ray Binaries and Recycled Pulsars, 75.
© *1992 Kluwer Academic Publishers.*

CHAPTER 2

Binary & Millisecond Pulsars

MILLISECOND PULSARS IN GLOBULAR CLUSTERS

A. G. Lyne
University of Manchester
Nuffield Radio Astronomy Laboratories
Jodrell Bank
Macclesfield
Cheshire SK11 9DL, United Kingdom

ABSTRACT. We review the various surveys of globular clusters for radio pulsars. During the four years since the first was discovered in 1987, a total of 30 pulsars have now been detected, of which 22 have periods of less than 20 milliseconds and 14 are in binary systems. The various problems and techniques used are discussed and we briefly consider the implications of the large number of detections for the formation processes and cluster evolution.

1. INTRODUCTION

Millisecond pulsars are thought to be the direct descendants of low-mass X-ray binary systems (LMXBs) (Alpar *et al.* 1982; Fabian *et al.* 1983; Webbink, Rappaport & Savonije 1983; Radhakrishnan & Srinivasan 1984) in which old neutron stars are spun up by accretion of material from a companion star as it overflows its Roche lobe during a giant phase (Smarr & Blandford 1976). Present statistics indicate that the probability of LMXB occurrence is a few orders of magnitude higher for stars in globular clusters than in the galactic disk (Verbunt & Hut 1987). This is underlined by the fact that globular clusters contain only $\sim 0.05\%$ of the mass of the Galaxy, but almost 20% of the LMXBs (Verbunt 1987). Globular clusters make particularly suitable birthplaces for LMXBs because of their high stellar concentrations and attendant higher rate of binary formation (Fabian, Pringle & Rees 1975). On the basis of these arguments, it might be expected that millisecond pulsars will exhibit the same bias (Fabian *et al.* 1983). However, until recently, no large-scale pulsed searches for millisecond pulsars exploited this possible association, mainly because of insufficient computing resources.

For this reason Hamilton, Helfand & Becker (1985) used the VLA in a survey of 12 globular clusters looking for unresolved sources which might be millisecond pulsars. Their only promising detection was a weak source deep within the core of M28. Later work on this object revealed a steep spectrum and a high degree of linear polarisation (Erickson *et al.* 1987), both of which are characteristic of pulsar radiation. Subsequently, with the aid of the 76-m Lovell telescope at Jodrell Bank and a Cray X-MP supercomputer, this object was found to be a pulsar with a period of 3.01 ms (Lyne *et al.* 1987). Somewhat mysteriously, the pulsar, PSR 1821-24, was solitary, any companion responsible for its spin-up having disappeared. This discovery led to a great increase in activity in this area and within 6 months a second millisecond pulsar was discovered, PSR 1620-26, with a period of 11 ms (Lyne *et al.* 1988), this time in the core of the cluster M4 (Goss, Kulkarni & Lyne 1988). This one was indeed in a binary system, supporting the formation hypothesis.

Subsequent searches have involved two research groups, one using the 1000ft Arecibo

79

E. P. J. van den Heuvel and S. A. Rappaport (eds.), X-Ray Binaries and Recycled Pulsars, 79–86.
© 1992 *Kluwer Academic Publishers.*

antenna, and the other the Jodrell Bank 76m and Parkes 64m telescopes. These have been augmented by continuum observations with the VLA. A total of 30 pulsars have now been discovered in the globular cluster system, this despite the large distance of most of the clusters from the Sun.

2. SEARCH TECHNIQUES

The main problem with searches for such pulsars is that they are very weak, so that only the largest radio-telescopes can be used. Long integration times and large total bandwidths are also required in order to achieve the required high sensitivity. Interstellar dispersion requires the sampling of many narrow frequency channels to prevent smearing of the pulses. The result is that a typical one-hour observation of a cluster requires the recording of more than 10^9 data samples for processing. Large, powerful computers are then required to dedisperse the data for many trial values of dispersion measure (DM), and to study the resulting time sequences for periodicity. Further, if the pulsar is a member of a close binary system, it is likely to experience an acceleration which will cause the received periodicity to vary during the observation, making detection difficult. A search in acceleration involves a series of quadratic "stretching" operations on the data prior to the standard periodicity search. For time sequences of a few (N_6) million points in length, a pulsed search on a Cray X-MP, Cyber 205, Amdahl VP or similar supercomputer requires between N_6 and $10N_6$ seconds of CPU time. A complete search over N_{DM} values of dispersion measure and N_{ACC} values of acceleration will thus take about

$$T_{CPU} \sim 3 \times N_6 \times N_{DM} \times N_{ACC} \text{ seconds.}$$

For instance, for a one-hour observation on a single cluster, $N_6 \sim 10$ and searches over 300 values of DM and 300 values of acceleration might be appropriate, requiring $3 * 10^6$ seconds or one month of CPU time. This is clearly not practical and more economical strategies need to be employed.

One of these is to perform "non-acceleration" searches just over DM in order to detect a first pulsar which will provide the DM of the cluster. Of course, once the dispersion measure of a cluster has been found in this way, one need search for other pulsars over only a small range of dispersion measure, since there is little ionised gas within globular clusters. If such a search is successful, then the required computational resources for further searches are reduced by perhaps two orders of magnitude, releasing them for acceleration searches. There are two dangers in adopting this approach: firstly it relies on the cluster having at least one solitary or long-period binary pulsar of sufficient luminosity. Secondly, there is a finite chance of such deep searches detecting a field pulsar, resulting in the subsequent search at the wrong DM.

An alternative strategy is to use a synthesis instrument, such as the VLA, to map clusters in order to detect pulsars as continuum radio sources, in the way that Hamilton, Helfand & Becker (1985) did, leading to the discovery of the pulsar in M28. If successful, then pulsed searches can be performed only on those clusters containing sources. The main advantages of this method are that it is sensitive to rapidly accelerating pulsars in close binary systems and it uses little CPU resource in the first phase. One of the problems is that some clusters contain radio-emitting LMXBs and may trigger CPU-intensive pulsed

TABLE 1: Globular Cluster Pulsars Discovered
in the Arecibo Surveys

Cluster	Pulsar	P_{bc} (ms)	DM	P_{orb}	Reference
M53	1310+18	33.163	25	255 days	Anderson et al. (1989b)
M5	1516+02A	5.553	29.5	Single	Wolszczan et al. (1989a)
	1516+02B	7.947	29.5	6.9 days	"
M13	1639+36A	10.378	30.4	Single	Anderson et al. (1989a)
	1639+36B	3.528	29.5	1.26 days	Anderson et al. (1991)
NGC 6760	1908+00	3.6	200	Single	Anderson et al. (1990a)
M15	2127+11A	110.665	67.25	Single	Wolszczan et al. (1989b)
	2127+11B	56.133	67.3	Single	Anderson et al. (1990b)
	2127+11C	30.529	67.1	8.0 hrs	"
	2127+11D	4.65	67.3	Single	Anderson et al. (1991)
	2127+11E	4.80	67.3	Single	"

searches unnecessarily. Since most pulsars have very steep spectral indices and often have high degrees of linear polarisation, spectral and polarisation information can be used to distinguish between LMXBs and pulsars. On the whole, this method is not as sensitive as lower frequency pulsed searches.

3. ARECIBO SEARCHES

The searches using the Arecibo telescope have resulted in the discovery of the 11 short period pulsars listed in Table 1. Of these, about half are true "millisecond" pulsars and 5 are in the one cluster M15. The first of the M15 pulsars to be discovered, PSR 2127+11A, has a period of 0.110 seconds and provided quite a surprise. Although it was clearly not in a binary system, it was the first of 500 known pulsars to have a negative period derivative (Wolszczan et al. 1989b). In other words it is apparently spinning up. This is clearly due to the acceleration of the pulsar towards the Earth as it moves in the gravitational potential well of the cluster as a whole. The pulsar PSR 2127+11C, also in this cluster, was subsequently found in an acceleration search conducted at the same DM as PSR 2127+11A (Anderson et al. 1990b). It is a member of an 8-hour eccentric binary system, very similar to the well known Hulse-Taylor system containing PSR 1913+16 which has proved to be such an exciting test-bed for theories of relativity. This also clearly contains two neutron stars, but unfortunately it will not be possible to use this pulsar for such precise studies because it is obviously not in an inertial frame. It is also somewhat unusual in that, while it is clearly associated with the cluster, it lies many cluster core radii away on the sky. The other four pulsars lie within the core of the cluster as does an LMXB, AC211, in which presumably another neutron star is in the process of being spun up. Some violence at some phase in the formation of the system must have kicked PSR 2127+11C out of the core. The small variations in DM amongst the pulsars is attributable to small-scale irregularities in the dispersing electrons across the line-of sight within the galactic disk.

TABLE 2: Globular Cluster Pulsars Discovered
in the Jodrell Bank/Parkes Surveys

Cluster	Pulsar	P_{bc} (ms)	DM	P_{orb}	Reference
47 Tuc	0021–72C	5.757	24.4	Single	Manchester et al. (1990)
	0021–72D	5.358	24.7	Single	Manchester et al. (1991)
	0021–72E	3.536	24.2	2.2 days	"
	0021–72F	2.624	24.4	Single	"
	0021–72G	4.040	24.2	Single	"
	0021–72H	3.210	24.3	~ days ?	"
	0021–72I	3.485	23.7	~ days ?	"
	0021–72J*	2.101	24.6	2.9 hours	"
	0021–72K	1.786	24.9	Binary	"
	0021–72L	4.346	24.5	?	"
	0021–72M	3.677	24.4	Binary	"
M4	1620–26	11.08	62.9	191 days	Lyne et al. (1988)
NGC 6342	1718–19	1004.0	70	6 hours	Biggs & Lyne (1991)
NGC 6440	1745–20	288.6	220	Single	Manchester et al. (1989b)
Terzan 5	1744–24A	11.56	242	1.8 hrs	Lyne et al. (1990)
	1744–24B**	442.8	210	Single	"
NGC 6539	1802–07	23.10	187	2.6 days	D'Amico et al. (1990)
NGC 6624	1820–30A	5.440	86.8	Single	Biggs et al. (1990)
	1820–30B	378.6	86.7	Single	"
M28	1821–24	3.054	120	Single	Lyne et al. (1987)

* *Possibly 4.201 ms pulsar with strong interpulse.*
** *Possibly a foregound object not associated with the cluster.*
Note that PSR 0021–72A (Ables et al. 1989) and PSR 0021–72B (Ables et al. 1988) are not thought to be located within 47 Tuc and efforts to detect them have proved unsuccessful (Manchester et al. 1990).

The other 6 of the Arecibo pulsars lie in the clusters M5, M13, M53 and NGC 6760, all of which are clusters of rather modest density, in which the collision rate and hence formation rate of binary systems is expected to be very low. This suggests that many of the LMXBs and hence millisecond pulsars may be formed in primordial binary systems.

4. JODRELL BANK/PARKES SEARCHES

The surveys conducted at Jodrell Bank and Parkes have resulted in the discovery of a total of 19 pulsars, including those in M28 and M4. These are listed in Table 2. One of these is a most exciting source, in the cluster Terzan 5 (Lyne et al. 1990). This is a pulsar with a period of 11.1 milliseconds in a binary system with an orbital period of only 1.8 hours. Moreover, it is occulted by its companion, often for more than half the orbit, and occasionally for a complete orbit. The pulsar is embbedded within the atmosphere of the

TABLE 3: Continuum Sources in Globular Clusters

Cluster	α_{1950}	δ_{1950}	$S_{1400}(mJy)$	Sp Index	Identification	Ref
M3	$13^h39^m53.1^s$	$+28°37'45''$	0.18 ± 0.05		?	K
M4	$16^h20^m34.1^s$	$-26°24'58''$	1.5 ± 0.2		PSR 1620-26	K
Terzan 5	$17^h44^m59.9^s$	$-24°45'39''$	1.9 ± 0.2	-1.6	?	F
NGC 6440	$17^h45^m54.2^s$	$-20°20'45''$	1.45 ± 0.10	< -2.0	PSR 1745-20	F
NGC 6539	$18^h02^m07.4^s$	$-07°35'39''$	0.90 ± 0.15		PSR 1802-07	F
NGC 6544	$18^h04^m15.3^s$	$-25°00'18''$	0.9 ± 0.3		?	F
NGC 6624	$18^h20^m27.9^s$	$-30°23'20''$	0.95 ± 0.10	-0.7	LMXB	F
M28	$18^h21^m27.4^s$	$-24°53'51''$	1.8 ± 0.2		PSR 1824-21	H,K
M15	$21^h27^m33.3^s$	$+11°56'51''$	1.8 ± 0.2		LMXB	K
M15	$21^h27^m33.2^s$	$+11°56'49''$	0.34 ± 0.09		PSR 2127+11A	K

REFERENCES: F: Fruchter & Goss (1990) H: Hamilton, Helfand & Becker (1985) K: Kulkarni et al. (1991)

companion which is less than 0.1 solar masses and seems to be in the process of evaporation by the intense pulsar radiation (Lyne *et al.* 1990; Nice *et al.* 1990). This pulsar does not lie in the core of the cluster but about 30 arcsec or 10 core radii away.

Within the last year, we have conducted a deep search of the globular cluster 47 Tucanae with the Parkes radiotelescope. We have previously discovered a 5.75 millisecond pulsar with a dispersion measure of 24.5 pc cm^{-3} (Manchester *et al.* 1990). With a local search around this dispersion measure, a further 10 millisecond pulsars have been discovered in the cluster, all with period of less than 6 milliseconds (Manchester *et al.* 1991). More than half of the 11 are in binary systems, one with an orbital period of only about 3 hours and which is probably eclipsed by its companion. More than half the known millisecond pulsars with such short period and more than a quarter of the known binary pulsars lie within this one globular cluster. This preponderance of millisecond pulsars contrasts strongly with M15 in which three of the 5 pulsars have period in excess of 30 milliseconds. What is so special about this cluster to produce such an abundance of millisecond pulsars is not clear, but the discovery does reveal that such clusters must have had many massive stars in their youth in order to produce such an abundance of neutron stars. There is clearly much to be learned about these pulsars and their binary systems in the future. Moreover, with such a large number in this one cluster, measurement of their positions and period derivatives will give an interesting insight into the distribution of mass through the cluster.

5. CONTINUUM OBSERVATIONS

The main continuum searches for pulsars in globular clusters have been conducted by Hamilton, Helfand & Becker (1985), Fruchter & Goss (1990) and Kulkarni *et al.* (1991) all at frequencies around 1400 MHz using the VLA. Out of a total of about two dozen clusters which have been studied in these programs, 10 sources have been detected and these are listed in Table 3. Five of these are almost certainly identified with known pulsars and include the original detection of PSR 1821-24 in M28. Two, PSR 1620-26 in M4

(Lyne *et al.* 1988) and PSR 2127+11A (Wolszczan *et al.* 1989b) in M15, were originally discovered in pulsed searches. No other pulsars are seen in these maps, despite the fact that there are 4 others present in M15. The remaining two detections, of PSR 1745-20 in NGC 6440 (Manchester *et al.* 1989b) and PSR 1802-07 in NGC 6539 (D'Amico *et al.* 1990) pre-dated the pulsed detections and assisted in their discovery.

Of the remaining 5 sources, two are identified with LMXBs and 3 so far have no identification. The most notable of these is the source near the core of Terzan 5. It is almost certainly pulsar radiation, because of its steep spectrum and variability. This source is not PSR 1744-24A (Lyne *et al.* 1990) since timing measurements show that the position of this pulsar is coincident with another source, which was subsequently found in the VLA maps and which lies 30 arcsec or ten core radii from the centre. Neither is it PSR 1744-24B (Lyne *et al.* 1990) whose flux is only 0.5 mJy. Even though it has a relatively high flux density, all efforts to detect the pulsations have failed so far. The most likely reasons for this are either that it has a very short period, is accelerating rapidly in a short-period massive binary system or that the source is in fact a composite of several pulsars each of much smaller flux density.

The lack of pulsations found in the remaining two sources in M3 and NGC 6544 may be for the same reasons if they are pulsars, or they may be accretion sources with low X-ray luminosity. Spectral information is required to distinguish these possibilities.

6. CONCLUSION

The advent of larger and more powerful computers, together with improved data reduction techniques has allowed the development of very sensitive pulsed searches for millisecond pulsars, covering a wide range of parameter space. The continuum observations suggest that there is still a region of this space not yet penetrated by these searches, in either period or acceleration, and efforts will be made to remedy this deficiency as resources become available.

These pulsars and future discoveries will provide information on the evolution of millisecond pulsars in globular clusters and their probable relation to low-mass x-ray binaries. The success of these searches in finding pulsars in globular clusters is somewhat surprising in view of the large distances to most of them and indicates a large neutron star population in the clusters. This presumably results from a large initial population of massive stars and provides an insight into the initial mass functions of the clusters. However a good quantitative assessment of the neutron star population requires an accurate knowledge of the sensitivities of the various surveys and the clusters searched, together with a good knowledge of the millisecond pulsar luminosity function and beaming factor. We have very poor knowledge of the last two because of the very small numbers of millisecond pulsars so far known, particularly low luminosity ones. It is to be hoped that this situation will improve before long as new, all-sky surveys are completed.

The clusters containing multiple pulsars are particularly interesting, as we have the possibility of studying the mass distribution through the cluster from measurements of the spatial distribution and accelerations of the pulsars. It will be interesting to see how much dark matter exists in the clusters. Apart from their value in studying globular clusters, the new pulsars are also a valuable collection to provide a much improved study of the physics of millisecond pulsars.

References

Ables, J. G., Jacka, C. E., McConnell, D., Hamilton, P. A., McCulloch, P. M. & Hall, P. J., 1988. *IAU Circ. No. 4602.*

Ables, J. G., McConnell, D., Jacka, C. E., McCulloch, P. M., Hall, P. J. & Hamilton, P. A., 1989. *Nature*, **342**, 158.

Alpar, M. A., Cheng, A. F., Ruderman, M. A. & Shaham, J., 1982. *Nature*, **300**, 728.

Anderson, S., Kulkarni, S., Prince, T. & Wolszczan, A., 1989a. *IAU Circ. No. 4819.*

Anderson, S., Kulkarni, S., Prince, T. & Wolszczan, A., 1989b. *IAU Circ. No. 4853.*

Anderson, S., Kulkarni, S., Prince, T. & Wolszczan, A., 1990a. *IAU Circ. No. 5013.*

Anderson, S. B., Gorham, P. W., Kulkarni, S. R., Prince, T. A. & Wolszczan, A., 1990b. *Nature*, **346**, 42.

Anderson, S., Kulkarni, S., Prince, T. & Wolszczan, A. 1991. Unpublished.

Biggs, J. D. & Lyne, A. G. 1991. Unpublished work.

Biggs, J. D., Lyne, A. G., Manchester, R. N. & Ashworth, M., 1990. *IAU Circ. No. 4988.*

D'Amico, N., Lyne, A. G., Bailes, M., Johnston, S., Manchester, R. N., Staveley-Smith, L., Lim, J., Fruchter, A. S. & Goss, W. M., 1990. *IAU Circ. No. 5013.*

Erickson, W. C., Mahoney, M. J., Becker, R. H. & Helfand, D. J., 1987. *Astrophys. J.*, **314**, L45.

Fabian, A. C., Pringle, J. E., Verbunt, F. & Wade, R. A., 1983. *Nature*, **301**, 222.

Fabian, A. C., Pringle, J. E. & Rees, M. J., 1975. *Mon. Not. R. astr. Soc.*, **172**, 15P.

Fruchter, A. S. & Goss, W. M., 1990. *Astrophys. J. Lett.*, **365**, L63.

Goss, W. M., Kulkarni, S. R. & Lyne, A. G., 1988. *Nature*, **332**, 47.

Hamilton, T. T., Helfand, D. J. & Becker, R. H., 1985. *Astron. J.*, **90**, 606.

Kulkarni, S., Goss, W., Wolszczan, A. & Middleditch, J., 1991. *Astrophys. J. Lett.*, . submitted.

Lyne, A. G., Brinklow, A., Middleditch, J., Kulkarni, S. R., Backer, D. C. & Clifton, T. R., 1987. *Nature*, **328**, 399.

Lyne, A. G., Biggs, J. D., Brinklow, A., Ashworth, M. & McKenna, J., 1988. *Nature*, **332**, 45.

Lyne, A. G., Manchester, R. N., D'Amico, N., Staveley-Smith, L., Johnston, S., Lim, J., Fruchter, A. S., Goss, W. M. & Frail, D., 1990. *Nature*, **347**, 650.

Manchester, R. N., Lyne, A. G., Johnston, S., D'Amico, N., Lim, J., Kniffen, D. A., Fruchter, A. S. & Goss, W. M., 1989. *IAU Circ. No. 4905*.

Manchester, R. N., Lyne, A. G., D'Amico, N., Johnston, S., Lim, J. & Kniffen, D. A., 1990. *Nature*, **345**, 598.

Manchester, R. N., Lyne, A. G., Robinson, C., D'Amico, N., Bailes, M. & Lim, J. 1991. Unpublished work.

Nice, D. J., Thorsett, S. E., Taylor, J. H. & Fruchter, A. S., 1990. *Astrophys. J. Lett.*, **361**, L61.

Radhakrishnan, V. & Srinivasan, G., 1984. In: *Proc. 2nd Asian-Pacific Regional Meeting of the IAU*, p. 423, eds Hidayat, B. & Feast, M. W., Tira Pustaka, Jakarta.

Smarr, L. L. & Blandford, R., 1976. *Astrophys. J.*, **207**, 574.

Verbunt, F. & Hut, P., 1987. In: *The Origin and Evolution of Neutron Stars, IAU Symposium no. 125*, p. 187, eds Helfand, D. J. & Huang, J., Reidel, Dordrecht.

Verbunt, F., 1987. *Astrophys. J. Lett.*, **312**, L23.

Webbink, R. F., Rappaport, S. & Savonije, G. J., 1983. *Astrophys. J.*, **270**, 678.

Wolszczan, A., Anderson, S., Kulkarni, S. & Prince, T., 1989a. *IAU Circ. No. 4880*.

Wolszczan, A., Kulkarni, S. R., Middleditch, J., Backer, D. C., Fruchter, A. S. & Dewey, R. J., 1989b. *Nature*, **337**, 531.

RECENT OBSERVATIONS OF RECYCLED PULSARS[1]

J. H. TAYLOR
Joseph Henry Laboratories and Physics Department
Princeton University
Princeton, NJ 08544 USA

ABSTRACT. More than 30 of the 520+ known radio pulsars show evidence of "recycling" during a spin-up phase in which mass and angular momentum were accreted from an evolving companion star. Recycled pulsars spin faster and live longer than the ordinary variety; they are also considerably more stable as clocks, and most of them are found in gravitationally bound binary systems. For all of these reasons, they make exquisite experimental tools for investigations ranging from fundamental physics to stellar evolution to cosmology. My colleagues and I have put much recent effort into observing recycled pulsars; I summarize here some recent work, much of it still in progress, involving especially PSRs 1744-24A, 1855+09, 1913+16, and 1957+20.

1. Introduction

A surprisingly large fraction of the most useful observations of pulsars involve measurements of their pulse times of arrival (TOA's) on Earth. If repeated over many months, timing observations give access to very high precision astrometry, effectively using the Earth's orbit as the baseline of an extremely stable, phase-coherent interferometer. They also permit accurate characterization of the deterministic spin-down behavior of the rotating neutron stars. Typical accuracies for measured TOA's are around 10^{-3} periods, and with special effort accuracies of 10^{-4} periods can be achieved. These numbers make possible milli-arcsecond astrometry, 15-significant-digit pulsar periods, and highly accurate orbital elements of binary pulsars. Such high precision results can lead to important results in cosmology, fundamental physics, and astrophysics. In this paper I will concentrate on some very recent results on four recycled pulsars, PSRs 1744-24A, 1855+09, 1913+16, and 1957+20. In some cases the new results represent updates of work published at an earlier stage; in other cases the new information is a preview of results slated for publication in more complete form elsewhere.

2. PSR 1913+16

The first recycled radio pulsar to be discovered [1] was found in 1974, about 7 years after the discovery of pulsars. PSR 1913+16 has a 59 ms period and moves in an eccentric, 8 hour orbit. Suggestions were made soon afterward that its evolutionary history involved an accretion phase [2], although a full appreciation of the significance of recycling took several more years and benefited from both observational and theoretical breakthoughs

---

[1]Based on work in collaboration with D. J. Nice, M. F. Ryba, S. E. Thorsett, and J. M. Weisberg

E. P. J. van den Heuvel and S. A. Rappaport (eds.), X-Ray Binaries and Recycled Pulsars, 87–92.

[3, 4]. The experimental challenge of measuring general relativistic effects in the orbital motion of this pulsar motivated special efforts aimed at improving the accuracy of its TOA's [5–8]. As it turned out, these efforts proved timely and fortuitous when further examples of recycled pulsars were found and their accurate timing became highly desirable [9–12].

PSR 1913+16 has now been observed regularly for more than 16 years, using the Arecibo Observatory's 305 m telescope. The most recent session was completed just six weeks ago, in December 1990. Early hopes that new tests of relativistic gravity might be obtained from timing observations of this pulsar have been more than fulfilled. In a binary system having seven important unknown parameters, ten distinct phenomenological orbital parameters have been measured—six of them with fractional accuracies in the parts-per-million range, or better, and two more with accuracies around a part in a thousand. The astrophysical parameters of this orbiting system are therefore over-determined, and the excess of observational constraints provide access to explicit tests of gravitation theories. The best known test involves comparison of the measured secular decrease of orbital period with the value predicted from the general relativistic "quadrupole formula" for gravitational radiation damping. Damour and Taylor [13] have shown that the present level of accuracy requires small corrections for the galactic accelerations of the pulsar and the Sun, and the proper motion of the pulsar. An updated summary of their error budget for the experiment is contained in the following table:

Table 1. Error budget for orbital decay measurements of PSR 1913+16.

Quantity	Value (10^{-12})
Observed $\dot{P}_b$	-2.4252 ± 0.0099
Galactic contribution ...	-0.0167 ± 0.0055
Observed $-$ Galaxy	-2.4085 ± 0.0113
Theory (GR)	-2.4026 ± 0.0001
(Obs$-$Gal)/Theory	1.0023 ± 0.0047

As shown in the last line of the table, this quantitative proof of the existence of gravitational radiation now agrees with the general relativistic prediction to better than 0.5%.

3. PSR 1855+09

Motivated in part by the rich results produced by the timing measurements of PSR 1913+16, observers have been eager to accept the new challenges posed by the discovery millisecond pulsars. The first such object, PSR 1937+21, remains the prototype of the class; it provides excellent examples of the interesting results obtainable from accurate timing observations even of non-binary pulsars. Stinebring et al. [14] recently published a summary of results from a series of bi-weekly timing observations extending over more than 7 years for PSR 1937+21, and nearly 4 years for PSR 1855+09—providing, among other things, a cosmologically significant limit on the energy density of low-frequency gravitational waves in the universe.

PSR 1855+09 is a pulsar with 5.3 ms period in a nearly circular, 12.3 d orbit [15]. The orbit is much less relativistic than that of PSR 1913+16: the maximum pulsar velocity is only $v/c \approx 6 \times 10^{-5}$, as opposed to 10^{-3}, and the small orbital eccentricity ($e \approx 2 \times 10^{-5}$)

and much longer orbital period preclude any hope of being able to measure relativistic effects seen easily in the PSR 1913+16 system, including perihelion advance, variations in time dilation and gravitational redshift, and orbital period decay. However, Nature kindly arranged the orbital plane of PSR 1855+09 to make a very small angle with our line of sight, thereby rendering measurable the general relativistic "Shapiro delay" of the pulsar signals. Fig. 1 illustrates the observed excess delays as a function of orbital phase, together with the fitted theoretical curve [16].

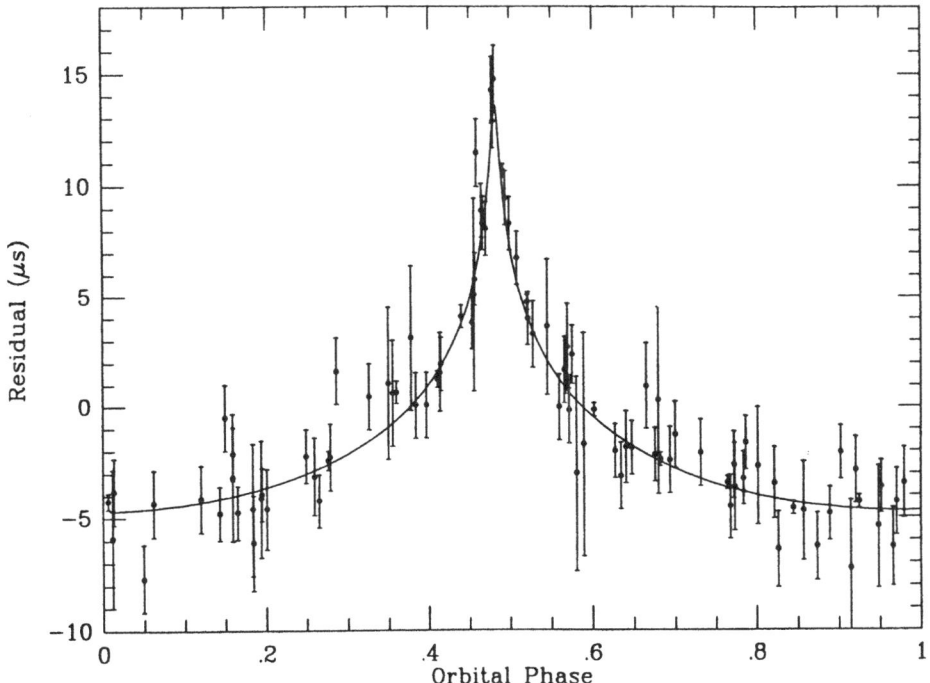

Figure 1. Observed excess delays in the PSR 1855+09 system, superimposed on a theoretical curve giving the relativistic "Shapiro delay" for a companion mass $m_2 = 0.23\,M_\odot$. The observations were made at frequencies near 1400 MHz.

Two parameters determine the shape and amplitude of the theoretical curve plotted in Fig. 1. Their values can be estimated from the data, and in turn they determine the mass m_2 of the companion star and the inclination i between the plane of the orbit and the plane of the sky [17, 8]. With the additional information provided by the Keplerian orbital elements, the Shapiro-delay parameters show that in the PSR 1855+09 system the pulsar and companion masses are, respectively, $m_1 = 1.27^{+0.23}_{-0.15}\,M_\odot$ and $m_2 = 0.23^{+0.026}_{-0.017}M_\odot$. The timing data are also precise enough, and extensive enough, to provide an accurate measurement of the proper motion of this binary system, $\mu_\alpha = -2.92 \pm 0.12$ and $\mu_\delta = -5.32 \pm 0.19\,\mathrm{mas\,yr^{-1}}$, and, for the first time, a significant timing measurement of a pulsar's annual parallax. We find the parallax to be $\pi = 1.2 \pm 0.5$ mas, corresponding to a distance of about 1 kpc [16].

4. PSR 1957+20

It often turns out that Nature is more imaginative than even the most creative-thinking of astrophysicists. However, the bizarre class of eclipsing binary pulsars, first epitomized by PSR 1957+20 [18] and more recently joined by PSR 1744−24A [19], were foreseen in a remarkable paper by Ruderman, Shaham, and Tavani [20]. There is now good evidence to support a conclusion that such systems represent the previously missing link in an evolutionary process leading to single millisecond pulsars [18, 21, 22]. My colleagues and I have continued to observe this pulsar regularly with the Arecibo telescope. In a paper published a year ago [23], we presented a number of new facts and details about the system, including its extremely small pulsar spindown rate, the frequency dependence of the eclipse durations, time variability of phenomena at the eclipse edges, and an upper limit on the magnetic field in the eclipsing gas.

I have a few more new facts to report today [24]. Regular timing observations, extending now over a 2.7 yr interval, have provided a reliable measurement of the proper motion of PSR 1957+20, namely $\mu_\alpha = -16.9 \pm 1.9$ and $\mu_\delta = -23.0 \pm 2.0$ mas yr^{-1}. The direction of motion is in excellent agreement with the direction suggested by the shape of the cometary emission nebula found in the field by Kulkarni and Hester [25]. This agreement confirms and establishes without doubt that the luminous nebulosity is a "pulsar-wind nebula" created by motion of the pulsar through the ambient interstellar medium, and excited by its relativistic wind. In addition, the same timing data now establish that the orbital period of PSR 1957+20 is gradually decreasing at the rate $\dot{P}_b = (-3.9 \pm 0.6) \times 10^{-11}$, implying that significant orbital evolution must take place on a time scale $P_b/\dot{P}_b = 30$ Myr. It is interesting to note that the dynamics of this system, including ablation of the companion's atmosphere by a relativistic pulsar wind, are so complicated that it has not been possible to predict theoretically even the sign of the expected orbital period derivative. One may hope that with some experimental guidance on the subject, a more complete understanding will follow.

We have now accumulated 44 observations of eclipse disappearance/reappearance events for PSR 1957+20. One would hope that these glimpses through edges of the partially transparent wind of the companion would provide important clues about the evaporation process taking place. We find that the duration of eclipses varies with observing frequency as $f^{-0.4}$. Just outside the regions of eclipse, we find excess propagation delays proportional to f^{-2}, consistent with the cold plasma dispersion relation. The delays can be converted to column densities of free electrons in the transition layer of the eclipsing region; logarithms of this quantity are plotted as a function of orbital phase in Fig. 2. The eclipsing region is clearly asymmetric. The column density e-folding scale corresponds to a change of orbital phase $\Delta\phi = 0.0043$, or a transverse path length of some 40,000 km at eclipse entrance. The corresponding numbers at eclipse exit are about three times larger.

5. PSR 1744−24A

Discovery of a second example of a recycled pulsar in an eclipsing binary system [19] showed that these systems are perhaps not so very rare—and that they can also occur in the population of recycled pulsars in globular clusters. PSR 1744−24A is an 11.5 ms pulsar in a 109 min orbit, and early observations showed that it disappears behind its 0.1 $M_\odot$ companion for a variable portion (sometimes as much as 100%) of each orbit [19, 26]. Recently we have established several additional facts about the pulsar. Timing observations show that the period derivative is surprisingly small, $|\dot{P}| < 3 \times 10^{-20}$. The apparent spin-down timescale is thus $P/\dot{P} > 10^{10}$ yr; quite possibly we are witnessing the partial cancellation of a somewhat larger, intrinsically positive, $\dot{P}$ by a negative contribution from acceleration of the pulsar in the cluster potential. The timing data also yield a limit on the orbital period derivative, in this case $P_b/\dot{P}_b > 5\,\mathrm{Myr}$; further observations should be able to tighten this limit considerably within a few years.

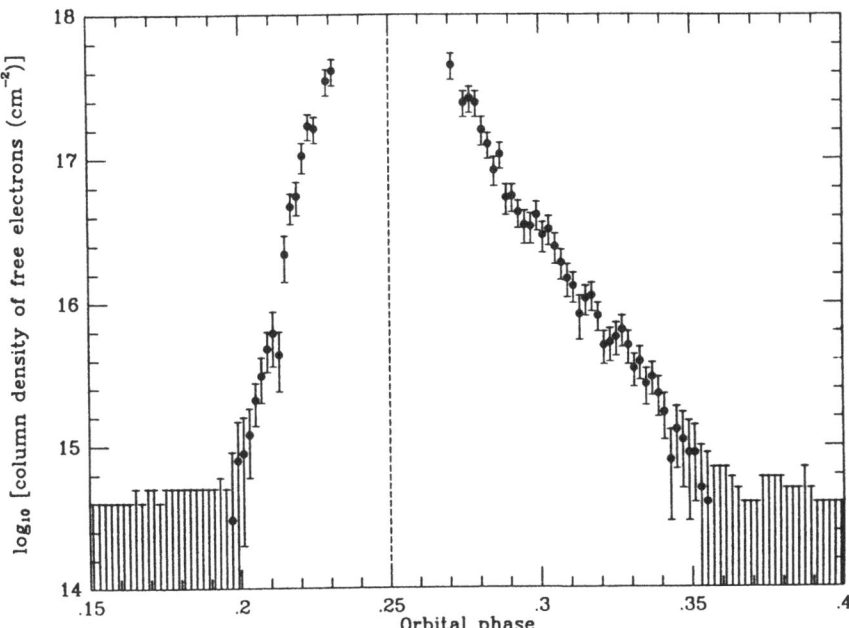

Figure 2. Profile of the free-electron column density in boundary regions of the eclipsing medium in the PSR 1957+20 system.

Perhaps the most interesting new result on PSR 1744−24A is that the total attenuation of pulsar signal by the eclipsing medium is sometimes no more than a factor of a few. In other words, the intervening medium, even at eclipse center, can be partially transparent. On the other hand, it has been known for some time that the pulsar signal sometimes disappears entirely for several entire orbits. It appears that in this system the evaporating gas flow from the companion is turbulent and variable.

References

[1] Hulse, R. A. and Taylor, J. H. *Astrophys. J.*, **195**, L51–L53 (1975).

[2] Smarr, L. L. and Blandford, R. *Astrophys. J.*, **207**, 574–588 (1976).

[3] Backer, D. C., Kulkarni, S. R., Heiles, C., Davis, M. M., and Goss, W. M. *Nature*, **300**, 615–618 (1982).

[4] Alpar, M. A., Cheng, A. F., Ruderman, M. A., and Shaham, J. *Nature*, **300**, 728–730 (1982).

[5] McCulloch, P. M., Taylor, J. H., and Weisberg, J. M. *Astrophys. J. (Letters)*, **227**, L133–L136 (1979).

[6] Taylor, J. H., Fowler, L. A., and McCulloch, P. M. *Nature*, **277**, 437–439 (1979).

[7] Taylor, J. H. and Weisberg, J. M. *Astrophys. J.*, **253**, 908–920 (1982).

[8] Taylor, J. H. and Weisberg, J. M. *Astrophys. J.*, **345**, 434–450 (1989).

[9] Backer, D. C., Kulkarni, S. R., and Taylor, J. H. *Nature*, **301**, 314–315 (1983).

[10] Davis, M. M., Taylor, J. H., Weisberg, J. M., and Backer, D. C. *Nature*, **315**, 547–550 (1985).

[11] Rawley, L. A., Taylor, J. H., and Davis, M. M. *Nature*, **319**, 383–384 (1986).

[12] Rawley, L. A., Taylor, J. H., and Davis, M. M. *Astrophys. J.*, **326**, 947–953 (1988).

[13] Damour, T. and Taylor, J. H. *Astrophys. J.*, **366**, 501–511 (1991).

[14] Stinebring, D. R., Ryba, M. F., Taylor, J. H., and Romani, R. W. *Phys. Rev. Lett.*, **65**, 285–288 (1990).

[15] Segelstein, D. J., Rawley, L. A., Stinebring, D. R., Fruchter, A. S., and Taylor, J. H. *Nature*, **322**, 714–717 (1986).

[16] Ryba, M. F. and Taylor, J. H. *Astrophys. J.*, **371**, 739–751 (1991).

[17] Damour, T. and Deruelle, N. *Ann. Inst. H. Poincaré (Physique Théorique)*, **44**, 263 (1986).

[18] Fruchter, A. S., Stinebring, D. R., and Taylor, J. H. *Nature*, **333**, 237–239 (1988).

[19] Lyne, A. G. *et al. Nature* **347**, 650–652 (1990).

[20] Ruderman, M., Shaham, J., and Tavani, M. *Astrophys. J.*, **336**, 507–518 (1989).

[21] Kluźniak, W., Ruderman, M., Shaham, J., and Tavani, M. *Nature*, **334**, 225–227 (1988).

[22] van den Heuvel, E. P. J. and van Paradijs, J. *Nature*, **334**, 227–228 (1988).

[23] Fruchter, A. S. *et al. Astrophys. J.*, **351**, 642–650 (1990).

[24] Ryba, M. F. and Taylor, J. H. *Astrophys. J.*, **380**, in press (1991).

[25] Kulkarni, S. R. and Hester, J. J. *Nature*, **335**, 801–803 (1988).

[26] Nice, D. J., Thorsett, S. E., Taylor, J. H., and Fruchter, A. S. *Astrophys. J.*, **361**, L61–L63 (1990).

DISCOVERY OF TWO MILLISECOND PULSARS AT HIGH GALACTIC LATITUDES

A. WOLSZCZAN
Arecibo Observatory
P.O. Box 995
Arecibo, Puerto Rico 00613

ABSTRACT. A 37.9–ms binary pulsar PSR1534+12 and a 6.2–ms pulsar PSR1257+12 have been discovered at high galactic latitudes with the Arecibo radiotelescope. PSR1534+12 is in a 10.1–hr, eccentric (e=0.274) orbit around a massive ($\sim$ 1.36 $M_\odot$), compact companion. High timing precision of this pulsar makes it a new, excellent tool to study gravity and test relativity theories. The new pulsars are nearby ($\sim$ 0.5 kpc) and have low intrinsic luminosities (8–10 mJy kpc^2 at 400 MHz). These discoveries indicate that the future all–sky surveys will have a good chance to detect a sizeable local population of millisecond pulsars.

1. Introduction

Among the pulsar surveys conducted in the past, which searched in the directions away from the galactic plane, only few were specifically designed to detect millisecond pulsars (Stokes et al. 1986; Backer, private communication). Until the time, when the search described in this paper was carried out, no detections of millisecond pulsars at high galactic latitudes had been reported. The undirected galactic plane surveys have discovered two nearby millisecond pulsars: PSR1855+09 (Segelstein et al. 1986) and PSR1957+20 (Fruchter, Stinebring and Taylor 1988). Despite the meager statistics, evolutionary studies (e.g. Kulkarni and Narayan 1988) and direct estimates (Fruchter 1989) suggest that the population of old, 'spun up' pulsars could be comparable in size to that of the so–called 'normal' pulsars. Motivated by these predictions, we have conducted a search for millisecond pulsars at high galactic latitudes using the Arecibo radiotelescope at 430 MHz. Here, we describe the two millisecond pulsars discovered during this search.

2. Observations

2.1 THE SURVEY

Observations were made in February 1990 with the Arecibo 305–m radiotelescope at 430 MHz using the 40 MHz correlation spectrometer and a 10 MHz receiver bandwidth. Details of the receiving system and the data acquisition and analysis procedures are described elsewhere (e.g. Wolszczan et al. 1989). The telescope was used as a transit instrument to drift–scan the sky observable from Arecibo at galactic latitudes $b \geq 30°$. The survey covered

E. P. J. van den Heuvel and S. A. Rappaport (eds.), X-Ray Binaries and Recycled Pulsars, 93–98.
© *1992 Kluwer Academic Publishers.*

$\sim$ 150 square degrees of the sky (Fig. 1) with a 6σ sensitivity of $\sim$1.5 mJy down to periods of about 4 ms and detected two new millisecond pulsars: PSR1257+12 and PSR1534+12 (Wolszczan 1991).

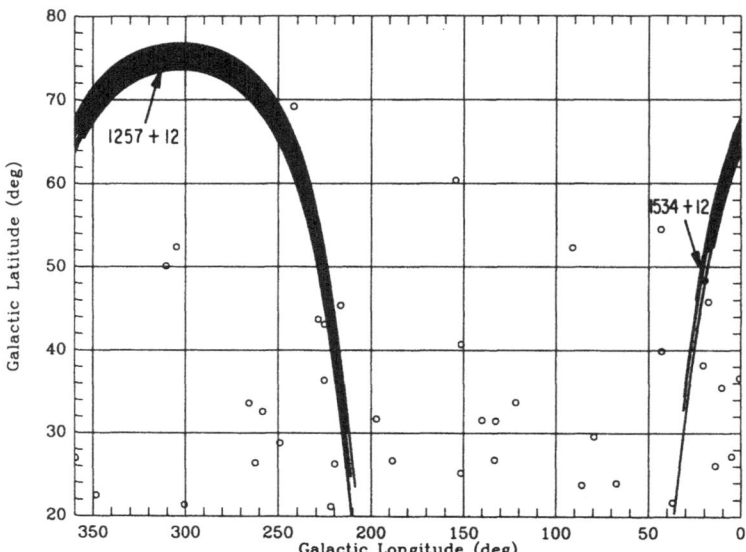

Figure 1. Sky coverage of the pulsar survey described in the text. Positions of new pulsars are indicated by arrows. Open circles denote locations of the previously known pulsars.

2.2 BASIC PROPERTIES OF THE NEW PULSARS

PSR1257+12 and PSR1534+12 were detected in the data acquired on February 5 and 9, respectively. PSR1257+12 has a period of 6.2 ms and its average flux density at 430 MHz is $\sim$ 20 mJy. The timing data accumulated so far suggest that the pulsar may be in a long period binary system, but more observations are necessary to verify this possibility. A 37.9-ms pulsar PSR1534+12, on the other hand, is in a 10.1–hr binary orbit around a massive companion. The average flux of this pulsar is $\sim$ 36 mJy. The integrated pulse profiles of both objects observed at 430 MHz are shown in Fig. 2. PSR1534+12 is of particular interest because of its extended off–pulse emission including a weak interpulse ($\sim$ 3% intensity level) separated from the main pulse by approximately half a period.

3. The 1534+12 Binary System

3.1 ANALYSIS OF PULSE ARRIVAL TIME DATA

Measurements of the barycentric period of PSR1534+12 as a function of orbital phase are shown in Fig. 3. The short, 10.1–hr orbital period and a $\sim$ 400 km/s radial velocity amplitude of the pulsar immediately suggest that its companion is another compact object.

Based on the measurements of pulse arrival times accumulated over the period of 10 months in 1990, a fairly complete timing model of PSR1534+12 has been derived. The analysis has been carried out with the aid of the model–fitting program TEMPO using an explicitly general–relativistic binary timing model, in which the post–Keplerian orbital parameters are expressed in terms of a total system mass M and a companion mass m_2 (DDGR model; Taylor and Weisberg 1989). The derived parameters of PSR1534+12 and its binary orbit are listed in Table 1. Post–fit residuals for the arrival times measured with the Princeton Mark III pulsar backend at 430 MHz did not exceed 5 μs for integration times of 2.5 minutes.

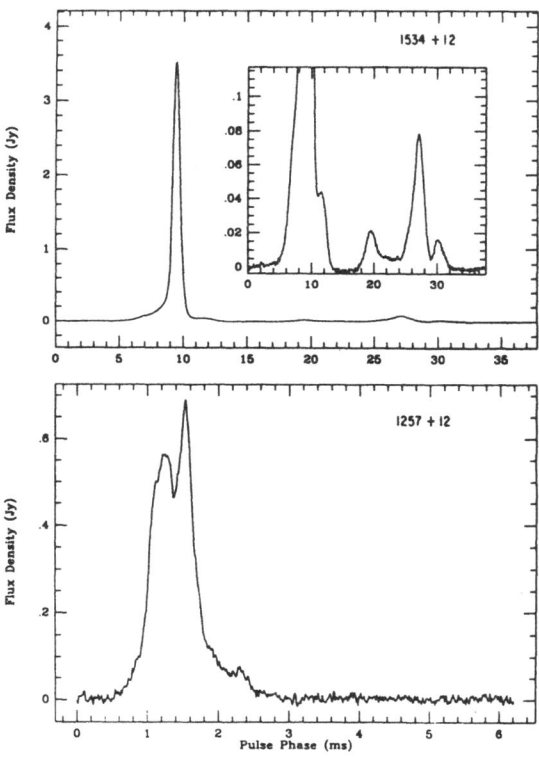

Figure 2. Average pulse profiles of PSR1257+12 and PSR1534+12 at 430 MHz. The inset shows low level emission of PSR1534+12 enhanced by a factor of 20.

3.2 MASSES OF THE COMPANION STARS

The small slowdown rate of PSR1534+12 implies that it is old and has a weak magnetic field (Table 1). These characteristics are typical of pulsars that underwent an accretion–induced spin–up and suggest that 1534+12 has evolved in a massive binary system (van den Heuvel and Taam 1984). Since the 1534+12 system is compact ($\sim 2R_\odot$) and eccentric and no eclipses of the pulsar are observed in spite of the high orbital inclination, the pulsar's companion is almost certainly another neutron star (Smarr and Blandford 1976). Consequently, the masses derived from the timing analysis of PSR1534+12 (Table 1), together with earlier measurements for the well-known 1913+16 binary system (Taylor and Weisberg 1989), represent the most accurate neutron star mass determinations to date and agree well with the theoretically predicted $1.3 - 1.35\ M_\odot$ range for the gravitational mass of a neutron star (Woosley 1986).

According to the standard model (van den Heuvel and Taam 1984), the companion of PSR1534+12 was formed as the result of a collapse of its helium star progenitor. Using the observed eccentricity and the pulsar mass and assuming a spherical explosion, it can be shown that the progenitor mass should be $\sim 2.1 M_\odot$, which appears to be at the lower limit to a helium core mass that would still collapse to a neutron star ($\geq 2.2 M_\odot$; e.g. Habets 1986). A more massive progenitor requires invoking an asymmetry in the supernova explosion that led to its formation (e.g. Flannery and van den Heuvel 1975). Interestingly, a preliminary modelling of the geometry of PSR1534+12 (Wolszczan and Phillips 1991) based on the observed polarization of the extended off–pulse emission (Fig. 1) indicates a significant spin–orbit misalignement ($\sim 17°$), which is precisely the expected consequence of such an asymmetry (e.g. Hills 1983).

TABLE 1. Parameters of the PSR 1534+12 System

Pulsar period, P	$0.0379044403665 \pm 0.0000000000004$ s
Period derivative, $\dot{P}$	$2.43 \times 10^{-18} \pm 8 \times 10^{-20}$ s s^{-1}
Right Ascension (B1950.0)	$15^h\ 34^m\ 47^s.686 \pm 0.003$
Declination (B1950.0)	$12°\ 05'\ 45''.23 \pm 0.03$
Epoch	JED 2448200.0
Dispersion measure	11.61 ± 0.01 pc cm^{-3}
Flux density (430 MHz)	36 ± 8 mJy
Projected semi–major axis, $a_1 \sin i$	3.72943 ± 0.00003 light s
Eccentricity, e	0.2736761 ± 0.0000007
Epoch of periastron, T_o	JED $2448199.8121643 \pm 0.0000003$
Orbital period, P_b	36351.7027 ± 0.0003 s
Longitude of periastron, ω	$264°.672 \pm 0.006$
DDGR masses and derived parameters	
Total mass, M	$2.679 \pm 0.003\ M_\odot$
Companion mass, m_2	$1.36 \pm 0.03 M_\odot$
Pulsar mass, $m_1 = M - m_2$	$1.32 \pm 0.03 M_\odot$
Mass function, $f(m_1, m_2)$	$0.3146 \pm 0.0002\ M_\odot$
Periastron advance, $\dot{\omega}$	1.7562 ± 0.0015 ° yr^{-1}
Time dilation, gravitational redshift, γ	0.0017 ± 0.0004 s
Orbital inclination, i	$\sim 74°$
Pulsar surface magnetic field, B	1.0×10^{10} G
Pulsar characteristic age, τ_c	2.4×10^8 yr

4. Discussion

The discovery of two millisecond pulsars described in this paper will have several interesting consequences. PSR1534+12 is the second binary pulsar after the well-known PSR1913+16 (Taylor and Weisberg 1989) to become a tool in the experimental studies of gravitation. In fact, a very high timing accuracy achievable with this pulsar (~ 2 μs as compared to ~ 20 μs for 1913+16) will most likely allow further tests of gravity theories to be conducted at an unprecedented level of accuracy. The existence of a significant spin–orbit misalignment in the 1534+12 system implies that the pulsar's spin axis should exhibit the relativistic geodetic precession at the rate of $\sim 0.5°$ yr^{-1}. The a priori knowledge of the spin–orbit angle will provide an important constraint for the interpretation of possible changes in the observed pulse morphology and polarization that may be caused by geodetic precession (e.g. PSR1913+16; Weisberg, Romani and Taylor 1989; Cordes, Wasserman and Blaskiewicz 1990).

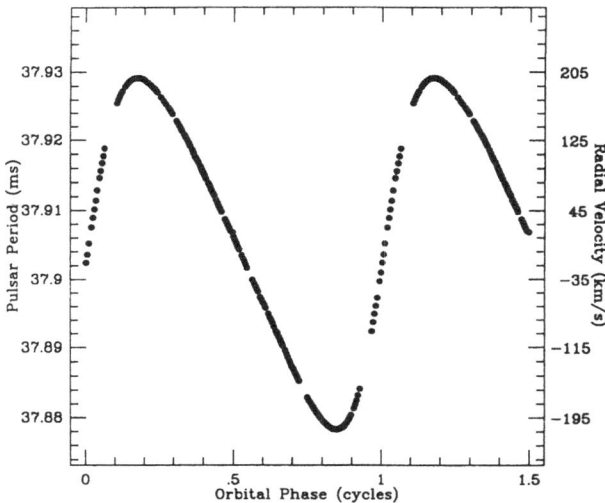

Figure 3. Radial velocity curve of the orbital motion of PSR1534+12.

A spin–orbit misalignment in massive binaries is obtained as the consequence of an asymmetry in the explosion of the helium star companion to the pulsar and the resulting off–plane 'velocity kick' imparted on the system (Flannery and van den Heuvel 1975). Consequently, a spin–orbit angle along with the systemic velocity of the pulsar are directly related to the details of its binary evolution (Hills 1983). A proximity of the 1534+12 system makes it highly probable that its spatial velocity will be measured within the next few years by means of the pulsar timing method. This, together with the observed spin–orbit angle will provide means to describe the evolution of this binary system before the time of explosion of the pulsar's stellar companion.

Millisecond pulsars are of great interest as potential detectors of the cosmic background of gravitational waves (Detweiler 1979). To detect the signature of a passing gravitational wave in the pulsar timing data, a sufficient number of the millisecond pulsar 'clocks', widely spaced over the sky should be available (Romani 1989; Foster and Backer 1990). PSR1257+12 and PSR1534+12 are the first galactic millisecond pulsars that are located away from the R.A. $= 19^h$ region, where all the other known millisecond pulsars in the galactic disk have been detected with the Arecibo radiotelescope. Therefore, adding PSR1257+12 and PSR1534+12 to the timing array of pulsars will significantly improve its angular characteristics and hence its capability to detect gravitational waves.

Detections of PSR1257+12 and PSR1534+12 are important for our understanding of the millisecond pulsar statistics (Bhattaharya and Srinivasan 1986; Kulkarni and Narayan 1989). Both pulsars are located at high galactic latitudes, have low intrinsic luminosities (8–10 mJy kpc^2 at 400 MHz) and are at distances not exceeding 0.5 kpc. Their discovery provides new suggestive evidence for the existence of a large population of old, 'spun up' pulsars in the galaxy and indicates that the future all–sky surveys may have a good chance to detect more millisecond pulsars in the solar neighbourhood.

REFERENCES

Bhattacharya, D. & Srinivasan, G. *Curr. Sci.* **55**, 327–330 (1986).
Cordes, J. M., Wasserman, I. & Blaskiewicz, M. *Astrophys. J.* **349**, 546–552 (1990).
Detweiler, S. *Astrophys. J.* **234**, 1100–1104 (1979).
Flannery, B. P. & van den Heuvel, E. P. J. *Astron. Astrophys.* **39**, 61–67 (1975).
Foster, R. S. & Backer, D. C. *Astrophys. J.* **361**, 300–308 (1990).
Fruchter, A. S., Stinebring, D. R. & Taylor, J. H. *Nature* **333**, 237–239 (1988).
Fruchter, A. S. *Ph. D. Thesis*, Princeton U. (1989).
Habets, G. M. H. J. *Astron. Astrophys.* **167**, 61–76 (1986).
Hills, J. G. *Astrophys. J.* **267**, 322–333 (1983).
Kulkarni, S. R. & Narayan, R. *Astrophys. J.* **335**, 755–769 (1988).
Romani, R. W. in *Timing Neutron Stars*, NATO/ASI Conf. Proc. No. 262, (eds H. Ogelman & E. P. J. van den Heuvel) 113–117 (Kluwer, Dordrecht 1989).
Segelstein, D. J. *et al. Nature* **322**, 714–717 (1986).
Smarr, L. L. & Blandford, R. *Astrophys. J.* **207**, 574–588 (1976).
Stokes, G. H. *et al. Astrophys. J.* **311**, 694–700 (1986).
Taylor, J. H. & Weisberg, J. M. *Astrophys. J.* **345**, 434–450 (1989).
van den Heuvel, E. P. J. & Taam, R. E. *Nature* **309**, 235–237 (1984).
Weisberg, J. M., Romani, R. W. & Taylor, J. H. *Astrophys. J.* **347**, 1030–1033 (1989).
Wolszczan, A. *et al. Nature* **337**, 531–533 (1989).
Wolszczan, A. *Nature* in press (1991).
Wolszczan, A. & Phillips, J. A., in preparation.
Woosley, S. E. in *The Origin and Evolution of Neutron Stars, IAU Symp. 125* (eds Helfand, D. J. & Huang, J. H.) 255–272 (Reidel, Dordrecht, 1986).

MILLISECOND PULSARS AND QUIESCENT LMXBS

S. R. Kulkarni, J. Navarro & G. Vasisht
California Institute of Technology, 105-24
Pasadena, CA 91125, USA

Y. Tanaka & F. Nagase
Institute of Space and Astronomical Science
1-1, Yoshinodai 3-chome, Sagamihara-shi, Kanagawa 229, Japan

ABSTRACT. The origin of millisecond pulsars still remains a matter of considerable debate. The conventional picture of spin up by accretion during an LMXB phase has been steadily challenged by a variety of statistical arguments. The most conclusive proof, the detection of coherent millisecond pulsations from LMXBs continues to elude us. We present a new approach: to search for radio pulsations from Soft X-ray Transients during quiescence. We argue that any reasonable millisecond pulsar would be able to expel matter during the low state and hence should be detectable in the radio window. Equivalently, in the low state there should be a cutoff in the X-ray luminosity accompanied by disappearance of optical emission lines. We present new radio, X-ray and optical observations and from the totality of the data we conclude that Cen X-4 does not contain a millisecond pulsar.

Introduction

Millisecond pulsars were discovered almost a decade ago. Despite this their origin remains cloaked in mystery. It is commonly assumed that millisecond pulsars (MSPs) are spun up by accretion of matter from a companion star (see van den Heuvel, this volume). During the accretion phase, the system is expected to be visible as a low mass X-ray binary (LMXB), of which there are about 100 in the Galaxy. This picture, the 'standard' picture, appears reasonable. Nonetheless, it is essentially an unproven idea. Indeed, various data (see below) go against this hypothesis. The most direct evidence, the detection of millisecond periods at X-ray wavelengths, has yet to be demonstrated despite several efforts (eg. Wood et al. 1991, Hertz et al. 1990). This embarrassing result (or lack of it) has been explained by appealing to a variety of propagation effects that might quench the pulsations.

Other lines of evidence also cast some doubt on the LMXB-MSP evolutionary link. In steady state, one expects the birthrate of MSPs and LMXBs to be approximately equal. However, early studies (Kulkarni & Narayan 1988; Narayan et al. 1989) indicated a large discrepancy, ~ 100 between the birthrates of MSPs and LMXBs. With the recent revision of distances to some of the key pulsars (Johnston & Bailes 1991), the birthrates of low-luminosity MSPs like 1855+09 and 1957+20 are reduced by almost by an order of magnitude to $\sim 11 \times 10^{-7}$ y^{-1} and $\sim 15 \times 10^{-7}$ y^{-1}. The birthrate of the LMXBs is $100/\tau_X$, where τ_X is the mean mass-transfer time or X-ray lifetime. Equality of the two birthrates requires $\tau_X \sim 4 \times 10^7$ y.

The X-ray lifetime is constrained by the amount of matter that has to be accreted in order to spin up the neutron star to millisecond periods,

$$\Delta M \sim (0.1 \text{ M}_\odot) I_{45} (P/1.5\text{ms})^{-4/3} \qquad (1)$$

99

E. P. J. van den Heuvel and S. A. Rappaport (eds.), X-Ray Binaries and Recycled Pulsars, 99–104.
© 1992 *Kluwer Academic Publishers.*

where P is the final period in milliseconds (Alpar et al. 1982). At the Eddington rate, the required mass transfer time-scale for spin-up is $\sim 10^7$ y, a similar value to τ_X. This would require that most of the progenitors of MSPs have undergone Eddington-limited mass transfer for most of their lifetime, whereas the mean luminosities of the observed LMXBs are an order of magnitude less. A discrepancy of a factor of at least 10 still appears to exist.

The problem may be even more acute than stated here if low luminosity pulsars like 1257+12 and 1534+12 that have been recently discovered are very common (see Wolszczan's contribution). In this regard, we note that the birthrate estimates derived from high frequency surveys (Johnston & Bailes 1991) may not be definitive since most MSPs have a steep spectral index. Another, difficulty is that the orbital periods of the *observed* LMXBs are considerably different from those of the *observed* MSPs. None of these reasons by itself is enough to reject the basic link between LMXBs and MSPs, but taken together they question the standard evolutionary model.

A New Approach

Detection of millisecond pulsations at radio wavelengths from LMXBs would also prove, beyond doubt, that LMXBs are MSPs in the making. However, the accretion process may suppress the radio emission, either by free-free absorption or by shorting the electric potential in the magnetosphere. Fortunately, there exists a group of transient X-ray sources, the so-called soft transients (Bradt & McClintock 1983) in which the accretion rate appears to vary by very large factors, eg. 4U2129+47 and Cen X-4. Our new approach is to look for radio pulsations from such systems in their low state (see Tanaka, this volume). We argue below that the putative millisecond pulsar should be able to disrupt the accretion disk and pulse in the radio window.

There are two groups of soft X-ray transients (White et al. 1984): soft ($kT \sim 5$ keV) and ultrasoft with a long hard tail (eg. A0620-00 and Cyg X-1). The latter are supposed to be black hole binaries and will not be discussed any further. The former, hereafter SXTs (soft X-ray transients) are supposed to be neutron star binaries, an assertion supported by dynamical studies and detection of type I bursts (attributed to thermonuclear burning instabilities on the surface of a neutron star). It can be proved by using simple but robust arguments based on energetics that energy released during the transient high state must come from gravitational energy released during accretion rather than thermonuclear burning (van Paradijs and Verbunt 1984). Since most of the accreted energy comes out in the X-ray window, we can accurately measure the accretion rate, $\dot{M}$. The inferred accretion rates vary from the Eddington limit ($\sim 10^{-8}$ $M_\odot$ y^{-1}, in the high or outburst state) to four or more orders smaller (van Paradijs & Verbunt 1984).

There are two models for SXTs (see Bailyn 1991 for a recent review). In one model, accretion from the secondary is essentially constant but the viscosity of the disk abruptly increases leading to large increase in the accretion rate. In another, heating of the secondary star by X-ray emission from the accretion disk leads to unstable mass loss rate. The details of the mass transfer instability are of little direct concern to us. The changes in $\dot{M}$ are central to our discussion and there is no dispute about this point. Another issue of prime concern to us is the evolutionary status of the donor star or equivalently the amount of matter accreted by the neutron star. If the donor star is evolved then it stands to reason that mass transfer has been going on for quite some time in which case the neutron star will be spun up. Cen X-4 very much looks like a precursor of the ablating binary pulsar, 1957+20. Indeed, McClintock & Remillard (1990) argue, on the basis of optical observations, that the secondary in Cen X-4 is a 0.1 $M_\odot$ mass star, the remains of a star that has lost most of its mass. We will henceforth assume that all SXTs are LMXBs with a long history of mass transfer.

Matter close to a rapidly spinning neutron star is subject to two opposing forces: disk viscosity which drives it inwards and the neutron star magnetosphere (or the pulsar wind; see below) which expels it outwards. This issue has been treated in detail by Illarionov & Sunyaev (1975) and more recently by Shaham & Tavani (1991). The discussion that follows is from the latter reference.

Consider an accreting neutron star rotating with period P ms and magnetic moment $\mu = \mu_{27} \times 10^{27}$ G cm^3. Let us assume that the light cylinder radius $r_c = 4.8 \times 10^6 P$ cm is considerably smaller than r_L, the Roche lobe radius of the neutron star. We assume that accretion suppresses pulsar activity, in which case the magnetic pressure at any radius is given by the static dipole formula, $\propto \mu^2 r^{-6}$. Equating the ram pressure of the infalling material to the magnetic pressure, both evaluated at the light cylinder radius, yields a critical accretion rate

$$\dot{m}_{exp} = 7.3 \times 10^{-11} \mu_{27}^2 (P/3 \text{ ms})^{-7/2} M_\odot \text{y}^{-1}. \tag{2}$$

When $\dot{M}$ falls below $\dot{m}_{exp}$ expulsion is certain and the spun up neutron starts functioning as a pulsar. Radio emission from this pulsar should be detectable for nearby SXTs (Figure 1).

Now consider a functioning millisecond pulsar (i.e. not accreting) in a binary system with a mass-losing donor. Since the pulsar is "on", the magnetic field pressure outside r_c scales as r^{-2}. The maximum accretion rate above which the pulsar will be forced to accrete is then given by

$$\dot{m}_{acc} = 6 \times 10^{-9} \mu_{27}^2 (P/3 \text{ ms})^{-4} (r_L/10^{11} \text{cm})^{1/2} M_\odot \text{y}^{-1}. \tag{3},$$

Note that accretion is possible but not necessary for $\dot{M}$ below $\dot{M}_{acc}$.

For most LMXBs, the secondary is of sufficiently low mass that $r_L \sim a$, the orbital separation. For accretion rates in between these two limits, a bow shock is formed somewhere between the neutron star and the donor but the details will depend upon the history of the mass loss rate (Shaham & Tavani 1991).

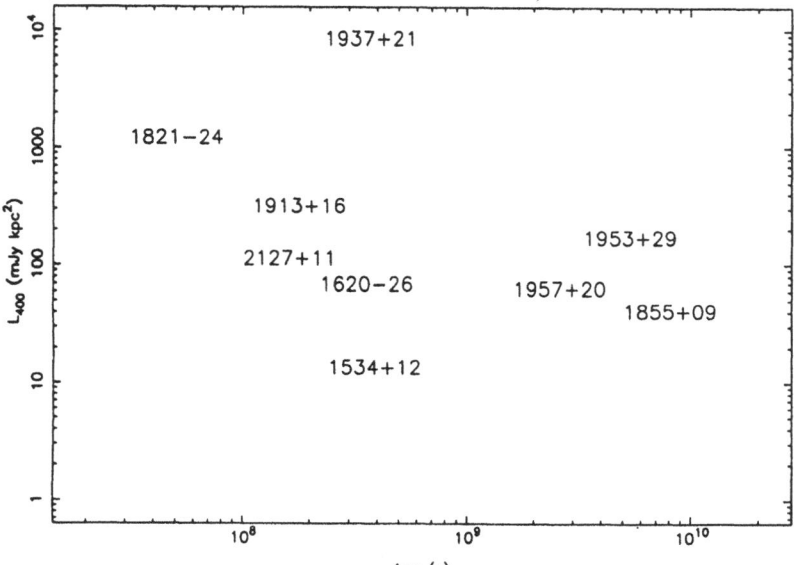

Figure 1. 400 MHz luminosity of millisecond and binary pulsars as a function of their characteristic age.

Soft X-ray transients provide a convenient laboratory to test out these ideas. The inferred $\dot{M}$ range over four or more orders of magnitude, spanning the range of critical accretion rates (Eqs. 2-3). Given this fortuitous situation we made multi-wavelength observations of Cen X-4 and 4U2129+47.

Cen X-4 is an SXT situated at a distance of 1.2 kpc (McClintock and Remillard 1990) with a recurrence time scale of about 10 y. It was last known to have undergone an outburst in 1979 (Kaluzienski et al. 1980). Detection of a type I burst during this period sets a firm upper limit of 1.3 kpc for the distance besides establishing that Cen X-4 is a genuine neutron star binary (Matsuoka et al. 1980). Quiescent emission was detected by *Einstein* and *Exosat* during 1981 and 1983 (van Paradijs et al. 1987). Assuming a thermal bremsstrahlung spectrum and $kT = 5$ keV these limits translate to dereddened luminosities of $(3\text{-}10) \times 10^{32}$ erg s^{-1}.

4U2129+47, an SXT at a distance of 1.6 kpc (Thorstensen et al. 1979), has been in low state at least since 1983 when *Exosat* observations failed to detect it (Pietsch et al. 1986). The dereddened flux limit in the 2-6 keV window is 10^{-12} erg cm^{-2} s^{-1} corresponding to an X-ray luminosity limit of 3×10^{32} erg s^{-1}.

Observations

VLA. Cen X-4 was observed on June 5, 1991 with the VLA in D/A array in the frequency range 1450-1478 MHz. We performed two types of observations: regular imaging and pulse searching. For the latter, the output of the phased VLA was sampled at a rate of 806 Hz with an analog filter bank (14×2 MHz, each polarization; see Frail & Kulkarni, this volume for additional details) and recorded on magnetic tapes for further processing. A pulse search has the advantage of being direct, but multi-path propagation within the local electron corona might quench the pulsations. On the other hand, imaging observations are immune to such propagation effects. (Here we note that free-free absorption at GHz frequencies is not expected to be important, F. Lamb, pers. comm.)

Cen X-4 was not detected to a 4σ limit of 0.4 mJy in the CLEANed image. Assuming a spectral index of -2.5 (typical of MSPs), this amounts to a 400 MHz flux density, $S \lesssim 9$ mJy and an upper limit to the radio luminosity (defined the usual way as the product of the flux density and the square of the distance) of the putative pulsar of less than 48 mJy kpc^2.

4U2129+47 was observed on 19 August, 1991 with the VLA in A array at the standard 1.4-GHz frequency. In the CLEANed image there is no emission to 0.25 mJy (4σ) at the position of the source. The corresponding limit on the radio luminosity is an impressive 15 mJy kpc^2 at 400 MHz.

GINGA. Prior to the VLA observations, an X-ray observation was carried out from the Ginga satellite on March 19, 1991. We performed scanning observations across Cen X-4 to find out any excess counts above the background at the source position. The result is negative with a 90% confidence upper limit of 0.5 counts/s or $\sim 1.3 \times 10^{-12}$ erg cm^{-2} s^{-1}, in the range 2-7 keV for which the S/N ratio is expected to be the highest. This corresponds to an upper limit of 5×10^{32} erg s^{-1} or $\dot{M} < 10^{-13} M_\odot$ y^{-1} for an assumed thermal bremsstrahlung spectrum with $kT = 5$ keV, typical of nearby LMXBs (eg. Matsuoka et al. 1980).

Interpretation

We now interpret two related issues: (i) the absence of radio emission (pulsations or otherwise) in Cen X-4 and 4U2129+47 during their off state and (ii) the detection of X-ray emission from Cen X-4 in quiescence (van Paradijs et al. 1987). In Table 1 we present the two critical accretion rates for these two objects as well as a number of interesting binary pulsar systems. For Cen X-4 and 4U2129+47 we assume $\mu_{27} = 0.5$ and $P = 3$ ms, values perfectly reasonable for young millisecond pulsars.

Table 1. Critical Accretion Rates

Pulsar Name	μ_{27} 10^{27}G cm^3	P ms	P_{orb} d	M_c M$_\odot$	r_L 10^{11} cm	$\dot{m}_{exp}$ M$_\odot$ y^{-1}	$\dot{m}_{acc}$ M$_\odot$ y^{-1}
1620-26	3.0	11.1	191.4	0.35	96	6.7×10^{-12}	2.8×10^{-9}
1744-24A	(0.5)	11.6	0.07	0.1	0.6	7.9×10^{-13}	5.0×10^{-12}
1855+09	0.33	5.4	12.3	0.22	11	1.1×10^{-12}	2.1×10^{-10}
1953+29	0.43	6.1	117.3	0.3	54	1.0×10^{-12}	4.8×10^{-10}
1957+20	0.18	1.6	0.38	0.01	1.6	2.1×10^{-11}	3.0×10^{-9}
Cen X-4	(0.5)	(3.)	0.63	0.1	1.6	1.8×10^{-11}	1.9×10^{-9}
4U2129+47	(0.5)	(3.)	0.22	0.6	0.6	1.8×10^{-11}	1.2×10^{-9}

From Table 1, we see that for the SXTs expulsion is certain when $\dot{M}$ typically falls below $\sim 10^{-11}$ M$_\odot$ y^{-1}. For both Cen X-4 and 4U2129+47 the upper limits to $\dot{M}$ are well below this value and thus we expect to detect radio pulsations. *Yet, we do not detect any radio emission.*

One of the following conclusions can be drawn from our failure to detect radio emission:

1. A millisecond pulsar exists but is not beamed towards us. This is certainly a possibility. However, we note that it is now widely accepted that millisecond pulsars have a beaming fraction close to unity (Narayan 1987).
2. A millisecond pulsar of low radio luminosity exists. This is certainly possible in the case of Cen X-4 but the limits in the case of 4U2129+47 are quite good and comparable to the lowest luminosity millisecond pulsars (see Figure 1). In any case, we intend to improve on these limits in the very near future.
3. The underlying neutron star is not rotating rapidly. All the disk millisecond pulsars have $\mu_{27} \sim$ 0.5-1 and were born with initial periods of about a few milliseconds. Adopting this and a typical upper limit of one solar luminosity for the quiescent emission we find from Eq. (2) that $P \gtrsim 13$ ms. From Eq. (1) we see most of the LMXB phase is spent in getting spun up to the last few milliseconds. Thus our constraint that $P \gtrsim 13$ ms does not fit the standard picture.

[Parenthetically, we note that 4U2129+47 has been argued to be an ADC source. The same material could thus hide both the radio and optical emission. Nonetheless do note that if, as Garcia & Grindlay (1987) argue, we are seeing only 10^{-3} of the total emission, then the *Exosat* quiescent limits are close to the predicted $\dot{M}_{exp}$. In this context, the detection of a full strength type I burst from Cen X-4 is very important since that implies the accretion disk is not hidden to our line of sight.]

The detection or even a stringent upper limit of X-ray emission from SXTs in their quiescence provides a clue as important as the search for radio pulsations. The detection of X-ray emission in quiescence means that *the underlying neutron star is not spinning rapidly enough to clear the accretion disk even when $\dot{M}$ is low.* This simple but powerful argument seems to have not been noted in the literature. Thus for at least Cen X-4 we can conclude that the neutron star is not spinning at millisecond periods.

Continuing in the same vein, the presence of emission lines in the off state of an SXT also implies that the putative pulsar has been unable to disrupt the disk. This argument is not as secure as the detection of X-ray emission since it is possible to have the outer disk left intact with the inner disk cleared by the active pulsar. Nonetheless, it is an extremely useful diagnostic. All this discussion highlights the importance of measuring the X-ray and optical observations of SXTs during their off states.

It is clear that the next step is to do simultaneous observations at radio, X-ray and optical wavelengths. If, during quiescent phases, no radio pulsations are detected or *equivalently* evidence for an accretion disk is seen then we must conclude that at least SXTs are not the progenitors of millisecond pulsars. We conclude the need for very good upper limits for the quiescent emission from SXTs. Stringent upper limits allow us to conclude on the nature of the putative pulsar without the uncertainty introduced by pulsar radio beaming.

Acknowledgements. We are grateful to J. Shaham, J. McClintock and J. van Paradijs for their enthusiastic encouragement and discussions. We gratefully acknowledge the help rendered by Dale Frail at the VLA. Our pulsar work at the VLA is supported by the Perkin Fund and support for other activities is from the Packard Foundation and the National Science Foundation.

References

Alpar, M. A., Cheng, A. F., Ruderman, M. A. & Shaham, J. 1982, *Nature*, **300**, 728.

Bailyn, C. D. 1991, *Astrophys. J.* (submitted).

Blair, W. P., Raymond, J. C., Dupree, A. K., Wu, C. C., Holm, A. V. & Swank, J. H. 1984, *Astrophys. J.*, **278**, 270.

Bradt, H. V. D. & McClintock, J. E. 1983, *Annu. Rev. Astron. Astrophys.*, **21**, 13.

Garcia, M. R. & Grindlay, J. E. 1987, *Astrophys. J.*, **313**, L59.

Hertz et al. 1990, *Astrophys. J.*, **354**, 267.

Illarionov, A.F. & Sunyaev, R. A. 1975, *Astron. Astrophys.*, **39**, 185.

Johnston, S. & Bailes, M. 1991, *Mon. Not. R. Astr. Soc.* (in press).

Kaluzienski, L. J., Holt, S. S. & Swank, J. H. 1980,, *Astrophys. J.*, **241**, 779.

Kulkarni, S. R. & Narayan, R. 1989, *Astrophys. J.*, **335**, 755.

Matsuoka et al. 1980,, *Astrophys. J.*, **240**, L137.

McClintock, J. E. & Remillard, R. A. 1990, *Astrophys. J.*, **350**, 386.

Narayan, R. 1987, *Astrophys. J.*, **319**, 162.

Narayan, R., Fruchter, A.S., Kulkarni, S.R. & Romani, R. W. 1989, in *Accretion-Powered Compact Binaries*, ed. C.V. Mauche, Cambridge University Press, p. 451.

Pietsch, W., Steinle, H., Gottwald, M. & Graser, U. 1986, *Astron. Astrophys.*, **157**, 23.

Shaham, J. & Tavani, M. 1991, *Astrophys. J.*, **377**, 588.

Thorstensen, J. et al. 1979, *Astrophys. J.*, **233**, L57.

van Paradijs, J. & Verbunt, F. 1984, in *High Energy Transients in Astrophysics*, A. I. P. Conf. Proc., **115**, 49.

van Paradijs, J., Verbunt, F., Shafer, R.A. & Arnaud, K. A. 1987, *Astron. Astrophys.*, **182**, 47.

White, N. E., Kaluzienski, J. J. & Swank, J. H. 1984, in High Energy Transients in Astrophysics, A. I. P. Conf. Proc., **115**, 31.

Wood, K. S. et al. 1991, *Astrophys. J.* (in press).

CONTINUUM OBSERVATIONS OF RECYCLED PULSARS

A. S. FRUCHTER[1]
Department of Terrestrial Magnetism
Carnegie Institution of Washington
and
Department of Astronomy
University of California
Berkeley, CA 94720

W. M. GOSS
National Radio Astronomy Observatory
P.O. Box 0
Socorro, NM 87801

ABSTRACT: Continuum observations are used to investigate the nature of the eclipse of PSR 1957+20 and the distribution of pulsars in globular clusters. The frequency dependence of continuum and pulsed eclipse are found to differ dramatically. While a substantial loss of continuum power occurs during pulsed eclipse at an observing frequency of 90 cm, the continuum flux density of PSR 1957+20 is largely undiminished at 20 cm, with the possible exception of a short asymmetric eclipse. This observation severely limits the density of the companion's wind and implies that the 20 cm pulsed eclipse is largely due to the loss of coherent pulsations. Continuum studies are also shown to be useful for examining the pulsar population of globular clusters. From observations of a sample of clusters it is found that the number of pulsars in a cluster scales approximately as the square root of the stellar collision rate. Deep continuum observations of the rich globular cluster Terzan 5 reveal steep spectrum sources distributed throughout the cluster core, implying the presence of a large population of pulsars.

1. Introduction

The distinguishing characteristic of radio pulsars is the periodic variation of their radio flux. Continuum, or time averaged, observations of pulsars thus appear to be wasteful, for they ignore the most informative feature of these objects. There are, however, instances where continuum observations of pulsars are particularly valuable. In this talk, continuum observations will be shown to afford a view of the eclipse of PSR 1957+20 that differs substantially from that provided by pulsed studies. In addition, continuum studies will be seen to powerfully constrain the numbers and distribution of pulsars in globular clusters.

---

[1] Hubble Fellow

E. P. J. van den Heuvel and S. A. Rappaport (eds.), X-Ray Binaries and Recycled Pulsars, 105–113.
© 1992 Kluwer Academic Publishers.

2. Studies of Pulsar Eclipse

An important question that has gone unanswered since the discovery of the first eclipsing pulsar, PSR 1957+20 (Fruchter, Stinebring and Taylor 1988), is whether the eclipses of this object are due to absorption of the pulsar's radio emission, or merely scattering or smearing of the pulsed radiation. (In general pulsed observations are sensitive only to the time variation of the signal, and cannot detect the small increase in system temperature caused by the presence of a weak continuum source.) The response to this question is crucial for deciding between the many eclipse mechanisms that have been proposed for this pulsar (see, for instance, Phinney *et al.* 1988, Kluzniak *et al.* 1988, Wasserman and Cordes 1988, Rasio *et al.* 1988, Michel 1988, London and Emmering 1989, Fruchter *et al.* 1990, Rasio *et al.* 1991, Eichler 1991).

We, therefore, undertook a program of interferometric observations at the VLA[2] in order to investigate the continuum eclipse. PSR 1957+20 was observed at 20 cm (1465 MHz) for about two hours on each of nine days while the VLA was in its A (or largest) configuration, and at 90 cm (327 MHz) for several hours on each of four days while the VLA was in the B/C configuration. Due to limitations of space, readers interested in the details of these observations are requested to examine Fruchter and Goss (1991).

At 90 cm, PSR 1957+20 is a $\approx$ 40 mJy source out of pulsed eclipse. However, an image made from data taken during the central 45 minutes of two pulsed eclipses (which at this frequency typically last 56 minutes) shows a flux density of only 2 $\pm$ 3 mJy at the position of the pulsar. Because it is extremely unlikely that this loss of signal can be due to scintillation, the dramatic drop in flux density observed here implies that the eclipsing medium must absorb or deflect the pulsar's radio flux during most of the pulsed eclipse. Our present observations do not allow us to accurately determine the length of the continuum eclipse at 90 cm, however further observations, which are already underway, should allow us to make this measurement.

Our 20 cm data, on the other hand, allow a more detailed look at the continuum eclipse. The 20 cm data were first combined to obtain an average flux density and accurate pulsar position. The data were then divided into orbital phase bins 2.5° wide using the Software Development Environment (SDE), written by Tim Cornwall. The forty-one bins which contained significant numbers of uv records were imaged, and the flux at the position of the pulsar determined. Figure 1A displays the results of this procedure. For each orbital phase bin the observed flux is shown, as is an error bar representing the r.m.s. noise of the image. As a result of scintillation the source variability is far greater than the noise in the images.

Earlier in this conference, Joe Taylor showed that PSR 1957+20 is eclipsed as a pulsed source at 1400 MHz between orbital phase $\sim 0.22 - 0.28$. The most surprising result of the continuum observations is therefore immediately obvious: for much of the time that the object is eclipsed as a pulsed source at 20 cm, it is still observable in the continuum.

While a statistically significant drop in average flux does not occur during the pulsed eclipse, an F test indicates with greater than 98% confidence that the variance of the

[2]The Very Large Array of the National Radio Astronomy Observatory is operated by Associated Universities, Inc., under co-operative agreement with the National Science Foundation.

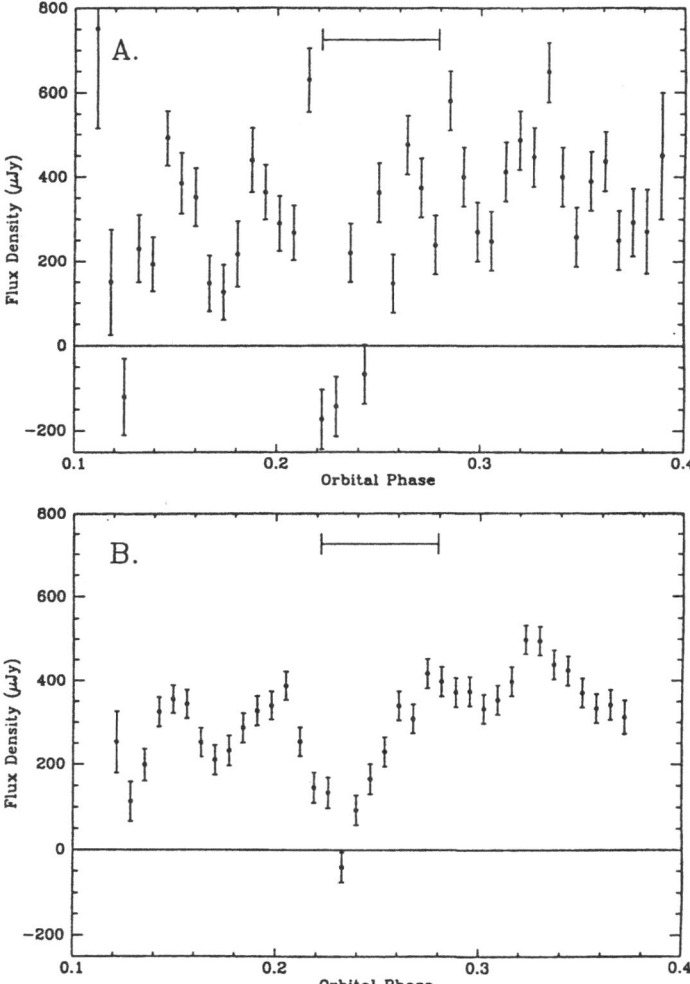

Figure 1: The observed 20 cm flux density of PSR 1957+20 as a function of orbital phase. Phase bins are 2.5° wide. The graph is the result of summing 9 eclipses observed on separate days. The error bars show the noise of the images. The upper graph displays the raw data; the lower graph the data smoothed with a boxcar average 4 bins wide. The barred horizontal lines show the average extent of pulsed eclipse at 20 cm.

continuum flux is larger in this region of orbital phase. The observed variability is apparently caused by a diminution of flux at the onset of pulsed eclipse. This effect is easily visible in Figure 1B, where we display the same data as in Figure 1A, but smoothed using a boxcar average of four data points. Bootstrap statistical analysis suggests that the probablility of this loss of signal being due to scintillation, rather than a true eclipse, is less than two percent, in agreement with the results of the F test.

While a short asymetric loss of 20 cm continuum flux probably occurs, and may be indicative of a bow compression of the ablated wind, throughout most of the pulsed eclipse the continuum flux is undiminished. Therefore, the wind that passes through our line of sight cannot have a density substantially greater than 10^9 cm^{-3}, and cannot be responsible for the orbital period derivative of PSR 1957+20 reported here by Marty Ryba and Joe Taylor, unless the specific angular momentum of the wind is more than an order of magnitude greater than that of the companion, or unless the wind is largely neutral. It seems likely then that either the companion's mass loss is largely restricted to a region away from our line of sight, or a mechanism other than mass loss, such as the interaction of the pulsar wind with the magnetic moment of the companion (C. Thompson, personal communication), is responsible for the orbital period derivative.

3. Globular Cluster Pulsars

As many speakers at this conference have noted, it has long been supposed that millisecond pulsars are spun up to their rapid rotation rates in low mass X-ray binary systems (LMXB's) (see for example van den Heuvel, this conference). Yet when the birthrate of millisecond pulsars is compared with that of LMXB's, a dramatic inconsistency appears: millisecond pulsars appear to be born 10 – 100 times more frequently than the standard theory of LMXB formation would allow (Kulkarni and Narayan 1988, Fruchter 1989, Coté & Pylyser 1990, Narayan et al. 1990), a point which Shri Kulkarni has stressed in his lecture at this conference.

Globular clusters provide a valuable system for probing the relationship between LMXB's and millisecond pulsars: while less than 10^{-3} of the stars in the galaxy reside in globular clusters, between one tenth and one fifth of all LMXB's in the galaxy are in globular clusters (Bradt and McClintock 1983). If millisecond pulsars are formed from LMXB's, then they should be similarly overpopulous in globular clusters.

Although a typical millisecond pulsar is too weak to detect at a distance of a few kiloparsecs, it seemed possible to observe the integrated radio luminosity from the many faint millisecond pulsars in a globular cluster. Of course, such an observation precludes the use of pulse detection to show that one has indeed found pulsars. However, the spectral index of the source can serve as a distinguishing marker. Pulsars have perhaps the steepest spectral indices of any objects in the sky; the flux density of a millisecond pulsar typically falls with observing frequency faster than $\nu^{-1.5}$ (see for instance Foster et al. 1991). The other sources of radio emission found in globular clusters, LMXB's, tend to have flat spectra.

Searchers have often taken advantage of the steep spectral indices of pulsars. Indeed, the first millisecond pulsar discovered in a globular cluster, PSR 1821-24, was located at the VLA in a search for steep spectrum point sources (Hamilton, Helfand and Becker

TABLE 1
PREDICTED AND OBSERVED CLUSTER PROPERTIES

Source	Coll. Rate	Radius "	Contam.	Flux Density mJy
Ter 5	1.000	2.9	0.004	1.9 ± 0.3
Lil 1	0.667	3.7	0.004	< 0.6
N6440	0.501	7.0	0.004	1.45 ± 0.15
N6441	0.341	8.3	0.004	< 0.8
N6266	0.282	14.5	0.009	< 1.2
N1851	0.233	3.9	0.004	< 0.6
N6626	0.117	22.7	0.022	1.6 ± 0.3
N6522	0.094	10.6	0.005	< 0.7
N6624	0.083	5.2	0.004	0.95 ± 0.15
N6541	0.066	16.8	0.012	< 0.6
N6273	0.057	30.1	0.040	< 0.6
N6293	0.038	9.9	0.004	< 0.7
N6544	0.034	27.7	0.034	0.9 ± 0.3
N6656	0.030	73.1	0.233	< 0.7
N6380	0.030	20.6	0.018	< 0.4
HP1	0.023	12.6	0.007	< 0.6
N6539	0.012	31.9	0.044	0.9 ± 0.15

Table 1: List of the observed clusters, showing the source name, predicted collision rate, core radius of the cluster in arcseconds, probability of detecting a contaminating background source within the cluster radius, and flux density at 20 cm. The sources are sorted by their predicted collision rates, which are scaled to Terzan 5. The errors given for the flux density reflect source variability. Cluster parameters are taken from Webbink (1984).

1985), a method of search later employed by Kulkarni et al. (1990). In contrast to these searches, we have attempted to insure that our VLA beamsize is as large as, or larger than, the core size of the observed clusters, so that our images would show the integrated emission of the cluster pulsars.

Seventeen globular clusters were observed at 20 cm. The clusters were chosen for their proximity (< 5kpc), richness, or collapsed cores (Djorgovski and King 1986), as well as for their propinquity on the sky for ease of observing.

Each cluster was observed in the C configuration of the VLA at 20 cm for 35 minutes. In addition, detected clusters, as well as clusters whose core radii exceeded the C beam size (typically $18'' \times 10''$), were reobserved in the more compact C/D and D arrays. Details of the observing procedure can be found in Fruchter and Goss (1990).

Of the seventeen clusters examined in this program, six show detectable emission at 20 cm either within the core radius of the cluster, or, in cases where the core diameter is smaller the resolution of the image, within one beam width of the optical center of the cluster. Flux densities of the detected clusters as well as 3σ upper limits on the flux density from clusters without detectable emission are provided in Table 1.

We have detected new sources in the clusters Terzan 5, NGC 6440, NGC 6539 and NGC 6544 and confirmed the existence of previously controversial source in NGC 6624 (Geldzahler 1983, Grindlay and Seaquist 1986). NCG 6266 contains PSR 1821-24, and was observed because it fit the criteria of the search and acted as a check on our method. Follow-up observations in several arrays at 6 cm have shown that all of these sources, with the possible exception of NGC 6624, possess spectra steeper than -1.5. NGC 6624 contains a highly variable 6 cm source that may be related to the LMXB in that cluster. On the basis of spectral indices, source variability and the subsequent discovery of pulsars in some of these clusters (described at this conference by Andrew Lyne), we conclude that all of the 20 cm flux observed, again with the possible exception of NGC 6624, is due to pulsars. (See Fruchter and Goss 1990 for a more detailed discussion of the nature of the sources.)

In order to analyze the cluster flux densities we have written a Monte Carlo simulation program which randomly places pulsars in clusters according to a variable distribution function, and assigns pulsar luminosities consistent with the well-studied field luminosity function (Manchester and Taylor 1977).

Present theoretical belief appears to generally favor a distribution of pulsars in globular clusters based upon the stellar collision rate. This implies that the number of pulsars in a globular cluster should scale approximately as the probability of forming a binary through tidal capture in the cluster (Verbunt, Lewin, and van Paradijs 1989). However, our Monte Carlo studies show that the stellar collision rate model fits the data no better than a model which assumes that all clusters contain equal numbers of pulsars. The equal number model underestimates and the collision rate model overestimates the difference in pulsar population between low and high collision rate clusters.

We have therefore considered a distribution which is proportional to the square root of the collision rate, i.e. up to a constant of proportionality is equal to the geometric mean of the collision rate and equal cluster models . *We find that the likelihood of the geometric mean distribution is more than an order of magnitude greater than that of the collision rate or equal number models.* (A similar result is presented at this conference in a poster by Johnston et al.)

We can fit the distribution of flux density discovered in this survey equally well by assuming that pulsars in globular clusters come from two populations, one distributed uniformly among clusters and the other distributed as the cluster collision rate. As noted in Fruchter and Goss (1990) and discussed in detail here by Phinney, this most likely implies that primordial binaries are important in the formation of cluster pulsars.

Preliminary analysis suggests that the relative ability of the various distributions to fit the observed total flux density is largely independent of the choice of pulsar luminosity function. As noted by Wijers and van Paradijs in a conference poster, the estimated number of pulsars in a cluster depends more sensitively on the luminosity function. Using the field luminosity function, either of the two preferred models predicts that the galactic globular cluster system contains 500 to 2000 pulsars which beam toward earth.

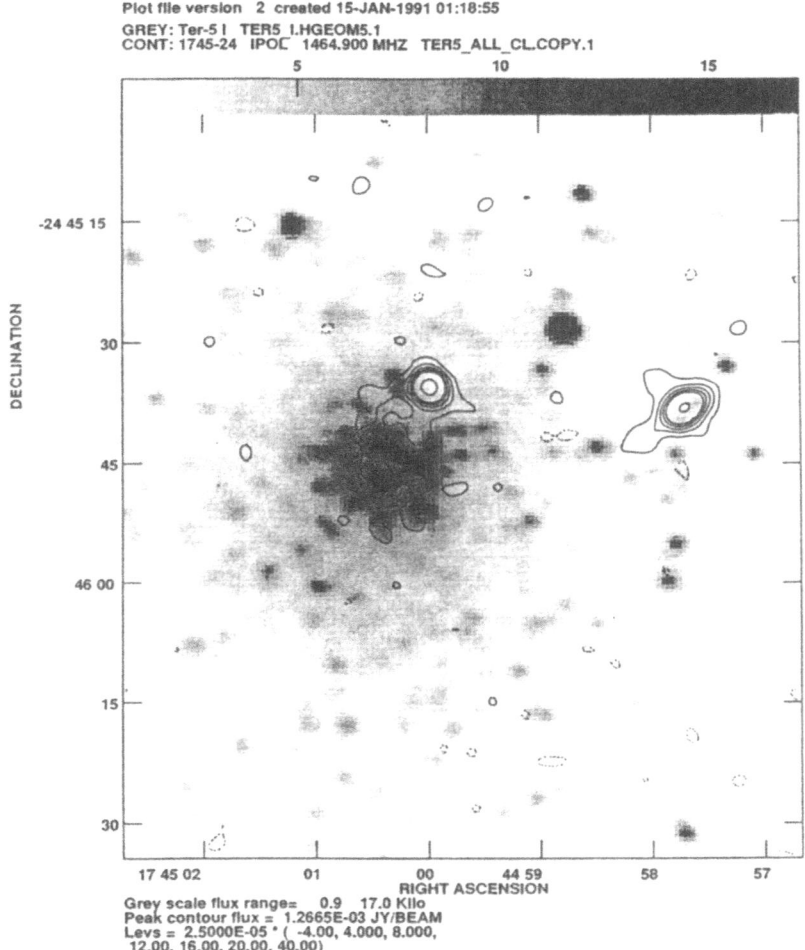

Figure 2: Radio contours are shown superposed on an I band image of the rich cluster Terzan 5. The 20 cm radio image has a beamsize of $3'' \times 3''$.

Our best fit model predicts that Terzan 5 contains the most pulsars of any globular cluster in the Galactic system, around 50 pulsars. In order to investigate this cluster, we have obtained a deep image using about 16 hours of data taken at 20 cm in the A, A/B, B and B/C configurations of the VLA. Figure 2 shows the contours of the resulting radio image superposed on an I band image of the cluster taken by Taft Armandroff with the CTIO 1-m telescope. The VLA map was made using uniform weighting and a 40 kilolambda taper. This weighting produces a small beam size (3″ × 3″) which sharpens the details of the image with the loss of some sensitivity to extended structure. The bright radio source about 30″ to the west of the cluster is the eclipsing pulsar, PSR 1744-24A. The sidelobes on this source are due to its time variability. The most striking feature of the radio image, however, is the diffuse radiation centered on the optical cluster. The flux density of the diffuse component is ≈ 2 mJy: stronger than the two prominent point sources in the image. Observations at 6 cm show that both the diffuse emission and the the ≈ 1.5 mJy source about ten arc seconds to the northwest of the cluster center possess spectral indices steeper than −1.4. The only astrophysical sources likely to possess spectra this steep are pulsars. It is therefore probable that the vast majority of the 20 cm radiation visible in this image is due to the presence of pulsars. Both the magnitude of the observed flux and its distribution suggest that the number of pulsars is very large, possibly reaching or exceeding the 50 predicted by the statistical model.

In this short paper we have briefly discussed two of the applications of continuum observations to pulsar studies. We hope that the reader will agree that pulsars need not pulse to fascinate.

We thank Tim Cornwell for use of his powerful Software Development Environment and Taft Armandroff for providing us with images of the cluster Terzan 5.

References

Bradt, H.V.D. & McClintock, J.E., *Ann. Rev. Ast. Ap.* **21**, 13 (1983).

Coté, J. & Pylyser, E.H.P., *Astron. Astrophys.* **218**, 131 (1989).

Djorgovski, S. & King, I.R., *Ap. J.* **305**, L61 (1986).

Emmering, R.T. & London, R.A., *Ap. J.*, **336**, 589 (1990).

Eichler, D., *Ap. J.*, **336**, L27 (1991).

Foster, R., Backer, D & Fairhead, L., *Ap. J.* (in press).

Fruchter, A.S., Ph.D. Thesis, Princeton University (1989).

Fruchter, A.S. & Goss, W.M., *Ap. J.* **365**, L5 (1990).

Fruchter, A.S. & Goss, W.M., *Ap. J.* submitted (1991).

Fruchter, A.S., Stinebring, D.R. & Taylor, J.H., *Nature* **333**, 237 (1988).

Geldzahler, B.J., *Ap. J.* **264**, L49 (1983).

Grindlay, J.E. & Seaquist, E.R., *Ap. J.* **310**, 172 (1986).

Hamilton, T.T., Helfand, D.J. & Becker, R.H., *A. J.* **90**, 606 (1985).

Kluzniak, W., Ruderman, M., Shaham, J., and Tavani, M. *Nature*, **334**, 225 (1988).

Kulkarni, S.R., Goss, W.M., Wolszczan, A. & Middleditch, J., *Ap. J.*, **363**, L5 (1990).

Kulkarni, S.R. & Narayan, R., *Ap. J.* **335**, 755 (1988).

Kulkarni, S.R., Narayan, R. & Romani, R., *Ap. J.*, **356**, 174 (1990).

Manchester, R.N. & Taylor, J.H., *Pulsars*, W.H. Freeman and Company (1977).

Michel, F. C., *Nature*, **337**, 236 (1988).

Narayan, R., Fruchter, A.S., Kulkarni, S.R. & Romani, R.W., *Proceedings of the 11th North American Workshop on CV's and LMXRB's*, ed. Manche, C.W. (1990).

Phinney, E. S., Evans, C. R., Blandford, R. D., and Kulkarni, S. R., *Nature*, **333**, 832 (1988).

Rasio, F., Shapiro, S., and Teukolsky, S. A., *Ap. J.*, **342**, 934 (1989).

Rasio, F.A., Shapiro S.L. & Teukolsky, S.A., *Astron. Astrophys.*, **241**, 25 (1991).

Verbunt, F., Lewin, W.H.G & van Paradijs, J., *M.N.R.A.S.* **241**, 51 (1989).

Webbink, R.F., *Dynamics of Star Clusters, I.A.U. Symposium No. 113*, Dordrecht, 541 (1984).

MILLISECOND PULSAR SPECTRA

R. S. FOSTER
National Research Council/NRL Cooperative Research Associate
Center for Advanced Space Sensing
Naval Research Laboratory
Washington, DC 20375 USA

D. C. BACKER and L. FAIRHEAD
Astronomy Department and Radio Astronomy Laboratory
University of California at Berkeley
Berkeley, CA 94720 USA

ABSTRACT. A study of the flux density and component emission of four millisecond radio pulsars shows that on average millisecond radio pulsars have steeper spectra than slow period pulsars. The results of flux density measurements of four millisecond pulsars, PSRs 1620−26, 1821−24, 1855+09, and 1937+21, have been made to determine their spectral indices in the range between 425 MHz to 3 GHz. The four objects are shown to have indices that range from −1.3 to −2.6. The flux densities are compared with other measurements down to 10 MHz. None of the objects shows any indication of a sharp spectral break at low frequencies. An analytic pulse component model is developed for each object. A more detailed presentation of this work can be found in Foster, Fairhead, and Backer (1991).

1. Observations

The observational data were collected as part of a continuing pulsar timing experiment using the 43 m radio telescope at the National Radio Astronomy Observatory (NRAO) in Green Bank, West Virginia (Foster and Backer 1990). Observations were conducted over 16 days between 1989 August 8 and 1989 August 27 at frequencies between 425 MHz and 3 GHz. Supplemental observations were obtained at 800 MHz and 1330 MHz during four days in both 1989 October and 1990 January. Data for this experiment were collected using the NRAO spectral processor in its pulsar timing configuration. The spectral processor is capable of sampling a 20 MHz wide, single side band, baseband voltage signal at ~40 MHz and performing 512-point Fast Fourier Transforms (FFT) every ~ 12.8 μs on each of two polarization channels.

Dispersion removal is done in software as a post processing procedure after filtering the bandpass for interference. The dedispersed pulse profile is accumulated by shifting each frequency channel by the appropriate dispersion delay first by an integer bin shift and then by a fractional bin shift using a cubic spline interpolation. Calibration measurements were made against the extragalactic continuum sources 3C 48 and 3C 286 at least once per night at each frequency.

2. Data Analysis

The results from flux density measurements are shown in Figure 1. The mean logarithmic spectral indices for the four pulsars in this study were determined from unweighted fits and range from −1.3 to −2.6 (Table 1). Table 1 also gives the best distance estimate to

E. P. J. van den Heuvel and S. A. Rappaport (eds.), X-Ray Binaries and Recycled Pulsars, 115–119.

each pulsar. Figure 2 shows average pulse profiles for three millisecond pulsars at various radio frequencies.

TABLE 1

MILLISECOND PULSAR MODEL SPECTRA

Pulsar	Flux Density (mJy)	Index	$\Delta\nu_d$ (MHz)	D (kpc)	$V_\perp$ (km s^{-1})	T_{DISS} (seconds)	T_{RISS} (days)
1620−26	3.3 ± 0.3	-2.5 ± 0.2	(0.45)	2.1	(200)	62	2.0
1821−24	3.9 ± 0.4	-2.3 ± 0.2	0.016	5.8	(200)	20	18
1855+09	10.7 ± 1.1	-1.3 ± 0.2	4.1	0.5	10	1830	6.6
1937+21	25.9 ± 2.6	-2.60 ± 0.05	0.3	5.0	50	310	15

In order to assess the contributions of diffractive and refractive scintillation on the uncertainties in these measurements, we compare their time scales with the duration of the observations. The diffractive and refractive scintillation time scales can be estimated numerically from the known pulsar distance, diffraction bandwidth, observing frequency, and transverse velocity (e.g., Stinebring and Condon 1990). The estimated time scales for the pulsars in this study are given in columns 7 and 8 of Table 1. The scintillation bandwidth at 1 GHz, $\Delta\nu_d$ and estimated distance and transverse velocity are also given. PSR 1855+09 has the longest diffractive time scale, while PSR 1821−24 has the longest refractive time scale at the nominal frequency of 1 GHz. Values in parenthesis in Table 1 are estimated.

Individual pulse component ratios vary as a function of frequency, making the spectral indices of the individual components different from the mean spectral index of the pulsar. We devised a pulse component model for each of the four objects based on the multifrequency data collected in 1989 August and published data. The pulse profile P(t) is decomposed into the sum of L Gaussian components. The model used to generate the analytic profile I(t) is given as

$$I(t) = [III(\Delta t) \star \, \sqcap \,] \cdot [P \star h_{DM} \star h_{ISS}], \tag{1}$$

where the analytic profile P(t) has been convolved with the two instrumental functions $h_{DM}(t)$ and $h_{ISS}(t)$, and sampled uniformly over N bins. The former smearing term comes from the dispersion of the signal within each passband, and the latter term comes from interstellar scattering. The uniform sampling time Δt is represented by the dot product of the convolved profile with the Shah sampling function, $III(\Delta t)$, convolved with the bin sample size $\sqcap$. This analytic description was necessary to deconvolve the pulse components with the limited resolution of the spectral processor data.

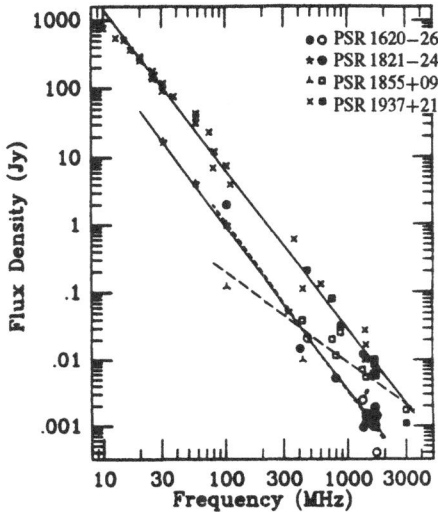

Figure 1 The calibrated flux density measurements from four millisecond pulsars over the frequency interval 10 MHz to 3 GHz are plotted. The best fit spectral index through all the data points are shown for each object. All the data represented by solid circles were collected using the NRAO 43 m telescope in Green Bank during the month of August 1989. Additional data from the literature are added with open circles. The primary source for the low frequency data is Erickson and Mahoney (1985).

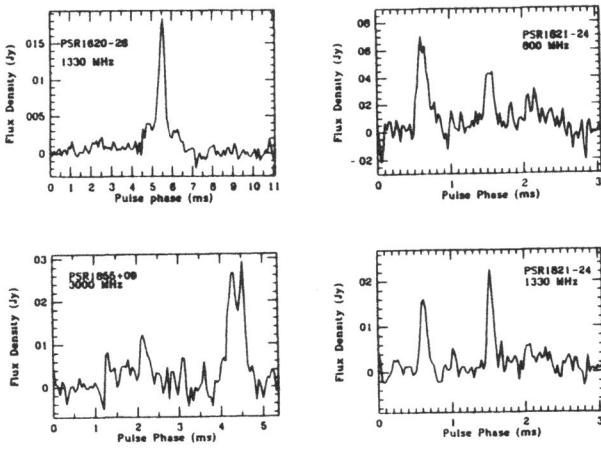

Figure 2 The integrated profiles for PSR 1620−26, PSR 1821−24, and PSR 1855+09 at various frequencies are shown. Each profile represents several hours of accumulated data collected with a bandwidth of 20 MHz.

TABLE 2

COMPONENT GAUSSIAN MODELS

Pulsar	Component	Peak Flux Density (mJy)	Index α	Pulse Width FWHM (μs)
1620−26	1	60 ± 6	−2.5 ± 0.2	400 ± 40
1821−24	1	31 ± 4	−1.4 ± 0.2	95 ± 15
	2	34 ± 4	−3.1 ± 0.2	130 ± 25
1855+09	1	20 ± 5.6	−1.3 ± 0.2	183 ± 56
	2	6.4 ± 3.2	−1.3 ± 0.2	144 ± 45
	1 and 2**	—	—	—
	3	80 ± 20	−1.3 ± 0.2	223 ± 31
	4	69 ± 30	−1.3 ± 0.2	123 ± 19
	3 and 4**	—	—	—
1937+21	1	540 ± 50	−2.6 ± 0.1	49 ± 5
	2	300 ± 30	−3.7 ± 0.2	17 ± 1
	3	290 ± 30	−3.2 ± 0.2	40 ± 2

The fit results for the four pulsars are given in Table 2 and Table 3. The model includes pulse component width(s) (Gaussian) as a function of frequency, component separation(s), and the spectral index of each component referenced to a fictitious 1 GHz profile. Gaussian models are specified as $S_0 \nu^\alpha exp\left[-(t/\sigma_w)^2\right]$ where the amplitudes S_0 are given in column 3, the spectral indices α are given in column 4, and the widths σ_w are given in column 5. The fits shown assumed a constant pulse width although some dependence on frequency was noted. The analytic models are used to determine arrival times which are in turn used to solve for an accurate dispersion measure toward each pulsar (see Foster, Fairhead, and Backer 1991).

TABLE 3

COMPONENT SEPARATION

Pulsar	Separation (μs)	(degrees)
1821−24		
(component 2 − component 1)	913 ± 12	107.6 ± 1.4
1855+09		
(main pulse [3] − inter pulse [2])	2146 ± 105	144.1 ± 7.0
(main pulse [4] − main pulse [3])	283 ± 24	19.0 ± 1.6
(inter pulse [2] − inter pulse [1])	406 ± 197	27.3 ± 13.2
1937+21		
(main pulse [1] − inter pulse [3])	744.9 ± 1.3	172.1 ± 0.3
(notch pulse [2] − main pulse [1])	37.0 ± 3.3	8.6 ± 0.8

3. Summary and Conclusions

Multifrequency observations collected over a two week period between radio frequencies of 425 and 3 GHz have been used to determine analytic profiles for four millisecond pulsars. The range of spectral indices seen in the individual pulse components from the four millisecond pulsars indicate that there is a distribution of spectral indices ranging from $\alpha \sim -1$ to -3. The mean spectral index of these four millisecond period pulsars is steeper than the average spectral index ($\alpha = -1.5$) of the slow period pulsar population (*e.g.* Manchester and Taylor 1977). The three fastest known pulsars, PSR 1821−24, PSR 1937+21, and PSR 1957+20 (see Fruchter *et al.* 1990) all have at least one very steep ($\alpha \leq -3$) component. Each of the four pulsars have pulse component widths at 1 GHz that are less than ~ 4 % of the pulse period.

We have used the average profiles obtained at each frequency to determine Gaussian component pulse widths, positions, and flux densities after removing instrumental and interstellar broadening functions appropriate for each pulsar. These results are then used to derive a model template which, in turn, was required to determine accurate dispersion measures.

4. Acknowledgements

The authors of this paper would like to express our thanks to M. Clark, J. M. Cordes, J. R. Fisher, and A. Wolszczan. The National Radio Astronomy Observatory is operated by Associated Universities, Inc. through support from the National Science Foundation. Our research was partially supported by the Center for Particle Astrophysics, a National Science Foundation Science and Technology Center operated by the University of California at Berkeley under Cooperative Agreement No. AST-8809616, and by grant AST-8719094 from the National Science Foundation.

References

Cordes, J. M. Wolszczan, A., Dewey, R. J., Blaskiewicz, M., and Stinebring. D. R. 1990, *Ap. J.*, **349**, 245.

Erickson, W. C. and Mahoney, M. J. 1985, *Ap. J. (Letters)*, **299**, L29.

Foster, R. S. and Backer, D. C. 1990, *Ap. J.*, **361**, 300.

Foster, R. S., Fairhead, L., and Backer, D. C. 1991, submitted to *Ap. J.*.

Fruchter, A. S., *et al.* 1990, *Ap. J.*, **351**, 642.

Manchester, R.N. and Taylor, J. H. 1977, *Pulsars*, (San Francisco: W. H. Freeman and Company), p. 21-24.

McKenna, J. and Lyne, A. G. 1988, *Nature*, **336**,226. Erratum, **336**, 698.

Stinebring, D. R. and Condon, J. J. 1990, *Ap. J.*, **352**, 207.

Thorsett, S. and Stinebring, D. R. 1990, submitted to *The Astrophysical Journal*.

SEARCHING FOR PULSARS IN GLOBULAR CLUSTERS : THE PARKES/JODRELL BANK SURVEYS

C. ROBINSON, A. G. LYNE, M. BAILES & J. D. BIGGS
University of Manchester
Nuffield Radio Astronomy Laboratories
Jodrell Bank
Macclesfield
Cheshire
U.K.

R. N. MANCHESTER
Australia Telescope National Facility
CSIRO
Epping
New South Wales
Australia

N. D'AMICO
Istituto di Fisica dell'Universita
Palermo
Italy

ABSTRACT. Recent searches have established a substantial population of millisecond pulsars in globular clusters. The surveys conducted with radio telescopes at Parkes, Australia and Jodrell Bank, England have been responsible for the discovery of 18 of the 29 globular cluster pulsars known to date. Here, we describe the general features of searching for pulsars in globular clusters and give particular details of the Parkes/Jodrell Bank surveys.

1. Introduction

The pulsar discoveries in M28 [1] and M4 [2] were the earliest successes of the searches for globular cluster pulsars conducted with the 76 m Lovell radio telescope at Jodrell Bank. In a collaborative effort between Jodrell Bank and the Australia Telescope National Facility (ATNF), recent globular cluster surveys have also been carried out using the 64 m radio telescope at Parkes in Australia, with the subsequent data processing being performed at the ATNF and at Jodrell Bank. These searches have led to further pulsar discoveries in the globular clusters 47 Tucanae [3,4], Terzan 5 [5], NGC 6440 [6], NGC 6539 [7] and NGC 6624 [8].

The globular cluster pulsars whose detection have resulted from these efforts include:

- Eleven millisecond pulsars in 47 Tucanae.

- An eclipsing binary pulsar in Terzan 5.

- A binary pulsar in an eccentric orbit in NGC 6539.

Further details of the above discoveries can be found in the review by A. G. Lyne contained in this volume.

121

E. P. J. van den Heuvel and S. A. Rappaport (eds.), X-Ray Binaries and Recycled Pulsars, 121–129.
© 1992 *Kluwer Academic Publishers.*

2.   Radio Observations of Pulsars in Globular Clusters

Globular clusters are relatively distant places in which to look for pulsars, but their well defined locations in the sky provide a significant advantage in searching for apparently weak pulsars, as compared with pulsar surveys of the galactic disk. The disk surveys cover extended areas of sky, which need to be scanned, and this limits the integration time that can be achieved for any particular direction in the sky. However, in the globular cluster surveys, a cluster core is small enough to fit within the main beam of the radio telescope, and the integration times are therefore not so constrained. Since the sensitivity of a pulsar search increases with the square root of the data length, the longer data sets that can be obtained for globular clusters enhance the prospects of pulsar detection.

The success of a pulsar survey depends critically on the observational system employed. The system design must take into account certain observational problems in detecting pulsars. These are now briefly described and this is followed by details of the particular observational systems used in the Parkes and Jodrell Bank surveys.

2.1. OBSERVATIONAL CONSIDERATIONS

Pulsars are steep spectrum sources; their flux density, S, is related to frequency, ν, by a power-law of the form $S(\nu) \propto \nu^{\alpha}$, where α is typically -2 for millisecond pulsars. This suggests that pulsar surveys should be more sensitive at lower radio frequencies. However, there are a number of adverse effects which offset this spectral advantage at low frequencies. One is the reduction in sensitivity caused by galactic background noise, a problem which is more critical when making observations towards the galactic centre. Others are the effects of dispersion and scattering, due to the ionised component of the interstellar medium, each of which contributes towards pulse broadening; this reduces the survey sensitivity in accordance with the dependence of sensitivity on the square root of the observed pulse duty cycle.

Dispersion is a consequence of the different phase velocities in the interstellar medium of the various frequency components of a pulse. It results in pulse broadening and puts a lower limit on the detectable periods of a pulsar survey, equal to $2\delta t$, where δt is the time delay between the components of a pulse at the extremes of the receiver bandwidth. This restricts the bandwidth that can be used, but the problem can be largely overcome by employing a number of frequency channels with reduced bandwidths. This allows dedispersion of the observed signal to be implemented (see section 3.1.) and reduces the minimum detectable period to that determined by the dispersion across the individual frequency channels. Interstellar scattering produces further pulse broadening, as a result of the multiple paths taken by the scattered waves from a pulsar through the interstellar medium. The degree of scattering increases for larger dispersion measures.

The frequency dependencies of dispersion and scattering are ν^{-3} and ν^{-4} respectively, and their effects are therefore reduced at higher observing frequencies. Thus, in searching for short period pulsars with high dispersion measures, the limitations introduced by dispersion and scattering can be partially overcome by using an increased observing frequency. This is the case when searching for pulsars in globular clusters of low galactic latitude, for which the line-of-sight makes a significant passage through the electron layer of the galactic disk.

Further limitations on a pulsar survey are inherent in the data acquisition process. In addition to the period restrictions imposed by dispersion and scattering, another lower period limit is dictated by the sampling rate used, due to the application of post-detection anti-aliasing filters. The minimum detectable flux of a pulsar survey depends on the system temperature of the radio receiver. Improved sensitivity can be achieved by the cryogenic cooling of the receiver system.

2.2. THE PARKES AND JODRELL BANK OBSERVING SYSTEMS

The Parkes and Jodrell Bank surveys have so far involved observations in both the λ50cm and λ20cm wavebands, corresponding to frequencies of 640 MHz and 1500 MHz respectively. The systems receive orthogonal polarizations which are summed after detection and low-pass filtering. Multi-channel sampling is employed and the recorded signals are encoded by means of one-bit digitization in each channel. All of the Parkes receiver systems are now cryogenically cooled. Details of the receiver systems used are listed in Table 1.

Receiver System	Channels	Channel BW	Sampling	S_{sys}(Jy)
Parkes : λ50cm	128	0.25 MHz	0.3 ms	90
Parkes : λ20cm System 1	80	1 MHz	0.3 ms	70
Parkes : λ20cm System 2	64	5 MHz	1.2 ms	70
Jodrell Bank : λ50cm	32	0.25 MHz	0.3 ms	100
Jodrell Bank : λ20cm	32	1 MHz	0.3 ms	45

Table 1: Details of the Parkes and Jodrell Bank Observing Systems

The higher frequency of 1500 MHz is used primarily to survey those clusters with large expected dispersion measures which reside at low galactic latitudes. In the λ20cm waveband, increased total bandwidths are used to compensate for the reduced pulsar fluxes at the higher frequency. However, the diminished effects of dispersion at this frequency allow larger channel bandwidths and so an increase in the number of frequency channels is not required.

3. Data Processing

The data obtained from Parkes are processed using an Alliant FX-8 computer system and a Sun 330 Sparcstation at Jodrell Bank and a Convex C220 computer system at the ATNF. The Jodrell Bank data were mainly processed using Amdahl 1100 and 1200 computer systems of the Manchester Computing Centre. The processing of data from our pulsar surveys consists of two main phases, as described below.

3.1. SPECTRAL ANALYSIS

The data, first of all, goes through the process of dedispersion; the time series in the individual frequency channels are realigned to compensate for the dispersive effects of the interstellar medium, according to the value of dispersion measure at which the search is being performed. This is followed by the summation of the data across the frequency channels to obtain a single channel dedispersed time series. A spectrum of periodicities is then obtained by performing a fast Fourier transform on the dedispersed data. The resulting spectrum undergoes harmonic summing, a process in which the spectral component for each periodicity is combined with a number of its harmonics. Since the narrow pulse of a pulsar should give rise to strong harmonics in its Fourier analysis, the addition of these harmonics can produce an increase in signal power by a factor greater than that for the noise, resulting in an improved signal-to-noise ratio. Lists of candidate pulsar periods are then produced, based on these signal-to-noise ratios. An example of the output which results from this phase of processing is given in Figure 1. In searching for pulsars in a globular cluster with no previously known pulsars, the dispersion measure is unknown and so this process needs to be repeated for a range of dispersion measures.

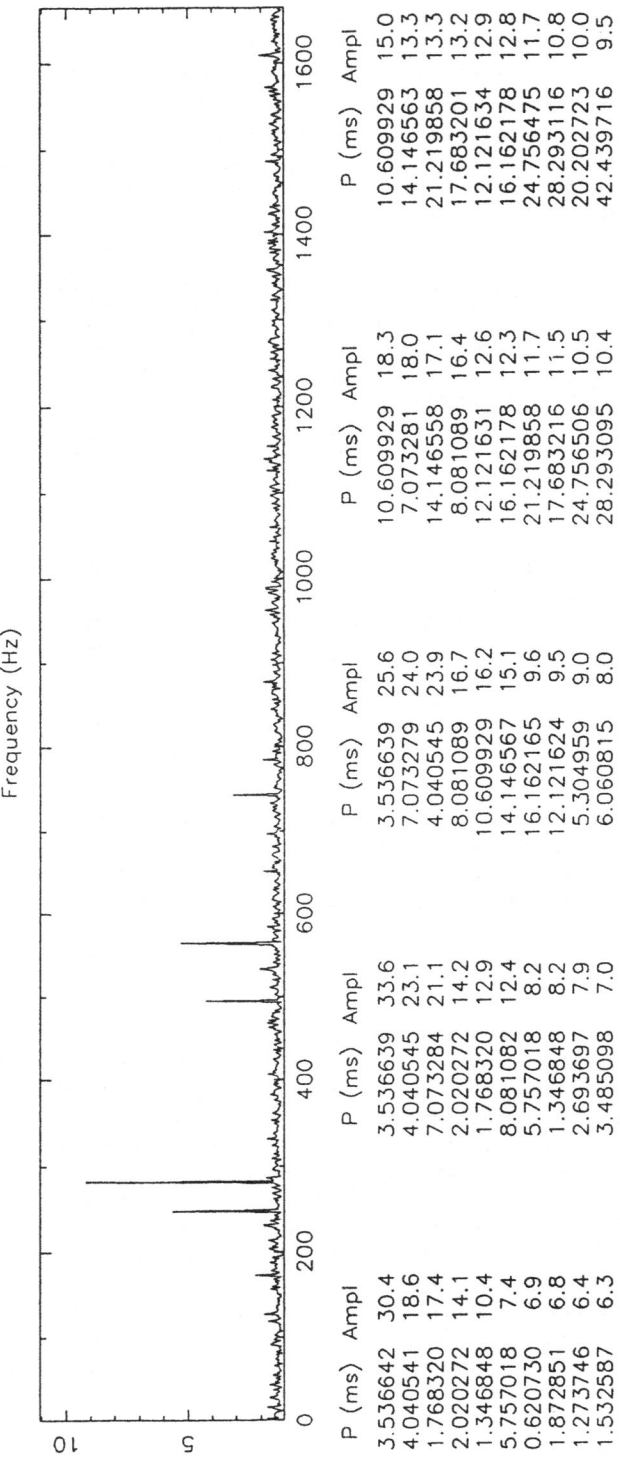

Figure 1: The above spectrum results from an 8M (2^{23}) point fast Fourier transform and features four of the 47 Tuc pulsars (C, E, G and I) in one observation. The five columns of figures give the ten best candidate periods, on the basis of signal-to-noise ratio, after harmonic summations which accumulate one, two, four, eight and sixteen harmonics, respectively.

3.2. PERIOD/DISPERSION MEASURE SEARCH

Searches in restricted ranges of period and dispersion measure are performed around the more promising candidate periods and associated dispersion measures obtained from the spectral analysis. Here, they are analysed more closely for pulsar characteristics on the basis of four plots. Examples of these plots for the solitary pulsar PSR 0021–72F and the binary pulsar PSR 0021–72E, both in 47 Tuc, are shown in Figures 2 and 3. The plots are explained below along with the features expected for a real pulsar.

DM/Period Plot (top) : A grey scale plot showing signal-to-noise ratios obtained over limited ranges of period and dispersion measure. A single pulsar will produce a tight and intense 'island', whilst for a fast binary pulsar this may be smeared in period.

Sub-integration/Phase Plot (middle left) : A grey scale plot showing signal-to-noise ratios resulting from sub-integrations in which the data, dedispersed at the nominal DM, is folded at the candidate period input. This results in narrow, straight tracks for a solitary pulsar, whilst curvature may be apparent for a fast binary pulsar.

Frequency/Phase Plot (middle right) : A grey scale plot showing signal-to-noise ratios in a number of frequency 'bands', as a result of integrating in these bands according to the best period obtained. A pulsar should give rise to a line of points across the frequency bands, the slope of which indicates an error in the assumed DM.

Amplitude/Phase Plot (bottom) : A pulse profile is obtained by integrating along the line corresponding to the best period in the sub-integration/phase plot.

4. Further Search Techniques

The pulsar search methods described above are biased in favour of bright solitary and long orbital period binary pulsars. More elaborate search techniques have, however, been developed which are designed to overcome this selection effect. They deal, separately, with two mutually exclusive classes of pulsar; strong fast binary pulsars and weak solitary pulsars. A further 'technique' takes advantage of a natural phenomenon in the interstellar medium which can aid the detection of weak pulsars. The respective techniques are outlined below.

4.1. DOPPLER ACCELERATION SEARCHES

A pulsar in a fast binary orbit may go undetected in the above search procedure due to the Doppler acceleration caused by its orbital motion; the changing line-of-sight component of its velocity produces changes in the observed period which results in a smearing of the spectral peaks in the pulsar's power spectrum. The search procedure can attempt to counter this effect by incorporating an algorithm which assumes and compensates for a constant period derivative. In such a search, the spectral analysis needs to be repeated for many different values of period derivative and the search in this additional parameter space adds greatly to the computational effort required. This type of search is, consequently, only usually feasible for clusters with known dispersion measures. The technique uncovered the fast binary pulsar PSR 2127+11C in M15 [9]. However, to date, all of the binary pulsars discovered in our surveys have been found in conventional non-acceleration searches. The detectability of fast binary pulsars is discussed in detail by Johnston and Kulkarni [10].

PSR 0021-72F

File: pg111011 RA: 00:21:53.0 Dec: −72:21:29 Gl: 305.8945 Gb: −44.8897 Date: 89/11/28
Centre freq. (Hz):   381.14229871   Centre period (ms):   2.62369200 Centre DM: 24.50
File start (blks):   1 Blks skipped: 0 Blks read: 4880 Blk length (s) 0.92160
Tsamp (ms): 0.3000 Nprd: 115 Ndm: 15 DM factor: 1.0
Start MJD   47858.49604

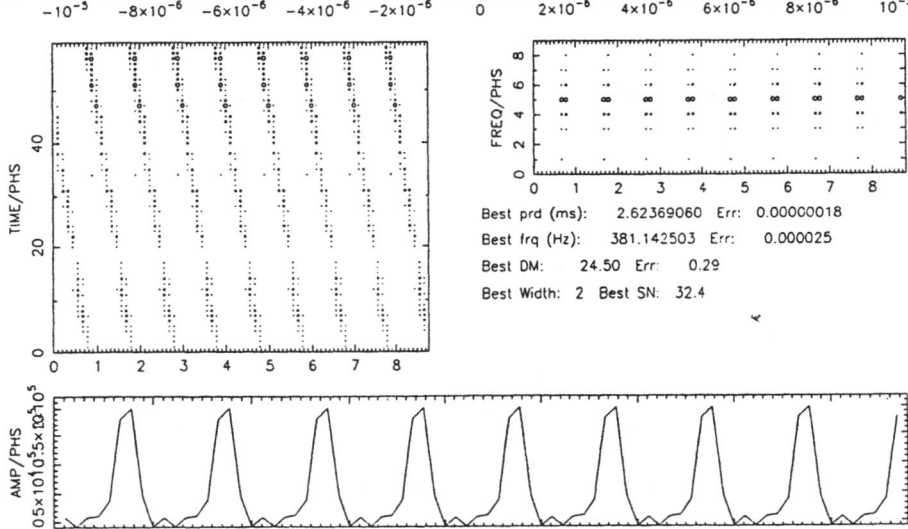

Best prd (ms):   2.62369060 Err: 0.00000018

Best frq (Hz):   381.142503 Err: 0.000025

Best DM: 24.50 Err: 0.29

Best Width: 2 Best SN: 32.4

Figure 2: Plot of the solitary pulsar PSR 0021-72F.

PSR 0021-72E

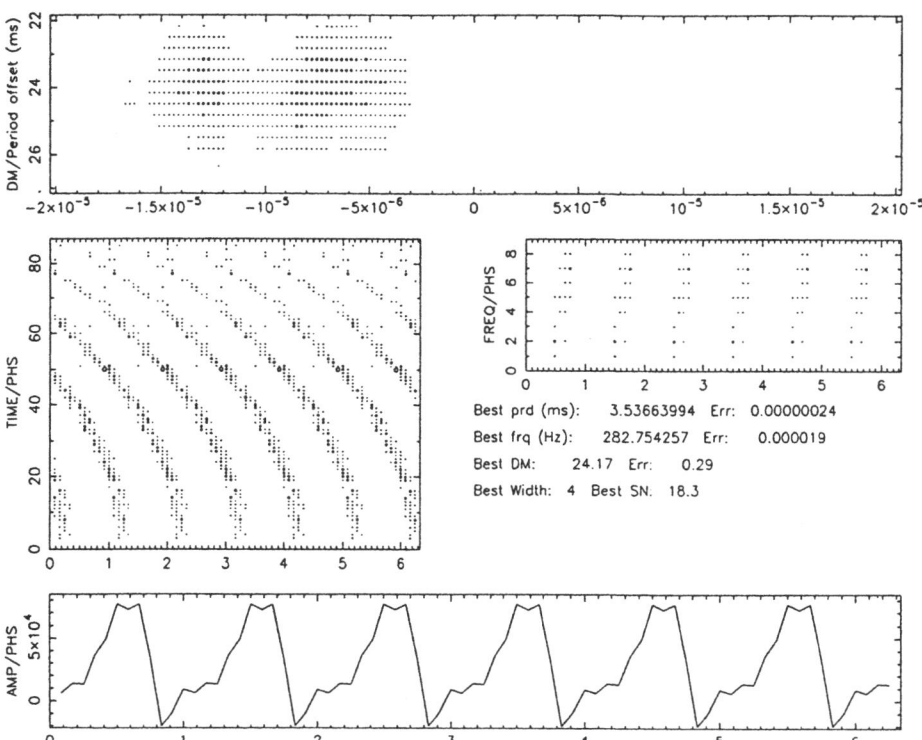

File: pg159011 RA: 00:21:53.0 Dec: −72:21:29 Gl: 305.8945 Gb: −44.8897 Date: 89/12/01
Centre freq. (Hz):   282.75369298   Centre period (ms):   3.53664700   Centre DM: 24.50
File start (blks):   1 Blks skipped: 0 Blks read: 4880 Blk length (s) 0.92160
Tsamp (ms): 0.3000 Nprd: 171 Ndm: 15 DM factor: 1.0
Start MJD   47861.40057

Best prd (ms):   3.53663994 Err: 0.00000024

Best frq (Hz):   282.754257 Err: 0.000019

Best DM: 24.17 Err: 0.29

Best Width: 4 Best SN. 18.3

Figure 3: Plot of the binary pulsar PSR 0021-72E.

4.2. SPECTRAL AVERAGING

A significant factor affecting the sensitivity of a pulsar survey is the integration time used. The sensitivity can be improved by increasing the integration time, although this improvement is restricted to those pulsars whose observed periods do not change appreciably, due to their orbital motion, during the observation. An increased integration time can be effectively achieved by combining the results from separate observations. Here, the periods in the spectrum resulting from each analysis need to be corrected to barycentric periods, after which the spectra can be added together and averaged. This technique will only be successful in detecting pulsars whose barycentric periods do not change between observations, i.e. solitary pulsars. Its successes, to date, are the single pulsars PSR 2127+11D and PSR 2127+11E in M15.

4.3. FLUX AMPLIFICATION BY THE INTERSTELLAR MEDIUM

Whilst the interstellar medium is in many respects something of a hindrance in searching for pulsars, it can also be of assistance. Pulsar fluxes are generally observed to fluctuate due to interstellar scintillation, of which there are thought to be two co-existing regimes; diffractive scintillation and refractive scintillation. The former can be understood in terms of the diffraction of the spatially coherent radiation of a pulsar by irregularities in the distribution of free electrons between the pulsar and the Earth. The pulsar is observed to scintillate as the Earth passes through the resulting diffraction pattern. Refractive scintillation is characterized by longer period fluctuations than the conventional diffractive type, and arises from the focussing and defocussing of radio waves by large scale irregularities. The observed fluctuations in a pulsar signal are then the combined effect of both types of scintillation. This can result in the amplification of the observed pulsar flux, and so, although it would seem that most globular clusters are too remote for our searches to reveal anything but the brightest of the cluster pulsars, scintillation effects may sometimes enhance the fluxes of pulsars which would normally be beyond our sensitivity. It is thought that this mechanism was responsible for the discovery of many of the 47 Tuc pulsars, which are actually not detectable in the majority of our observations. Indeed, our success with 47 Tuc may suggest that the interstellar medium could have a very effective and important role in searching for pulsars in globular clusters.

REFERENCES

1. Lyne, A. G., Brinklow, A., Middleditch, J., Kulkarni, S. R., Backer, D. C. and Clifton, T. R. 1987, *Nature*, **328**, 399.

2. Lyne, A. G., Biggs, J. D., Brinklow, A., Ashworth, M. and McKenna, J. 1988, *Nature*, **332**, 45.

3. Manchester, R. N., Lyne, A. G., D'Amico, N., Johnston, S., Lim, J. and Kniffen, D. A. 1990, *Nature*, **345**, 598.

4. Manchester, R. N., Lyne, A. G., Robinson, C., D'Amico, N., Bailes, M. and Lim, J. 1991, *Nature*, submitted.

5. Lyne, A. G., Manchester, R. N., D'Amico, N., Staveley-Smith, L., Johnston, S., Lim, J., Fruchter, A. S., Goss, W. M. and Frail, D. 1990, *Nature*, **347**, 650.

6. Manchester, R. N., Lyne, A. G., Johnston, S., D'Amico, N., Lim. J., Kniffen, D. A., Fruchter, A. S. and Goss, W. M. 1989, *IAU Circ.* No. 4905.

7. D'Amico, N., Lyne, A. G., Bailes, M., Johnston, S., Manchester, R. N., Staveley-Smith, L., Lim, J., Fruchter, A. S. and Goss, W. M. 1990, *IAU Circ.* No. 5013.

8. Biggs, J. D., Lyne, A. G., Manchester, R. N. and Ashworth, M. 1990, *IAU Circ.* No. 4988.

9. Anderson, S. B., Gorham, P. W., Kulkarni, S. R., Prince, T. A. and Wolszczan, A. 1990, *Nature*, **346**, 42.

10. Johnston, H. M. and Kulkarni, S. R. 1991, *Ap. J.*, **368**, 504.

DISCOVERY AND TIMING OF RADIO PULSARS IN GLOBULAR CLUSTERS M13 AND M15

S.B. Anderson, S.R. Kulkarni, and T.A. Prince
California Institute of Technology, Pasadena, CA 91125 USA

A. Wolszczan
Arecibo Observatory, Arecibo, Puerto Rico 00613

ABSTRACT. We announce the discovery of PSR1639+36B (M13B), a 3.5 ms binary pulsar in the globular cluster M13 (NGC6205). In addition, we report the timing positions of PSR1639+36A (M13A), PSR2127+11D&E (M15D&E), as well as a second general relativistic parameter for the 8-hr binary pulsar PSR2127+11C (M15C). Initial doppler shift measurements of M13B indicate that it is in a highly circular orbit of ~one day. Timing positions of M13A and M15D indicate that they are within the central cores of their respective clusters, while M15E is out a few core radii and M15C has been ejected to near the half-mass radius of the cluster. Further observations of M15C have determined the mass of the pulsar and companion to be $m_p = 1.33 \pm .22\ M_\odot$ and $m_c = 1.38 \pm .22\ M_\odot$, indicating that the companion star is either a massive white dwarf or another neutron star.

INTRODUCTION

The work presented here is part of an ongoing survey of globular clusters observable from Arecibo. Observations were made using the 305-m Arecibo radio telescope at a central frequency of 430 MHz and a 10 MHz receiver bandwidth. The data were sampled at an effective rate of 1.974 kHz using the Arecibo 40 MHz, 3–level correlation spectrometer.

The pulse search was carried out on the Caltech NCUBE/10, a concurrent computer having 512 Central Processing Units (CPUs) with 256 Mbytes of memory and hypercube interconnection topology. For the two clusters discussed here the observations were de-dispersed (i.e. collapsed to a fiducial radio frequency) based on known dispersion measures (DM's) from previously detected pulsars (PSR1639+36A in the case of M13 and PSR2127+11A for M15).

In the case of M13, a search for additional pulsars in tight binary orbits was performed by approximating the time varying Doppler shift due to the orbital motion of the pulsar as a linear change in the observed pulse frequency with time. The success of this technique of first detecting an isolated pulsar in a globular cluster to determine the DM and then searching over orthogonal orbital parameters has resulted in the discovery of three binary pulsars, including M15C.

For M15 an attempt to detect faint pulsars, and hence probe the luminosity function of radio pulsars in globular clusters, was attempted by incoherently adding multiple observations. This technique resulted in the discovery of M15D&E.

131

E. P. J. van den Heuvel and S. A. Rappaport (eds.), X-Ray Binaries and Recycled Pulsars, 131–145.
© *1992 Kluwer Academic Publishers.*

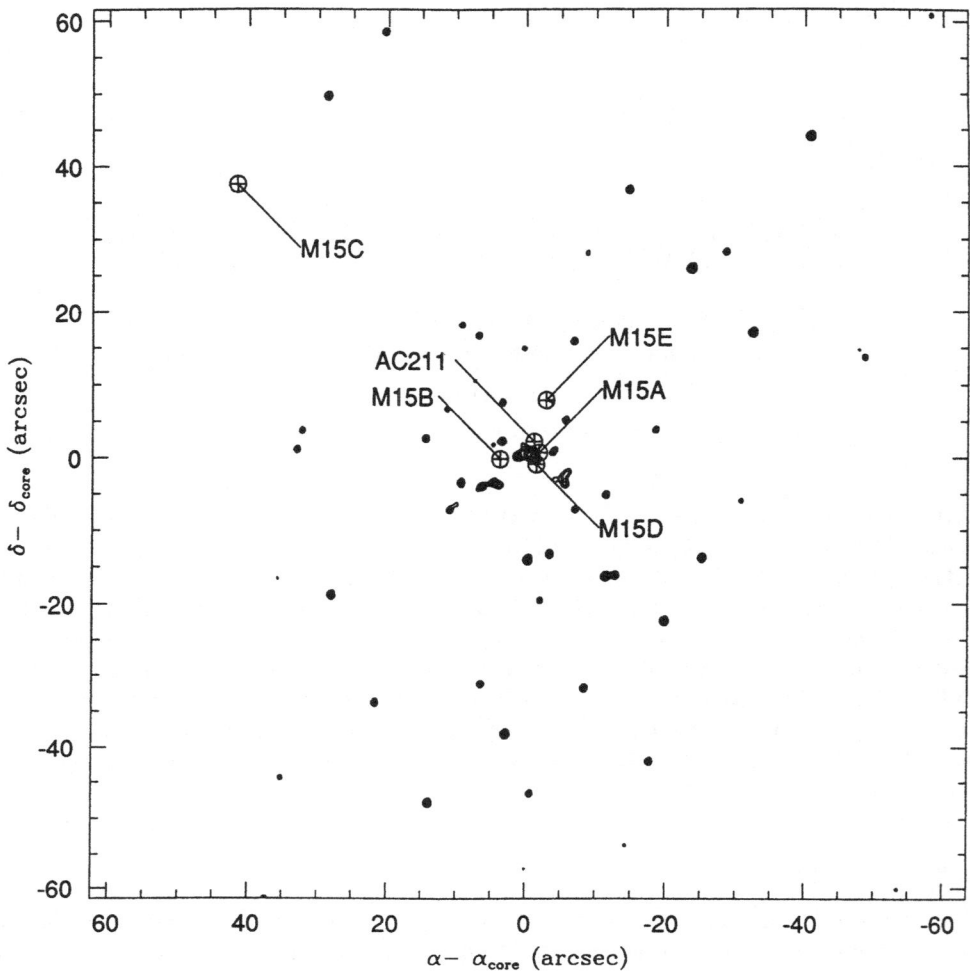

Figure 1: Optical image of Globular Cluster M15 with the positions of the 5 known radio pulsars and the 1 LMXB superimposed.

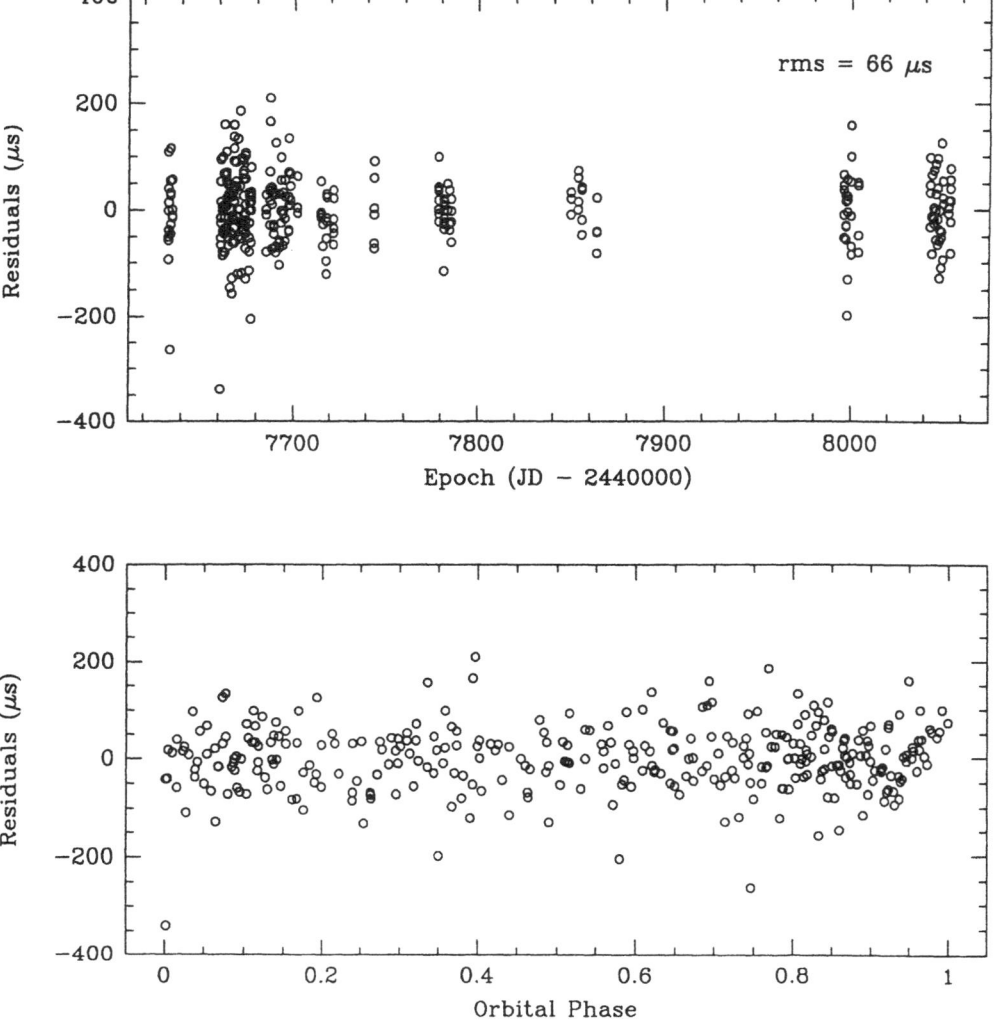

Figure 2: Timing residuals for the binary pulsar PSR2127+11C; (top) as a function of observation epoch, (bottom) as a function of orbital phase.

PSR 2127+11C

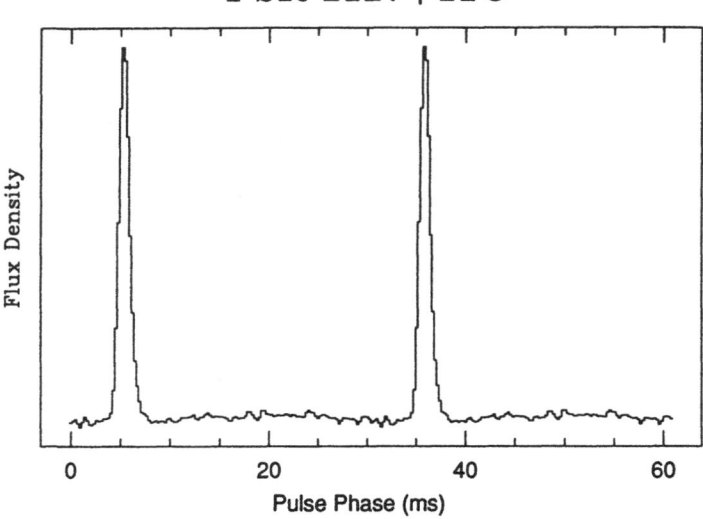

Pulse Phase (ms)

Timing Parameters

Pulsar period	P	$30.5292951287(6)$ ms
Dispersion measure	DM	67.25 ± 1.5 pc cm^{-3}
Right Ascension †	$\alpha - \alpha_{core}$	$+41''.65(5)$
Declination†	$\delta - \delta_{core}$	$+37''.47(4)$

Orbital Parameters

Orbital period	P_b	$28968.36974(20)$ s
Projected semi-major axis	$a_1 sini$	$2.518(1)$ ls
Eccentricity	e	$0.681405(15)$
Longitude of Periastron	ω_0	$316.36(3)°$
Apsidal Motion	$\dot{\omega}$	$4.465(5)°$ yr^{-1}
Gravitational Redshift	γ	4.9 ± 1.0 ms
Epoch of Periastron	T_0	$2447632.4672053(15)\ JD$

Derived Parameters

Pulsar Mass	m_p	$1.33 \pm .22 M_\odot$
Companion Mass	m_c	$1.38 \pm .22 M_\odot$

† Cluster Center[1]: $\alpha_{core} = 21^h 27^m 33^s.35$, $\delta_{core} = 11° 56' 48''.8$

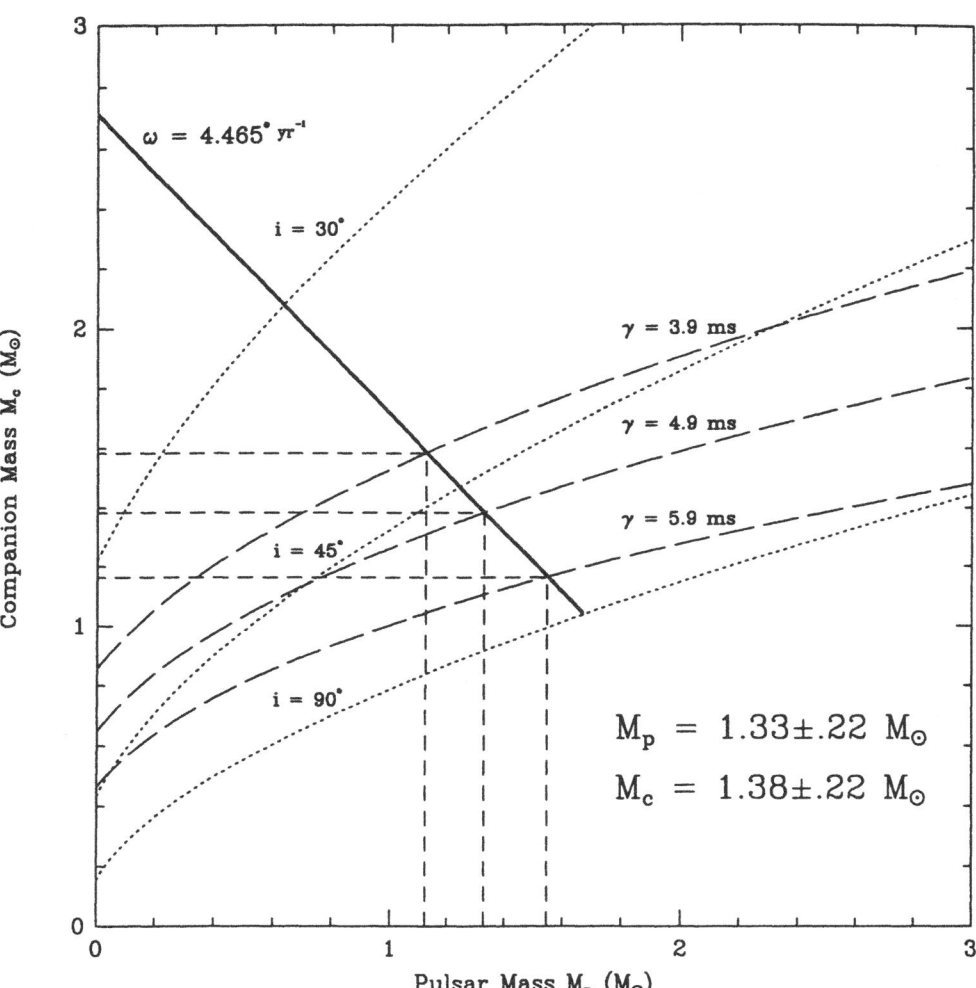

Figure 3: Constraints on the masses of the two components of the binary system 2127+11C. Errors on the $\dot{\omega}$ constraint lie within the thickness of the line.

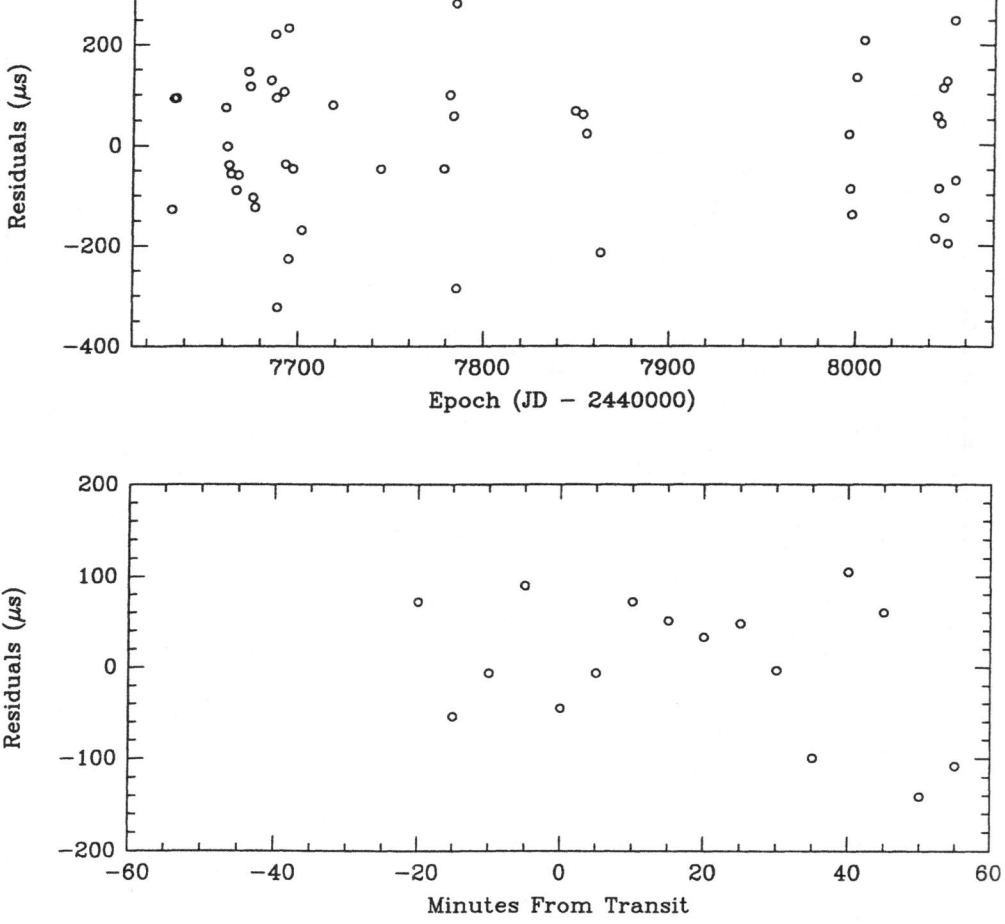

Figure 4: Timing residuals for PSR2127+11D; (top) as a function of observation epoch, (bottom) as a function of local sidereal time. An error in pulse counting between consecutive days of one pulse corresponds to a residual drift rate of $200\mu s$ hr^{-1}.

PSR 2127+11D

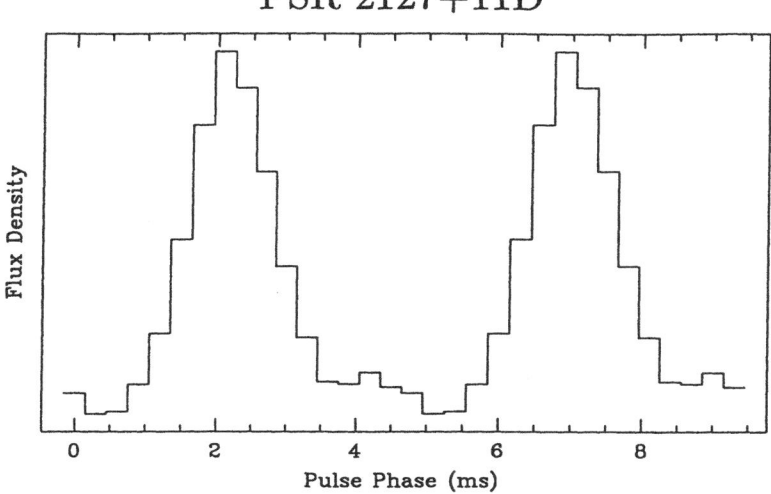

Flux Density

Pulse Phase (ms)

Timing Parameters

Pulsar period	P	$4.8028043458(3)\,\text{ms}$
Right Ascension †	$\alpha - \alpha_{core}$	$-1''.4(1)$
Declination†	$\delta - \delta_{core}$	$-0''.9(2)$
Epoch	T_0	$2447633.02\,\text{JD}$
Dispersion measure	DM	$67.25 \pm 1.5\,\text{pc cm}^{-3}$

† Cluster Center[1]: $\alpha_{core} = 21^h\,27^m\,33^s.35$, $\delta_{core} = 11°\,56'\,48''.8$

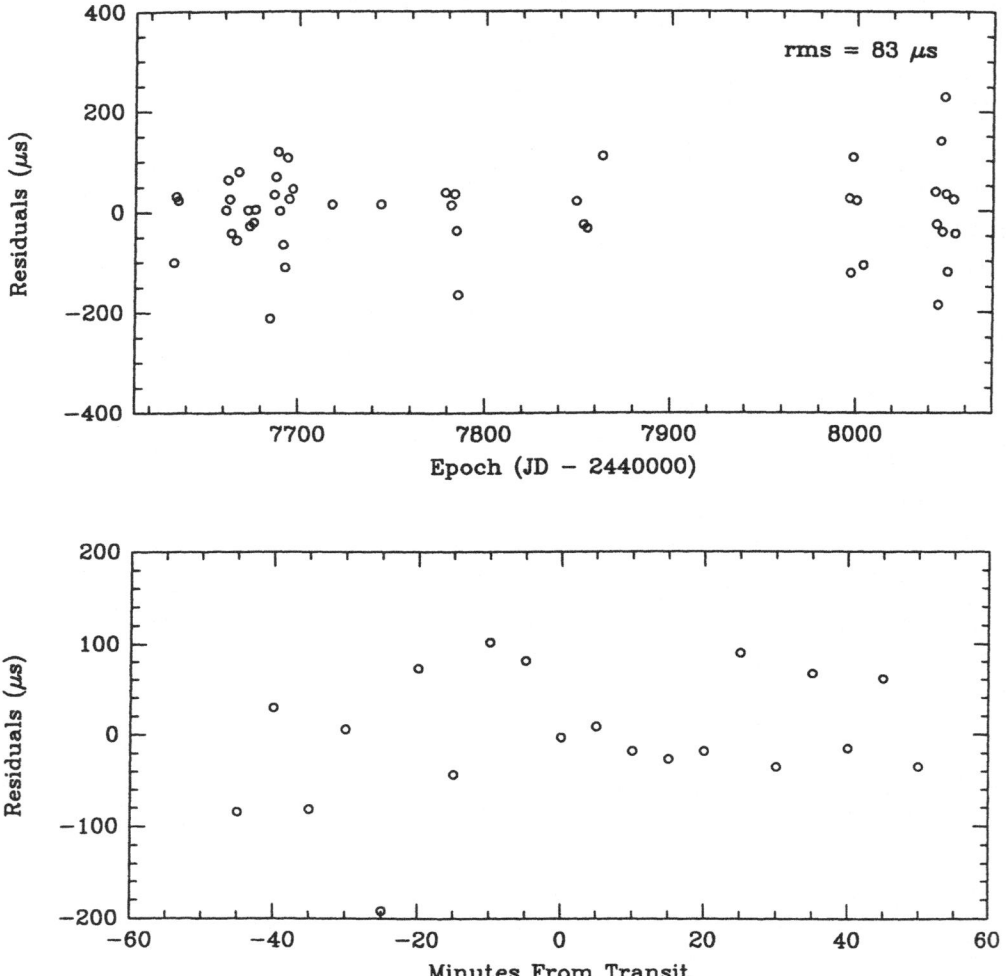

Figure 5: Timing residuals for PSR2127+11E; (top) as a function of observation epoch, (bottom) as a function of local sidereal time. An error in pulse counting between consecutive days of one pulse corresponds to a residual drift rate of $194\mu s\ hr^{-1}$.

PSR 2127+11E

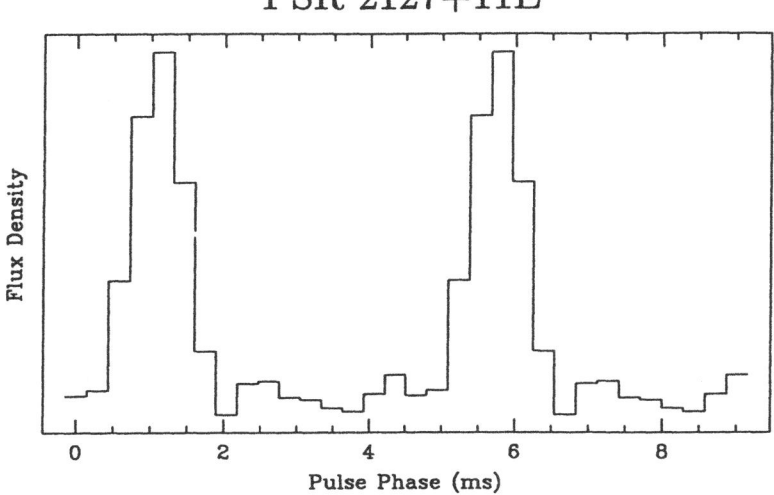

Pulse Phase (ms)

Timing Parameters

Pulsar period	P	4.6514352154(2) ms
Right Ascension †	$\alpha - \alpha_{core}$	$-2''.69(7)$
Declination†	$\delta - \delta_{core}$	$+8''.0(1)$
Epoch	T_0	2447633.02 JD
Dispersion measure	DM	$67.25 \pm 1.5\, \text{pc cm}^{-3}$

† Cluster Center[1]: $\alpha_{core} = 21^h\, 27^m\, 33^s.35$, $\delta_{core} = 11°\, 56'\, 48''.8$

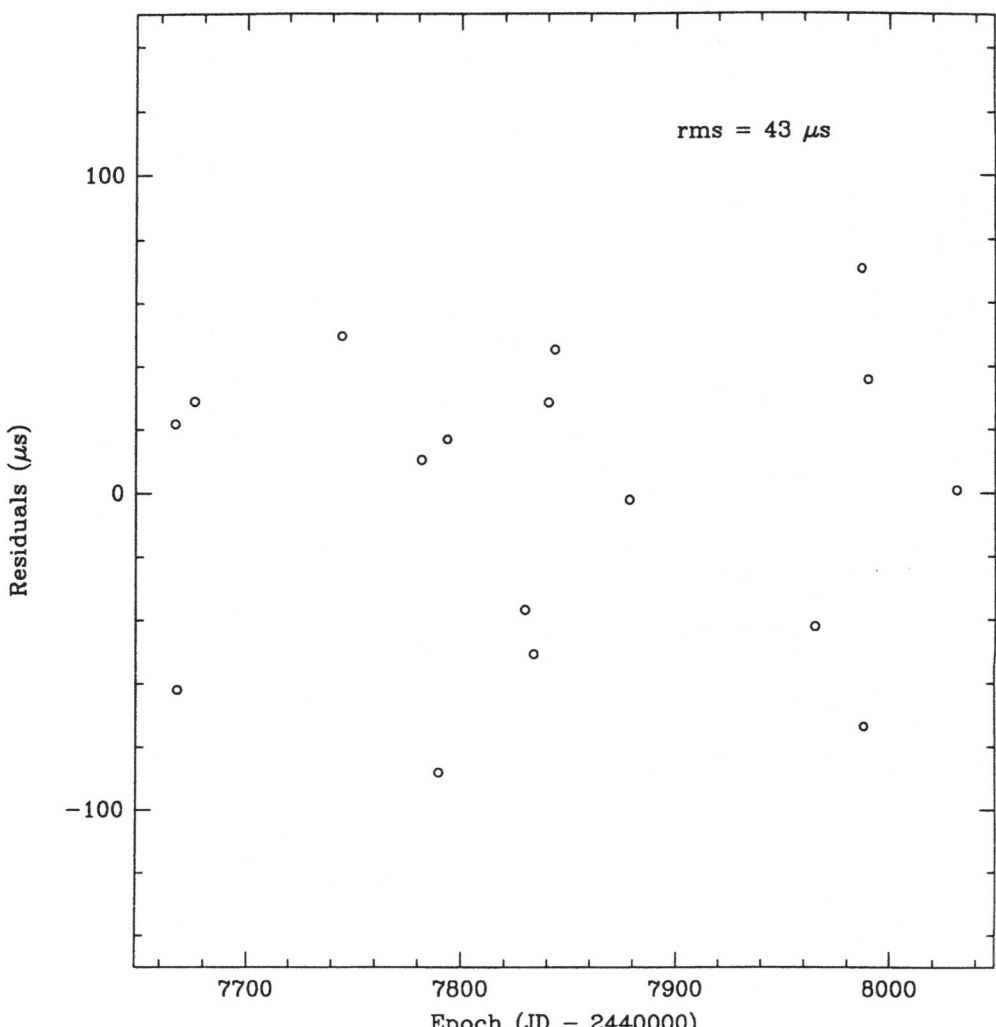

Figure 6: Timing residuals for PSR1639+36A.

PSR 1639+36A

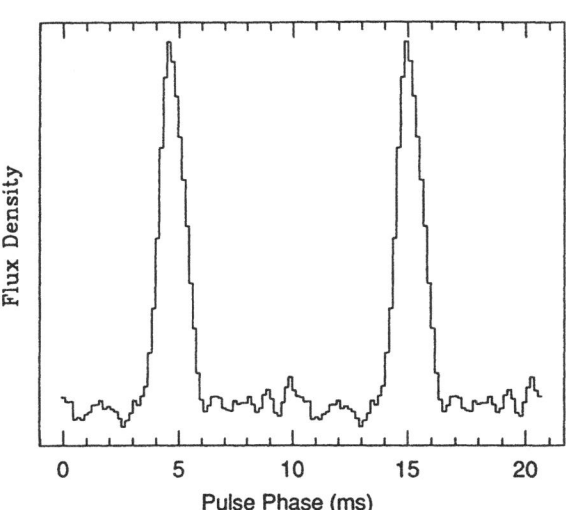

Timing Parameters

Pulsar Period	P	$10.3775094520(6)\,\text{ms}$
Pulsar Period Derivative	$\dot{P}$	$< 4.5 \times 10^{-20}\,\text{s s}^{-1}$
Epoch	T_0	$2447666.71\,\text{JD}$
Dispersion Measure	DM	$30.36(4)\,\text{pc cm}^{-3}$
Right Ascension †	$\alpha - \alpha_{core}$	$-6''.80(5)$
Declination†	$\delta - \delta_{core}$	$-21''.36(9)$

Derived Parameters

Acceleration	$|a_l|$	$< 4 \times 10^7\,\text{cm s}^{-2}$
Magnetic Field	B	$< 1.2 \times 10^9\,\text{G}$
Characteristic age	τ_c	$> 0.9 \times 10^9\,\text{yr}$

---

† Cluster Center[1]: $\alpha_{core} = 16^h\,39^m\,54^s.19$, $\delta_{core} = 36°\,33'\,16''.3$

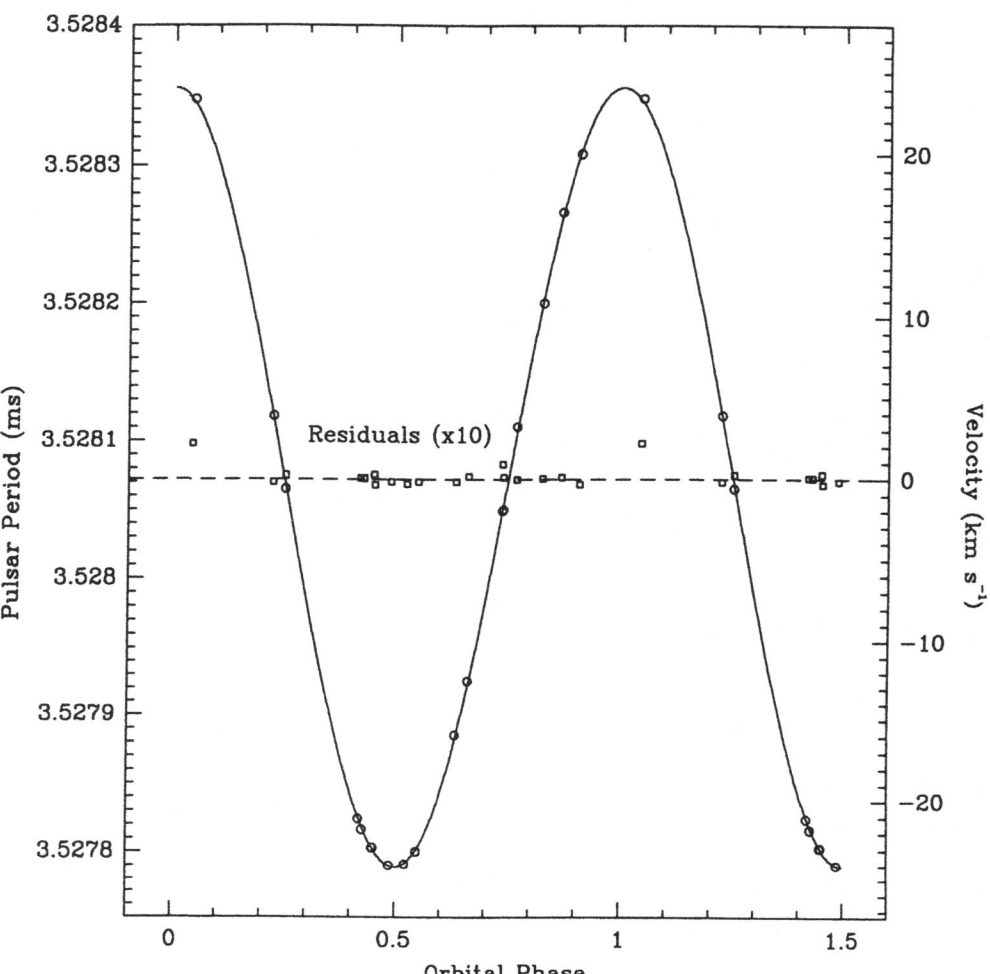

Figure 7: Pulse period residuals for the binary pulsar PSR1639+36B.

PSR 1639+36B

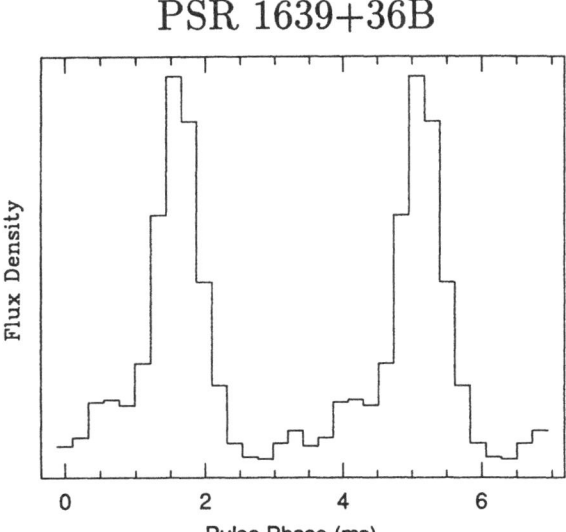

Timing Parameters

Pulsar period	P	3.528(7) ms
Dispersion measure	DM	29.5 ± 1.5 pc cm^{-3}
Right Ascension †	α	$16^h\, 39^m\, 54^s.19 \pm 10'$ (B1950.0)
Declination†	δ	$36° \, 33' \, 16''.3 \pm 10'$ (B1950.0)

Keplerian Orbital Parameters

Orbital period	P_b	1.259113(3) d
Projected semi-major axis	$a_1 sini$	1.389(2) ls
Eccentricity	e	< 0.001
Epoch	T_0	2447149.6835(18) JD

Derived Parameters

Mass function	f	$0.001815(8) M_\odot$

† Assumed position is coincident with cluster center.[1] Errors correspond to beam size at 430MHz.

PSR 2127+11C

As can been seen in figure 1, M15C has an unexpectedly large offset from the center of the cluster of 56″. With the exception of M15C, all the cluster pulsars known to date and the dozen cluster low mass X-ray binaries are located within the core of clusters, or within a few core radii. However, it is highly unlikely that M15C is on an escape trajectory from the cluster because of the very short time to be ejected to its current radius ($< 10^5$ yr), rather it is almost certainly in a highly eccentric orbit with its pericenter in the cluster core (see ref. 3 for a discussion of the distribution and evolution of orbits of ejected pulsars). The low stellar density at the radius of M15C implies a lifetime of $\mathcal{O}$ (10^{13}yr) for either tidal capture or exchange reactions. Since at least one such event occurred within $\tau_c \sim \tau_{GR} \simeq 10^8$ yr to form the currently observed binary system, it is almost certain that the system was not formed at its current location, but rather in the highly concentrated core of this post-core collapse globular cluster.

A plausible explanation for ejection is the heating of the cluster core through the release of binary binding energy[4]. Core collapse of dense globular clusters is considered to be an inevitable consequence of dynamical evolution. The collapse is halted by the formation of tidal binaries and the release of energy through additional close interactions which cause a decrease in the binary semi-major axis ('hardening') and therefore a release of binding energy. For close binaries, exothermic interactions are kinematically favored over endothermic reactions which tend to disrupt the binary. The highly eccentric M15C system is likely to have gone through such an exothermic collision with the potential energy released in the collision being converted into kinetic energy resulting in ejection from the core.

Given the small size of the orbit and the environment of a globular cluster with a turn-off mass $\sim .8$ $M_\odot$ it can be argued that there are no classical contributions to the observed rate of orbital precession[5]. Given this and the measurement of the gravitational redshift and time dilation presented here, the mass of the pulsar and its companion are determined to be, $m_p = 1.33 \pm .22$ $M_\odot$ and $m_c = 1.38 \pm .22$ $M_\odot$ (figure 3).

PSR 2127+11D&E

The faintness of M15D&E necessitates the integration of a full transit observation to measure a single pulse arrival time. This results in a pulse numbering ambiguity from one day to the next. However, if one records the raw data stream for $\mathcal{O}$ (25) transits it is possible to coherently fold back the data to a fiducial transit, thereby increasing the signal-to-noise sufficiently to measure the pulse period accurately enough to remove the day-to-day pulse numbering ambiguity. Figures 4 and 5 show the resulting residuals for M15D&E.

PSR 1639+36A

Timing measurements of M13A indicate that $\mid \dot{P} \mid < 4.5 \times 10^{-20}$ s s^{-1} (figure 6), however, the mean gravitational field of the cluster can be expected to generate a line-of-site acceleration of the pulsar $\mid a_l \mid \lesssim 4 \times 10^{-7}$ cm s^{-2} with a corresponding corruption of the intrinsic $\dot{P}$ by $c \mid a_l \mid P^{-1} \lesssim 14 \times 10^{-20}$ s s^{-1} (ref. 2). Consequently, a lower limit of 0.9×10^9 yr may be placed on the characteristic age, τ_c, making M13A the oldest known cluster pulsar. The corresponding magnetic field may be estimated to be $\lesssim 1.2 \times 10^9$ G.

PSR 1639+36B

M13B is a 3.5 ms binary pulsar in a 30 hr orbit (figure 7). With a mass function, $f = 0.001815$ M$_\odot$, the mass of the companion $> .16$ M$_\odot$ (assuming $m_p = 1.4$ M$_\odot$), indicating a white dwarf as the probable companion. The discovery of a second pulsar in the globular cluster M13 provides additional evidence that the birthrate of radio pulsars in low-density globular clusters is higher than allowed from two-body tidal interactions. This lends support to other formation scenerios such as three-body interactions involving primordial binaries[6].

REFERENCES

1. Shawl, S. J. & White, R. E. *Astron. J.* **91**, 312 − 316 (1986).

2. Kulkarni, S.R., Anderson, S.B., Prince, T.A. & Wolszczan, A. *Nature* **349**, 47 − 49 (1991).

3. Phinney, E.S., *Mon. Not. R. Astro. Soc.*, submitted.

4. Elson, R., Hut, P. & Inagaki, S. *Ann. Rev. Astron. Astrophys.*, **25**, 565-601 (1987).

5. Prince, T.A., Anderson, S.B., Kulkarni, S.R. & Wolszczan, A. *APJL*, accepted.

6. Phinney, E.S. & Kulkarni, S.R. *Nature* (submitted).

CHAPTER 3

Supernovae, Pulsar Proper Motions

A NEW PULSAR - SUPERNOVA REMNANT ASSOCIATION

D. A. FRAIL
National Radio Astronomy Observatory*, Socorro, NM, 87801, USA.

S. R. KULKARNI
California Institute of Technology, Pasadena, California, 91125, USA.

ABSTRACT. We report high-resolution VLA observations which have revealed a compact, highly-polarized, flat-spectrum radio nebula on the western periphery of the galactic supernova remnant G 5.4–1.2. We argue that this radio emission is powered by the 125 msec pulsar PSR 1758–24 and that the pulsar is physically associated with G 5.4–1.2. We interpret the peculiar morphology of G 5.4–1.2 in terms of the "rejuvenation" hypothesis of Shull et al. and make several testable predictions of this model.

1. Introduction

Over the years there has been much speculation about the nature and origin of the peculiar galactic radio source G 5.4–1.2. It was first classified by Milne and Dickel (1971) as a galactic supernova remnant (SNR) on the basis of its negative spectral index ($\alpha = -0.2$) and the detection of strong linear polarization. Higher resolution images at 6 and 20 cm by Becker and Helfand (1985a) reveal an unusual morphology for this object. G 5.4–1.2 is fan-shaped with edge-brightened emission along a western arc becoming fainter and more diffuse towards the east. Strong linear polarization is seen along the entire western arc, indicating the presence of a well-ordered magnetic field. Just on the periphery of the western arc there is an extended, center-filled radio source (G 5.27–0.90) that is joined to G 5.4–1.2 by a short bridge of emission.

To account for these unusual characteristics Becker and Helfand (1985b) proposed that G 5.4–1.2 is one of a new class of non-thermal sources formed from the emission of relativistic particles produced by an accreting binary system. This interpretation was disputed when a wide-field image at 843 MHz by Caswell *et al.* (1987) revealed two arcs of emission: the bright western arc mentioned above, and a fainter eastern arc missed by Becker and Helfand because it was outside their field of view. Caswell *et al.* (1987) interpret these arcs as forming a shell-type SNR and the strong brightness contrast between the arcs is explained as being due to large-scale gradients in the interstellar gas density and/or magnetic field.

---

* The National Radio Astronomy Observatory is operated by Associated Universities Inc., under contract with the National Science Foundation.

149

E. P. J. van den Heuvel and S. A. Rappaport (eds.), X-Ray Binaries and Recycled Pulsars, 149–153.
© 1992 *Kluwer Academic Publishers.*

A short period pulsar PSR 1758–24 (P=125 msec) is known to be nearby G 5.4–1.2 (Manchester *et al.* 1985). Its dispersion measure-derived distance of 7 kpc is nearly identical to that derived to the SNR from the Σ-D relation (7.5 kpc) (Clark and Caswell 1976), strongly suggesting a physical association. However, because the pulsar is weak (S_{mean}=1 mJy at 20 cm), no follow-up work had been done and the position was uncertain by ±6 arcmin. For this reason we decided to undertake sensitive, high resolution radio observations of the region to search for the characteristic signature of a pulsar, with the aim of clarifying the nature and origin of G 5.4–1.2 and its (possible) association with PSR 1758–24.

2. Observations

In August 1990 we used the Very Large Array (VLA) in its B-array configuration to observe a field at 20 cm and 6 cm, centered on the nominal position of the pulsar. In this configuration the telescope is an effective spatial filter, attenuating large-scale emission from the remnant and the rest of the Galaxy, thus giving greater contrast to aid in the detection of point sources. The bandwidth and total integration time were 25 MHz and 160 minutes at 20 cm, and 50 MHz and 25 minutes at 6 cm. The AIPS package was used in the reduction of these data following standard procedures.

3. Discussion

Only a subsection of the full 20-cm field is shown in Figure 1. The western arc of G 5.4–1.2 is visible on the left edge of Figure 1, linked to G 5.27–0.9 (center) by the bridge of emission. The most striking aspect about this image is the presence of a bright, partially resolved, compact object just to the west of G 5.27–0.9, joined to it by another bridge of emission. This source is located at 17^h 57^m 55.4^s and –24° 51′ 25″, and has a peak flux of 3.5 mJy. Linear polarization is detected over most of the region in Figure 1 and is brightest at the compact source. The measured degree of polarization for the compact source is 20%. However, since we see evidence of significant Faraday rotation in our data this is likely only a lower limit. There are only 4 other point sources within the error bars of the pulsar's nominal position. Two of these appear to form the components of an extragalactic double. We detect no polarized flux from the remaining two point sources, but they are weak (<1 mJy).

A 6-cm observation, shown in Figure 2, is of shorter duration than the 20-cm but has a resolution 3 times better than that at 20 cm (i.e. 1.5″ versus 5″). The increased resolution reveals that the compact source is a ring-like structure with a radius of 1.3″). The emission bridge resolves into several blobs, suggesting a flow of material from the ring in a westward direction. The integrated flux over this region at 6 cm is 9±2 mJy, while at 20 cm it is 10±1 mJy. Within the errors, this corresponds to a flat ($\alpha \sim 0$) spectrum source.

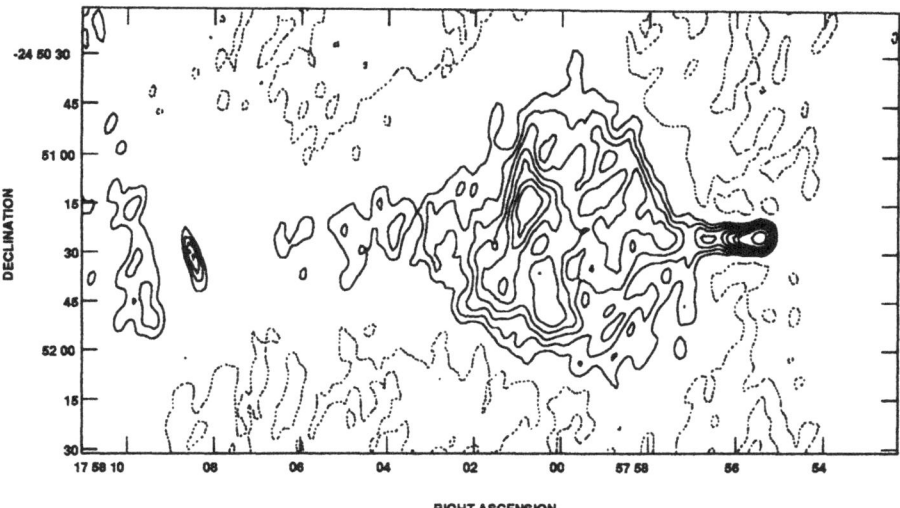

Fig. 1. A 20-cm image with G 5.4–1.2 visible on the eastern edge, G 5.27–0.9 in the center, and the newly discovered compact nebula to the west. Contour levels are –3, 3, 5, 7, 9, 11, 13, 15, 20 times the rms noise of 150 μJy ba^{-1}.

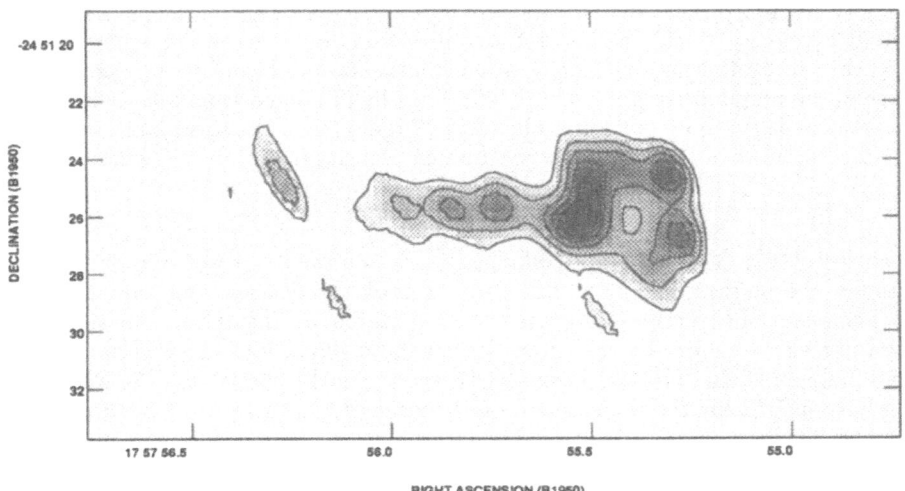

Fig. 2. A 6-cm image of the compact nebula. Contour levels are –3, 3, 4, 5, 6, 7, 8 times the rms noise of 86 μJy ba^{-1}.

4. Interpretation

What is the compact nebula and what is its relation to either G 5.4−1.2 or PSR 1758−24? We note several points:

• The flat-spectrum, highly polarized emission from the compact nebula is characteristic of a pulsar-powered nebula (cf. CTB 80/PSR 1951+32, Hester and Kulkarni 1988).
• The extended synchrotron emission that joins G 5.4−1.2 with the compact nebula is not likely due to a line-of-sight coincidence. A more plausible explanation is that there is a real association between these two objects.
• PSR 1758−24 is also likely to be physically associated with G 5.4−1.2 because they both have nearly identical distances as determined from two different methods.

When taken together the evidence is persuasive enough to argue that the compact nebula harbors the active pulsar PSR 1758−24 (now more appropriately called PSR 1757−24). We suggest that the distinct cometary shape of the nebular emission arises due to the interaction of the relativistic wind from the pulsar as it moves through the interstellar medium. The head of the "comet" is formed where the pulsar wind is confined by the ram pressure of the interstellar medium and the tail is the downstream flow of relativistic particles from the wind.

5. The Rejuvenation Model

The interaction of pulsars with the shells of supernova remnants such as CTB 80 and PSR 1951+32, has been considered in some detail by Shull *et al.* (1989). They term this process the "rejuvenation" hypothesis. Interpreting our results in this framework we speculate that some t_{age} years ago there was a type II supernova event that produced a rapidly expanding supernova shell and the pulsar was given a velocity "kick" in the westward direction. In time, as the supernova evolved into its isothermal or radiative snowplow phase it began to slow down. The age of the remnant (in units of 10^5 years) can be predicted from a knowledge of its distance d_6 (in units of 6 kpc), diameter θ (in degrees), the total mechanical energy E_{51} (in units of 10^{51} ergs) and the ambient density n_o (in units of cm^{-3}) by

$$t_{age} = 8.2 \, (\theta \, d_6)^{3.33} \, E_{51}^{-0.737} \, n_o^{0.857}.$$

For G 5.4−1.2 $\theta = 35'$, $d_6=1$ and taking $n_o \sim E_{51}=1$ then $t_{age} = 1.4 \times 10^5$ yrs. Eventually the pulsar caught up with and overtook the decelerating supernova shell. The result of this interaction was a "rejuvenation" of the aging SNR as the pulsar injected energetic particles from its magnetosphere into the compressed gas and magnetic field of the isothermal SNR shock. In this model it is the pulsar that is responsible for the increased radio brightness in the western portion of G 5.4−1.2. The transverse velocity required for a pulsar to overtake the shell is given by

$$V_{PSR} = 514 \, \text{km s}^{-1} \, \beta \, \theta \, d_6 \, t_{age}^{-1},$$

where β is the fractional diameter of the supernova remnant that the pulsar has traveled. PSR 1757−24 is located 3′ *outside* G 5.4−1.2 and so $\beta=1.17$, for which we derive $V_{PSR}=250$ km s^{-1}.

6. Predictions

Several predictions, all testable, emerge from this model.

(1) PSR 1757–24 should be precisely located in the cavity of the compact nebula as defined by Figure 2.

(2) A value for the period derivative $\dot{P}$ of PSR 1757–24 can be estimated by setting its characteristic age ($\tau_c \equiv P/2\dot{P}$) equal to the age of G 5.4–1.2 $t_{age} = 1.4 \times 10^5$ yrs. We predict that $\dot{P} = 14 \times 10^{-15}$ s s^{-1} and that the rate of rotational energy loss by the pulsar is $\dot{E} = 2.9 \times 10^{35}$ erg s^{-1}, exactly the same as that for PSR 1957+20 (Kulkarni and Hester 1988).

(3) The proper motion of the pulsar is expected to be about 9 mas yr^{-1} due west.

Since the distance is rather uncertain these estimates could be in error by as much as a factor of 2. We conclude that the rejuvenation hypothesis offers an attractive but not a unique explanation for the peculiar morphology of G 5.4–1.2. In the future the most important task is to confirm that PSR 1757–24 is embedded within the compact nebula by detecting pulsed emission. A timing program is also needed both to get an independent position for the pulsar and to determine its $\dot{P}$. The proper motion will be somewhat more difficult to measure but it would be a valuable check on the rejuvenation hypothesis. One additional test would be to look for a steepening of the spectral index (due to synchrotron losses) away from the pulsar and towards the SNR.

SRK's pulsar research at the VLA is supported by a grant from the Perkin Fund to which he is very grateful.

7. References

Becker, R. H. & Helfand, D. J. *Nature* **313**, 115 (1985a).
Becker, R. H. & Helfand, D. J. *Nature* **313**, 118 (1985b).
Caswell, J. L., Kesteven, M. J., Komesaroff, M. M., Haynes, R. F., Milne, D. K., Stewart, R. T. & Wilson, S. G. *M. N. R. A. S.* **225**, 329 (1987).
Clark, D. H. & Caswell, J. L. *M. N. R. A. S.* **174**, 267 (1976).
Hester, J. J. & Kulkarni, S. R. *Ap. J.* **331**, L121 (1988).
Kulkarni, S. R. & Hester, J. J. *Nature* **335**, 801 (1988).
Manchester, R. N., D'Amico, N. & Tuohy, I. R. *M. N. R. A. S.* **212**, 975 (1985).
Milne, D. K. & Dickel, J. R. *Nature. Phys. Sci.*, **231**, 425 (1971).
Shull, J. M., Fesen, R. A. & Saken, J. M. *Ap. J.* **346**, 860 (1989).

Added Note: On January 30, 1991 we successfully detected pulsed emission using the VLA in the phased-array mode with a ~ 20″ beam. Pulsations were detected with the VLA phased towards the cavity in Figure 2. This immediately confirms our model. The pulsar appears to have a period 40 μsec longer than its discovery period (Manchester *et al.* 1985). If this slowdown is attributed purely to the rotational spindown then $\dot{P} = 140 \times 10^{-15}$ s s^{-1}, a factor of 10 larger than our prediction. Thus the pulsar must be moving at a speed close to 1% of the speed of light. If this is indeed the case, the proper motion of the nebula should be detectable within a year.

NEW PULSAR PROPER MOTION RESULTS FROM JODRELL BANK

P. A. Harrison, A. G. Lyne & B. Anderson
University of Manchester
Nuffield Radio Astronomy laboratories
Jodrell Bank
Macclesfield SK11 9DL
UK

ABSTRACT. A programme of observations to determine the proper motions of 44 pulsars has been in progress at Jodrell Bank since October 1984. This programme is now complete, and publication of the results is in preparation. The programme yielded 26 new determinations of proper motion and 17 upper limits. This paper presents some highlights of the study.

1. Observational Details

Observations have been made at 408MHz using two baselines of the MERLIN system, namely the 127km Lovell–Defford baseline and the 68km Lovell–Knockin baseline. Using wide–field phase referencing to multiple reference sources within the primary beam of the interferometer, accurate differential positions between the pulsar and its reference sources have been determined for a number of epochs. The difference between the differential positions found at each epoch allows a proper motion to be determined. The method is similar to that described by Lyne, Anderson & Salter (1982) (hereafter LAS).

The main improvements in the new observations over those of LAS are;

- The use of two baselines and two hands of circular polarization, resulted in the observations being more sensitive, allowing weaker, more distant pulsars to be observed.

- An increased number of reference sources was identified for each pulsar, an average of 2.9 reference sources per pulsar, compared with 1.2 for the LAS experiment. This improved the accuracy of measurement and allowed a better assessment of the contribution of any apparent proper motion of the reference sources.

This experiment, along with that of LAS and Bailes *et al.* (1990) bring the total number of pulsars with well determined interferometric proper motions to 54, with a further 17 upper limits. This paper utilises the measurements from these three major proper motion surveys.

2. The Velocity Distribution

In figure 1 the velocity histogram of all the pulsars with interferometrically measured proper motions is compared with the sample measured via the scintillation characteristics of the pulsar emission by Cordes (1986). The lower plot is a histogram of the scintillation speeds

155

E. P. J. van den Heuvel and S. A. Rappaport (eds.), X-Ray Binaries and Recycled Pulsars, 155–160.
© 1992 *Kluwer Academic Publishers.*

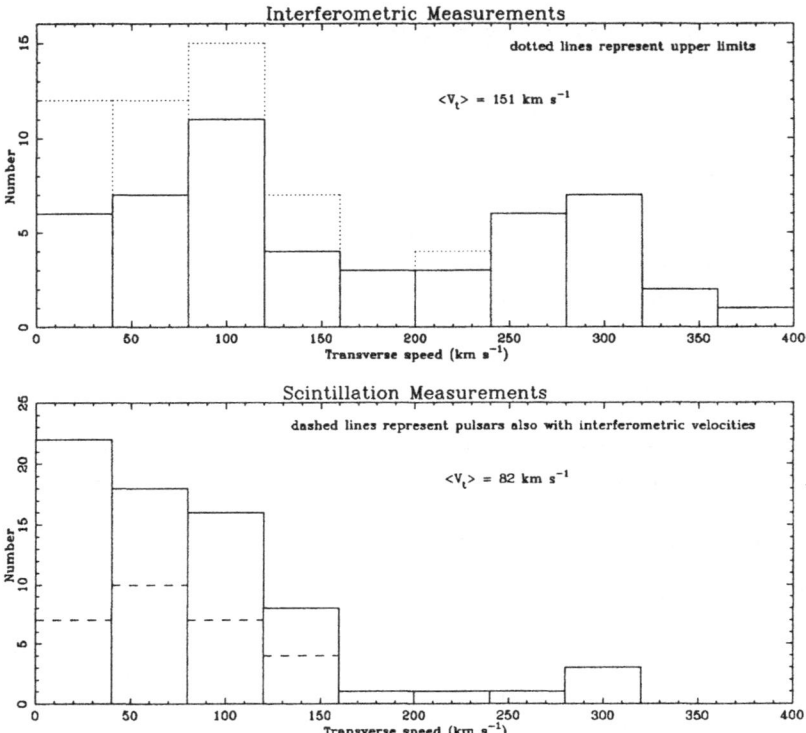

Figure 1: A comparison of the transverse speed distributions of pulsars as measured using interferometry and scintillation observations.

with the dotted line marking those pulsars for which an interferometric proper motion has also been measured. The upper plot is a histogram of the transverse speeds determined from the interferometric proper motions through multiplication by the pulsar distance. The histograms have been truncated at 400 km s⁻¹ for clarity, but there are three pulsars with speeds between 400–550 km s⁻¹ and one, PSR2224+65 with a transverse speed of $\simeq 1000 \, \text{km s}^{-1}$.

The distributions are strikingly dissimilar, with the scintillation speeds more strongly biased towards low values, as is illustrated by the average transverse speed of the interferometer sample being nearly twice that of the scintillation sample.

This systematic difference can be explained by either;

- An incorrect distance scale. The proper motion speed is proportional to the distance, whereas the scintillation speed is proportional to the square root of the distance.

- There being a "leverage" effect in the scintillation speed determination. The calcula-

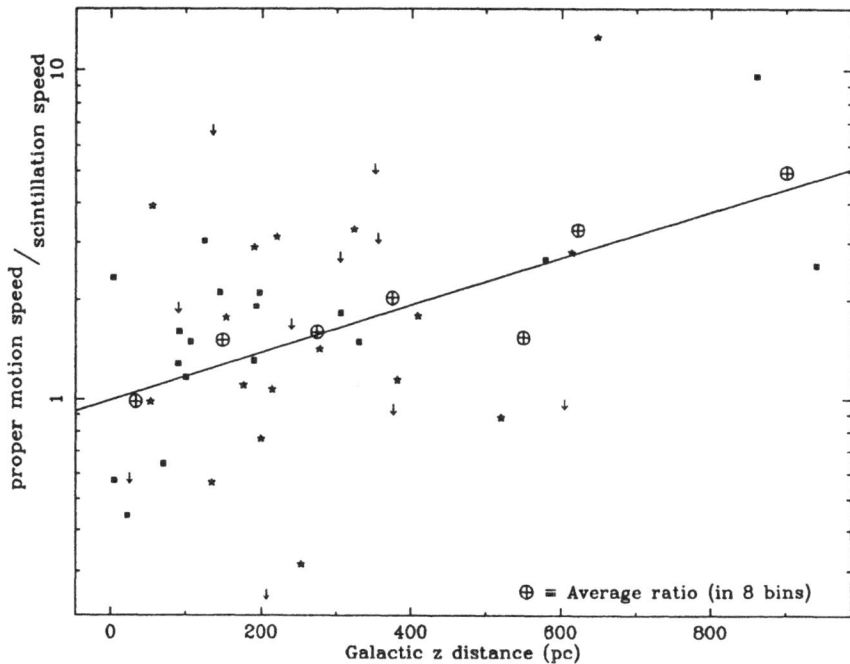

Figure 2: Ratio of proper motion and scintillation speeds as a function of galactic Z height.

tion of the scintillation speed from the observed parameters usually assumes that the scattering material is distributed evenly between the observer and the pulsar. If the scattering material is situated closer to the observer then the derived speed will be lower than the true speed, and if the scattering material is nearer to the pulsar then the true speed will be magnified.

Figure 2 plots the ratio of the two speed determinations as a function of galactic z distance. If the scattering material is concentrated about the galactic plane then pulsars at high z should have their velocities diminished by the leverage effect. The circles containing a cross on the plot are binned averages of the ratio as a function of z. These averages do indeed show an increase from a value of about 1 at $z = 0$ pc to about 5 at $z = 1000$ pc, which strongly suggests that the scattering material is concentrated relatively close to the plane.

158

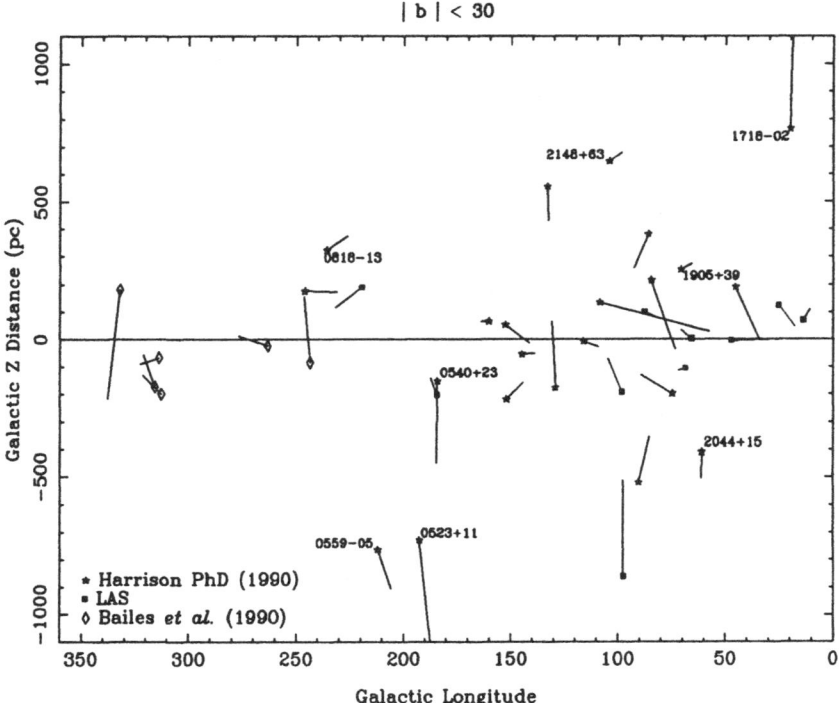

Figure 3: Distances travelled by pulsars in the last million years.

3. Migration from the Plane

LAS concluded that, taking into account projection effects, the data were consistent with all of their pulsars further than 100pc from the plane moving away from the plane. This fitted in well with the hypothesis (Gunn & Ostriker 1970) that pulsars are born from population I stars which have a scale height of about 70pc. The large space velocity that the pulsar acquires allows it to migrate far from the plane so that the observed population has a scale height of about 350pc (Lyne, Manchester & Taylor 1985).

Figure 3 shows the present position in galactic longitude and z distance of all the proper motion pulsars at $|b| < 30°$ (meaning that the z velocity is relatively uncontaminated by the unknown radial component of the velocity) and with errors of less than 25 degrees in their direction of proper motion. Each pulsar has a small vector attached to it which represents the approximate distance that the pulsar has travelled in the last million years.

The current experiment shows that a significant number of the pulsars are moving towards the plane. All of the pulsars at $z > 100$pc which are returning to the plane are

Pulsar	l	b	μ_l	μ_b	dist	z	V_t	V_z	T_c	T_r
	°	°	mas yr^{-1}	mas yr^{-1}	Kpc	pc	km s^{-1}	km s^{-1}	Myr	Myr
0523+11	192.7	−13.3	19±6	24±6	3.2	−729	454±105	365	79.1	8.7
0540+23	184.4	−3.3	−1±8	23±7	2.7	−153	288±92	288	0.3	0.2
0559−05	212.2	−13.5	23±9	9±6	3.3	−764	384±118	139	4.8	2.7
0818−13	235.9	12.6	35±10	−7±8	1.5	323	247±51	−51	9.3	3.9
1718−02	20.1	18.9	3±7	−28±8	2.4	765	316±95	−332	87.0	8.9
1905+39	71.0	14.2	19±2	−5±3	1.0	253	93±14	−23	36.9	6.9
2044+15	61.1	−16.8	−1±4	13±6	1.4	−409	89±40	93	97.5	9.2
2148+63	104.3	7.4	21±4	−1±3	5.0	648	489±80	−30	35.8	6.8

Table 1: Properties of pulsars apparently returning to the plane. T_c is the characteristic age of the pulsar and T_r is the 'true' age of the pulsar assuming an exponential magnetic field decay timescale of 5 million years

named on the plot, and are shown in table 1.

The possible origins of these pulsars are interesting. It should be noted that several of the pulsars have large characteristic ages, opening up the possibility that they might have undergone more than a quarter of a cycle of gravitational oscillation about the galactic plane. Using the z force law of Oort (1960), a quarter cycle of gravitational oscillation about the galactic plane takes about 40 million years for a pulsar with an initial z velocity of 100 km s^{-1}. A natural explanation of the position and velocity of a pulsar such as PSR2044+15 would then be that its true age is greater than 40 million years, which in turn would indicate that magnetic field decay (if any occured) would have had to have been halted at some point. A pulsar similar to 2044+15 which had undergone just more than half a cycle of oscillation would, of course, have important implications for the kinetic age–characteristic age diagram (see LAS).

A problem with the gravitational oscillation interpretation is that some of the pulsars appear to be returning to the plane with large velocities ($\simeq 300$ km s^{-1}), rather close to the galactic escape velocity. This would make them very old indeed if they were born on the plane. Also there are some pulsars which have small characteristic ages which appear to be returning. An explanation for the origin of these pulsars could be that they were born a large distance from the plane and the velocity imparted at birth directed them back towards the plane. It is then a problem in understanding the large z distance of the pulsar's progenitor system. Likely systems would be extreme OB runaway stars, or globular clusters, but it seems unlikely that such systems would contribute many pulsars to the general population.

4. Conclusions

The results presented in this paper are to be published in more detail (Harrison, Lyne & Anderson 1991), but even from this brief exposition it is apparent that there is much to be learnt from analysis of pulsar proper motions. Further research is required in at least two areas to realise their full value.

New scintillation measurements are under way at Jodrell Bank to attempt to measure scintillation speeds for all of the pulsars with interferometrically measured speeds. This enlarged dataset will allow a determination of the distribution of scattering material. In turn, this will permit better determination of pulsar speeds from scintillation observations.

A more detailed analysis of how the velocities of a population of pulsars born near the galactic plane evolve with time, and of how the observed population is biased by selection effects is needed before the observed dynamical behaviour can be understood in terms of the initial space distribution, and any magnetic field evolution.

References

Bailes, M., Manchester, R. M., Kesteven, M. J., Norris, R. P. & Reynolds, J. E., 1990. *Mon. Not. R. astr. Soc.*, **247**, 322.

Cordes, J. M., 1986. *Astrophys. J.*, **311**, 183.

Gunn, J. E. & Ostriker, J. P., 1970. *Astrophys. J.*, **160**, 979.

Harrison, P. A., 1990. Manchester University Ph.D. thesis.

Harrison, P. A., Lyne, A. G. & Anderson, B., 1991. In preparation.

Lyne, A. G., Anderson, B. & Salter, M. J., 1982. *Mon. Not. R. astr. Soc.*, **201**, 503.

Lyne, A. G., Manchester, R. N. & Taylor, J. H., 1985. *Mon. Not. R. astr. Soc.*, **213**, 613.

Oort, J. H., 1960. *Bull. Astr. Inst. Netherlands*, **15**, 45.

THE EVOLUTION OF THE OPTICAL REMNANT OF KEPLER'S SN

R. BANDIERA
Osservatorio Astrofisico di Arcetri
Luigi E. Fermi 5
I–50125 Firenze
Italy
and
S. VAN DEN BERGH
Herzberg Inst./Natl. Research Council
Dominion Astrophys. Obs.
5071 W. Saanich Road
Victoria BC V8X 4M6
Canada

ABSTRACT. We present the results of an extensive work of image restoration on several plates of Kepler's SNR, obtained at the Mt. Wilson Observatory in 1941–43 and at the Palomar Observatory in 1950–83. Using CCD frames taken on La Silla in 1989 we produced images of the nebular emission as well as a reference image of the stellar emission. The latter has then been used for restoring the old plates by subtracting the stellar continuum, after linearization of the plates characteristic curves, and equalization of the respective point spread functions.

This allowed us to follow in detail the evolution of 50 optical knots during the last half century, for which we obtained accurate estimates of positions and proper motions, as well as the H-α light curves. By the astrometric analysis we confirmed, at a greater level of accuracy, the results of Van den Bergh and Kamper (1977), namely that the expansion of the knots pattern is small, but that there is evidence for a common transversal motion (respectively about 70 km/s and 160 km/s, for a distance of 4.5 kpc).

Considerable changes took place in the last decades. New knots appeared along the outer part of the remnant northern edge, while some inner knots are seen to fade gradually. In a group of knots located near the remnant center, instead, new components brightened to the south of those already existing. This behaviour is consistent with that predicted by Bandiera (1987), assuming that the blast wave is moving through an anisotropic circumstellar medium, formed by the interaction of the interstellar medium with the wind of the progenitor, that was presumably a runaway star.

E. P. J. van den Heuvel and S. A. Rappaport (eds.), X-Ray Binaries and Recycled Pulsars, 161.
© 1992 *Kluwer Academic Publishers.*

II THEORETICAL ASPECTS

CHAPTER 4

Stellar Collapse, Supernovae, Supernova Remnants

ACCRETION INDUCED COLLAPSE

S. E. WOOSLEY,   F. X. TIMMES
Board of Studies in Astronomy and Astrophysics
University of California at Santa Cruz
Santa Cruz CA 95064

and

E. BARON
Department of Physics and Astronomy
University of Oklahoma
Norman OK 73019

ABSTRACT: As an accreting white dwarf approaches the Chandrasekhar mass, nuclear reactions ignite in or near the center and, under certain circumstances, the dwarf collapses directly to a neutron star. We consider in some detail the speed and stability of the nuclear combustion front that propagates through the star and determine critical ignition densities above which collapse will ensue. The speed of the flame is severely restricted by the maximum wavelength of Rayleigh-Taylor instability that can develop before it is suppressed by electron capture. We also examine the continued collapse of a white dwarf to nuclear density and the subsequent neutrino energized wind that blows from its surface. This wind obscures any potential burst of hard radiation that might have accompanied the collapse. Its nucleosynthesis, however, places severe constraints upon the frequency with which such events can occur.

1. Introduction

A number of groups have shown that, under certain circumstances, a slowly accreting white dwarf composed either of carbon and oxygen or of neon and oxygen can collapse directly to a neutron star with little or no accompanying optical emission (see, for example, Canal, Isern, and Labay 1990; Kawai, Saio, and Nomoto 1987; Nomoto and Kondo 1991; Isern, Canal, and Labay 1991). Concurrently other groups have suggested

E. P. J. van den Heuvel and S. A. Rappaport (eds.), X-Ray Binaries and Recycled Pulsars, 167–187.

that such collapses could give rise to brilliant γ-ray bursts visible even at cosmological distances (Goodman, Dar, and Nussinov 1987; Paczynski 1990; Ramaty and Dar 1990; although see Shemi and Piran 1990) or at least to certain types of x-ray binaries (Canal, Isern, and Labay 1990; Van den Heuvel 1984; 1987) and the formation of millisecond pulsars (Bailyn and Grindlay 1990).

Here we investigate some of the interesting physics associated with this kind of event. Important details of just how the white dwarf grows to the Chandrasekhar mass are glossed over, as is the brief convective phase that precedes the localization of the nuclear flame to a sharp surface. We pay attention however, to two important areas of present uncertainty: 1) the speed with which a localized nuclear burning front, hereafter referred to simply as "the flame", propagates through the degenerate matter that composes the white dwarf, and 2) the possible observational manifestation of the event. The former is critical in determining whether the white dwarf collapses to a neutron star or explodes as a supernova. We shall find a critical ignition density above which the "effective flame speed", the speed of the flame after its potential Rayleigh-Taylor instability is included, is sufficiently slow compared to the sound speed that collapse is assured. Since present day models suggest ignition, at least of neon-oxygen white dwarfs, at a density exceeding this value, we go on to consider the hydrodynamical collapse, shock propagation, and neutrino transport that ensues. A "wind" is found to blow from the surface of the young neutron star which greatly limits its observability in hard wavebands.

Both sets of calculations are complex and are only tersely summarized in this proceedings. For further detail see Timmes and Woosley (1991) and Woosley and Baron (1991).

2. The Laminar Conductive Flame Speed

For a laminar front which propagates as a result of radiative diffusion or conduction, it is easy to obtain an order of magnitude estimate for the steady state width and speed of the subsonic wave front. The width of the flame δ, can be approximated by setting the diffusion time scale,

$$\tau_{diff} \approx \frac{\delta^2}{A} \approx \frac{\delta^2}{\lambda c}, \tag{1}$$

equal to the nuclear burning time scale,

$$\tau_{burn} \approx \frac{E}{\dot{S}}, \tag{2}$$

where A is a characteristic thermal diffusion constant, λ is an electron mean free path, E is a characteristic energy per unit mass, and $\dot{S}$ is a typical energy generation rate.

Solving for the width,

$$\delta \approx \sqrt{\frac{\lambda c E}{\dot{S}}}, \tag{3}$$

allows an estimate of the speed of the deflagration front (see also Landau and Lifshitz 1959)

$$v_{cond} \approx \frac{\delta}{\tau_{burn}} \approx \frac{\delta \dot{S}}{E} \approx \sqrt{\frac{\lambda c \dot{S}}{E}}. \tag{4}$$

If the diffusion time scale is much shorter than that for nuclear burning, any thermal perturbation will diffuse away to the ambient conditions. If the converse is true, the disturbance will be overdriven and the two time scales will approach equality in the steady state. Thus equality of the diffusion and nuclear burning time scales is a necessary condition for the flame to propagate in a steady state.

Unfortunately evaluation of the above expression requires the choice of a temperature which is not known *a priori* and to which the answer is very sensitive. Physically this temperature corresponds to the point on the flame where conductive diffusion is balanced by nuclear energy generation. Interior to that point energy generation dominates and exterior to it conduction dominates. Looking ahead to the results of the following section, a reasonable choice for this temperature is 5×10^9 K for either carbon-oxygen or neon-oxygen mixtures. Then eqs. (3) and (4) imply (for $\rho \sim$ a few times 10^9 g cm^{-3}) flame speeds on the order of $10 - 100$ km s^{-1} and flame widths of $\sim 10^{-4}$ cm. But, without knowing the temperature beforehand, this method is, at best, an order of magnitude estimate.

2.1 NUMERICAL DETERMINATIONS OF THE FLAME SPEED

The most general line of attack in determining the physical properties of conductive nuclear flames is to solve exactly the continuity, momentum, energy and nuclear reaction equations. This has been done (Timmes and Woosley 1991; see also Woosley, 1986 and Woosley and Weaver, 1986) using the implicit hydrodynamics code KEPLER (Weaver, Zimmerman and Woosley 1978; hereafter WZW). Small spheres of carbon and oxygen (in variable proportions) and of neon, oxygen, and magnesium were generated under conditions of constant pressure (accomplished by use of a surface boundary pressure). These samples, typically 10 to 100 grams, were much larger than what was ultimately determined to be the flame width, but small enough that the temperature, density, and composition could be smoothly resolved in the flame. They represented small samples of the actual conditions that would exist for degenerate matter at the densities that were examined, $10^8 - 10^{10}$ g cm^{-3}.

A flame was initiated in the inner few per cent of the mass of each sphere and its propagation observed. Because the flame speed is so subsonic, pressure equilibrium is easily maintained across the flame, both in nature and in the code. This represents our first and, for a given set of nuclear and thermodynamic physics, most accurate calculation of the flame speed (Table 1).

Considerable simplification can be achieved however, with no loss of accuracy, if one assumes from the beginning that pressure is constant. An isobaric thermodynamic condition reduces the momentum equation employed in the hydrodynamic code to simply the acceleration due to gravity. Terms arising from the pressure gradient in, for example, eq. (1) of WZW, can be neglected, and of course there are no shocks. Thus, under isobaric conditions, one need solve only the energy equation,

$$\frac{dE}{dt} + P\frac{\partial(1/\rho)}{\partial t} = \frac{\partial}{\partial x}\left(\sigma\frac{\partial T}{\partial x}\right) + \dot{S}, \tag{5}$$

where

$$\sigma = 4acT^3/(\kappa\rho), \tag{6}$$

S is the nuclear energy generation rate, and σ is the conductivity.

For this calculation we employ a moving, adaptive mesh. That is, zones can be added, deleted, and refined as the flame propagates. Such a strategy alleviates the need to specify a complete region for simulation before the computation even commences and has the added benefit of a substantial savings in computer time and storage. The equations may also be expressed in spherical or cylindrical coordinates, but this turns out not to matter so long as the flame moved many times its thickness, which is certainly the case in the situation we are considering. A Dirichlet boundary condition, $T = T_{cold}$, is imposed at the outer boundary and a Neumann boundary condition, $dT/dx = 0$, representing zero heat flux, is enforced at the inner boundary. As before, a perturbing step function in temperature is taken as the initial condition and we seek the steady state properties of the wave.

The temporal evolution of the temperature for a typical case, obtained with the adaptive mesh diffusion code and verified with the implicit hydrodynamics code KEPLER, is given in Fig. 1. The isobaric, laminar flame is propagating to the right into a degenerate O+Ne+Mg composition which has a density of $\rho_9 = 10$. The grid used in the calculation is represented by the solid circles. Also given in Fig. 1 is the density profile corresponding to the last temperature profile calculated. As expected for a subsonic deflagration wave, the density decreases behind the flame front. The region shown comprises about 10 grams of material. The burning front in the steady state is traveling at a speed of $v_{cond} = 84.1$ km s^{-1} while maintaining a constant width $\delta = 3.6 \times 10^{-5}$ cm. The is the second and our most favored approach. The results are summarized in Table 1.

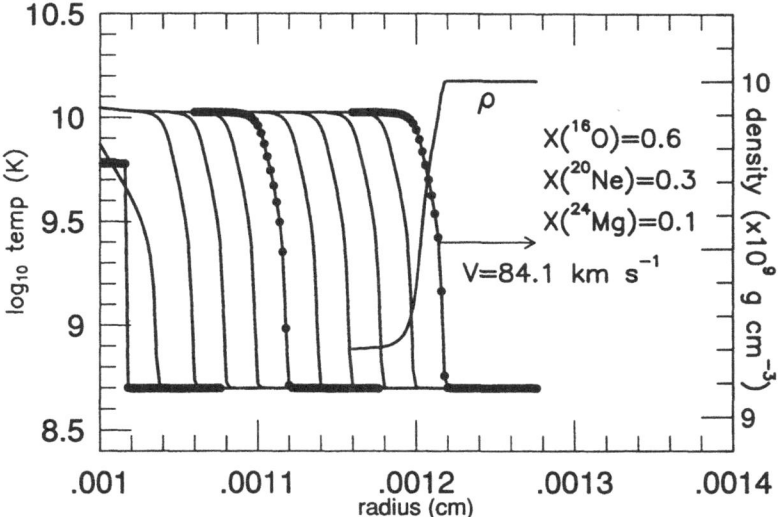

Figure 1. Temperature evolution of a laminar flame in degenerate O+Ne+Mg matter initially at a density of $\rho_9 = 10$. Dots are the mesh points.

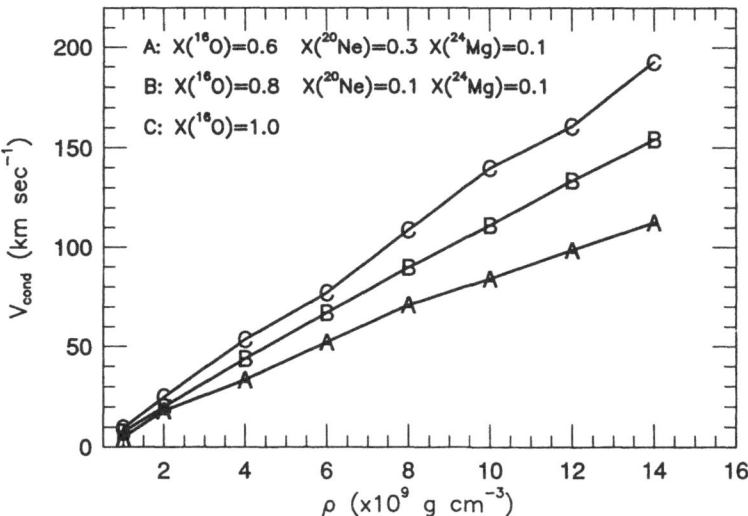

Figure 2. The conductive flame speed as a function of density for three different O+Ne+Mg mixtures.

TABLE 1

Conductive Nuclear Flame Speeds

Composition		ρ_9	microzone hydrocode	diffusion equation	eigenvalue equation	minimum integral
X(^{16}O)=0.6	X(^{20}Ne)=0.3					
		1.0	4.9	5.0	7.8	4.6
		2.0	18.2	18.5	22.2	14.7
		4.0	33.4	34.2	39.3	25.8
		6.0	52.3	52.6	54.4	35.0
		8.0	71.0	69.2	67.5	21.2
		10.0	84.1	83.4	82.7	50.4
		12.0	98.7	98.3	96.9	57.5
		14.0	112.6	109.7	110.1	63.7
X(^{12}C)=0.5	X(^{16}O)=0.5					
		0.2	6.0	6.2	6.1	5.1
		0.5	20.2	19.5	22.7	11.6
		1.0	42.5	42.6	45.7	18.2
		2.0	71.6	69.7	69.9	31.1
		4.0	115.4	112.9	112.4	52.8
		6.0	172.4	170.7	165.9	70.2
		8.0	210.3	206.7	201.8	84.9
		10.0	259.1	259.4	236.2	98.0

Two other means of determining flame speeds exist and are simpler to evaluate, but unfortunately less accurate as well. They are both discussed in detail by Timmes and Woosley (1991). The first assumes the existence of a steady state flame propagating into the cold unburned fuel with a constant speed v_{cond} and drops all time dependence. This transforms the energy diffusion eq. (5) to

$$v_{cond}\left(\frac{dE}{dx} + P\frac{d(1/\rho)}{dx}\right) = \frac{d}{dx}\left(\sigma\frac{dT}{dx}\right) + \dot{S}. \qquad (7)$$

This is an ordinary differential equation having v_{cond} as its eigenvalue. Unfortunately with the elimination of the time derivative, hence the explicit need for specifying some initial condition, an additional boundary condition must be imposed to determine a unique solution. The obvious choice of a Dirichlet condition $T = T_{burn}$ at the inner boundary leaves one with the undesirable situation of having to specify the final temperature T_{burn} of the material. This is not so bad as having to guess the *critical* temperature where conduction balances burning but still adds an extra (and unnecessary) degree of uncertainty to the calculation. The values from this method were given along with those determined by the other two methods in Table 1.

A fourth method, described by Zeldovich *et al.* (1985) goes even one step further and assumes, in addition to a steady state, that all the combustion takes place at the final temperature. This definitely is not the case here. Typically the final temperature after all burning is about 9×10^9 K whereas the critical temperature where burning becomes important is about 5×10^9 K. Still because of its ease of evaluation, the Zeldovich approach remains of interest and is tabulated in Table 1. Note that this method always gives a *lower bound* to the actual flame speed.

2.2 SENSITIVITY TO REACTION NETWORK AND OTHER PHYSICS

In the first part of our study the nuclear energy generation rate $\dot{S}$ and the resulting nucleosynthesis was obtained by integrating a 19 isotope reaction network that coupled 80 nuclear reaction rates (see WZW). The tabulation of Caughlan and Fowler (1988) was employed for all critical nuclear reaction rates. Screening was implemented as discussed in Wallace, Woosley, and Weaver (1982). For consistency, the same network was employed initially to determine $\dot{S}$ in all four formalisms. To obtain a more detailed and presumably more accurate picture of the flame's dynamics and nucleosynthesis, an arbitrary reaction network was implemented in the diffusion code. The network is capable of including reactions among about 500 isotopes, ranging from neutrons, protons and alpha particles to ^{96}Tc. We then repeated some of the calculations of the previous section using different reaction networks to test the sensitivity (Table 2). A network substantially smaller than our standard 19 isotopes is clearly inadequate to give better than factor of two accuracy (the 9 isotopes were nucleons, treated as one species, α-particles, the α-particle nuclei, $Z = N = 2n$, from ^{12}C though ^{28}Si and ^{56}Ni, with quasiequilibrium assumed between ^{28}Si and ^{56}Ni as specified in Woosley, 1986). Of course the heavy ion reactions between ^{12}C and ^{16}O in various permutations were included in all calculations. On the other hand substantially larger networks lead to energy generation rates and flame speeds that are only marginally faster. The values in Table 1 are thus considered adequate.

The equation of state and conductivity employed in the calculation are also important. Here we assumed Planckian photons; nonrelativistic but possibly degenerate ions; electrons having all degrees of relativity and degeneracy; and the possibility of electron pair production at elevated temperatures. In addition to these basic features we included electrostatic Coulomb corrections to the chemical potential, pressure and internal energy in the low temperature, high density regime as formulated by Salpeter (1957). The functional form of the expressions for the radiative and conductive opacities, and their coefficients, were adopted from the works of Hubbard and Lampe (1969), Christy (1966), Iben (1975), Chin (1965) and Canuto (1970).

TABLE 2

Flame Speed (km s^{-1}) for Different Nuclear Reaction Networks

Number of Isotopes	O+Ne+Mg at $\rho_9=10$	C+O at $\rho_9=6$
9	58	112
19	84	172
32	100	185
89	110	190
134	112	193

2.3 FITS TO THE CONDUCTIVE FLAME SPEED AS A FUNCTION OF DENSITY AND COMPOSITION

The conductive flame speed was shown as a function of density for neon-oxygen compositions in Fig. 2 for O+Ne+Mg compositions (see Timmes and Woosley, 1991, for carbon-oxygen compositions). The formula

$$v_{cond} = 80.2 \left(\frac{\rho}{2\text{x}10^9}\right)^{0.839} \left(\frac{X(^{12}C)}{0.5}\right)^{1.12} \text{ km s}^{-1} \tag{8}$$

fits the computed C+O flame speeds to about 10% in the density range $0.2 \leq \rho_9 \leq 10$ while the formula

$$v_{cond} = 87.7 \left(\frac{\rho}{8\text{x}10^9}\right)^{1.02} \left(\frac{X(^{16}O)}{0.8}\right)^{0.961} \text{ km s}^{-1} \tag{9}$$

fits the O+Ne+Mg compositions to within about 12% in the density range $1 \leq \rho_9 \leq 14$.

3. Stability of the Burning Front and the Condition for Collapse

As the laminar, conductive wave travels radially outward in the white dwarf, the temperature behind the wave front increases, the composition is burned into nuclear statistical equilibrium, the density behind the flame front decreases, and the strength of the gravitational field increases. The crossed gravitational and density gradients cause a Rayleigh-Taylor instability to develop. The normal conductive propagation of the flame inhibits the growth of deformations that have small length scales while larger deformations will not exist until they are comparable to the size of the white dwarf's burned out region. The minimum length scale, λ_{min}, that can begin to deform the conductive wave front and grow turbulent before being eroded by the burning is given by equating the time scales for the Rayleigh-Taylor instability to the crossing time of the conductive flame (Woosley, 1990)

$$\lambda_{min} = 4\pi v_{cond}^2 \left(g \frac{\Delta\rho}{\rho} \right)^{-1} \approx 3 v_{cond}^2 \left(Gr\rho \frac{\Delta\rho}{\rho} \right)^{-1}. \tag{10}$$

As the flame propagates radially outward, the minimum length scale for turbulent motion becomes smaller.

At densities appropriate to C+O and O+Ne+Mg white dwarfs, electron capture under isobaric conditions and in nuclear statistical equilibrium causes the number of electrons per baryon $Y_e = (n_e/\rho N_A)$ to decrease and the baryon density to increase behind the flame. The temporal evolution of the density for an initial O+Ne+Mg composition is shown in Fig. 3. The dashed lines represent the initial density while the solid lines represent the density evolution. The intersection of these two curves defines a "recovery time" t_{recov}, for the density to return to its value ahead of the flame. In order to follow the evolution of Y_e, it was necessary to use a larger nuclear reaction network. Consequently, following oxygen depletion of the burning material, we switched to a 125 isotope nuclear reaction network in the hydrodynamical code to follow the nucleosynthesis. The weak interaction rates and network have been discussed by WZW and by Weaver, Woosley, and Fuller (1984).

The fact that the density recovers its initial value implies the existence of a maximum spatial extent of density inversion. This in turn suggests the existence of a maximum deformation length scale λ_{max}

$$\lambda_{max} = v_{cond} t_{recov}. \tag{11}$$

Only wavelengths smaller than the thickness of the density inversion layer can grow and deform the flame front. On the other hand only wavelengths larger than given by eq. (10) can grow before the conductive flame passes over them.

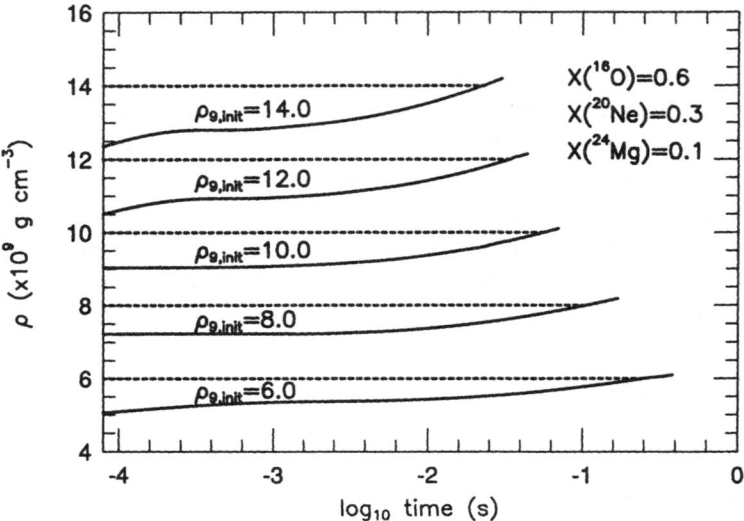

Figure 3. The temporal evolution of the density behind the deflagration front for isobaric nuclear statistical equilibrium. The increase in density is due to electron capture which also decreases Y_e. The intersection of the lines defines $t_{recover}$.

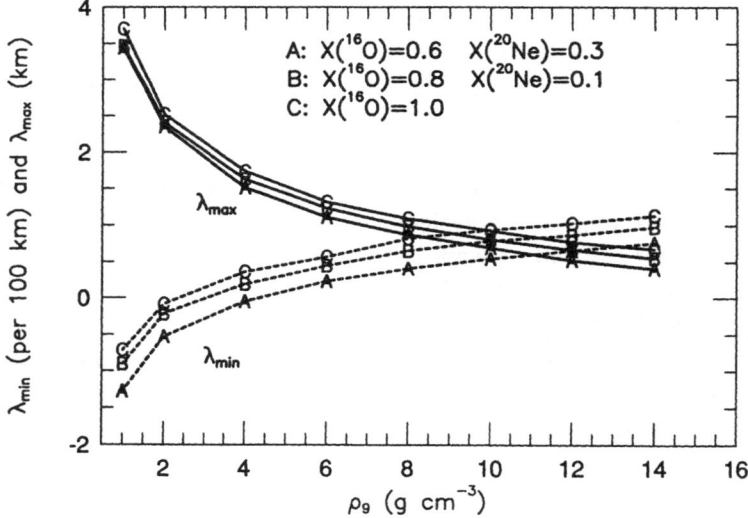

Figure 4. The thickness of the density inversion λ_{max}, behind the subsonic conduction front and the minimum wavelength instability λ_{min} per 100 km as a function of the initial density for various O+Ne+Mg mixtures.

TABLE 3

Oxygen-Neon-Magnesium Conductive Wave Properties

Composition		ρ_9	V_{cond}	Width	$\Delta\rho/\rho$	T_{recov}	λ_{max}	λ_{min}
$X(^{16}O)=0.6$	$X(^{20}Ne)=0.3$							
		1.0	4.9	2.1(-3)	0.20	6.94(+2)	2776.0	0.053
		2.0	18.2	5.2(-4)	0.15	1.25(+1)	225.0	0.496
		4.0	33.4	1.8(-4)	0.14	9.81(-1)	32.4	0.895
		6.0	52.3	7.8(-5)	0.12	2.49(-1)	12.9	1.71
		8.0	71.0	4.7(-5)	0.11	1.06(-1)	7.52	2.58
		10.0	84.1	3.6(-5)	0.09	5.95(-2)	4.99	3.53
		12.0	98.7	2.8(-5)	0.08	3.42(-2)	3.35	4.56
		14.0	112.6	2.4(-5)	0.07	2.29(-2)	2.56	5.81
$X(^{16}O)=0.8$	$X(^{20}Ne)=0.1$							
		1.0	9.4	1.4(-3)	0.21	3.85(+2)	3619.0	0.189
		2.0	24.6	4.8(-4)	0.15	1.27(+1)	304.0	0.907
		4.0	36.1	1.2(-4)	0.14	9.67(-1)	34.8	1.05
		6.0	67.1	5.8(-5)	0.12	2.54(-1)	17.0	2.81
		8.0	89.8	3.5(-5)	0.10	1.08(-1)	9.61	4.53
		10.0	111.3	2.6(-5)	0.09	5.81(-2)	6.44	6.18
		12.0	133.4	2.1(-5)	0.09	3.50(-2)	4.66	7.41
		14.0	154.1	1.6(-5)	0.08	2.32(-2)	3.57	9.53
$X(^{16}O)=1.0$	$X(^{20}Ne)=0.0$							
		1.0	9.5	1.4(-3)	0.21	5.31(+2)	5044.0	0.193
		2.0	24.9	4.7(-4)	0.17	1.33(+1)	331.0	0.820
		4.0	53.6	1.5(-4)	0.14	1.03(+0)	55.2	2.31
		6.0	77.1	5.2(-5)	0.12	2.67(-1)	20.6	3.72
		8.0	108.9	3.0(-5)	0.10	1.14(-1)	12.4	6.66
		10.0	139.8	2.5(-5)	0.10	6.12(-2)	8.55	8.78
		12.0	160.6	1.9(-5)	0.09	3.70(-2)	5.94	10.7
		14.0	192.7	1.4(-5)	0.09	2.47(-2)	4.76	13.6

The width of the density inversion behind the subsonic conduction front, λ_{max}, and the minimum instability wavelength λ_{min} per 100 km (of burned material) are shown as a function of the initial density for three different neon-oxygen compositions in Fig. 4 (see Timmes and Woosley 1991 for carbon-oxygen compositions). Previous calculations by the groups cited in §1.1 have shown that the decision to explode or implode is made sometime during the first few hundred kilometers of flame propagation. At initial C+O densities above about $\rho_9 = 6$ and initial O+Ne+Mg densities above about $\rho_9 = 10$, with a fairly strong dependence on composition, the maximum length scale of deformation λ_{max} that can grow begins to be smaller than the minimum length scale λ_{min}. *This means that the conductive flame will remain absolutely stable.* This case corresponds to an undeformed surface, a plane whose fractal dimension is, by definition, $D = 2$. The flame speed remains at its conductive value and the analysis in §2 is directly applicable.

On the other hand, at slightly lower densities the thickness of the front is large compared to the smaller unstable wavelengths. Then the front will become unstable and consequently wrinkled, deformed and perhaps even quasi-connected. The flame surface is characterized by a range of deformations from λ_{min} to λ_{max}. The velocity of the front is increased compared to the diffusively propagated deflagration, primarily due an increase in the surface area of the front and subsequently by an increase in the fractal dimension. Woosley (1990) showed that one may treat the expanding area of the turbulent burning region as a fractal whose tile size is identical to the minimum unstable Rayleigh-Taylor wavelength λ_{min}. This choice ensures that the prerequisite Mandelbrot (1983) scaling relation between area and volume is preserved. Woosley (1990) further demonstrated that the effective turbulent velocity v_{eff} is given by

$$v_{eff} = v_{cond} \left(\frac{\lambda_{min}}{\lambda_{max}} \right)^{D-2} \tag{12}$$

where D is the fractal dimension. The effective velocity flame speed as a function of distance, density and fractal dimension for the standard neon-oxygen composition is given in Fig. 5 (see Timmes and Woosley, 1991, for carbon-oxygen compositions).

Various calculations (Nomoto and Kondo 1991; Canal, Garcia, Isern and Labay 1990; Isern, Canal, and Labay 1991) have found a bifurcation between collapse and explosion of the white dwarf at ratios of the effective flame speed to the sound speed around 3% (larger in the study by Isern *et al*). Figures 4 and 5 indicate that, for all reasonable values of the fractal dimension and for initial C+O densities above 6×10^9 g cm^{-3} and initial O+Ne+Mg densities above 8×10^9 g cm^{-3}, the white dwarf probably collapses to a neutron star. While the fractal dimension $2 \lesssim D \lesssim 3$ remains a free parameter, and is quite likely itself a function of distance, the physical conditions that lead to accretion induced collapse of white dwarfs are apparently not particularly sensitive to the choice of the fractal dimension, given the small range between λ_{min} and λ_{max}. Also one does not expect values for D greater than about 2.7; values of 2.6 commonly typify fully developed turbulence or a highly convoluted surface (Mandelbrot, 1983, and references therein).

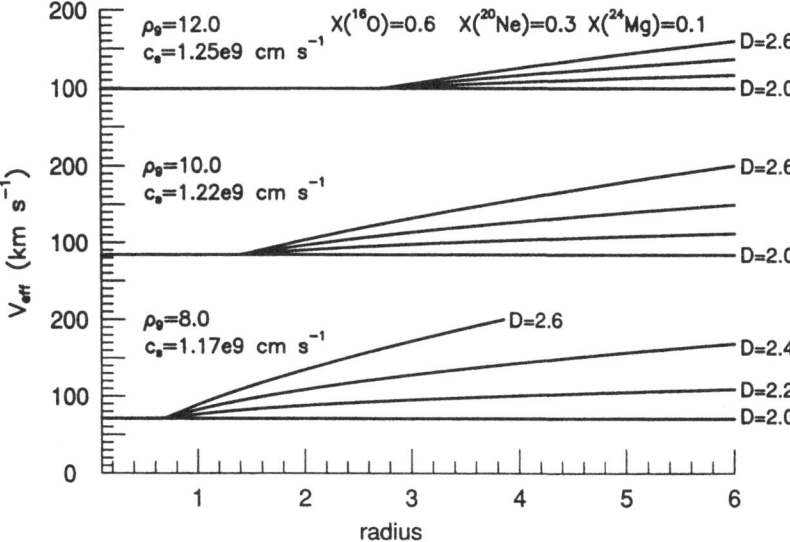

Figure 5. The effective velocity as a function of the average radius of burned material and the fractal dimension, D, O+Ne+Mg ignition densities of $\rho_9 = 8$, $\rho_9 = 10$ and $\rho_9 = 12$. Radius is in units of 100 km.

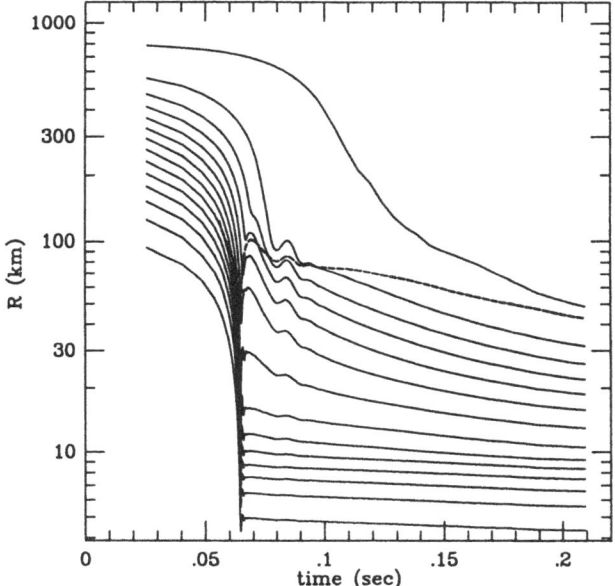

Figure 6. Radius vs. time for the mass points M=0.1 to 1.3 $M_\odot$ in steps of 0.1 $M_\odot$. The dashed line is the neutrinosphere. The outermost line is the surface.

Since the ignition densities cited in the literature for neon-oxygen white dwarfs (and some carbon oxygen white dwarfs) are larger than this, we conclude that collapse to a neutron star is likely.

4. The Collapse of the White Dwarf

Once the decision to collapse has been made, negative velocities, whose origin may be traced to the pressure deficit from electron capture, begin to accelerate. Electron capture is augmented by photodisintegration and the subsequent evolution of the core resembles closely that of the collapsing iron core of a massive star. A steady burning front persists briefly at a radius of several hundred kilometers ultimately to be swept inwards as the rest of the white dwarf falls through the front, but the energy release by nuclear reactions is of little consequence. Collapse continues to well beyond nuclear density.

Such a situation has already been studied by Baron et $al.$ (1987) and others ($e.g.$ Mayle and Wilson 1988). Here we re-examine this model with improvements in the physics and finer surface zoning (see Woosley and Baron, 1991, for additional detail). It turns out that the optical appearance of and nucleosynthesis ejected from these events is critically dependent upon the use of very fine surface zoning. Zoning of 10^{-5} $M_{\odot}$ is needed to see the phenomena we shall describe and 10^{-7} $M_{\odot}$ is needed for its adequate resolution.

In the calculation we employ fully general relativistic non-LTE neutrino transport including all neutrino flavors (Cooperstein, van den Horn, and Baron 1986; Baron et $al.$ 1989). The specific initial model consisted of a white dwarf of mass 1.39998 $M_{\odot}$ composed in its deep interior of carbon and oxygen (Nomoto 1984; Baron, et $al.$ 1987). The outer 0.22 $M_{\odot}$ consisted of accreted helium, but that played no important role in what followed. At the time our calculation began, collapse had already become hydrodynamic (collapse speed equal to 3000 km s^{-1} at 1.16 $M_{\odot}$) and the radius and central density were 9.6×10^7 cm and 6.5×10^{10} g cm^{-3} respectively. The central temperature was 1.04 MeV (1 MeV = 11.604 billion K). A total of 141 zones was employed in the study (approximately constant mass zoning inside of 1.35 $M_{\odot}$ and logarithmic zoning at the surface). The calculation was dynamically rezoned after bounce. The outer zone in this part of the study was 10^{-5} $M_{\odot}$ and initially located at a radius of 963 km with density of 2.2×10^5 g cm^{-3}.

Fifty-six ms after the calculation was begun, the center had collapsed to a density of 5.5×10^{11} at which point neutrinos from electron capture had begun to be trapped. Here the central temperature was 1.39 MeV. By that time the outer zone had moved in only slightly, to 860 km. Heating was almost entirely due to gravitational compression and the flame had not moved from its location at 0.78 $M_{\odot}$.

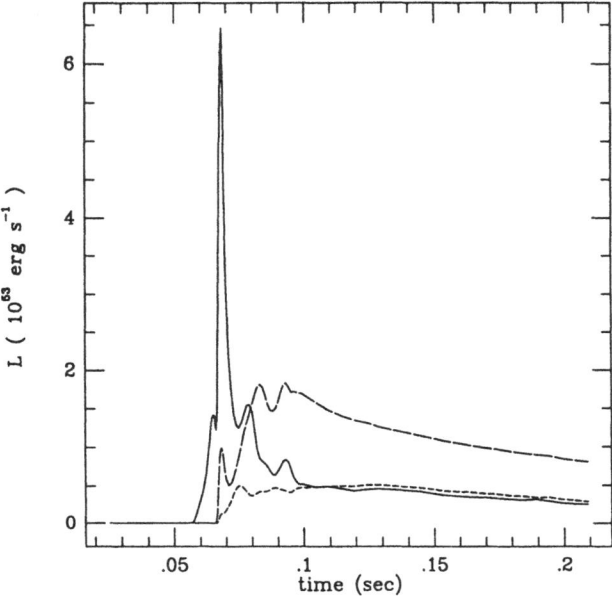

Figure 7. Neutrino luminosities as a function of time. The electron neutrino (solid line), anti-electron neutrino (short dashed line), and μ- and τ-neutrinos (summed together as the long dashed line).

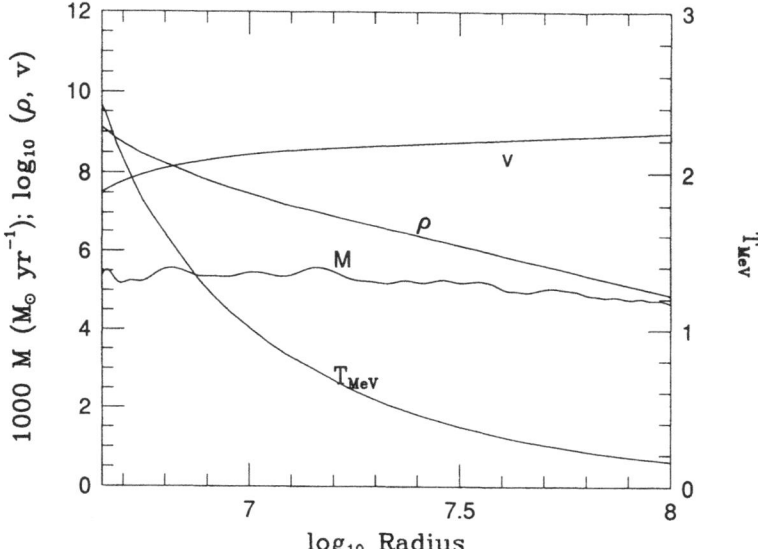

Figure 8. Temperature, density, velocity, and mass loss rate as a function of radius in a fine zoned calculation of the region of the protoneutron star external to the neutrinosphere at 400 ms.

Next came maximum central compression at t = 64.8 ms, a central density of 6.1×10^{14} g cm^{-3}, and a central temperature of 19.33 MeV. The homologous core mass here was about 0.6 M$_\odot$ and other conditions were as in Baron et al. (1987). A shock was born, located at 0.75 M$_\odot$ (12.8 km). The flame was still at 0.78 M$_\odot$, but by then had moved inwards in radius to 19 km. The outer zone was at 907 km and the neutrino luminosity was 1.03×10^{53} erg s^{-1} due almost entirely to electron capture, and hence consisting of mostly electron neutrinos.

The shock stalled (lost all positive velocity) at 70.8 ms when it was at 152 km and 1.25 M$_\odot$. The remaining part of the white dwarf then fell through this accretion shock which reached the surface at 137 ms at a radius of 700 km. By then neutrino heating had become effective behind the shock and 5×10^{-4} M$_\odot$ of material was beginning to move out with positive velocity. The neutrino luminosity was 1.97×10^{53} erg s^{-1} (electrons 0.40, antielectrons 0.46, and pairs 1.12) and the neutrinosphere was at 65 km.

At the last point calculated with the neutrino-transport code, t = 262 ms, the shock had left the grid and 1.3×10^{-4} M$_\odot$ of material was being ejected from the star. The neutrinosphere was at 37 km and had a temperature of 6.15 MeV for μ- and τ-neutrinos, 3.47 MeV for electron neutrinos, and 4.4 MeV for antielectron neutrinos. The neutrino luminosity is 0.80×10^{53} erg s^{-1} (0.15 for electrons, 0.17 for antielectrons, and 0.48 for pairs). Figures 6 and 7 show the radius and neutrino luminosity history of the model.

5. The Wind

As noted above, in order to estimate the mass loss correctly including the coupled effects of energy deposition, expansion, and work against the gravitational field, one must carry out a numerical calculation using much finer zoning than traditionally employed by those studying core collapse. It did not prove feasible to use such fine zoning in the (explicitly coded) neutrino-transport/hydrodynamics code used for the core collapse calculation. Instead a finely zoned calculation of the outer 0.01 M$_\odot$ of the protoneutron star at age 209 ms was carried out using the KEPLER implicit hydrodynamics code (WZW). Initially the 0.01 M$_\odot$ was ruled into 100 zones of logarithmically decreasing mass such that the outer zone had a mass of 10^{-6} M$_\odot$. The neutron star was represented by an inner boundary condition at 30 km and a mass of 1.39 M$_\odot$. The temperature of the inner zone was 2.8×10^{10} K (though this did not turn out to be critical) and the density, in hydrostatic equilibrium, 2.1×10^{11} g cm^{-3}. Note that the entire region studied is optically thin to neutrinos. The model was then evolved for roughly one day (obviously much longer than the stellar evolutionary time) to allow the material to relax into near thermal and hydrostatic equilibrium (with the neutrino luminosity still at zero).

Neutrino energy deposition was then initiated. The neutrino luminosity was 1×10^{53} erg s^{-1} with appropriate partition into the various flavors. Several neutrino processes appropriate to the "optically thin" limit were employed: electron and positron capture on nucleons, pair production by $\nu - \bar{\nu}$ annihilation and losses due to the inverse process — pair annihilation; and neutrino electron scattering (of all flavors). Approximation formulae from Woosley and Weaver (1991) were used for these processes. The equation of state, as usual, included ions, radiation, and electrons and positrons of arbitrary degeneracy and relativicity. As the atmosphere began to expand, zones were continually added wherever needed, while conserving momentum and energy, to keep the temperature and density profiles smooth (see also Wallace, Woosley, and Weaver 1982). Eventually there were close to 200 zones, with zone masses of 10^{-7} M$_\odot$ being typical, and the smallest (surface) zone, 10^{-9} M$_\odot$. Figure 8 shows the thermodynamic structure and velocity distribution in the envelope after 0.2 s (*i.e.*, $t_{total} = 0.4$ s) when the flows seem to have reached steady state. From the density, radius, and velocity we estimate a mass loss rate, $4\pi r^2 \rho v \approx 10^{30}$ g s^{-1}.

At this time the total energy being deposited by neutrino annihilation is about 5×10^{50} erg s^{-1} most of which is going to driving mass loss (the terminal velocity of the surface is roughly 2/3 c, not relativistic). The mass loss rate is about 0.005 M$_\odot$ s^{-1}, but this will decline with time and one expects a total of about a hundredth of a solar mass to be lost during the event.

6. Gamma-Ray Bursts?

Paczynski (1990) has considered the observable consequences of a highly super-Eddington wind pumped by neutrino energy deposition at its base. He finds a critical mass loss rate, $\sim 10^{27}$ g s^{-1} for an energy deposition rate of 10^{50} erg s^{-1}. Above this value radiation at the photosphere changes its qualitative properties. For smaller values the radiation dominates over the matter producing a relativistic outflow and a hard x-ray or γ-ray transient. Below this value, expansion of the matter dilutes and cools the radiation to comparatively faint emission. Though sensitivity and parameter studies remain to be done, the high mass loss rate we calculate here would certainly rule out the production of γ-ray bursts by collapsing white dwarfs at cosmological distances. For a mass loss rate of 10^{30} g s^{-1} and an energy deposition rate of 5×10^{50} erg s^{-1} as the present study implies, Paczynski's eq (26)

$$L_{out} \approx 3 \times 10^{48} \left(\frac{\dot{Q}_{\nu\bar{\nu}}}{10^{50}} \right)^3 \left(\frac{\dot{M}}{10^{27}} \right)^{-8/3} \quad \text{erg s} \tag{13}$$

would imply a luminosity of only $\sim 10^{42}$ erg s^{-1} lasting for a few seconds. The effective temperature would not be in the gamma-ray range. Further, since our mass loss estimate scales with $\dot{Q}_{\nu\bar{\nu}}$, one could not increase the luminosity simply by increasing the energy input at the base.

7. Nucleosynthetic Limits

The amount of mass ejected during the total outburst of several seconds will be $\sim 10^{-2}$ $M_\odot$. The value of Y_e in the ejecta comes from our calculation and is roughly 0.40 in about 0.002 $M_\odot$ of the ejecta (smoothly joined to material having $Y_e \approx 0.50$ above and below; see also Mayle and Wilson 1988). The low value of Y_e is a consequence of electron capture on protons during the time the matter spends at high density. Woosley and Hoffman (1991) have recently studied the nucleosynthesis that occurs as this sort of neutron-rich matter, initially composed of nucleons in nuclear statistical equilibrium, is rapidly cooled. For Y_e near 0.40, the principal nucleosynthetic products are ^{90}Kr and ^{86}Se which later decay to ^{90}Zr and ^{86}Kr respectively. Though the actual identity of the most abundant species is sensitive to the entropy and time scale for the ejected material to cool down, it is clear that neutron rich species well above the iron group will be produced. In the calculation of Woosley and Hoffman, ^{90}Kr had a mass fraction in the ejecta of 0.33. In the sun, the mass fraction of ^{90}Zr is 1.6×10^{-8}, which is relatively large for species in this mass range.

A simple minded approach to Galactic chemical evolution would be as follows. Suppose core collapse supernovae (Type II and Ib) occur twice a century and, on the average, eject 1 $M_\odot$ of ^{16}O. The ratio of ^{16}O to ^{90}Zr in the sun is about 500,000 by mass. Thus events of the sort we are describing could happen no more frequently than once every $(500,000)(.002/1)(0.3)(50) = 15,000$ years, even if there were no other source of ^{90}Zr in nature. Nor is one able to hide the material in other heavy nuclei since the mass fractions of all isotopes heavier than germanium ($Z = 32$) are individually less than a few times 10^{-8} in the sun. Our limit is probably a tight one, at least to one order of magnitude, which means that these must be very rare events indeed (compared to supernovae, the major source of neutron stars). Indeed a generic problem with the delayed supernova explosion mechanism may be keeping the ejection of this neutron rich material to a tolerable level.

It is also true that during its decay to ^{90}Zr, ^{90}Kr passes through ^{90}Sr, a gamma-ray line emitter with a half-life of 29 years. Some ^{56}Ni will also be produced in the region where $Y_e \approx 0.5$, and this may provide a detectable γ-ray line signal from AIC were such a (rare) event ever to occur in our Galaxy in modern times.

8. Postscript – Neutron Star Kick Velocities

It is often presupposed that a white dwarf undergoing accretion induced collapse will have its orbit altered very little by the explosion since, after all, the amount of mass lost from the system is very small. Actually such neutron stars are subject to the same sort of kicks from an asymmetric explosion mechanism that are thought by many to occur in massive stars (*e.g.*, Burrows and Woosley 1986; Woosley 1987). First it should be noted that the neutron star loses about 15% of its mass energy as neutrinos. In the case of accretion induced collapse this value far exceeds the actual baryonic mass that is ejected and has interesting, though possibly minor implications for the orbit and the continuation of accretion upon the neutron star.

More importantly, this neutrino emission may not be perfectly isotropic. A region emitting neutrinos on one side of the neutron star is out of touch with the other side. Of some importance is also the fact that the neutrino emitting region may be convective (*e.g.* Bethe 1990) and even subject to macroscopic overturn (Smarr *et al.* 1981). It is thus not difficult to imagine asymmetries of say a few tenths of a per cent in L_ν from one side to the other. Since the neutrinos account for about 15% of the mass energy of the neutron star and are emitted at the speed of light, conservation of momentum and a 0.3% asymmetry in emitted neutrino energy imply a recoil velocity on the order of 100 km s^{-1}. Such a value is typical of radio pulsars and may be enough to unbind the binary.

This work has been supported by the NASA Theory Program (NAGW-1273), the National Science Foundation (AST 88-13649), by the California Space Institute, and the Oklahoma Research Council. The calculations of EB reported herein were performed at the National Energy Research Computer Center, supported by the U.S. Dept. of Energy; we thank them for a generous allocation of computer time. EB also thanks Jerry Cooperstein for helpful discussions.

REFERENCES

Bailyn, C. D., and Grindlay, J. E. 1990, *Ap. J.*, **353**, 159.

Baron, E., Cooperstein, J., Kahana, S., and Nomoto, K. 1987, *Ap. J.*, **320**, 304.

Baron, E., Myra, E. S., Cooperstein, J., and van den Horn, L. J. 1989, *Ap. J.*, **339**, 978.

Bethe, H. A. 1990, *Rev. Mod. Phys.*, **62**, No. 4, 801.

Burrows, A., and Woosley, S. E. 1986, *Ap. J.*, **308**, 680.

Canal, R., Garcia, D., Isern, J., and Labay, J. 1990, *Ap. J.*, **356**, L51.

Canal, R., Isern, J., and Labay J. 1990, *Ann. Rev. Astron. and Ap.*, **28**, 183.

186

Canuto, V. 1970, *Ap.J.*, **159**, 641.

Caughlan, G. R., Fowler, W. A. 1988, *Atomic Data and Nuclear Data Tables*, **40**, 283.

Chin, C. 1965, *Ap.J.*, **142**, 1481.

Christy, R. F. 1966, *Ap.J.*, **144**, 108.

Cooperstein, J., van den Horn, L. J., and Baron, E. 1986, *Ap. J. (Letters)*, **321**, L129.

Goodman, J., Dar, A., and Nussinov, S. 1987, *Ap. J. Lettr.*, **314**, L7.

Hubbard, W. B. and Lampe, M. 1969, *Ap.J. Suppl.*, **18**, 279.

Iben, I., Jr. 1975, *Ap.J.*, **196**, 525.

Isern, J., Canal, R., and Labay, J. 1991, *Ap.J. Lettr.*, in press.

Kawai, Y., Saio, H., and Nomoto, K. 1987, *Ap. J.*, **315**, 229.

Landau, L. D., and Lifshitz, E. M. 1959, *Fluid Mechanics: Volume 6 of Course in Theoretical Physics*, (Pergamon Press: Oxford).

Mandelbrot, B. B. 1983, *The Fractal Geometry of Nature*, (W. H. Freeman: New York)

Mayle, R. W., and Wilson, J. R. 1988, *Ap. J.*, **334**, 909.

Nomoto, K. 1984, *Ap. J.*, **277**, 791.

Nomoto, K., Kondo, Y. 1991, *Ap. J. Lettr.*, **367**, L19.

Paczynski, B. 1990, *Ap. J.*, in press.

Ramaty, R., and Dar, A. 1990, preprint: Proc. of the COSPAR.

Salpeter, E. 1957, *Comp. Phys.*, **88**, 2.

Shemi, A., and Piran, T. 1990, *Ap.J.*, **365**, L55.

Smarr, L. L., Wilson, J. R., Barton, R. T., and Bowers, R. L. 1981, *Ap. J.*, **246**, 515.

Timmes, F. X., and Woosley, S. E. 1991, *Ap. J.*, submitted.

Van den Heuvel, E. P. J. 1984, *J. Ap. Astr.*, **5**, 209.

Van den Heuvel, E. P. J. 1987, in *IAU Symp. 125 - The Origin and Evolution of Neutron Stars*, ed. D. Helfand and J. Huang, (Dordrecht: Reidel), p. 393.

Wallace, R. K., Woosley, S. E., and Weaver, T. A. 1982, *Ap. J.*, **258**, 696.

Weaver, T. A., Zimmerman, G. B., and Woosley, S. E. 1978, *Ap. J.*, **225**, 1021, (WZW).

Weaver, T. A., Woosley, S. E., and Fuller, G. M. 1984, in *Numerical Astrophysics*, eds, J. Centrella, J. LeBlanc, and R. Bowers, (Boston: Jones and Bartlett), p. 374.

Woosley, S. E. 1986, in *Nucleosynthesis and Chemical Evolution* eds. Hauck, B., Meader, A. and Meynet, G., (Geneva: Swiss Society of Astrophysics and Astronomy, Geneva Observatory, p. 1.

Woosley, S. E. 1987, in *The Origin and Evolution of Neutron Stars*, Proc. IAU Colloq. 125, eds. D. Helfand and J. H. Huang, (D. Reidel), p. 255.

Woosley, S. E. 1990, in *Supernova*, ed. Petschek, A. G., (D. Reidel:Dordrecht, p. 182.

Woosley, S. E., and Weaver, T. A. 1986, in *Radiation Hydrodynamics in Stars and Compact Objects*, ed. D. Mihalas and K.-H. Winkler, (Springer Verlag: Berlin), p. 91.

Woosley, S. E., and Baron, E. 1991, *Ap. J.*, in preparation.

Woosley, S. E., and Weaver, T. A. 1991, *Proceedings of Session LIV, Les Houches*, eds. J. Audouze, S. Bludman, R. Mochovitch, and J. Zinn-Justin, (Elsevier Science Publishers), in press.

Woosley, S. E., and Hoffman, R. D. 1991, *Ap. J.*, in press.

Zeldovich, Ya. B., Barenblatt, G. I., Librovich, V. B., and Makhiladze, G. M. 1985, in *The Mathematical Theory of Combustion and Explosions*, (Plenum: New York).

ACCRETION-INDUCED COLLAPSE OF WHITE DWARFS

K. NOMOTO and H. YAMAOKA
Department of Astronomy, Faculty of Science, University of Tokyo, Tokyo

ABSTRACT. Evolutionary scenarios for the formation of some binary pulsars are discussed. (1) In relation to the origin of low mass X-ray binaries and binary millisecond pulsars, conditions for the occurrence of accretion-induced collapse of white dwarfs are examined. The outcome of the evolution of accreting white dwarfs is summarized as a function of accretion rate and the initial mass of the white dwarf. (2) In relation to the binary pulsar with a neutron star companion, some constraints on the progenitor system and on the asymmetry of the explosion are presented. The relation to Type Ib/Ic supernovae is discussed.

1. Introduction

We present possible evolutionary scenarios for the formation of some binary pulsars, in particular, low mass binary pulsars and double neutron star systems. Theoretically, the final forms of the stars in interacting binary systems are predicted to be either white dwarfs or helium stars of mass $\gtrsim 2.5\ M_\odot$ after the loss of their hydrogen-rich envelope by Roche lobe overflow. The helium stars evolve through the Fe core collapse and would undergo supernova explosions, which would be observed as Type Ib/Ic supernovae. White dwarfs, if accreting matter from companion stars, would either explode or collapse depending on the conditions of binary systems. These scenarios of neutron star formation in binary stars have been suggested to be related to the origin of some binary pulsars as follows:

(1) An unexpectedly large number of low mass binary pulsars (LMBPs) have recently been discovered. The birth rate of LMBPs is now estimated to be about 100 times higher than that of low mass X-ray binaries (LMXBs) in both the Galactic disk (Kulkarni and Narayan 1988; Narayan *et al.* 1990) and the globular clusters (Kulkarni *et al.* 1990; Romani 1990). Since LMXBs have been thought to be the progenitors of LMBPs, this birth rate discrepancy has raised a serious question about the evolutionary origin of LBMPs. Two scenarios have been proposed to resolve this problem: (a) accretion-induced collapse (AIC) of white dwarfs in close binaries (Michel 1987; Chanmugam and Brecher 1987; Bailyn and Grindlay 1990; Romani 1990; Ray and Kluzniak 1990), and (b) shortening of the LMXB phase due to the evaporation of the companion star (e.g., Tavani 1991 and references therein). Further,

E. P. J. van den Heuvel and S. A. Rappaport (eds.), X-Ray Binaries and Recycled Pulsars, 189–205.
© 1992 *Kluwer Academic Publishers.*

combinations of AIC and the tidal capture of neutron stars have been suggested as an explanation for the very high incidence of LMBPs in globular clusters (Romani 1990; Ray and Kluzniak 1990).

(2) Recently the masses of the component stars in the binary pulsar system 1532+12 has been determined, which strongly suggests a neutron star companion (Wolszczan 1991). The observed binary parameters raise some interesting questions about the evolutionary origin and explosion mechanism of such a double neutron star system.

2. Evolution of Accreting White Dwarfs

The scenario that possibly brings a close binary system to AIC or Type Ia supernovae (SNe Ia) is as follows (although the exact binary system has not been identified): Initially the close binary system consists of two intermediate mass stars ($M \lesssim 8\ M_\odot$). As a result of Roche lobe overflow, the primary star of this system becomes a white dwarf composed of C+O or O+Ne+Mg. When the secondary star evolves, it begins to transfer hydrogen-rich matter over to the white dwarf.

The mass accretion onto the white dwarf releases gravitational energy at the white dwarf surface. Most of the released energy is radiated away from the shocked region as UV and does not contribute much to heating the white dwarf interior. The continuing accretion compresses the previously accreted matter and releases gravitational energy in the interior. A part of this energy is transported to the surface and radiated away from the surface (*radiative cooling*) but the rest goes into thermal energy of the interior matter (*compressional heating*). Thus the interior temperature of the white dwarf is determined by the competition between compressional heating and radiative cooling; that is the white dwarf is hotter if the mass accretion rate $\dot{M}$ is larger, and vice versa (e.g., Nomoto 1982a).

2.1. Hydrogen Shell Flashes

When a certain amount of hydrogen $\Delta M_{\rm H}$ is accumulated on the white dwarf surface, hydrogen shell burning is ignited. Figure 1 shows $\Delta M_{\rm H}$ at the ignition as a function of $\dot{M}$ and M; $\Delta M_{\rm H}$ is smaller for larger $\dot{M}$ and larger M because of higher temperatures and higher pressures (eq. 2 below) in the accreted envelope. Here the compressional heating due to accretion is just balanced with the cooling due to heat conduction (Nariai and Nomoto 1979; Nomoto 1982a). The type and the strength of the hydrogen shell burning depend sensitively on $\dot{M}$ as follows:

(1) *Rapid accretion forming a red-giant-like envelope.* When the accretion is as rapid as $\dot{M} \gtrsim \dot{M}_{\rm RH}$, the accreted matter is too hot to be swallowed by the white dwarf, thereby forming a red-giant-like envelope (Nomoto et al. 1979b). This critical rate, $\dot{M}_{\rm RH}$, corresponds to the growth rate of the degenerate core in red giant stars due to hydrogen shell burning. Paczynski's (1970) relation of core mass to luminosity, assuming a hydrogen abundance of $X = 0.7$, gives

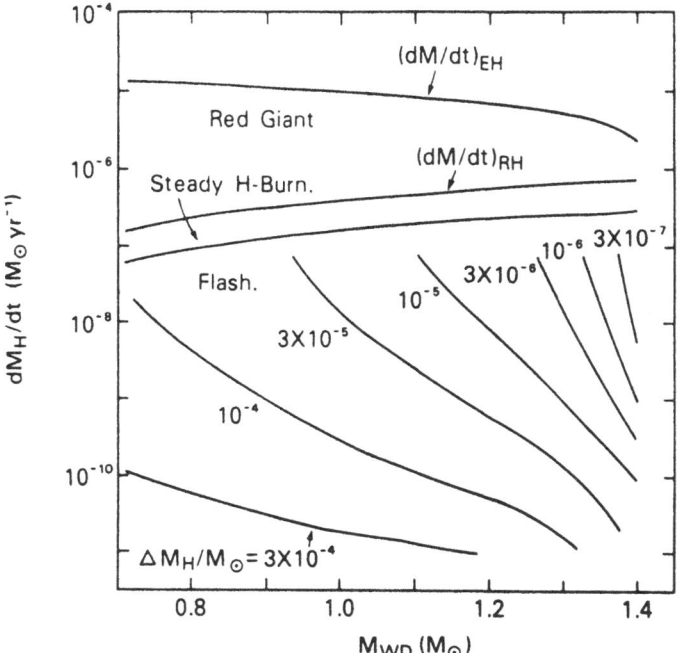

Fig. 1: Types and strength of hydrogen shell burning as a function of accretion rate and the white dwarf mass (Nomoto 1982a).

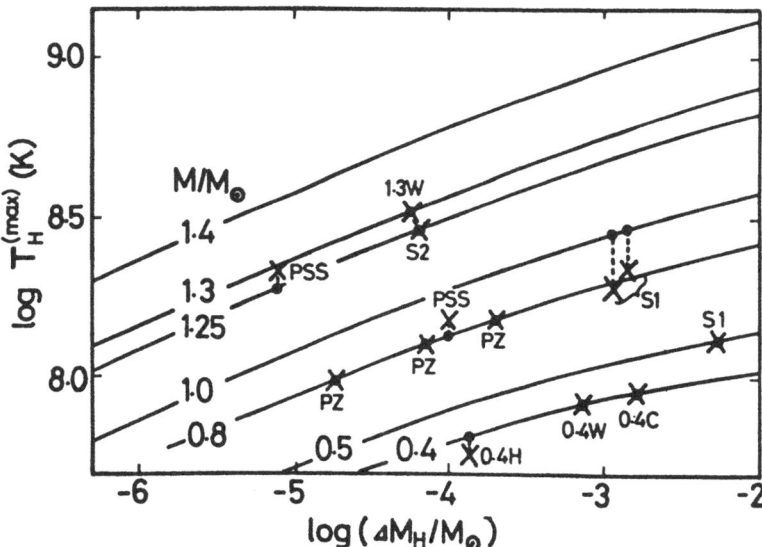

Fig. 2: Maximum temperature attained during the hydrogen shell flash as a function of the accreted mass ΔM_H and the white dwarf mass M. Analytical values (solid curves) are compared with those obtained by hydrodynamical calculations (X-mark) (see Sugimoto et al. 1979 for details).

$$(\dot{M})_{\text{RH}} = 8.5 \times 10^{-7}(M/M_\odot - 0.52) \quad M_\odot \text{yr}^{-1} \tag{1}$$

for $0.60 \leq M/M_\odot \leq 1.39$, where the core mass is replaced by the white dwarf mass M.

The matter would form a common envelope, which is eventually lost from the system. As a result of mass and angular momentum losses from the system, some binaries would form a pair of white dwarfs. Further evolution of such a double white dwarf system is driven by gravitational wave radiation and leads to a Roche lobe overflow of the smaller mass white dwarf (Iben and Tutukov 1984; Webbink 1984).

(2) *Steady burning*. For an accretion rate in the range of $0.4\,\dot{M}_{\text{RH}} \lesssim \dot{M} \lesssim \dot{M}_{\text{RH}}$, the hydrogen shell burning is stable (e.g., Sienkiewicz 1980). The radius depends very sensitively on $\dot{M}$ ranging from red-supergiant size to white dwarf size. Such hydrogen burning processes the accreted matter into helium at a rate of $\dot{M}$.

(3) *Recurrence of flashes*. When the accretion rate is lower than $0.4\,\dot{M}_{\text{RH}}$, hydrogen shell burning becomes unstable and flashes. The progress and the strength of the flashes is determined by two parameters (P^*, Ω^*),

$$P^* = \frac{GM\Delta M_{\text{H}}}{4\pi r^4}, \qquad \Omega^* = \frac{GM}{r}, \tag{2}$$

i.e., by the pressure and potential at the burning shell corresponding to completely flat configuration (Sugimoto and Fujimoto 1978; Sugimoto et al. 1979). A set of (P^*, Ω^*) can be transformed into $(M, \Delta M_{\text{H}})$ since the radius r at the burning shell is well approximated by the radius of the white dwarf which is determined only by M.

Progress of the shell flash can be treated semi-analytically. Initially the temperature at the burning shell increases along $P = P^*$. As the nuclear energy is released, the pressure decreases as a result of expansion, which is described by $P = fP^*$. Here the flatness parameter f is unity for plane-parallel configuration and $f < 1$ for more spherical configuration. This is expressed as

$$\frac{1}{f(V,N)} = \sum_{k=0}^{\infty} b_k, \qquad b_0 = 1, \quad b_k = b_{k-1}\frac{k+3}{N+k+1}\frac{N+1}{V} \tag{3}$$

$$V = r/H_{\text{p}}$$

where N and H_{p} denote the polytropic index for the convective envelope and the scale height of pressure, respectively (Sugimoto and Fujimoto 1978). As the specific entropy s in the hydrogen-burning shell increases, f decreases because of increasing H_{p}, i.e., expansion of the accreted envelope. (This corresponds to the change in the configuration of the burning shell from plane parallel to spherical.)

Then the temperature reaches its maximum $T_{\text{H}}^{\text{max}}$, which is higher for higher P^* and thus for higher M (smaller r) and larger ΔM_{H} (eq. 2). Such a relation between T_H^{max} and ΔM_H for several M obtained from Equation (3) is shown in Figure 2 (Sugimoto et al. 1979). Results of some hydrodynamical calculations (X-mark) are in excellent agreement with the corresponding analytical predictions (filled circles). (Some discrepancies are likely to be caused by a coarse zoning in numerical calculations).

The results in Figures 1 and 2 show that generally smaller M and higher $\dot{M}$ lead to a weaker flash because of the lower pressure at the flashing shell. The flash would induce little mass ejection so that a large portion of the accreted matter could be processed into helium for $\dot{M} \approx 2 \times 10^{-7} - 10^{-8}$ $M_\odot$ yr^{-1}.

For slow accretion ($\dot{M} \lesssim 1 \times 10^{-9}$ $M_\odot$ yr^{-1}), hydrogen shell flash is strong enough to grow into a nova explosion, which leads to the ejection of most of the accreted matter from the white dwarf (e.g, Nariai et al. 1980 and references therein). For these cases, the white dwarf does not become a supernova since its mass hardly grows. However, if the white dwarfs are close to the Chandrasekhar mass, novae could grow into AIC of SN Ia because the ejected mass from nova explosion is found to be significantly smaller than the accreted mass (Starrfield et al. 1991).

2.2. Helium Shell Flashes

Suppose that a helium layer grows on the white dwarf as a result of hydrogen shell burning or direct transfer of helium if the companion is a helium star (e.g., Iben et al. 1987). When a certain mass ΔM_{He} is accumulated, helium shell burning is ignited; the solid lines in Figure 3 show ΔM_{He} as a function of $\dot{M}$ and the white dwarf masses M (Kawai et al. 1987). The maximum temperature attained during the helium shell flash depends on ΔM_{He} and M (Figure 4; Fujimoto and Sugimoto 1982) as discussed for hydrogen in §2.1. The strength of the helium flash depends mainly on $\dot{M}$ as follows:

If the accretion of helium is as slow as $\dot{M} \lesssim 1 \times 10^{-9}$ $M_\odot$ yr^{-1}, the material is too cold to ignite helium burning, so that the white dwarf mass is increased. Exception is the case with $M_{CO} \lesssim 1.1$ $M_\odot$ where pycnonuclear helium burning is ignited (Nomoto 1982a,b).

For $\dot{M}_{det} \gtrsim \dot{M} \gtrsim 10^{-9}$ $M_\odot$ yr^{-1}, helium shell flash is strong enough to initiate an off-center helium detonation, which prevents the white dwarf mass from growing (e.g., Nomoto 1982b; Woosley et al. 1986; Iben and Tutukov 1991). Here we adopt $\dot{M}_{det} \sim 1 \times 10^{-8}$ $M_\odot$ yr^{-1}, since the $^{14}N(e^-, \nu)^{14}C(\alpha, \gamma)^{18}O$ (NCO) reaction ignites weak helium flashes (Hashimoto et al. 1986; Limongi and Tornambe 1991) if the mass fraction of CNO elements in the accreting material exceeds 0.005. For smaller CNO abundances, the NCO reaction is not effective and thus $\dot{M}_{det} \sim 4 \times 10^{-8}$ $M_\odot$ yr^{-1} (Nomoto 1982a).

For intermediate accretion rates (3×10^{-6} $M_\odot$ yr$^{-1} \gtrsim \dot{M} \gtrsim 1 \times 10^{-8}$ $M_\odot$ yr^{-1}), helium flashes are of moderate strength, thereby recurring many times to increase the white dwarf mass to the Chandrasekhar mass (e.g., Fujimoto and Sugimoto 1982).

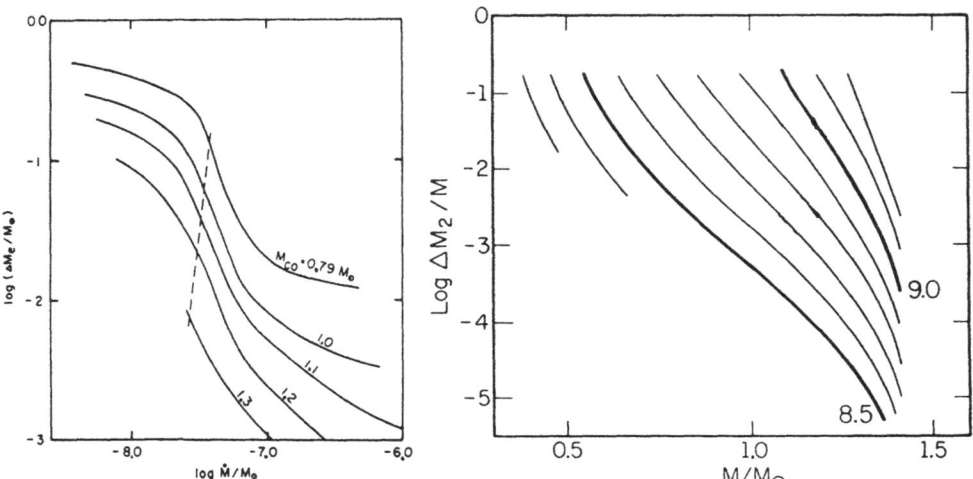

Fig. 3: *left*: The mass of accreted helium envelope, ΔM_e, at the helium ignition as a function of $\dot{M}$ and the mass of underlying C+O core, M_{CO} (Kawai *et al.* 1987).

Fig. 4: *right*: Maximum temperature log T (K) attained during the helium shell flash as a function of the accreted mass ΔM_2 and the white dwarf mass M (Fujimoto and Sugimoto 1982).

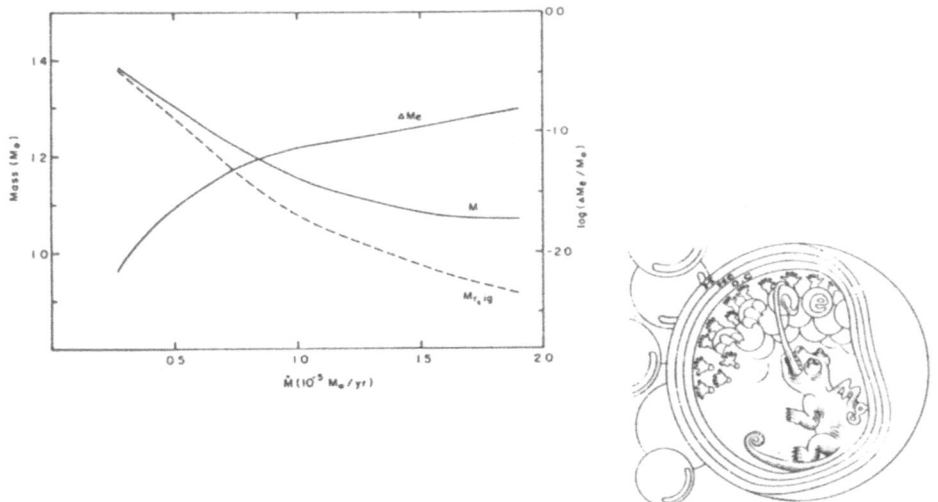

Fig. 5: *left*: The white dwarf mass (M), the location of the carbon ignition shell $(M_{r,ig})$, and $\Delta M_e = M - M_{r,ig}$ at the carbon ignition as a function of $\dot{M}$ (10^{-5} $M_\odot$ yr^{-1}) (Kawai *et al.* 1987).

Fig. 6: *right*: Collapse of an O+Ne+Mg white dwarf is triggered by electron capture on ^{24}Mg (H. Nomoto 1989).

2.3. Merging C+O White Dwarfs

Merging of double C+O white dwarfs is expected to take place as frequently as SN Ia (Iben and Tutukov 1984; Webbink 1984). After the smaller mass white dwarf filles its Roche lobe, mass transfer of carbon on to the more massive white dwarf is expected to be rapid (Iben 1988), which ignites off-center carbon burning for $\dot{M}$ $\gtrsim 2.7 \times 10^{-6}$ $M_\odot$ yr^{-1} in spherical accretion models (Nomoto and Iben 1985). In a steady accretion approximation (Kawai et al. 1987), the mass of C+O white dwarf M and the mass overlying carbon burning shell ΔM_e are determined by $\dot{M}$ as seen in Figure 5. As far as $\dot{M} \lesssim \dot{M}_{Edd}$, M should be larger than 1.07 $M_\odot$ at the carbon ignition.

Actual merging process would be more complicated and its fate is not clear yet. As an optimistic view for AIC, we adopt the following scenario for $\dot{M} \gtrsim 3 \times 10^{-6}$ $M_\odot$ yr^{-1}: First merging of double C+O white dwarfs forms a thick disk around more massive component (Benz et al. 1990). Subsequent heat generation at the boundary layer ignites off-center carbon burning (Mochkovitch and Livio 1990), which burns the entire C+O white dwarf into O+Ne+Mg quietly (Nomoto and Iben 1985; Saio and Nomoto 1985). Eventually the O+Ne+Mg white dwarf collapses as illustrated in Figure 6 (H. Nomoto 1989).

2.4. Conditions for Accretion-Induced Collapse

For C+O white dwarfs, the outcome of accretion depends not only on $\dot{M}$ but also on the initial mass M_{CO}. For $M_{CO} < 1.2 M_\odot$, substantial heat inflow from the surface layer into the central region ignites carbon at relatively low central density ($\rho_c \sim 3 \times 10^9$ g cm^{-3}), which make SNe Ia (Nomoto et al. 1984). On the other hand, if the white dwarf is sufficiently massive and cold at the onset of accretion, the central region is compressed only adiabatically, thereby being cold (and solid) when carbon is ignited in the center of density as high as 10^{10} g cm^{-3}.

The fate of the white dwarf after the ignition at high densities will be discussed in §3. With these results, we draw boundaries for AIC in a diagram of mass accretion rate ($\dot{M}$) versus mass of the white dwarf at the onset of accretion (M_{CO} and M_{ONeMg}) in Figures 7 and 8 (Nomoto 1986; Nomoto and Kondo 1991). We note that the boundaries must be regarded as relatively optimistic ones for the growth of white dwarfs since wind-type mass loss associated with shell flashes of hydrogen and helium is not fully taken into account (e.g., Kato and Hachisu 1989).

Figure 8 shows that for a relatively wide parameter range the O+Ne+Mg white dwarf can increase its mass. Since M_{ONM} can be very close to the Chandrasekhar mass, only a small increase in mass is enough to trigger a collapse. Such very massive O+Ne+Mg white dwarfs would give rise to recurrent novae (Nariai and Nomoto 1979).

Figures 7 and 8 clearly show that close binaries with relatively high $\dot{M}$ and high initial white dwarf mass are favored for AIC. This leads to the possibility that LMBPs with relatively long orbital period may originate from AIC (Nomoto 1987b;

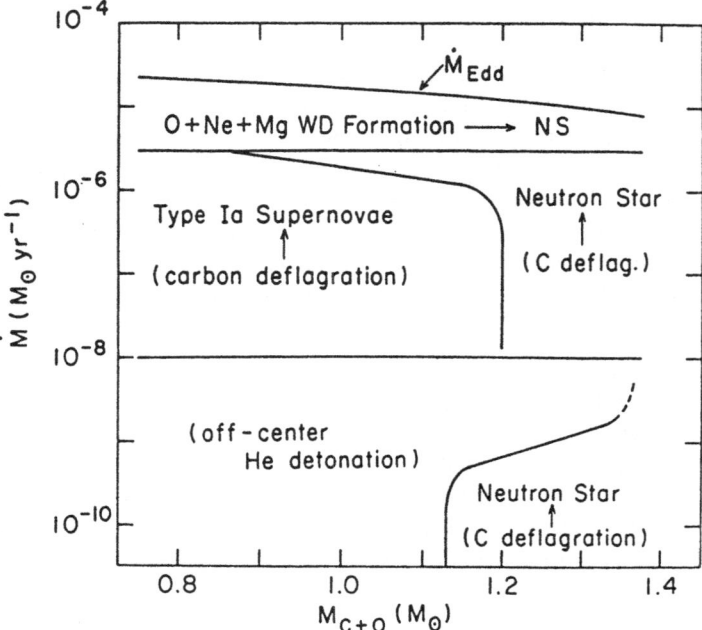

Fig. 7: The final fate of accreting C+O white dwarfs expected for their initial mass and accretion rate $\dot{M}$ (Nomoto and Kondo 1991).

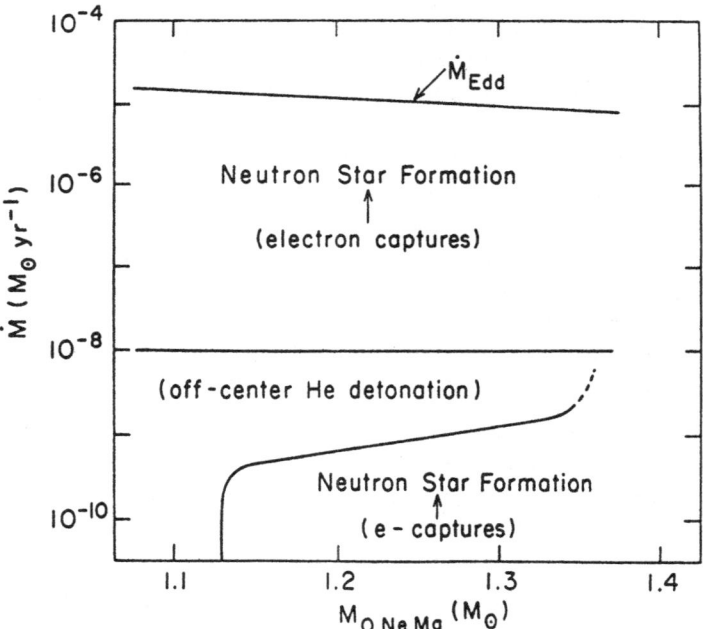

Fig. 8: Same as Figure 7 but for O+Ne+Mg white dwarfs. Collapse is triggered by electron capture on ^{24}Mg and ^{20}Ne (Nomoto and Kondo 1991).

Romani 1990; Ray and Kluzniak 1990) since mass transfer rate from giant stars may be relatively high. On the other hand, Wheeler (1990) suggested that many of the white dwarfs in cataclysmic variables could possibly increase their masses toward the Chandrasekhar mass ending up with $\dot{M} \sim 10^{-9}$ $M_\odot$ yr^{-1} at the lower-right hand corners of Figures 7 and 8; resultant systems could be short orbital period LMBPs. Also if we consider various types of helium star companions (helium main-sequence, helium subgiants, helium white dwarfs, etc.) as well as companions surrounded by a common envelope (Hachisu et al. 1989), LMBPs with relatively short orbital period could be formed from AIC.

3. Collapse Triggered by Deflagration at High Densities

Possible models for AIC previously advanced include solid C+O white dwarfs (Canal et al. 1980; Isern et al. 1983) and O+Ne+Mg white dwarfs (Nomoto et al. 1979a), whose masses could grow to the Chandrasekhar's mass limit for a white dwarf. In the AIC models, collapse of the white dwarf is induced by electron capture that effectively reduces the Chandrasekhar mass limit.

However, since the white dwarf contains nuclear fuel, whether the white dwarf undergoes collapse or explosion depends on which is faster behind the deflagration wave, nuclear energy release or electron capture. The energy generation rate is determined mainly by the propagation velocity of the deflagration wave, v_{def}, while the electron capture rate depends on the density. If v_{def} is lower than a certain critical speed, electron capture induces collapse. If on the other hand v_{def} is sufficiently high, complete disruption results. It is important to determine the critical velocity that divides collapse and explosion (Nomoto 1986).

3.1. Solid C+O White Dwarfs

It is possible that the accreting C+O white dwarfs could collapse rather than explode, depending on the conditions of the white dwarfs. As described in §2 (7), compression of the white dwarf by the accreted matter first heats up a surface layer because of the small pressure scale height there. Later, heat diffuses inward (Nomoto et al. 1984). The diffusion timescale depends on $\dot{M}$ and is short for larger $\dot{M}$ because of the large heat flux and steep temperature gradient generated by rapid accretion. For example, the time it takes for the heat wave to reach the central region is $\sim 2 \times 10^5$ yr for $\dot{M} \sim 10^{-6}$ $M_\odot$ yr^{-1} (Nomoto and Iben 1985) and 5×10^6 yr for $\dot{M} \sim 4 \times 10^{-8}$ $M_\odot$ yr^{-1} (Nomoto et al. 1984).

If the initial mass of the white dwarf, M_{CO}, is smaller than 1.2 $M_\odot$, the entropy in the center increases substantially due to the heat inflow and thus carbon ignites at relatively low central density ($\rho_c \sim 3 \times 10^9$ g cm^{-3}). On the other hand, if the white dwarf is initially more massive than 1.2 $M_\odot$ and cold at the onset of accretion, the central region is compressed only adiabatically and thus is cold when carbon is ignited in the center. In the latter case, the ignition density is as high as 10^{10} g cm^{-3} (e.g., Isern et al. 1983) and the white dwarf may well have a solid core. For such a case, it is necessary to determine the critical condition for which a carbon

198

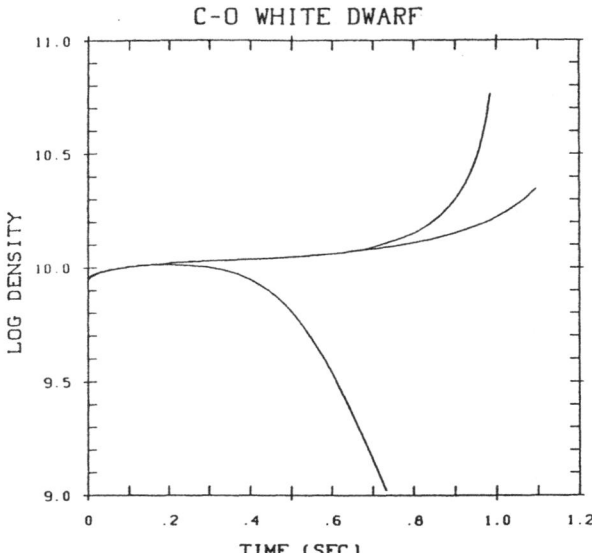

Fig. 9: Change in the central density of the C+O white dwarfs following the propagation of the conductive carbon deflagration wave in the initially solid core. Three cases with $v_{\mathrm{def}}/v_{\mathrm{s}} = 0.05$, 0.03, and 0.01 are shown and the latter two undergo collapse (Nomoto and Kondo 1991).

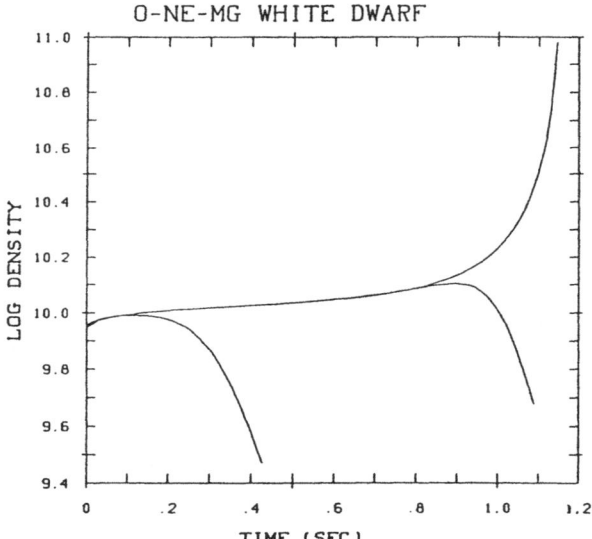

Fig. 10: Same as Figure 9 but for the O+Ne+Mg white dwarfs following the propagation of the oxygen deflagration wave for three cases with $\ell/\min(H_p, r) =$ 2.0, 1.4, and 0.7. For the slowest case of $\ell/H_p = 0.7$, the central density increases, i.e., white dwarf undergoes collapse. Faster propagation induces an explosion of the white dwarf (Nomoto and Kondo 1991).

deflagration induces collapse rather than explosion.

For solid C+O white dwarfs, recent work find that no significant separation between carbon and oxygen occurs during solidification (Barrata et al. 1988; Ichimaru et al. 1988). This in turn leads to the ignition of explosive carbon burning at much lower densities than in models which postulate chemical separation (Isern et al. 1983).

In solid cores, carbon ignition takes place in the pycnonuclear reaction regime (Ogata et al. 1991) and develops into explosive burning at $\rho_c \sim 1 \times 10^{10}$ g cm^{-3}. After thermal runaway of carbon burning, it is likely that a conductive deflagration wave propagates in the solid core. A detonation wave would not form because of steep temperature gradient in the central solid region. Convection would not be effective unless the solid core is melted by the heating from nuclear burning or neutrinos (Canal et al. 1990a).

The conductive deflagration in the solid C+O white dwarf is calculated assuming a constant ratio of v_{def}/v_s for conductive deflagration wave (Nomoto 1986, 1987b; Nomoto and Kondo 1991; Canal et al. 1990b).

Figure 9 shows the change in the central density associated with the propagation of the deflagration wave. Three cases with $v_{def}/v_s = 0.05, 0.03$, and 0.01 are calculated and the latter two slow cases undergo collapse. This implies that the critical velocity, v_{crit}, that divides collapse and explosion is $v_{crit} \sim 0.04 \, v_s$ for $\rho_c \sim 10^{10}$ g cm^{-3}. Since the realistic value of conductive deflagration speed is $v_{def} \sim 0.01 \, v_s$ (Woosley and Weaver 1986), the collapse is the most likely outcome for the solid white dwarf.

3.2. O+Ne+Mg White Dwarfs

The O+Ne+Mg white dwarfs are formed from stars of main-sequence masses of 8 - 12 $M_\odot$ in close binaries with the initial masses as large as 1.1 - 1.37 $M_\odot$ (Nomoto et al. 1979a; Nomoto 1984a). After mass accretion from the companion star, the mass of the white dwarf increases toward the Chandrasekhar mass for a certain range of accretion rate (Fig. 8). When ρ_c exceeds 4×10^9 g cm^{-3}, the O+Ne+Mg white dwarf undergoes electron captures ^{24}Mg (e$^-$, ν) ^{24}Na (e$^-$, ν) ^{20}Ne and ^{20}Ne (e$^-$, ν) ^{20}F (e$^-$, ν) ^{20}O. Electron capture not only reduces the effective Chandrasekhar mass but also releases heat due to γ-ray emission which eventually ignites oxygen deflagration at a certain central density.

In the previous AIC models (Nomoto et al. 1979a), oxygen is ignited at $\rho_{ig} \sim 2.5 \times 10^{10}$ g cm^{-3} after the initiation of collapse (Miyaji et al. 1980; Nomoto 1987a). At such central densities, electron capture is much faster than oxygen burning, thus promoting further collapse. However, ρ_{ig} has been found to depend on the timescale of semiconvective mixing in the electron capture region (Nomoto 1984b; Mochkovitch 1984; Miyaji and Nomoto 1987). If semiconvective mixing is negligible and the heating due to γ-ray emission is confined to the very central region, oxygen burning is ignited at $\rho_{ig} \sim 9.95 \times 10^9$ g cm^{-3} before collapsing (Miyaji and Nomoto 1987). Hydrodynamical calculation is carried out to see whether this model leads to collapse or explosion (Nomoto and Kondo 1991; Isern et al. 1991).

The heat released by electron capture on ^{24}Mg results in the formation of a liquid core even if the white dwarf had initially a solid core (Mochkovitch 1984; Miyaji and Nomoto 1987; Canal et al. 1990a). When γ-rays resulting from electron capture on ^{20}Ne ignite explosive oxygen burning, there exist several possible modes of the subsequent propagation of explosive burning front. Among them the formation of a detonation wave is very unlikely in the present case because negligible semiconvective mixing forms a very steep temperature gradient when explosive oxygen burning starts.

Therefore it is likely that an oxygen deflagration wave forms to propagate at a subsonic velocity. The propagation velocity, v_{def}, depends on which mode of heat transport is faster, conductive or convective. For conductive deflagration, we apply $v_{def} = 0.01\, v_s \sim 100$ km s^{-1} where v_s denotes the local sound velocity, which is a good approximation of v_{def} obtained by numerical calculations (Woosley and Weaver 1986). For the convective deflagration wave, we apply a time dependent mixing length prescription using the ratio between the mixing length and the pressure scale height (or radial distance) $\alpha = \ell/\min(H_p, r) = 0.7, 1.4$, and 2 (Nomoto et al. 1984). For small α, the deflagration speed in the very central region is slower than the conductive deflagration because of small buoyancy force across the burning front; then the minimum v_{def} is set to be 0.01 v_s.

Whether this will lead to collapse or explosion depends on which is faster behind the deflagration wave, nuclear energy release or electron capture. The energy generation rate is determined mainly by the propagation velocity of the deflagration wave, v_{def}, while the electron capture rate depends on the density. If v_{def} is lower than a certain critical speed, electron capture induces collapse. If on the other hand v_{def} is sufficiently high, complete disruption results.

Figure 10 shows the changes in the central density associated with the propagating deflagration front for three cases. It is seen that the slowest case of $\alpha = 0.7$ leads to increasing ρ_c, i.e., collapse, while the propagation with $\alpha = 2$ results in explosion; the intermediate case with $\alpha = 1.4$ is marginal. The fate of the convective deflagration wave depends mainly on whether v_{def} exceeds $\sim 0.04\, v_s$ in the central region at $M_r < \sim 0.1$ - $0.2\, M_\odot$.

Though the determination of v_{def} may require multi-dimensional calculations, the carbon deflagration model for Type Ia supernovae favors $\alpha = 0.7$; the model with $\alpha = 0.7$ can nicely account for many of the observed features of Type Ia supernovae, while the propagation with $\alpha = 0.8$ is a little too fast to be consistent with spectral features of SN Ia (Nomoto et al. 1984). For oxygen deflagration, $\alpha < 1$ may also be the case for the same prescription of deflagration; then the collapse of O+Ne+Mg white dwarfs would be the most likely outcome since $\alpha \sim 1.4$ is marginal between collapse and explosion.

If total disruption results from the white dwarf with central density of $\sim 10^{10}$ g cm^{-3} (Isern et al. 1991), such an explosion should be rare since the ejection of too much neutron-rich iron-peak elements would not be compatible with solar isotopic

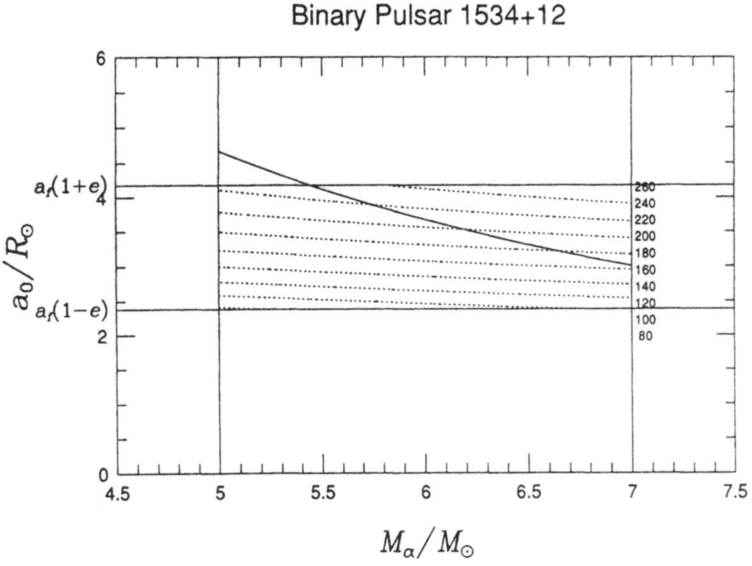

Binary Pulsar 1534+12

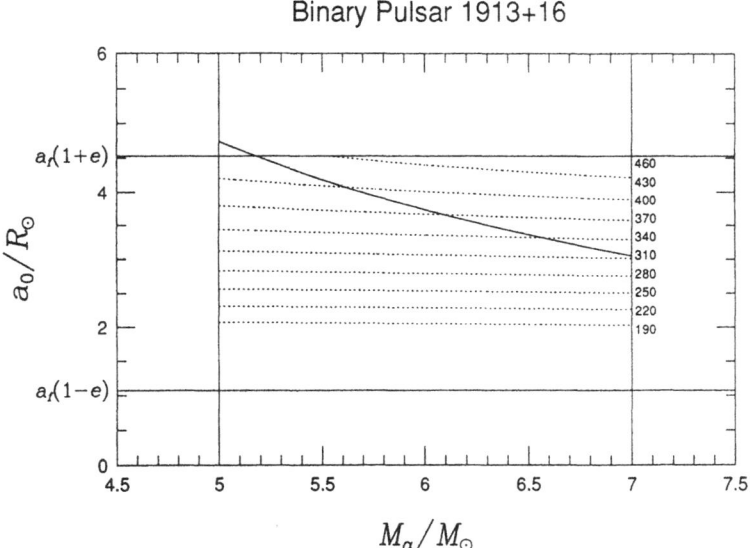

Fig. 11: Kick velocity imparted to the neutron star at the explosion of the helium star 2 of mass M_α as a function of M_α and the initial orbital radius a_0 before the explosion (Yamaoka *et al.* 1991). Here $a_f(1 - e) < a_0 < a_f(1 + e)$. For the solid line, the radius of the helium star is equal to its Roche lobe radius, so that only the upper-right part of the parameter space is allowed. For PSR 1534+12, $v_{\rm kick} \sim 180$ – 240 km s^{-1} is necessary to avoid the disruption of the binary system.

Binary Pulsar 1913+16

Fig. 12: Same as Figure 11 but for PSR 1913 + 16, where $v_{\rm kick} \sim 340 - 420$ km s^{-1} is necessary (Yamaoka *et al.* 1991).

ratios. In addition, the explosion with such low energy as $\sim$ a few times 10^{50} ergs due to large neutrino losses does not match any subclass of SN I frequently observed.

4. Formation of Double Neutron Stars

Recently the masses of the component stars in the binary pulsar system 1532+12 has been determined, which strongly suggests a neutron star companion (Wolszczan 1991). Their masses, eccentricity, semi-major axis, and orbital period are summarized in Table 1 together with those of the first binary pulsar 1913+16 (Taylor 1991). A possible evolutionary scenario for such systems are as follows (e.g., van den Heuvel 1991): (1) Two main-sequence stars 1 and 2, (2) Roche lobe overflow of star 1, which becomes a helium star 1, (3) the first supernova explosion of the helium star 1 to form a neutron star 1, (4) Roche lobe overflow of star 2, which leads to a spiral-in of the neutron star 1 into star 2 and thus to a considerable shrink of the system due to the losses of angular momentum and mass from the system; the system now consists of the recycled neutron star 1 and a helium star 2, (5) the second supernova explosion of the helium star 2; this forms a two neutron star system in an eccentric orbit.

Given the observed orbital parameters in Table 1 and the assumption of a circular orbit for the pre-explosion helium star 2–neutron star 1 system, the mass of the helium star 2 M_α and the possible kick velocity v_{kick} at the explosion can be calculated. If the explosion is spherical (i.e., $v_{kick} = 0$), $M_\alpha \sim 2.1$ $M_\odot$ (Wolszczan 1991). This is smaller than the minimum mass of the helium star that can form a neutron star [~ 2.5 $M_\odot$ (Nomoto 1984a) – ~ 2.2 $M_\odot$ (Habets 1986) depending on the treatment of overshooting and semi-convection]. This suggests that either the explosion is not spherical or the exploding star had lost even its helium layer before the explosion.

If we introduce a finite v_{kick} for the explosion of the helium star 2, the kick velocity and its direction can be calculated as functions of assumed M_α and the initial orbital radius a_0 before the explosion where $a_f(1 - e) < a_0 < a_f(1 + e)$ (Fig. 11; Yamaoka et al. 1991). It should be noted that smaller helium stars have larger radii at the collapse which is ~ 3 $R_\odot$ for $M_\alpha \sim 3.3$ $M_\odot$ (Nomoto and Hashimoto 1988). For the solid line in Figures 11 and 12, the radius of helium star is equal to its Roche lobe radius. As far as the star 2 is a helium star, therefore, M_α should be larger than 5 $M_\odot$ to underfill the Roche lobe. If this is the case, a kick velocity of $v_{kick} \sim 180 - 240$ km s^{-1} is necessary to avoid the disruption of the binary system. The same relation is obtained for 1913 + 16 in Figure 12, where $v_{kick} \sim 340 - 420$ km s^{-1} is necessary (see also Burrows and Woosley 1986).

Two possible extreme cases are: (1) the star 2 is a helium star more massive than ~ 5 $M_\odot$, for which the explosion produces a large kick velocity, and (2) the star 2, being initially a helium star of smaller than 5 $M_\odot$, loses its helium envelope to become an almost bare C+O star. The masses of C+O stars are: 6.0, 3.8, 2.1, and 1.8 $M_\odot$ for $M_\alpha = 8, 6, 4, 3.3$ $M_\odot$, respectively (Nomoto and Hashimoto 1988). For $M_\alpha \lesssim 4$ $M_\odot$, therefore, the explosion of star 2 could be spherical with $v_{kick} = 0$.

The supernova explosions of stars 1 and 2 might be observed as Type Ib/Ic supernovae (SNe Ib/Ic). From the light curve models, the low mass helium stars are the most plausible progenitors of SNe Ib/Ic (Shigeyama *et al.* 1990; Nomoto *et al.* 1990). Such helum stars would form mostly in close binaries (Yamaoka and Nomoto 1991), because the single 12 – 18 $M_\odot$ stars would not lose their entire hydrogen-rich envelope by wind mass loss. Meurs and van den Heuvel (1989) predicted that more than 70 percent of massive star explosions would occur in close binaries. This estimate predicts that the occurrence frequencies of SNe Ib/Ic are higher than SN II, which might be consistent with the increasing number of SNe Ic recently discovered.

The binary scenario suggests that SNe Ib/Ic might be closely related to the formation of binary pulsars and X-ray binaries. If the binary system is not disrupted by the supernova mass ejection, a neutron star is left to orbit around a various types of a companion star (a main-sequence star or a helium star). Many of them would become Be X-ray binaries.

If the main difference between SNe Ib and Ic are the absence and presence of hydrogen-rich envelope (Jeffery *et al.* 1991; Yamaoka and Nomoto 1991), the first explosion could be SN Ic because of possible presence of hydrogen, while the second explosion might be SN Ib. if star 2 loses even its helium envelope. These binary scenario also suggests that SNe Ic occur more frequently than SNe Ib.

Table 1. Binary Pulsars

PSR	M_p ($M_\odot$)	M_c ($M_\odot$)	e	a_f ($R_\odot$)	P_b (hours)	$P_b/\dot{P_b}$ (years)
1913+16	1.4421±0.0012	1.3875±0.0012	0.617127	2.8	7.752	3×10^8
1532+12	1.32±0.03	1.36±0.03	0.274	3.28	10.1	$\sim 1\times 10^9$

We would like to thank Toshikazu Shigeyama for collaborative work on the formation of double neutron star systems. K.N. thanks Ramon Canal, Jordi Isern, Shri Kulkarni, Marco Tavani, Ed van den Heuvel, and Stan Woosley for stimulating discussions during his visit to Barcelona (December 1990) and Santa Barbara (January 1991). This work has been supported in part by the grant-in-Aid for Scientific Research (01540216, 02234202, 02302024, 03218202) of the Ministry of Education, Science, and Culture in Japan, and by the Japan-U.S. Cooperative Science Program (EPAR-071/88-15999) operated by the JSPS and the NSF.

REFERENCES

Bailyn, C.D., and Grindlay, J.E. 1990, *Ap. J.*, **353**, 159.

Barrat, J.L., Hansen, J.P., and Mochkovitch, R. 1988, *Astr. Ap.*, **199**, L15.

Benz, W., Bowers, R.L., Cameron, A.G.W., and Press, W. 1990, *Ap. J.*, **348**, 647.

Burrows, A., and Woosley, S.E. 1986, *Ap. J.*, **308**, 680.

Canal, R., Garcia, D., Isern, J., and Labay, J. 1990b, *Ap. J. (Letters)*, **356**, L51.

Canal, R., Isern, J., and Labay, J. 1980, *Ap. J. (Letters)*, **241**, L33.

Canal, R., Isern, J., and Labay, J. 1990a, *Ann. Rev. Astr. Ap.*, **28**, 183.

Chanmugam, G., and Brecher, K. 1987, *Nature*, **329**, 696.

Fujimoto, M.Y., and Sugimoto, D. 1982, *Ap. J.*, **257**, 291.

Habets, G.M.H.J. 1986, *Astr. Ap.*, **167**, 61.

Hachisu, I., Kato, M., and Saio, H. 1989, *Ap. J. (Letters)*, **342**, L19.

Hashimoto, M., Nomoto, K., Arai, K., and Kaminisi, K. 1986, *Ap. J.*, **307**, 687.

Iben, I.Jr. 1988, *Ap. J.*, **324**, 355.

Iben, I.Jr., Nomoto, K., Tornambe, A., and Tutukov, A. 1987, *Ap. J.*, **317**, 717.

Iben, I.Jr., and Tutukov, A. 1984, *Ap. J. Suppl.*, **55**, 335.

Iben, I.Jr., and Tutukov, A. 1991, *Ap. J.*, **370**, 615.

Ichimaru, S., Iyetomi, H., and Ogata, S. 1988, *Ap. J. (Letters)*, **334**, L17.

Isern, J., Labay, J., Hernanz, M., and Canal, R. 1983, *Ap. J.*, **273**, 320.

Isern, J., Canal, R., and Labay, J. 1991, *Ap. J. (Letters)*, **372**, L83.

Jeffery, D., Branch, D., Filippenko, A.V., and Nomoto, K. 1991, *Ap. J.*, submitted.

Kato, M., and Hachisu, I. 1989, *Ap. J.*, **346**, 424.

Kawai, Y., Saio, H., and Nomoto, K. 1987, *Ap. J.*, **315**, 229.

Kulkarni, S.R., and Narayan, R. 1988, *Ap. J.*, **335**, 755.

Kulkarni, S.R., Narayan, R., and Romani, R.W. 1990, *Ap. J.*, **356**, 174.

Limongi, M., and Tornambe, A. 1991, *Ap. J.*, **371**, 317.

Meurs, E.J.A., and van den Heuvel, E.P.J. 1989, *Astr. Ap.*, **226**, 88.

Michel, F.C. 1987, *Nature*, **329**, 310.

Miyaji, S., and Nomoto, K. 1987, *Ap. J.*, **318**, 307.

Miyaji, S., Nomoto, K., Yokoi, K., and Sugimoto, D. 1980, *Pub. Astr. Soc. Japan*, **32**, 303.

Mochkovitch, R. 1984, in *Problems of Collapse and Numerical Relativity*, ed. D. Bancel and M. Signore (Dordrecht: Reidel), p. 125.

Mochkovitch, R., and Livio, M. 1990, *Astr. Ap.*, **236**, 378.

Narayan, R., Fruchter, A.S., Kulkarni, S.R., and Romani, R.W. 1990, in *Accretion-Powered Compact Binaries*, ed. C.W. Mauche (Cambridge University Press), p. 451.

Nariai, K., and Nomoto, K. 1979, in *IAU Colloquium 53, White Dwarfs and Variable Degenerate Stars*, ed. H.M. Van Horn and V. Weidemann (Rochester: Univ. of Rochester), p.525.

Nariai, K., Nomoto, K., and Sugimoto, D. 1980, *Pub. Astr. Soc. Japan*, **32**, 473.

Nomoto, H. 1989, *Exploring SN 1987A* (in Japanese), (Tokyo: Kodansha).

Nomoto, K. 1982a, *Ap. J.*, **253**, 798.

Nomoto, K. 1982b, *Ap. J.*, **257**, 780.

Nomoto, K. 1984a, *Ap. J.*, **277**, 791.

Nomoto, K. 1984b, in *Problems of Collapse and Numerical Relativity*, ed. D. Bancel and M. Signore (Dordrecht: Reidel), p. 89.

Nomoto, K. 1986, *Prog. Part. Nucl. Phys.*, **17**, 249.

Nomoto, K. 1987a, *Ap. J.*, **322**, 206.

Nomoto, K. 1987b, in *IAU Symposium 125, The Origin and Evolution of Neutron Stars*, ed. D.J. Helfand and J.-H. Huang (Dordrecht: Reidel), p. 281.

Nomoto, K., Filippenko, A.V., and Shigeyama, T. 1990, *Astr. Ap.*, **240**, L1.

Nomoto, K., and Hashimoto, M. 1988, *Physics Reports*, **163**, 13.

Nomoto, K., and Iben, I.Jr. 1985, *Ap. J.*, **297**, 531.

Nomoto, K., and Kondo, Y. 1991, *Ap. J. (Letters)*, **367**, L19.

Nomoto, K., Miyaji, S., Sugimoto, D., and Yokoi, K. 1979a, in *IAU Colloquium 53, White Dwarfs and Variable Degenerate Stars*, ed. H.M. Van Horn and V. Weidemann (Rochester: Univ. of Rochester), p. 56.

Nomoto, K., Nariai, K., and Sugimoto, D. 1979b, *Pub. Astr. Soc. Japan*, **31**, 287.

Nomoto, K., Thielemann, F.-K., and Yokoi, K. 1984, *Ap. J.*, **286**, 644.

Ogata, S., Iyetomi, H., and Ichimaru, S. 1990, *Ap. J.*, **372**, 259.

Paczynski, B. 1970, *Acta Astr.*, **21**, 1.

Ray, A., and Kluzniak, W. 1990, *Nature*, **344**, 415.

Romani, R.W. 1990, *Ap. J.*, **357**, 493.

Saio, H. and Nomoto, K. 1985, *Astr. Ap.*, **150**, L21.

Shigeyama, T., Nomoto, K., Tsujimoto, T., and Hashimoto, M. 1990, *Ap. J. (Letters)*, **361**, L23.

Sienkiewicz, R. 1980, *Astr. Ap.*, **85**, 295.

Starrfield, S., Sparks, W.M., Truran, J.W., and Shaviv, G. 1991, in *Supernovae*, ed. S.E. Woosley (Berlin: Springer), p. 602.

Sugimoto, D. and Fujimoto, M.Y. 1978, *Pub. Astr. Soc. Japan*, **30**, 467.

Sugimoto, D., Fujimoto, M.Y., Nariai, K., and Nomoto, K. 1979, in *IAU Colloquium 53, White Dwarfs and Variable Degenerate Stars*, ed. H.M. Van Horn and V. Weidemann (Rochester: Univ. of Rochester), p. 280.

Tavani, M. 1991, *Ap. J.*, in press.

Taylor, J. 1991, in *NATO ARW, X-Ray Binaries and the Formation of Binary and Millisecond Radio Pulsars*, ed. E.P.J. van den Heuvel (Kluwer), in press.

Van den Heuvel, E.P.J. 1991, in *NATO ASI, Neutron Stars: An Interdisciplinary Field*, ed. J. Ventura and D. Pines (Kluwer), in press.

Webbink, R. 1984, *Ap. J.*, **277**, 355.

Wheeler, J.C. 1990, in *Frontiers of Stellar Evolution*, ed. D.L. Lambert (San Francisco: Astron. Soc. of Pacific), in press.

Wolszczan, A. 1991, *Nature*, **350**, 688.

Woosley, S.E., Taam, R.E., and Weaver, T.A. 1986, *Ap. J.*, **301**, 601.

Woosley, S.E., and Weaver, T.A. 1986, *Lecture Notes in Physics*, **255**, 91.

Yamaoka, H., and Nomoto, K. 1991, in *SN 1987A and Other Supernovae*, ed. I.J. Danziger (Garching: ESO), in press.

Yamaoka, H., Shigeyama, T., and Nomoto, K. 1991, in preparation.

SOME PROBLEMS IN THE THEORY OF ISOLATED PULSARS

Franco Pacini
Arcetri Astrophysical Observatory
and
University of Florence
ITALY

Abstr. In this lecture I will briefly summarize some of the many problems which have confronted (or still puzzle) theorists since the discovery of pulsars, more than 20 years ago. I refer the reader interested in this subject to a recent paper by Srinivasan (1989), to a review by Ruderman (1987) and to the references contained in them. My main topics will be:

1. the magnetospheric model: basic theory versus real life
2. the radiation process, including the implications of the possible recent detection of the $\bar{e}$ - $\check{e}$ annihilation line from the Crab pulsar
3. relativistic winds and plerionic SN Remnants

1. *The magnetospheric model: energy loss*

The basic models developed for the energy loss from the magnetosphere of neutron stars involve the electrodynamics of a rotating magnet. In the first model (Pacini 1967, 1968) the loss is attributed to magnetic radiation from a dipole inclined with respect to the rotation axis. This radiation takes the form of a strong wave emitted at the basic rotation frequency Ω. Ostriker and Gunn (1969) have shown that this wave can accelerate particles up to ultrarelativistic energies.

On the other hand, even if the dipole is aligned with the rotation axis, an energy loss would occur. Indeed the neutron star would act as an unipolar inductor generating very large potential differences between the poles and the equator of the neutron star (Goldreich and Julian 1969). The consequent electric fields would overcome gravity and extract charges of both signs from the stellar surface. A dense magnetosphere would be created and forced to corotate up to a distance $R_c = \frac{c}{\Omega}$ (speed of light distance) where the corotation velocity becomes c. More precisely, one should distinguish between a "closed magnetosphere" where the magnetic field lines close before reaching a distance R_c (here the particles can actually corotate) and the region above the magnetic poles where the magnetic lines are "open". The strong electric fields along these open lines can accelerate particles to very high energy. In turn, this gives rise to a poloidal current flowing out of

207

E. P. J. van den Heuvel and S. A. Rappaport (eds.), X-Ray Binaries and Recycled Pulsars, 207–213.
© 1992 *Kluwer Academic Publishers.*

the pulsar and to a consequent, large scale, toroidal magnetic field permeating the space around the pulsar.

In essence, both models can (at least in principle) explain the energy loss from pulsars and account for the existence of relativistic particles and large scale magnetic fields in the Crab Nebula and similar SN Remnants.

In both cases the loss of rotational energy $\frac{1}{2} I \Omega^2$ (I is the moment of inertia of the star) is determined by the strength of the dipolar moment which is $\propto B R^3$ (B and R are the surface field and the stellar radius) and by the rotation frequency Ω

$$- \frac{d}{dt} \left(\frac{1}{2} I \Omega^2 \right) = \frac{2}{3c^3} B^2 R^6 \Omega^4$$

($B = B_0$ for a parallel field; $B = B_0 \sin \alpha$ for a dipole inclined with an angle α).

It is not surprising that both models give a similar expression for the energy loss. This loss corresponds to the energy outflow through a surface of order $4\pi \left(\frac{c}{\Omega} \right)^2$ and, as an order of magnitude, it is given by $B_c^2 \, 4\pi \left(\frac{c}{\Omega} \right)^2$ where B_c is the value of B computed at R = R_c . It is straightforward to derive from this expression the energy loss in a dipole field.

I would like to stress here two particular points:

1. the determination of the stellar magnetic field from the slowing down of pulsars actually provides its strength at $R_c = \frac{c}{\Omega}$. Only by assuming that the field is dipolar we can deduce the value of the surface field.

2. the above assumption that the field is dipolar can be checked by measuring the so-called braking index in the slowing down law $\dot{\Omega} \propto \Omega^n$ (n = 3 for a dipole). Unfortunately, up to now, this has been possible only for the Crab pulsar PSR 0531, PSR 1509 and for PSR 0540. The values found are in the range 2-3.

It is interesting, in the best studied case (Crab Pulsar), to compare the basic model with real life. This is done in Table 1.

TABLE 1: model versus real life in the Crab Nebula

Theory	Observation
$\dot{\Omega} \propto \Omega^3$	$\dot{\Omega} \propto \Omega^{2.5}$
charge outflow of order $10^{33}s^{-1}$	particle's outflow of order $10^{40}s^{-1}$
monocromatic energy distribution for the accelerated particles	power law energy distribution for the accelerated particles
nebular field around 10^{-4} gauss	nebular field 3×10^{-4} gauss

2.   *The radiation process.*

I come now to the origin of the pulsed radiation. We will only discuss the emission at radiofrequencies (a common feature of all pulsars) and at optical or X-ray frequencies (typical only of a few fast pulsars). Concerning the radioemission, apart from the sharpness of the pulses, the most significant characteristic of the radiation is the observed brightness temperature, typically much above $10^{20°}$ K (up to, say, $10^{33°}$ K). The radio emission is certainly coherent. Its origin is often ascribed to the acceleration of bunches of particles sliding along curved field lines in the open magnetosphere, possibly while these particles escape from above the magnetic poles (Radhakrishnan & Cooke 1969). If ρ is the radius of curvature of these lines and γ is the Lorentz factor of the moving particles, the characteristic emitted frequency v_c is given by

$$v_c = (c/\rho) \, \gamma^3 \tag{2}$$

Since $R_{star} < \rho < c/\Omega$, radiofrequencies would correspond to typical Lorentz factors around 10^2. Also, because of the thermodynamic limitation on the brightness temperature $kT_b \leq mc^2 \, \gamma F$, one can infer a lower limit to F, the number of particles radiating in phase. Just as an example, $T_b \sim 10^{26°}$ K corresponds to a degree of coherence (number of particles per bunch radiating in phase) $F \sim 10^{14}$.

The data available about spectral and polarization properties of pulsars also carry some information on the radiation process and the location of the emitting region, although without complete consensus among the researchers.

Concerning the optical emission, it must be stressed that this is a characteristic typical only of some very fast pulsars. Indeed, optical pulses have been detected only in the case of PSR 0531 (P = 33 ms), PSR 0833 (P = 89 ms) and the pulsar discovered in the Large Magellanic Cloud PSR 0540 (P = 50 ms). The optical radiation from pulsars can be interpreted as incoherent synchrotron radiation emitted by relativistic particles in the proximity of the speed of light distance.   An analysis of the high frequency emission from the Crab pulsar shows the presence of a low energy cut-off in the near infrared. If this is due to synchrotron reabsorption, one can deduce the strength of the perpendicular component for the magnetic field $B_\perp \sim 10^4$ - 10^5 gauss, a value compatible with the estimate (from the slowing down) $B \sim 10^6$ gauss. In order to produce the observed incoherent emission, the electrons must have $E > 10^8$ eV and their density should be close to 10^{14} cm^{-3}. Across the speed of light distance, this corresponds to an outflow about 10^{40} electrons s^{-1} injected into the Crab Nebula, in good agreement with the injection rate deduced on the basis of the spectral properties of this Remnant. In this case it is easy to show that the bolometric synchrotron radiation should scale roughly as $B_o^4 P^{-10}$ (Pacini 1971). The strong dependence on the period is largely due to the fact that, in a dipole field, slower pulsars have magnetic fields at $r \sim R_c$ much smaller than fast pulsars.

More elaborate calculations (Pacini & Salvati 1983) have taken into account the possibility of synchrotron reabsorption and the shape of the energy spectrum for the particles. These calculations lead to the possibility of predicting the optical emission of various pulsars as a function of their period and period derivative (alternatively, period and magnetic field). (It is worthwhile to stress that these predictions only hold as far as the duty cycle of the various sources is comparable to that of the Crab pulsar PSR 0531).

The excellent agreement between theory and observations for PSR 0833 and for the secular decrease of the optical luminosity of PSR 0531 (J. Kristian, private communication) lends strong support to the hypothesis that the optical pulses are synchrotron radiation from particles moving close to the speed of light distance. It also suggests a picture where the particles slide in bunches along the open field line, thus producing the radiowaves, while they also incoherently gyrate around the field lines.

We note that the same model could be used in order to understand the X-ray emission observed from PSR 0531, PSR 1509, PSR 0540. The main embarassment comes from the missed detection of PSR 0833: its X-ray emission cannot be very much lower than that of the surrounding nebula without forcing theorists to invoke an "ad hoc" high energy cut-off for the radiating electrons.

Although an analysis of the γ-ray emission detected from some pulsars (in particular PSR 0531 and PSR 0833) is outside the scope of the present lecture, we would like to mention very briefly the implications of the recently reported balloon measurement of an annihilation line from PSR 0531 (Massaro et al. 1991). The basic data involve a flux 0.86 ± 10^{-4} photons s^{-1} cm^{-2}, centered around an energy $E = 440 \pm 10$ KeV, corresponding to a luminosity around 3×10^{34} erg s^{-1} (assumed distance d = 2 Kpc). If this line corresponds to annihilation in the magnetosphere of PSR 0531, then its redshift suggests that the emission takes place at ≤ 2 stellar radii. The observed luminosity would imply, once more, a flux of about 10^{40} electrons s^{-1} moving outwards in the magnetosphere. The annihilation would occur because of collisions with a similar number of positrons photo-produced by γ-rays in the intense magnetospheric field (see, e.g., Ruderman 1987). If the existence of the annihilation line is confirmed by other experiments, it would provide an additional probe for the conditions in the magnetosphere of PSR 0531.

3.   *Pulsar winds and the evolution of SN Remnants.*

It is now recognized that several galactic supernova remnants, the so-called plerions, closely resemble the Crab Nebula and are probably driven by a central pulsar. Their expected evolution can be studied assuming that the pulsar injects continuously into the surrounding space magnetic energy and relativistic electrons. One can then derive, as a function of time, the field strength in the remnant, the energy distribution of the particles and the resulting nebular synchrotron luminosity at various frequencies (Pacini & Salvati 1973).

Concerning the evolution of the nebular field B, its average value is determined by the following equation

$$\frac{dU_B}{dt} = \frac{L_0}{(1+t/\tau)^2} - \frac{v}{R} U_B$$

where

$$U_B = \frac{1}{6} B^2 R^3$$

This equation expresses the balance between the change in the nebular magnetic energy due to the input from the pulsar (initial energy loss L_0, decay time τ), minus the rate of adiabatic energy losses determined by the expansion velocity v. In the case of the Crab Nebula one finds that at $\tau \sim 1$ day the average nebular field was of order 10^2 G. One expects a present day value 3×10^{-4} G, in agreement with the observations.

Once the time evolution of the field has been established, one can determine the history of the nonthermal nebular luminosity if one assumes that the pulsar injects continuously relativistic electrons with a power law energy distribution (these electrons evolve because of adiabatic and radiative losses).

Quite generally, one can expect that the early phases of the evolution are dominated by synchrotron reabsorption because of the very strong magnetic field. At later times (say, some months after the explosion) the newly born plerion should have a very strong radio luminosity which remains constant for a time comparable with the pulsar slowing-down timescale. Later, the flux decrease is due to the fact that the rate of injection has declined and the evolution of the source is dominated by adiabatic and radiation losses. This is confirmed by detailed calculations (Bandiera et al. 1984).

The expected very strong initial radio emission raises the question of a possible connection between the early evolution of plerions and the radiosupernovae. Concerning the latter, we recall here the main observational points:

1. the radio emission is generally observed in type 2 supernovae and it is usually delayed by several months with respect to the optical outburst.

2. the radio emission is nonthermal and may last for several years (possibly > 10 yr).

Two different models for the origin of the radio emission have been put forward. Chevalier (1982) has proposed that the radio flux originates outside the expanding remnant and the particles are accelerated when the shell interacts with pre-ejected circumstellar matter. The alternative model (Pacini & Salvati 1981; Bandiera et al. 1984) postulates the existence of a central pulsar which gives birth to a plerion.

The main objection raised against the latter model is that a uniform distribution of ionized matter resulting from the explosion would prevent the escape of radio emission for at least 10^2 years. This objection becomes invalid if the supernova shell fragments very early into filaments because of Rayleigh-Taylor instabilities or if particles can escape through "holes" and radiate outside the envelope. Radio observations of other supernovae which exploded during the last century will be crucial in establishing whether there is a continuity between the radiosupernovae and the remnant phase.

In any case, the observation of nonthermal phenomena from the site of stellar explosions provides a very important possibility of investigating the properties of supernovae in the post-outburst phase and the early life of pulsar, a stage which was completely obscure up to recent years.

Important evidence may obviously come from the close scrutiny of the behavior in the coming years of SN 1987A where, however, there is still only little evidence for the energy input associated with an energetic young pulsar.

References

Bandiera, R., Pacini, F., Salvati, M. 1983, *Astr. Astrophys.*, **126**, 7.

Bandiera, R., Pacini, F., Salvati, M. 1984, *Astrophys. J.*, **285**, 134.

Chevalier, R. 1982, *Astrophys. J.*, **259**, 302.

Goldreich, P., Julian, W. 1969, *Astrophys. J.*, **157**, 869.

Massaro, E. et al. 1991, *Astrophys. J.*, **376**, L11.

Ostriker, J., Gunn, J. 1969, *Astrophys. J.*, **157**, 1395.

Pacini, F. 1967, *Nature*, **216**, 567.

Pacini, F. 1968, *Nature*, **221**, 454.

Pacini, F. 1971, *Astrophys. J.*, **163**, L17.

Pacini, F., Salvati, M. 1983, *Astrophys J..*, **274**, 369.

Pacini, F., Salvati, M. 1981, *Astrophys. J.*, **245**, L107.

Pacini, F., Salvati, M. 1973, *Astrophys. J.*, **186**, 249.

Radhakrishnan, V., Cooke, D.J. 1969, *Astrophys. Letters*, **3**, 225.

Ruderman, M.A.V. 1987, in *High Energy Phenomena around collapsed Stars* (Editor Pacini, F.), Reidel Publishing Company.

Srinivasan, G. 1989, *Astronomy and Astrophysics Review*, **1**, 225.

RELATIVISTIC SHOCK WAVES AND THE EXCITATION OF PLERIONS

JONATHAN ARONS
Departments of Astronomy and of Physics,
University of California at Berkeley
and
Institute of Geophysics and Planetary Physics,
Lawrence Livermore National Laboratory

ABSTRACT. The shock termination of a relativistic magnetohydrodynamic wind from a pulsar is the most interesting and viable model for the excitation of the synchrotron sources observed in plerionic supernova remnants. I describe results on the structure of relativistic magnetosonic shock waves in plasmas composed purely of electrons and positrons, as well as those whose composition includes heavy ions as a minority constituent by number. Relativistic shocks in symmetric pair plasmas create fully thermalized distributions of particles and fields downstream. Therefore, such shocks are not good candidates for the mechanism which converts rotational energy lost from a pulsar into the nonthermal synchrotron emission observed in plerions. However, when the upstream wind contains heavy ions which are a minority constituent by number density, but carry the bulk of the energy density, much of the energy of the shock goes into a downstream, nonthermal power law distribution of *positrons* with energy distribution $N(E)dE \propto E^{-s}$. In a specific model presented in some detail, $s = 1.7$. These characteristics are close to those assumed for the pairs in macroscopic MHD wind models of plerion excitation. The essential acceleration mechanism is collective synchrotron emission of left-hand elliptically polarized extraordinary modes by the ions in the shock front at high harmonics of the ion cyclotron frequency, with the downstream positrons preferentially absorbing almost all of this radiation, mostly at their fundamental (relativistic) cyclotron frequencies. The spatial structure of the shock includes compressional overshoots of the magnetic field strength above what would be expected from the MHD jump conditions, at places where the ions are reflected. I outline a new model of the inner regions of the Crab Nebula, in which I propose that the shock structure is angularly resolved and the optical "wisps" are surface brightness enhancements created in the compressional overshoots of the magnetic field within and just down stream from the shock. I also briefly outline the application of these results to the relativistic shocks believed to terminate the winds from pulsars in compact binaries.

1. Introduction

The excitation of diffuse, nonthermal astrophysical synchrotron sources by energy lost from central compact objects has long been a puzzle in high energy astrophysics. Pulsars and their surrounding plerionic nebulae form the nearest at hand and best studied examples of this problem. Other examples include the excitation of extragalactic radio sources by jets, and possibly the emission from active galactic nuclei themselves. The Crab Nebula is the best studied of the plerions, and for my purpose here, I will take the physical problems it poses to be characteristic of the class. It is typical of what I shall call "diffuse" plerions, in which the synchrotron radiation

E. P. J. van den Heuvel and S. A. Rappaport (eds.), X-Ray Binaries and Recycled Pulsars, 215–228.

and adiabatic expansion dominate the energy losses from relativistic electrons and positrons. The termination of the outflow from a pulsar in a compact binary, as the relativistic wind encounters the mass lost from the companion star, may form an example of a "compact" plerion, the difference from the diffuse case being the possible importance of inverse Compton losses.

The X-ray and gamma-ray emission from the Crab requires continuous energy input, which is aptly explained by the energy lost from a central, rotating magnetized neutron star (Pacini 1967). Indeed, the identification of this idea with the subsequently discovered pulsars (Gold 1968) is one of the pillars of the reasoning by which we identify the rotating neutron star model with the observed pulsar phenomenon. However, the physics of the energetic link has remained an unsolved problem. The Crab has features similar to a wide variety of astrophysical synchrotron sources (power from a central compact object, a large scale magnetic field, diffuse, optically thin nonthermal synchrotron emission, dense filaments of thermal emission line plasma embedded in or adjacent to the diffuse nonthermally emitting gas). Therefore, an understanding of the physical mechanisms through which energy is carried, without emitting many photons, from the central compact object to the surrounding environment, then degraded into the observed nonthermal emission, has wide astrophysical significance. In addition, the nature of this conversion mechanism may provide constraints on models of the pulsar magnetosphere.

Relativistic, magnetohydrodynamic (MHD) models of the coupling between pulsar and nebula are the most successful at the present time. Building on earlier ideas (Piddington 1957, Rees and Gunn 1974, Kundt and Krotscheck 1980), Kennel and Coroniti (1984a,b) showed that such a wind from the pulsar, terminated by a standing, transverse magnetosonic shock wave, could give a good account of the nebular dynamics and of the nebular spectrum of high energy photons (near infrared, optical, X-rays and gamma rays), which are an instantaneous probe of the pulsar's outflow. The prominent new results of their model are several. 1) The post shock flow fits into the slowly expanding nebula only if the upstream flow is almost entirely dominated by kinetic energy, with

$$\sigma \equiv \frac{B_1^2}{4\pi m N c^2 \gamma_1} \simeq 0.003. \tag{1}$$

Subsequent work by Emmering and Chevalier (1987), who included the time dependence introduced by the nonrelativistic nebular expansion velocity, confirm this result, with an increase of σ to perhaps 0.005. The dynamics of the model provide a natural explanation of the apparent equipartition of the fields with the relativistic plasma in the main body of the nebula, since the decelerating flow behind the shock compresses the field until approximate equipartition is reached and no further deceleration is possible. 2) If the upstream flow is entirely composed of pairs (an assumption), and if the post-shock energy distributions are assumed to be power laws with $N(E) \propto E^{-s}$ (another assumption), then the synchrotron emission from the downstream flow can be fit to the observed X- and gamma ray spectra of the Nebula, taken as an unresolved object, if $s \approx 2.3$. Once this fit is made, the optical and near infrared spectra of the nebula (again taken as an unresolved object) are *predicted* (more or less) correctly by the model.

This theory is sufficiently successful as a macroscopic explanation of the phenomenology to provide a setting for investigation of the microscopic physics. Several questions of basic physics are implied by the macroscopic flow model. Does a relativistic, magnetosonic shock in a pair plasma actually create the assumed power law distributions downstream? If so, what is the mechanism of acceleration in such shocks? Are there other observable consequences of this mechanism, other than the desired particle acceleration, which allow one to test the shock hypothesis directly? I emphasize that one cannot appeal to the ever popular ideas of diffuse shock acceleration proposed for particle acceleration by interplanetary and nonrelativistic interstellar shocks (*e.g.*, Bell 1978, Quenby and Lieu 1989) even as extended to relativistic shocks with magnetic field parallel to the flow (*e.g.* Ellison *et al* 1990 and references therein), since the termination shocks of pulsar winds must be fairly close to having the magnetic field perpendicular to the flow upstream when the plerion surrounds the pulsar — indeed, for the Crab, a laminar Archimedean spiral in the outflow from the pulsar would have Θ_{Bn}, the angle between the flow direction and the magnetic field in the shock frame where we observe the system, differing from $90°$ only by one part in 10^9! In addition, in relativistic shocks, the particles can stream ahead of the shock and participate in the back and forth motion across the front required in diffusive shock acceleration only when $\Theta'_{Bn} < 1/\gamma_1$ (Begelman and Kirk 1990, Gallant *et al* 1991), where γ_1 is the Lorentz factor of the upstream flow and Θ'_{Bn} is the angle between the flow direction and the magnetic field in the proper frame of the upstream fluid. I will argue that $\gamma_1 \approx 10^7$ in the Crab Nebula. Then $(\Theta'_{Bn})_{Crab} \approx \pi/2 - 10^{-2} \gg \gamma_1^{-1}$; the magnetic field is still almost exactly transverse even in the upstream fluid frame. Therefore, some other mechanism of shock acceleration must apply.

In relativistic shocks, the mechanisms of high energy particle acceleration and of basic thermalization can be one and the same — accelerating relativistic particles can be part of the problem of shock structure itself. In fact, the highest energy electrons and positrons in the shock theory I outline below are those which have Larmor radii comparable to the gyro-radii of the heavy ions which control the shock thickness, and are precisely the $e^{\pm}$ which give rise to the highest energy synchrotron photons ($\varepsilon \sim 100$ MeV) observed in the *nebular* spectrum. Thus, the problems of shock structure and of nonthermal particle acceleration *are* one and the same in this case.

2. Shock Structure and Particle Acceleration

Pulsar outflows must have extremely high Mach number. The Alfven Mach number in the flow is $M_A = \beta_1 \gamma_1/\sqrt{\sigma} \cong 1.5 \times 10^8$! Under these circumstances, magnetic reflection of particles from the shock front plays an essential role in the shock dynamics. Relativistic, magnetosonic solitary waves with unidirectional flow exist only for $M_A < 1 + 1/\gamma$ (Kennel and Pellat 1976, Alsop and Arons 1988, Chiueh 1989). At higher Mach number, incoming particles reflect from the enhanced magnetic field in the wave front and are set into Larmor gyration, leading to a new kind of magnetosonic soliton with reflected particles self-consistently incorporated (Alsop and Arons 1988). This reflection process requires all the momentum of the incoming plasma to be temporarily stored in a magnetic overshoot, with $B_{peak}/B_1 = (1 + \sigma^{-1})^{1/2}$ in the wave frame of a soliton in a pair plasma. The dependence of the overshoot on σ alone is characteristic of relativistic magnetosonic shocks — once $\gamma_1 \gg 1$, the dependence on γ_1 itself is entirely subsumed in the scale of the flow being that of the Larmor radius based on the upstream parameters. σ provides a better parameterization than does M_A.

While it is possible to construct solitary wave models with magnetically reflected particles in pair plasmas (Alsop and Arons 1988), unraveling the instabilities and thermalization mechanisms in this highly inhomogeneous environment is a formidable task for analytical theory. In a collaboration with Y. Gallant, M. Hoshino, A.B. Langdon and C.E. Max, I have made use of fully non-linear, fully self-consistent particle-in-cell numerical simulations in order to "experimentally" uncover the relevant physics. The basic method is to integrate the equations of motion of many charged particles in their *self-consistent* fields. Suppose at some time t_n, one knows

218

the electromagnetic field on a spatial grid, and one knows the positions and momenta of all the particles $\{x_j(t_n), p_j(t_n)\}$ (they are not on the grid). One then finds the Lorentz force at each particle's position and uses Newton's law (in relativistic form) to advance the particles to new phase space positions $\{x_j(t_n + dt), p_j(t_n + dt)\}$. One then calculates new fields at time $t_n + (dt/2)$ using the charge densities and currents at time t_n, with the particles distributed on the grid with a suitable weight function. These new fields are used to advance the positions of the particles once again, with the time step set by various stability considerations. These techniques are quite well understood and have been in use for 20 years (Langdon and Lasinski 1976, Birdsall and Langdon 1985).

We modeled shock waves in pair plasmas by setting up a one-dimensional spatial grid in the computer, with length anywhere from 10 to 40 Larmor radii based on the magnetic field and particle energy in the upstream flow. The flow and field geometry is shown in Figure 1.

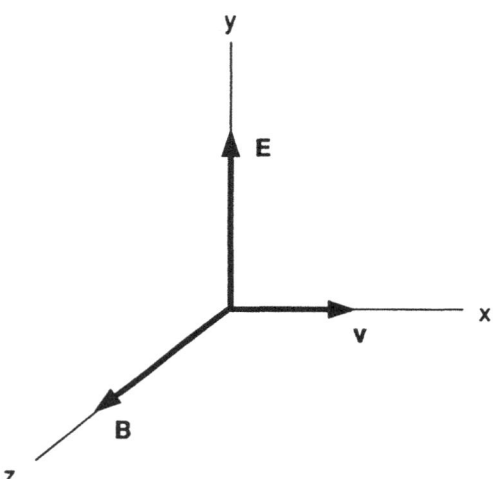

Figure 1: Upstream geometry of flow, magnetic field and electric field. The electrostatic field parallel to v is zero upstream, while the electric field has magnitude $E = (v/c)B$.

At the initial time, the cold plasma fills the computational box, with flow Lorentz factor $\gamma_1 \gg 1$ and a magnetic field polarized across the flow carried with the plasma at the fluid speed. The ratio of magnetic energy density to flow energy density upstream σ is the only parameter of significance. As expected from theoretical considerations, simulations with the same value of σ but differing γ_1 yield identical results, when $\gamma_1 \gg 1$. We have surveyed parameter space with these spatially 1D models from $\sigma = 13$ down to $\sigma = 0.001$.

At the injection point (the "left wall"), the particles are injected all with the same upstream velocity, and the magnetic field is carried with the fluid, by requiring the boundary electric field to be that of a perfectly conducting fluid. The opposite end of the grid is represented as a conducting rubber wall —

particles bounce off the wall, which acts as a perfect conductor as far as the fields are concerned. It would be better to have outflow boundary conditions on the plasma and radiation, but so far a useful algorithm to implement these in the dense downstream plasma has not been constructed. Because the magnetic field points across the flow, the wall has no influence on the shock structure, once the point of particle reflection is more than a few Larmor radii away from the wall. In the low σ shocks of interest here, the wave energy is never more than a few percent of the thermal and kinetic energy of the plasma, so the conducting wall, which forces the downstream waves to be standing modes, has little influence on the shock once it its well upstream of the wall.

The bouncing of the particles off the wall initiates the shock wave. After a confused initial adjustment, a well developed shock appears and travels upstream. Figure 2 shows the electromagnetic field structure of a shock located at $x = 21$ in a pure pair plasma, with $\gamma_1 = 10^6$ and $\sigma = 0.01$, as observed 34 cyclotron periods after the plasma bounced off the simulation wall. One finds an intense, quasi-coherent electromagnetic precursor travelling in front of the shock, while downstream, a few per cent of the total energy is turned into a linearly polarized Rayleigh-Jeans spectrum of electromagnetic waves. Figure 2 also shows the phase space of the electrons in the whole box at a fixed time in the flow. One sees the reflection of the particles from the shock front, followed by rapid thermalization of the Larmor gyration. The downstream spectra of electrons and positrons (not shown) are almost perfect relativistic Maxwellians. The Rayleigh-Jeans distribution (not shown) of down stream waves has a temperature close to the post-shock temperature of the plasma. Thus, transverse shocks in a pure pair plasma create a fully thermalized downstream medium, contrary to a basic assumption of the MHD model of plerionic supernova remnants.

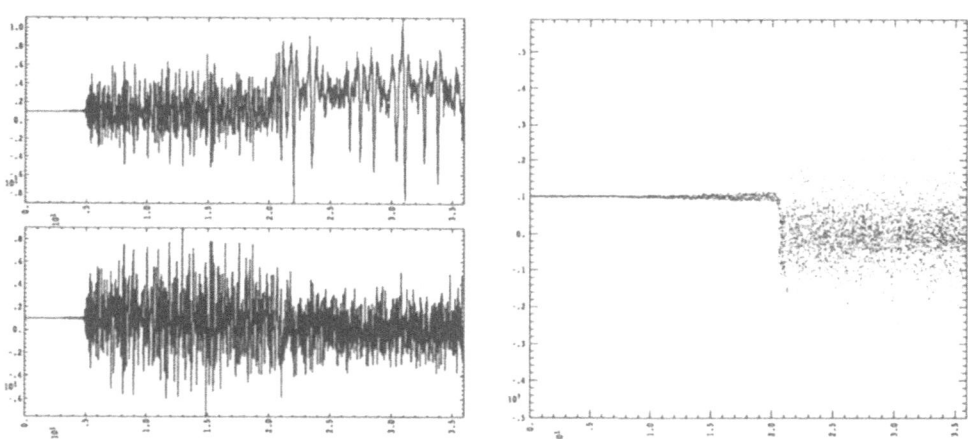

Figure 2: Structure of a relativistic shock in a symmetric pair plasma with $\sigma = 0.01$ and $\gamma_1 = 10^6$. Upper left panel: magnetic field. The shock occurs at $x = 21$, where the unit of length is the Larmor radius of the electrons and positrons based on the upstream flow parameters, and the unit of field strength is the upstream magnetic field. An intense electromagnetic precursor propagates ahead of the shock. Lower left panel: transverse electric field, showing the strong precursor and the weaker downstream extraordinary mode fluctuations. Upstream, the mean value of the electric field is $E_1 = (v_1/c)B_1 \approx B_1$, while downstream, the mean value of the electric field is zero, since the simulation is done in the frame where the downstream plasma is at rest. Right panel: The $p_x - x$ projection of the electrons' phase space, showing the cold plasma injected at $x = 0$. The large amplitude precursor increases the momentum "dispersion" upstream of the shock. The mechanical shock transition occurs where the particles are magnetically reflected from the shock front, at $x = 21$, with almost immediate thermalization of the plasma.

Some initial results of these calculations were reported by Langdon *et al* (1988), and full details are given by Gallant *et al* (1991). We have done a few preliminary calculations with the magnetic field oblique to the flow, using what is otherwise the same 1D model. So long as $\Theta'_{Bn} > 50°$, the results are semiquantitatively the same as for the purely transverse case, as one might suspect since the structure is dominated by the reflected particles and the rapid dissipation. Smaller obliquities may be different, but for technical reasons this conclusion is not yet warranted. We have also investigated some aspects of the two dimensional structure of these shocks, motivated by the observation that the 1D model leaves the plasma with a high temperature in the momenta orthogonal to the magnetic field, while it is still cold in the momenta along B. Such a medium is unstable, most prominently (at low σ) with respect to the Weibel instability. In one 2D shock calculation, we have found this effect does heat the plasma in the direction along B, as well as broaden the distribution of gyrational momenta, but the basic shock structure persists, with downstream anisotropy $T_\perp/T_\parallel \sim 2$. As this kind of anisotropy may be observationally relevant (Michel *et al* 1991) to the pattern of optical polarization in the Crab Nebula, further 2D work is of interest, for a different version of the shock model which shows a more interesting downstream particle spectrum.

The nature of the thermalization mechanism gives a clue to a shock model which does have nonthermal particle acceleration. As seen in the shock frame, the reflected particles form a gyrating ring with small momentum dispersion in the leading edge of the shock . Such rings are unstable — they form synchrotron masers (Zheleznyakov and Suvorov 1972, Yoon 1990, Hoshino and Arons 1991, Arons *et al* 1991). Calculations of the growth rate for a homogeneous pair plasma, along with estimates of the maximum harmonic number that can be generated, suggest that the dissipation observed in the pure pair plasmas is due to the formation of synchrotron masers, with consequent plasma heating as the extraordinary modes radiated are absorbed.

Now imagine what happens if the upstream flow contains heavy ions, as well as pairs. Since all the species flow into the shock with the same speed (the upstream Larmor radii are small compared to the overall flow scale in which the shock is embedded), $ZN_{1i} + N_{1+} = N_{1-}$. The ions have a rest mass large compared to that of the $e^\pm$. Therefore, they will contribute the dominant kinetic energy in the flow, if $N_{1i} > (m_\pm/m_i)(N_{1+} + N_{1-})$. As the plasma encounters the shock, the lighter pairs, which have Larmor radii a factor $m_\pm/m_i$ smaller than the ions, form a leading leptonic shock almost identical to the shock found in a symmetric plasma, if $N_{1i} \ll (N_{1+} + N_{1-})$. The more massive ions have no chance to respond, since the pair shock develops on the pair cyclotron time. They plow on through until they too are magnetically reflected in the enhanced magnetic field of the shock front. An electrostatic field in the direction of flow will be associated with this spatial separation of the ions from the leptons, with electrostatic potential having magnitude $Ze\Phi \sim m_i\gamma_1c^2$. Once the ions are set into gyration by the reflection from the shock front, they too are synchrotron maser unstable, as they gyrate in the much denser background of shock heated pairs.

Simulations of shock flows with ions (Hoshino, Arons, Gallant and Langdon 1991) support these conjectures. In Figure 3, I show the electromagnetic structure from a simulation done in the same manner as the simulations of the purely leptonic shocks, but now containing ions with a charge to mass ratio appropriate to protons, at $12\omega_{ci}^{-1}$ after the plasma bounced off the simulation wall. In order to keep the computer time finite, the mass ratio is small, $m_i/m_\pm = 20$, and the density ratio is $N_{1i}/(N_{1+} + N_{1-}) = 0.2$, chosen as a compromise to have $N_{1i} \ll (N_{1+} + N_{1-})$ but still have most of the kinetic energy density in the upstream flow. In this case, 80% of the plasma flow energy is in the ions. The upstream flow is almost entirely dominated by the plasma kinetic energy flux — $\sigma = 0.005$. While the choice of γ_1 matters only to the relative scale of the flow, in the interests of making the problem be of direct relevance to plerions and pulsars, we chose $\gamma_1 = 10^6$, corresponding to protons being accelerated through a few percent of the polar cap potential of the Crab pulsar. The presursor does not appear in this simulation, since the grid scale was equal to $0.2c/\omega_{c1\pm}$ based on upstream parameters, which leaves the precursor unresolved; in separate calculations we have shown that omitting the precursor does not affect the downstream particle acceleration. The electrostatic field in Figure 3 shows a strong double layer structure at the shock front, as the ions are first held back by the pairs frozen to the field, then begin gyrating and pull the pairs forward, while the potential reaches $\sim 0.5\gamma_1 m_i c^2/e$ in the double layer region. Behind the shock, the magnetic field settles down to a mean value of $B_2 \approx 3B_1$, as one would expect from the MHD jump conditions for such a plasma dominated flow, with intense extraordinary mode noise ($\delta B/B > 1$). In the simulation frame, the average transverse electric field is noise with a mean of zero, while in the shock frame (only approximately well defined, since the shock velocity is not strictly steady around its mean value of $c/3$), the mean transverse electric field is approximately constant. Note the large overshoot in the magnetic field and the long wavelength oscillations downstream. The first overshoot occurs because as the ions reflect from the shock front, their flow momentum is stored in an increase in the magnetic pressure and in the thermal pressure of the pairs.

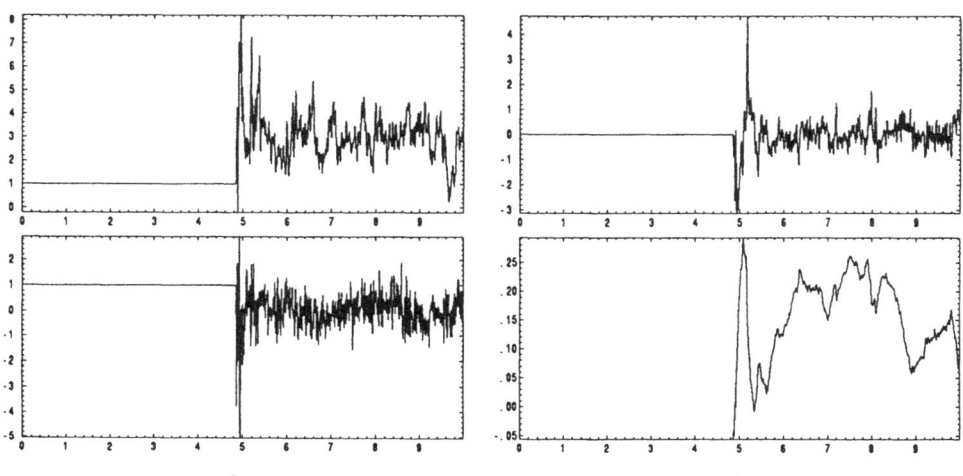

Figure 3: Electromagnetic structure of a relativistic magnetosonic shock in an electron-positron-proton plasma, at a time when the shock has crossed 50% of the simulation grid. Upper left panel: Magnetic field B_z, with upstream field amplitude equal to unity. Lower left panel: Transverse electric field E_y, with $E_y = (v_1/c)B_z$ at the injection wall. Upper right panel: Electrostatic field E_x. The downstream waves have electrostatic fields with amplitudes roughly one-half the magnitude of the electromagnetic field components, indicating strong elliptical polarization of the post-shock extraordinary modes. Lower right panel: Electrostatic potential in units of $\gamma_1 m_i c^2/e$. Since the potential never exceeds 0.3, the electrostatic field in the shock front acts to decelerate the ion stream, but the reflection is magnetic, as in the shocks in pure pair plasmas. The unit of length is the ion Larmor radius based on upstream prameters.

222

Behind the overshoot, the ions continue gyrating quasi-coherently, creating compressional enhancements at each turning point, until their gyration finally thermalizes. These oscillations are
spaced roughly one ion Larmor radius $r_{L2} = m_i c^2 \gamma_1 / e B_2$ apart.

In Figure 4 I show zoomed views of the $p_x - x$ projections of ion phase space for a region of
the simulation near the shock front itself, at 12 and 13 ion cyclotron times after the simulation
began. One sees the quasi-coherent reflection of the ions, followed by their rapid thermalization.
The figure also shows the unsteady character of the shock structure. On this scale, the structure
of the pair shock, with its initial thermalization to Maxwellian distributions, is infinitesimal,
achieved as the ions have barely begun to gyrate. Figure 5 shows the spectra of the

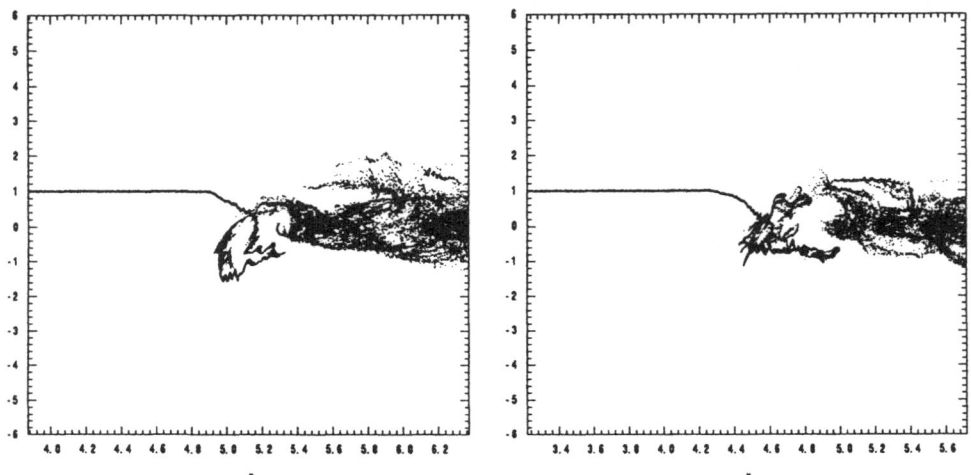

*Figure 4: Ion phase space near the shock front. Left panel: Flow momentum in units of $10^6 m_i c$
versus x for ions near the shock front at t = 12/ω_{cil}. The unit of lenght is the ion Larmor radius
based on upstream parameters. Note the initial deceleration of the ion stream in the electrostatic
field, followed by the rapidly dissolving loop in the magnetic field. The main body of an ion
Maxwellian distribution is produced within 2–3 Larmor times after the ions begin decelerating in
the shock front. Right panel: ion flow momentum versus x at t = 13/ω_{cil} near the shock front.
These snapshots illustrate the unsteady character of the shock.*

downstream electrons and positrons. The positrons are no longer everywhere Maxwellian in
energy space, but now have a power law distribution with $N(E) \propto E^{-1.7}$ at energy/particle
exceeding $\gamma_1 m_\pm c^2$, and have energy density comparable to the ions' kinetic energy density in the
upstream flow. In this simulation, the electrons also show some sign of power law behavior at
energies exceeding $\gamma_1 m_\pm c^2$. The positron spectrum is Maxwellian at energies below $\gamma_1 m_\pm c^2$, and
cuts off at positron energy equal to that of the upstream energy/particle of the protons, $\gamma_1 m_i c^2$.
Thus, a relativistic, low σ magnetosonic shock, composed of ions, electrons and positrons, with
$N_{1i} \ll N_{1+} + N_{1-}$ but with the upstream flow energy dominated by the ions, can accelerate
the downstream positrons into a power law distribution with high efficiency, with the slope of

the power law being very close to what is desired for the macroscopic model. This particular simulation is highly resolved, with 64 particles per grid cell per species in the initial state. In other calculations of the shock structure (Hoshino *et al* 1991) with various values of σ and of γ_1, and in simulations of the synchrotron maser process in isolation in a uniform medium (Hoshino and Arons 1991), we have found the slope to be between 1 and 2. It is likely that 2D effects (primarily pitch angle scattering due to Weibel "turbulence") will steepen the spectrum somewhat. In any case, it is clear that these electron-positron-proton shocks do produce downstream positron spectra whose characteristics are highly desirable for the modeling of plerions.

Downstream Particle Spectra

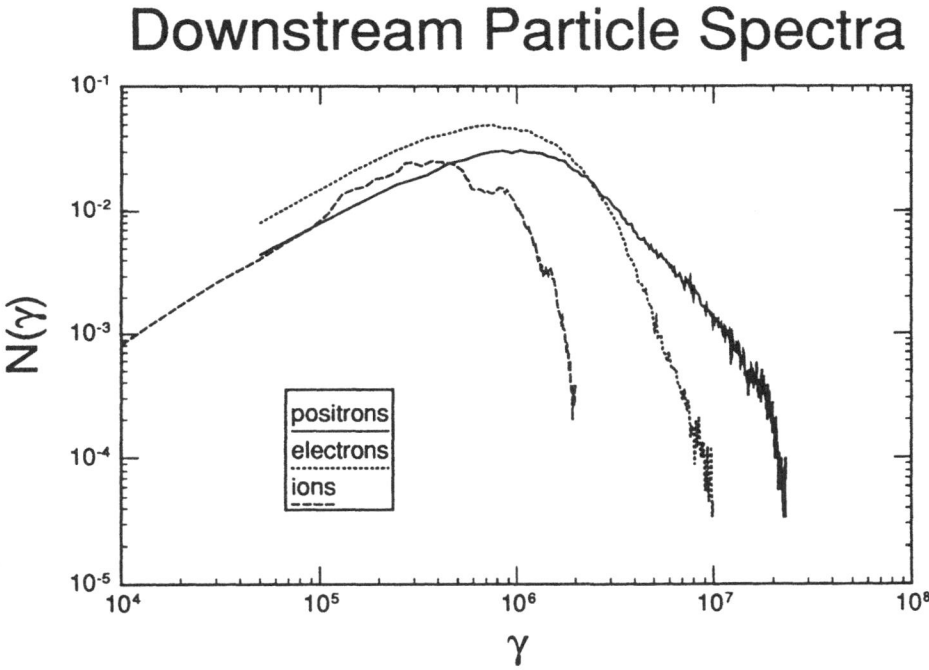

Figure 5: Post-shock distributions. The positrons have a high energy power law spectrum with slope s = 1.7, while the electrons show evidence for a power law supra-therma distribution with slope s = 3.7. The upper cutoff occurs where the energy/particle of the pairs equals the upstream energy/particle of the ions, as is expected from cyclotron resonant absorption of extraordinary modes emitted at and above the upstream ion cyclotron frequency.

The physics of this acceleration is simple. In this non-symmetric plasma, the extraordinary modes are no longer linearly polarized electromagnetic waves but have a noticeable electrostatic component as they propagate across the magnetic field. For frequencies between the relativistic analog of the lower hybrid frequency and the relativistic cyclotron frequency of the heated pairs as they emerge from the leading leptonic shock, the extraordinary modes behave like magnetosonic waves in the pair plasma, and have *elliptical* polarization with the electric vector rotating in the left handed sense with respect to the magnetic field. This is the same sense as the Larmor gyration of the positrons and of the ions. As a result, a positron can linearly absorb the magnetosonic modes at their cyclotron resonance (as well as at higher harmonics) and can gain energy. Because the polarization is elliptical, there is some power in right circularly polarized fields, so the electrons also can gain energy nonthermally from the ions. Because of the small mass ratio used in the simulations, the power in left handed circularly polarized fields emitted by the ions is much larger than in right handed circular polarization, so the positron acceleration is much stronger than the acceleration of electrons. Under realistic conditions, we believe the acceleration of electrons and positrons will be much more comparable (Hoshino *et al* 1991). It is possible to simulate this flow with mass ratios as high as 100, which gives much larger dynamic range in the particle spectra and should show more comparable electron and positron acceleration; such simulations are in progress.

This mechanism works efficiently only if the collective synchrotron emission of the ions yields extraordinary mode frequencies greater than or equal to the thermal cyclotron frequency of the positrons just behind the leading leptonic subshock. The linear theory shows this is possible only if the upstream thermal momentum dispersion of the ions is small compared to $m_\pm/m_i$. This is probably not a severe restriction, both because the acceleration of the ions is likely to be by an electromagnetic mechanism (Arons 1983), with the ion energy essentially unique on each streamline. In any case, adiabatic expansion in the outflow reduces any initial thermal dispersion drastically by the time the shock is reached. To the degree this shock acceleration model works successfully, this constraint on the acoustic Mach number of the flow should be regarded as a restriction on any theory of the wind. All of the simulations easily satisfy this requirement.

That this heating process can and should result in power law positron spectra can be shown by an elementary "quasi-linear" argument. If the ion gyration is modeled as a simple homogeneous ring in phase space, while the pairs are modeled as a hot homogeneous plasma, one finds an ion growth rate almost independent of frequency, up to an upper cutoff related to the momentum dispersion of the ion ring. If the positrons have a power law spectrum $N_+(\gamma) \propto \gamma^{-s}$, the linear damping rate at which positrons can cyclotron absorb the magnetosonic modes can be equal to the rate the ions emit the extraordinary modes at all frequencies only if $s = 1$. Such detailed balance cannot be achieved at all wave frequencies if the positrons have a Maxwellian distribution. This steady state, "on the spot" argument overdoes the efficiency of the acceleration process — in the "real" world of the simulations, the steady decline of the ions' energy needs to be incorporated in a theory. Nevertheless, the argument does support the appearance of power laws in the positrons as a consequence of the ion maser process. Furthermore, it does suggest that quasi-linear theory is applicable to the post-shock heating process, in spite of the large amplitude of the waves, since the total energy is almost entirely in the particles and even these large amplitude waves can make only moderate disturbances of the particles' orbits.

3. Application to the Crab Nebula and to Pulsars in Compact Binaries

We are in the middle of constructing models of astrophysical systems using these results. One can get the flavor of these from the following simple considerations. Suppose the magnetosphere of the Crab pulsar emits a dense flow of pairs, with $\dot{N}_\pm \sim 10^{38} \dot{N}_{\pm,38}$ coming out in the broad sector around the rotational equator where the pulsar must emit the wind required to feed the torus of X- and gamma-ray emission in the Nebula — for the nebular geometry and required particle injection rate, see Aschenbach and Brinkmann (1975) and Pelling *et al* (1987). A number of magnetospheric models, with varying quantitative credibility, suggest such a flow is possible (Arons 1983, Beskin *et al* 1983). In addition, suppose the magnetospheric electrodynamics causes an outflow of heavy ions to be pulled up from the stellar surface and accelerated to high energy in the outflow around the magnetic equator (Arons 1983 seems to have made the only suggestion of this sort, so far). The particle flux is constrained by the electrodynamics to be close to the Goldreich-Julian flux, $\sim 4 \times 10^{34} Z^{-1}$ for this pulsar, where Z is the charge on the ions. The resulting ratio of the mass densities in the wind is $m_i N_i / m_{\pm} n_{pair} \approx 0.7 A / Z \dot{N}_{\pm,38}$, where A is the atomic number of the ions. The most likely candidate is fully stripped iron, with $A/Z \approx 2.2$, yielding approximate equipartition between the components of the upstream pair energy. While our simulations so far have been done for somewhat higher ratios of the ion flow energy to that in the pairs, we do not expect major differences for these smaller values.

These ions might be accelerated through a fair fraction of the total polar cap potential $\Phi_{cap} \sim 10^{17}$ Volts, corresponding to $\gamma_1 \sim 10^8 (\Phi/\Phi_{cap})$. The equatorial wind with these accelerated ions carries a fraction Φ/Φ_{cap} of the spindown luminosity of the pulsar. If the shock really puts essentially all the energy into the accelerated positrons, as our simulations suggest, and if these positrons radiate essentially all of this in X-rays and gamma-rays, as is indicated by the macroscopic flow models, then one would conclude, from the hard photon luminosity of the nebula, that the ions experience an acceleration potential $\Phi/\Phi_{cap} \sim 0.3/I_{45}$, where I_{45} is the moment of inertia of the neutron star in units of 10^{45} cgs.

We can go much further. From momentum conservation, one expects the shock to occur roughly $10^{17.5}$ cm from the pulsar (Rees and Gunn 1974). Optical observations reveal the presence of the curious "wisps" in the surface brightness in this region (Scargle 1969), a series of enhancements in the surface brightness elongated in the direction perpendicular to the flow lying to the Northwest of the pulsar; there is also a fainter wisp to the Southeast (van den Bergh and Pritchett 1989). The broad band synchrotron emissivity in the magnetic overshoot and downstream magnetic compressions shown in Figure 3 is readily found to be proportional to the $B^{11/3}$, assuming the downstream pitch angle scattering has largely isotropized the pairs. The essential new astrophysical idea is to identify the enhanced surface brightness expected in these compressions with the observed wisps.

From this identification, one can extract a host of conclusions. The details of these will be presented elsewhere (Arons, Gallant, Hoshino and Langdon, in preparation). However, several qualitative statements can be made immediately. 1) A reasonable model, with $\sim 10^{38}$ pairs/s flowing out in the wind and perhaps 10^{34} iron ions/s, yields both a roughly correct optical surface brightness for the thin wisp and wisp 1, the observed spacing between the thin wisp and wisp 1, and the thickness of wisp 1 itself, if $\gamma_1 \sim 10^7$. This is a particle outflow rate and flow energy/particle *roughly* consistent with the values inferred above for the gross energetics of the nebular emission. 2) The series of downstream, large amplitude compressions of the plasma are spaced roughly one ion Larmor radius apart. This spacing is similar to that seen between the first few wisps in van den Bergh and Pritchett's image if $\gamma_1 \sim 10^7$. 3) The spectrum of wisp 1 should be close to that of Maxwellian synchrotron emission; see Jones and Hardee (1979) for a detailed computation of these spectra in a uniform medium. With $B \sim 10^{-4}$ Gauss and pair temperature $\sim 10^7 m_{e\pm} c^2$, the characteristic exponential decline of the spectrum from a relativistic Maxwellian should appear around ~ 100 eV. 4) Successive wisps further from the ion overshoot should show a progressively more nonthermal spectrum, as the ions transfer their energy to the positrons. 5) Because the pair Larmor radius is very small (~ 10 AU for $B \sim 10^{-4}$ Gauss and $\gamma_1 \sim 10^7$), we identify the observed thickness of the thin wisp [$\sim 10^{16.5}$ cm in the image obtained by van den Bergh and Pritchett (1989)] as being due to the projection of an infinitesimally thin sheet projected on the plane of the sky, as one expects from the flow geometry implied by the Aschenbach and Brinkmann (1975) model of the X-ray emission in the nebula. 6) The flow velocity of the pairs throughout this region is between 0.3c and 0.5c, which provides a Doppler asymmetry to the luminosity just about right to explain the brightness ratio seen between wisp 1 and the one faint wisp observable to the pulsar's southeast. Possible partial anisotropy of the pairs with respect to the local magnetic field enhances this conclusion. 7) The shocks observed in our simulations are unsteady, as is shown in Figure 4. Translated into the shock frame, which is where we observe the Crab Nebula, the unsteady shock velocity becomes oscillation of the position of the various ion overshoots on a time scale of several months, a phenomenon consistent with time variability of the wisps reported long ago by Scargle (1969). Such variability should be uncorrelated with pulsar glitches and other rotational anomalies. 8) Finally, the parameters derived from the wisp model seem appropriate to those expected in the diffuse X-ray source formed further downstream. The detailed modeling in progress is designed to test whether these rough numerical conclusions stand up to careful analysis of the flow and predicted surface brightness distribution.

Therefore, I suggest that the structure of a relativistic shock in a relativistic wind composed of pairs plus heavy ions has its spatial structure revealed to us in the detailed morphology of the wisp region in the Crab Nebula. That this may be so is a consequence of the very high rigidity of the outflows expected from pulsars, an expectation based on the enormous voltages developed in the magnetospheres of these objects combined with the negligible radiation drag on the flow near these photon poor compact objects.

The same ideas can be applied to the millisecond radio pulsars recently discovered in compact binaries (Fruchter et al 1988, Lyne et al 1990), whose relativistic winds interact with the mass lost from the "normal" stellar companion (Kluzniak et al 1988, Phinney et al 1988, Ruderman et al 1989). If one assumes the same fraction of the potential drop on open field lines goes into the energy/particle of these winds as appears in the wind from the Crab pulsar, one expects $\gamma_1 \sim 5 \times 10^4$, while if σ is as small as it appears to be in the Crab pulsar's wind, the magnetic field behind the termination shock of the relativistic wind is on the order of 1 Gauss, since the shock occurs less than one solar radius from these pulsars, at least along the line of centers between the pulsar and the star. If synchrotron losses were the only process competing with the cyclotron resonant acceleration of the positrons and the electrons, the particles would accelerate to energies on the order of $10^8 m_{\pm} c^2$ and much of the wind energy would be given up as 100 MeV synchrotron photons, whose flux at earth might be as high as 10^{-4} photons/cm^2-s (Hoshino et al 1991). However, these systems differ from a diffuse plerion like the Crab in a fundamental sense. Because the standing relativistic shock occurs near to a (small) star, the optical and infrared radiation from the star provide an energy density in the shock region perhaps 100 times that of the magnetic field in and behind the relativistic shock. Therefore, inverse Compton radiation forms the dominant incoherent energy loss, and may occur at a rate competetive with the resonant cyclotron acceleration of the pairs. If so, the pairs are limited to energies well below $\gamma_1 m_i c^2$, although the energy lost still is likely to be largely in the form of photons with energy above 10 MeV. The implications of this change in energy loss mechanism for the emission properties of these "compact plerions" is under investigation.

4. Acknowledgments

The ideas and results described in this paper come from close collaboration with Yves A. Gallant,

Masahiro Hoshino, A. Bruce Langdon and Claire E. Max, and include work done by all of us. Our research on relativistic shock waves was supported in part by NSF grant AST-8615816 and by IGPP-LLNL grant 90–14, both to the University of California at Berkeley. It was also supported by NASA astrophysical theory grant NAGW-1301. Part of the work was performed under the auspices of the U.S. Department of Energy at the Lawrence Livermore National Laboratory under contract W-7405–Eng-48.

5. References

Alsop, D., and Arons, J. 1988, "Relativistic Magnetosonic Solitons with Reflected Particles in Electron-Positron Plasmas", *Phys. Fluids*, **31**, 839

Arons, J. 1983, "Electron-Positron Pairs in Radio Pulsars", in *Proc. Workshop on Electron-Positron Pairs in Astrophysics*, M.L. Burns, A.K. Harding and R. Ramaty, eds. (New York: American Institute of Physics), 163

————, Hoshino, M. and Gallant, Y.A. 1991, "Synchrotron Instability, Absorption and Suprathermal Particle Acceleration in Relativistic, Magnetosonic Shock Waves", to be submitted to *Ap.J.*

Aschenbach, B., and Brinkmann, W, 1975, "A Model of the X-Ray Structure of the Crab Nebula", *Astron. and Ap.*, **41**, 147

Begelman, M.C., and Kirk, J. 1990, "Shock Drift Acceleration in Superluminal Shocks: A Model for Hotspots in Extragalactic Radio Sources", *Ap.J.*, **353**, 66

Bell, A.R. 1978, "The Acceleration of Cosmic Rays in Shock Fronts", *Mon. Not. Roy. Astron. Soc.*, **182**,147

Beskin, V.S., Gurevich, A.V., and Istomin, Ya. N. 1983, "Electrodynamics of Pulsar Magnetospheres", *Zh. Eksp. Teor. Fiz.*, **85**, 401 (*Soviet Physics — JETP*, **58**, 235)

Birdsall, C., and Langdon, A.B. 1985, *Plasma Physics via Computer Simulation* (New York: McGraw-Hill)

Chiueh, T. 1989, "Relativistic Solitons and Shocks in Magnetized $e^- - e^+ - p^+$ Fluids", *Phys. Rev. Lett.*, **63**, 113

Ellison, D.C., Jones, F.C., and Reynolds, S.P. 1990, "First-Order Fermi Particle Acceleration by Relativistic Shocks", *Ap.J.*, **360**, 702

Emmering, R.T, and Chevalier, R.A. 1987, "Shocked Relativistic Magnetohydrodynamic Flows with Application to Pulsar Winds", *Ap. J.*, **321**, 334

Fruchter, A. S., Stinebring, D. R., and Taylor, J. H., 1988, "A Millisecond Pulsar in an Eclipsing Binary", *Nature*, **333**, 237

Gallant, Y.A., Hoshino, M., Langdon, A.B., Arons, J., and Max, C.E. 1991, "Structure of Relativistic Magnetosonic Shock Waves in Electron-Positron Plasmas", submitted to *Ap. J.*

Gold, T. 1969, "Rotating Neutron Stars and the Nature of Pulsars", *Nature*, **221**, 25

Hoshino, M., and Arons, J. 1990, "Differential Heating and Acceleration of Positrons by Synchrotron Maser Instabilities", *Phys. Fluids B*, **3**, 818

————, Arons, J., Gallant, Y.A., and Langdon, A.B. 1991, "Relativistic, Magnetized Electron-Positron-Proton Shock Waves in Synchrotron Sources: Shock Structure and Non-Thermal Acceleration of Positrons", submitted to *Ap. J.*

Jones, T. W., and Hardee, P.E. 1979, "Maxwellian Synchrotron Radiation", *Ap.J.*, **228**, 268

Kennel, C.F., and Pellat, R. 1976, "Relativistic Nonlinear Waves in a Magnetic Field", *J. Plasma Phys.*, **15**, 335

Kennel, C.F., and Coroniti, F.V. 1984a, "Confinement of the Crab Pulsar's Wind by its Supernova Remnant", *Ap. J.*, **283**, 694.

———. 1984b, "Magnetohydrodynamic Model of Crab Nebula Radiation", *ibid.*, 710

Kluzniak, W., Ruderman, M., Shaham, J., and Tavani, M. 1988, "Nature and Evolution of the Eclipsing Millisecond Binary Pulsar PSR1957+20", *Nature*, **334**, 225

Kundt, W., and Krotscheck, E. 1980, "The Crab Nebula — a Model", *Astron. and Ap.*, **83**, 1

Langdon, A.B., and Lasinski, B.F. 1976, "Electromagnetic and Relativistic Plasma Simulation Models", in *Methods in Computational Physics*, **16**, B. Alder, S. Fernbach and M. Rotenberg, eds. (New York: Academic Press), 327

Langdon, A.B., Arons, J., and Max, C.E. 1988, "Structure of Relativistic Magnetosonic Shocks in Electron-Positron Plasmas", *Phys. Rev. Lett.*, **61**, 779

Lyne, A. G., Manchester, R. N., D'Amico, N., Stabeley-Smith, L., Johnston, S., Lim, J., Fruchter, A. S., Goss, W. M., and Frail, D. 1990, "An Eclipsing Millisecond Pulsar in the Globular Cluster Terzan 5", *Nature*, **347**, 650

Michel, F.C., Scowen, P.A., Dufour, R.J., and Hester, J.J. 1991, "Observations of a Pulsar Wind: CCD Polarimetry of the Crab Nebula", *ApJ.*, **368**, 463

Pacini, F. 1967, "Energy Emission from a Neutron Star", *Nature*, **216**, 567

Pelling, R.M., Paciesas, W.S., Peterson, L.E., Makashima, K., Oda, M., Ogawara, Y., and Miyamoto, S. 1987, "A Scanning Modulation Collimator Observation of the High Energy X-Ray Source in the Crab Nebula", *Ap. J.*, **319**, 416

Phinney, E. S., Evans, C. R., Blandford, R. D., Kulkarni, S. R. 1988, "Ablating Dwarf Model for Eclipsing Millisecond Pulsar 1957+20", *Nature*, **333**, 832

Piddington, J.H. 1957, "The Crab Nebula and the Origin of Interstellar Magnetic Fields", *Aust. J. Phys.*, **10**, 530

Quenby, J., and Lieu, R. 1989, "Enhanced Shock Acceleration in Relativistic Jets and Cosmic ray Origin in Active Galactic Nuclei", *Nature*, **342**, 654

Rees, M.J., and Gunn, J.E. 1974, "The Origin of the Magnetic Field and Relativistic Particles in the Crab Nebula", *Mon. Not. Roy. Astron. Soc.*, **167**, 1

Ruderman, M., Shaham, J., and Tavani, M. 1989, "Accretion Turnoff and Rapid Evaporation of Very Light Secondaries in Low Mass X-Ray Binaries", *ApJ.*, **336**, 507

Scargle, J. D., 1969, "Activity in the Crab Nebula", *Ap. J.*, **156**, 401

van den Bergh, S., and Pritchett, C.J. 1989, "The Crab Synchrotron Nebula at 0.5″ Resolution", *ApJ. (Lett.)*, **343**, L69

Yoon, P. 1990, "Amplification of a High-Frequency Electromagnetic Wave by a Relativistic Plasma", *Phys. Fluids B*, **2**, 867

Zheleznyakov, V.V., and Suvorov, E.V. 1972, "Results and Problems in the Investigation of the Synchrotron Instability", *Ap. and Space Sci.*, **15**, 24

Neutrino Driven Neutron Star Formation

J. Cernohorsky

Max–Planck Institut für Astrophysik, Karl Schwarzschild–Straße 1, D–8046 Garching, F.R.G.

Summary– Two neutron star evolution simulations are presented. The ν_e and $\bar{\nu}_e$ transport is calculated with the Flux–limited Neutrino Diffusion Theory (FNDT) transport code[1, 2, 3]. The transport is spectral and monochromatic. The evolution of two proto–neutron star models is simulated for 30 and 10 sec, respectively. In the first stellar model, without nuclei in its baryonic composition, the neutrino flows deposit a substantial amount of energy, about $3 \ 10^{49} \ erg$, during the first tenths of a second at matter densities around and below $5 \ 10^{10} g \ cm^{-3}$. Initially, this causes the star to expand. In the second model nuclei are added to the composition of the star in the relevant regions to investigate the qualitative effect of the modified opacities on the deposition. It is found that the inclusion of nuclei impedes the expansion. Nevertheless, the energy deposition slows down the contraction of the atmosphere in both models considerably, and over many seconds. After about 1 sec a Rayleigh–Taylor unstable density inversion develops at the edge of the dense inner core at $\rho \approx 10^{13} g \ cm^3$ in both models. This density inversion is unstable against convection as well. Both the energy deposition in the outer regions of the star and the density inversion at the edge of the inner core may have some effect on the working of the delayed–explosion mechanism [3].

References

[1] J.Cernohorsky, L.J. van den Horn and J. Cooperstein, *JQSRT*, **42**, Vol. 6, 603 (1989)

[2] J. Cernohorsky and L.J. van den Horn *JQSRT*, **43**, Vol. 1, 33 (1990)

[3] J.Cernohorsky PhD. thesis, University of Amsterdam, (1990)

E. P. J. van den Heuvel and S. A. Rappaport (eds.), X-Ray Binaries and Recycled Pulsars, 229.
© 1992 *Kluwer Academic Publishers.*

CHAPTER 5

Formation and Evolution of
Neutron Star Binaries & Millisecond Pulsars

FORMATION AND EVOLUTION OF NEUTRON STAR BINARIES

E.P.J. VAN DEN HEUVEL
Astronomical Institute
& Center for High Energy Astrophysics
University of Amsterdam
Kruislaan 403, 1098 SJ Amsterdam
The Netherlands
and
Institute for Theoretical Physics
University of California
Santa Barbara, CA 93106

ABSTRACT. The High Mass X-ray Binaries (HMXB) represent a normal stage in the evolution of high-mass binary systems, since due to the effects of mass transfer high-mass binaries in general are not expected to be disrupted by the supernova explosion of the initially most massive component. The formation rate of neutron stars in high-mass binaries in the Galaxy is estimated to be $(3-4,5) \times 10^{-3}$/yr. Roughly half of these systems will later on evolve into "standard" HMXB, the other half into Be/X-ray binaries. The number of Be/X-binaries in the Galaxy is estimated to be about $2 \cdot 10^4$.

During their later evolution all standard HMXB and about half of the Be/X-ray binaries are expected to merge and form Thorne-Zytkow stars. In the other half of the Be/X-ray binaries (systems with $P \gtrsim 100^d$) the neutron star will not completely spiral into its companion and a very close binary consisting of the neutron star and the heavy-element core of its companion will remain. Since the latter core is in most cases less massive than $4M_\odot$, its collapse to a neutron star, in the second supernova explosion, will in most cases not unbind the systems, and a close eccentric-orbit neutron star binary with an orbital period $1,5^H - 12^H$ will remain. These systems will later on merge due to the emission of gravitational radiation. Although most double neutron-star systems are expected to be probably formed at orbital periods $< 7^H$, the short orbital decay timescale of these systems will make that binary pulsars with $P > 7^H$ will outnumber those with $P < 7^H$ in a ratio of more than two to one.

The birthrate of these double neutron star binaries (which will all merge within a few billion years after their birth) in the Galaxy is conservatively estimated to be $(4-8) \times 10^{-5}$ yr^{-1}, with a lower limit of $(1-2) \times 10^{-5}$ yr^{-1} and an upper limit of $(2-4) \times 10^{-4}$ yr^{-1}. Although a sizeable number of neutron star-black hole binaries and some black hole-black hole binaries will be formed, the long orbital periods expected for these systems will prevent them from merging within a Hubble time, making them probably uninteresting as sources of bursts of gravitational radiation.

The evolution of Low-Mass Binaries containing neutron stars is discussed, with an emphasis on the mechanisms driving the mass transfer and on the final fate of these systems.

E. P. J. van den Heuvel and S. A. Rappaport (eds.), X-Ray Binaries and Recycled Pulsars, 233–256.
© 1992 *Kluwer Academic Publishers.*

1. Introduction

The binary neutron stars come basically in four varieties: the high- and low-mass X-ray binaries (here-after designated as HMXB and LMXB), schematically depicted in Figure 1, and the two types of binary radio pulsars, schematically depicted in Figure 2 (after van den Heuvel and Taam 1984).

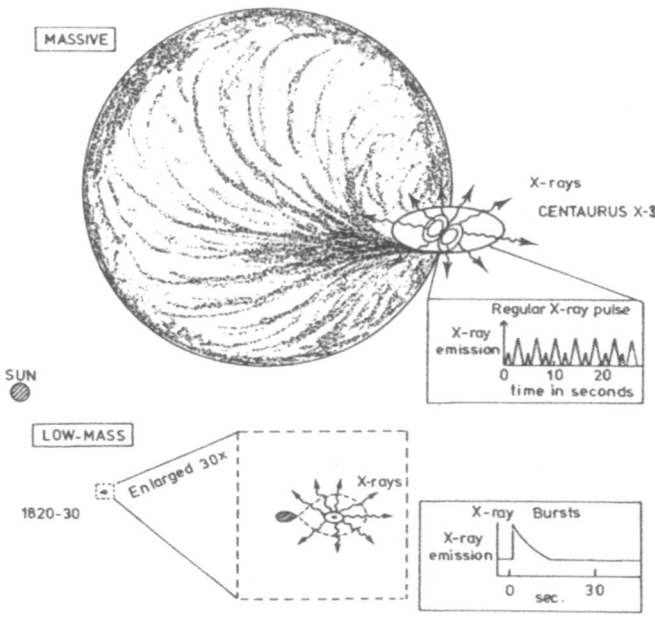

Figure 1. The two basic types of accreting binary neutron star systems in the Galaxy. The High Mass X-ray Binaries (top) consist of a massive (8 to $40M_\odot$) star together with a neutron star. In almost all systems the neutron star is a regular X-ray pulsar, indicating that it has a relatively strong surface dipole magnetic field ($B \gtrsim 10^{11}G$) which channels the inflowing matter towards the magnetic poles. In the Low Mass X-ray Binaries (lower panel) the neutron star has a companion of mass $\lesssim 1, 2M_\odot$. These sources almost never show regular X-ray pulsations, but many systems show thermonuclear X-ray bursts. Characteristic representatives of each type of system are Centaurus X-3 and XB 0820-30, respectively, depicted here to scale with the sun. See further the text and Table 1.

Although both the HMXB and LMXB appear to contain neutron stars, the two groups show a number of distinct differences, listed in Table 1, indicating that they belong to different populations, with presumably a different evolutionary history. The HMXB, having companion stars more massive than $\sim 8 - 10M_\odot$, belong to the youngest stellar population of our Galaxy (age $\lesssim 10^7$ yrs), and in most cases their neutron stars have strong magnetic fields ($> 10^{11}$ G) as evidenced by the presence of regularly periodic modulations of their X-ray emission, with periods ranging from 0,069 s to 835 s.

On the other hand the LMXB belong to a generally much older stellar population, as evidenced by their concentration in the Galactic bulge and in globular star clusters. The absence of regular X-ray pulsations in most of them and the occurrence of the thermonuclear type I X-ray bursts in many, shows that surface dipole magnetic fields, if present, should be weaker than $\sim 10^{11}$G (*cf.* Lewin and Joss 1983).

TABLE 1

The two groups of strong Galactic X-ray Sources.

Group I (HMXB)	Group II (LMXB)
–Optical counterparts massive and luminous early type stars, spectrum O and early B; $L_{opt}/L_x > 1$.	Faint blue optical counterparts $L_{opt}/L_x < 0,1$
–Concentrated in space towards the galactic plane: young stellar population, age $< 10^7$ years	Concentrated in space towards the galactic center; fairly wide spread around the galactic plane: old stellar population, age $(5 - 15) \times 10^9$ years
–Type of time variability: Regular X-ray pulsations; no X-ray bursts	Type of time-variability: often X-ray bursts; only in 3 cases regular X-ray pulsations.
–Relatively hard X-ray spectra: $kT \gtrsim 15$KeV	Softer X-ray spectra: $kT \lesssim 10$KeV

Also the binary radio pulsars appear to fall into two distinct classes, as depicted in Figure 2:

(i) The PSR 1913+16 class, consisting of systems with relatively narrow and in most cases very eccentric orbits, and relatively massive companion stars: $M \approx 0,8 - 1,45M_\odot$;

(ii) the PSR 1953+29 class, consisting of systems with, on average, much wider and practically circular orbits, with companions of low mass: $M \approx 0,2 - 0,3M_\odot$.

We would like in this paper to address the following topics:

- The formation and later evolution of the HMXB, and the incidence of their descendants (recycled single, and binary radio pulsars) in the galaxy.
- The mechanisms driving the mass transfer in LMXB systems and (briefly) the possible fate of such systems. The final fate of LMXB is dealt with in more detail in this volume

236

in the papers by Bhattacharya, Shaham, Kluzniak and Tavani.

As to the origins and formation mechanisms of the LMXB systems, I refer to the article in this volume by Webbink.

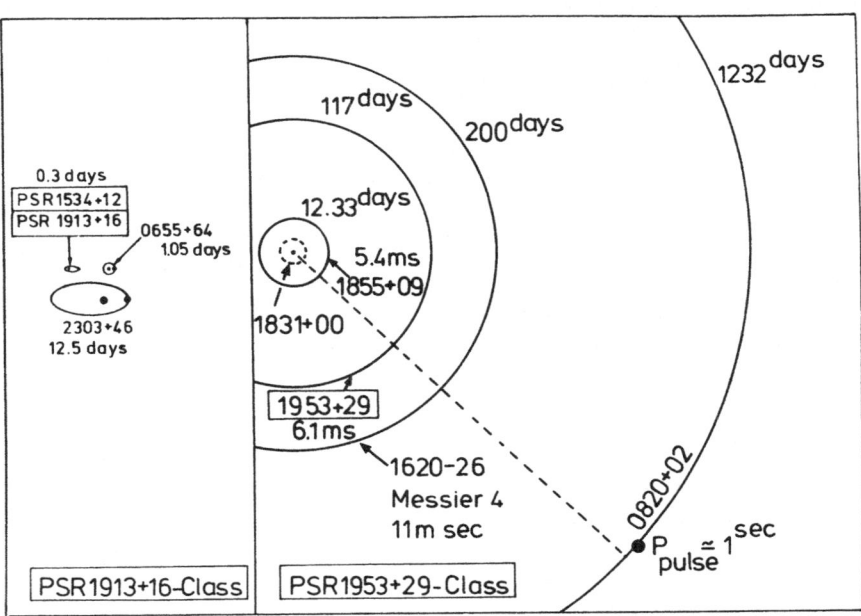

Figure 2. The two classes of binary radio pulsars. Left: The PSR 1913+16-class of systems tend to have narrow and very eccentric orbits; the companion of the pulsar is itself a neutron star or a massive white dwarf. Right: The PSR 1953+29-class systems tend to have wide and circular orbits; here the companion stars always have a low mass, in the range $0, 2 - 0, 4 M_\odot$, or even smaller – and most probably are helium white dwarfs (see text).

2. Origins and Fate of the HMXB

2.1. INTRODUCTION

The general course of the evolution of a HMXB up to the formation of the neutron star is well understood and has been extensively reviewed elsewhere (*e.g.*, see van den Heuvel 1976, 1977, 1978, 1983, 1989, Verbunt 1989). The reasons why the systems were not disrupted in the supernova explosion in which the neutron star was formed is (van den

Heuvel and Heise 1972): the large-scale mass transfer that takes place in a close binary system prior to the supernova explosion. This causes the more evolved component of the system to be less massive than its companion at the time of the explosion (see, for example, the evolution of a system initially consisting of stars of 13 and $6.5M_\odot$, depicted in Figure 4). In such a situation one does not expect the binary to be disrupted by the explosive mass ejection, even if the effects of the impact of the SN-shell on the companion, and asymmetries (random kicks with $v \simeq 100 - 150$km/s) in the explosion are taken into account (cf. Blaauw 1961; Sutantyo 1975; de Cuyper 1985; Fryxell and Arnett 1981; Bailes 1989).

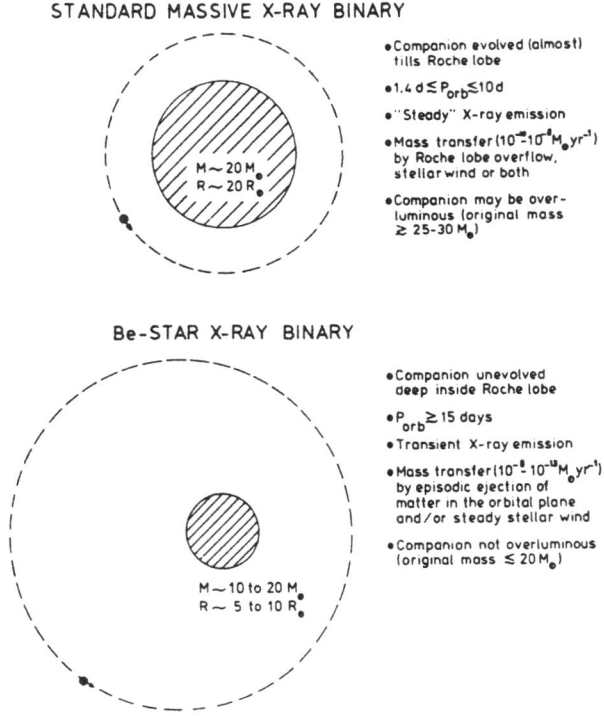

Figure 3. Differences between the standard and B-emission-type high-mass X-ray binaries. Explanation in the text.

2.2. THE TWO TYPES OF HMXB

Looking in more detail, the HMXB present a number of interesting characteristics that require an explanation. The most important one of these is the broad division of the HMXB into two groups, the so-called "standard" HMXB and the B-emission/X-ray binaries, which show a number of distinctively different characteristics summarized in Figure 3. In the standard systems the companion stars tend to be massive (in general $\gtrsim 18 - 20M_\odot$), the orbital periods tend to be short ($1.4^d - 10^d$ in all but one case), and the companions tend to be evolved stars, near the end of core-hydrogen burning or beyond (in the blue supergiant phase) and almost or completely filling their Roche lobes.

On the other hand, in the Be/X-ray binaries we are dealing with companion stars of lower mass ($8 - 15M_\odot$), the orbital periods are long (15^d up to several years) and the companion stars, which are relatively unevolved rapidly rotating main-sequence early B-type stars, are deep inside their Roche lobes. The mass transfer is not steady here, but is due to irregular outbursts of mass ejection from the equatorial regions of the rapidly rotating Be star. The driving mechanism of these outbursts is not known but is undoubtedly related to the rapid rotation of the Be star. During such an outburst the neutron star suddenly appears as a strong (often: regularly pulsating) transient X-ray source for a period of weeks to months (*cf.* Maraschi *et al.*1975; van den Heuvel and Rappaport 1987), and can be off for decades in between.

2.3. REASONS FOR THE DIFFERENCE IN ORBITAL CHARACTERISTICS BETWEEN THE "STANDARD" AND B-EMISSION HMXB: CONSERVATIVE EVOLUTION DURING THE FIRST PHASE OF MASS TRANSFER

The difference between the orbital characteristics of the "standard" and Be/X-ray binaries is most probably due to a combination of (i) conservation of mass and orbital angular momentum in the binary systems during the first phase of mass transfer, and (ii) the increase with stellar mass of the fractional mass of the hydrogen-burning core and thus: of the helium core produced. This can be shown as follows.

In the majority of the unevolved massive close binaries the orbital periods are so long ($\geq 1 - 2^d$) that the primary star (the more massive component) will fill its Roche lobe and begin to transfer matter to its companion not before it has terminated core-hydrogen burning. In such a case (so-called "case B" evolution, *cf.* Paczynski 1971) the primary star will transfer practically all of its hydrogen envelope to its companion and only its helium core will be left after the transfer (see Figure 4 for an example). If one assumes the total mass and orbital angular momentum of the system to be conserved during the mass transfer (so-called "conservative" evolution) the change in orbital period due to the transfer is simply given by (*cf.* Paczynski 1971):

$$P^f/P_0 = \left\{ M_1^0(M - M_1^0)/(M_1^f(M - M_1^f)) \right\}^3 \qquad (1)$$

where subscript and superscripts f and 0 denote the final and initial situation, respectively; M is the total mass of the system and subscript 1 denotes the original primary star.

The relative mass fraction of the helium core M_1^f/M_1^0 increases considerably with increasing stellar mass M_1^0. For example, if one includes convective overshooting (*cf.* Chiosi and Maeder 1986) a star of $30M_\odot$ develops a helium core of mass $12 - 15M_\odot$, whereas a star of $M_1^0 = 12M_\odot$ develops a core of only $2, 5 - 3, 0M_\odot$, which later on, during core carbon burning will transfer another $0, 3 - 0, 5M_\odot$ to its companion, leaving a $2, 2 - 2, 5M_\odot$ helium star to explode as a supernova. Inserting these values into Eq. (1) one observes that in systems with the same initial mass ratio $q_0 = M_2^0/M_1^0 = 0, 5$, if $M_1^0 = 30M_\odot$, after the mass transfer $P^f = P_0$ to $1, 46P_0$, whereas for $M_1^0 = 12M_\odot$ one finds: $P^f = 5, 8P_0$ to $8, 9P_0$.

Hence, due to the smaller mass fraction of the helium core in lower-mass systems the orbital periods increase systematically by a large factor due to the mass transfer, whereas in the very massive systems the transfer hardly changes the orbital period. (One may easily verify that the same will also be true for other choices of the value of q_0.)

Thus, the generally very much longer orbital periods ($\gtrsim 15^d$) of the Be/X-ray binaries relative to those of the "standard" massive systems are a natural consequence of the mass transfer that took place prior to the supernova explosion of their helium stars (van den Heuvel 1983).

The supernova explosions themselves will have introduced a further increase in orbital period (*i.e.*, from phase (c) to phase (d) in Figure 4), will have introduced an orbital eccentricity and will have imparted a mild runaway velocity to the systems, in general not more than 10 − 40 km/s (*cf.* Habets 1985, 1986). For a detailed discussion of the evolutionary history of the Be/X-ray binaries and their progenitors I refer to Habets (1985) and to van den Heuvel and Rappaport (1987).

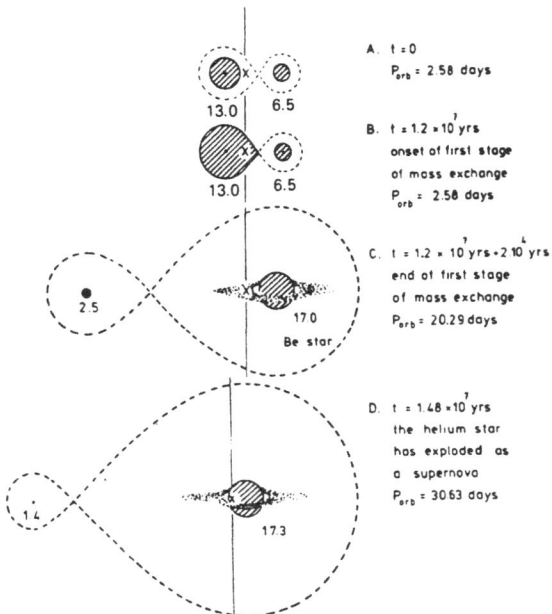

Figure 4. Conservative evolutionary scenario for the formation of a Be/X-ray binary out of a close pair of early B stars with masses $13, 0M_\odot$ and $6, 5M_\odot$. The numbers indicate mass (in units of $M_\odot$). After the end of the mass transfer the Be star presumably has a circumstellar disk or shell of matter associated with its rapid rotation (induced by the previous accretion of matter with high angular momentum; from Habets, 1985, 1986).

2.4. THE FINAL EVOLUTION AND FATE OF THE HIGH-MASS X-RAY BINARIES: SPIRAL-IN AND THE FORMATION OF BINARY PULSARS AND THORNE-ZYTKOW STARS

a) Introduction

The HMXB are all characterized by a mass ratio of secondary and primary (the normal star) very far from unity. The neutron star (secondary) has a mass of only about $1, 4M_\odot$, whereas its companion has a mass of between $8M_\odot$ and $\sim 40M_\odot$. When the companion is near or beyond finishing its core-hydrogen burning phase, its hydrogen-rich outer layers will expand and will, at a certain point, begin to overflow their Roche lobes, thus initiating a second (reverse) phase of mass transfer. In the narrow systems this phase will be preceded by a phase in which matter from the tenuous outer atmospheric layers (or from its stellar wind) is "trickling" over to the neutron star causing it – temporarily – to become observable as a strong High Mass Binary X-ray source.

This is the "standard" HMXB phase which, however, will be soon, within about $2,5 \times 10^4$ yrs, (cf. Meurs and van den Heuvel 1989) be followed by a phase of full-scale Roche-lobe overflow. Because of the very unequal masses of the components the onset of full-scale Roche-lobe overflow will cause the neutron star to be "dragged" into the envelope of the companion, initiating a so-called "common-envelope" (CE) phase of the system. For a variety of reasons, which are discussed in this volume by Taam and Webbink, the occurrence of such a CE-phase in the HMXB seems unavoidable. The helium core of the companion together with the neutron star will orbit inside the common envelope and, since this envelope cannot be kept in co-rotation with the binary system (see Taam, this volume and Webbink, this volume), large frictional drag will be generated which will cause the orbit of the two cores to rapidly shrink to very small dimensions.

If the initial orbit was sufficiently wide, the drop in orbital potential energy may be sufficiently large to exceed the binding energy of the common envelope, such that the released energy is sufficient to eject the envelope. In such a case a very close binary, consisting of the neutron star and the helium core of its companion will remain. This situation is depicted in Figure 5b. Cygnus X-3 has been suggested to have resulted from such a spiral-in evolution (van den Heuvel and De Loore 1973; van den Heuvel 1974).

On the other hand, if the initial X-ray binary has a short orbital period, the drop in orbital potential energy may not be sufficient to expel the entire hydrogen-rich envelope. In such a case the neutron star is expected to spiral down completely into the core of its companion, such that a massive star with a neutron-star core will result (Figure 5a). The structure of such a Thorne-Zytkow star (Thorne and Zytkow 1977) has recently been analyzed by Biehle (1991) and Cannon (1991), to which we refer for details. The limiting initial orbital separation for complete spiral-in vs. survival of a (very) close binary can be obtained by using the treatment of CE evolution given by Webbink (1984; see also Webbink, this volume). According to this treatment the ratio of the final and initial orbital separation a_2 and a_1, respectively, is given by:

$$\frac{a_2}{a_1} = \frac{M_1 M_{2f}}{M_{2f} + M_{2e}} / \left(M_1 + \frac{2 M_{2e}}{\eta \cdot \lambda \cdot r_{L_2}} \right) \qquad (2)$$

where M_1 is the mass of the low-mass companion (neutron star), M_{2f} and M_{2e} are the core mass and envelope mass, respectively, of the massive star, r_{L_2} is the ratio of the Roche lobe radius of the massive star and its orbital radius, λ is a weighting factor for the binding energy of core and envelope of the massive star, and η is the efficiency of the conversion of orbital potential energy into the kinetic energy that provides the outward motion of the envelope.

For situations such as the one described here, r_{L_2} and λ are about $0,6$, whereas the numerical calculation by Taam and Bodenheimer suggest $\eta \simeq 0,5$ (cf. Taam and Bodenheimer 1989; Bodenheimer and Taam 1984; Taam, this volume). In view of this we will assume here $\eta \cdot \lambda \cdot r_{L_2} = 0,2$.

In that case, for an HMXB with $M_2 = 15 M_\odot$, $M_1 = 1,4 M_\odot$ one has $M_{2e} = 11 M_\odot$, and

$$a_2/a_1 = 1/294.$$

Hence, in order to have $a_2 > R_\odot$, as is required to fit a $4 M_\odot$ helium star into the post-spiral-in binary, a_1 should have been $> 294 R_\odot \simeq 1,36 AU$. Such a separation corresponds to an orbital period of 0.39 yrs = 143 days. On the other hand, if one allows for some convective overshoot such that the $15M_\odot$ star produced a $5M_\odot$ helium core, the spiral-in survival condition becomes $a_1 \gtrsim 1AU$, $P_1 \gtrsim 90^d$ (the same would hold if prior to spiral-in a few solar masses had been lost from the envelope by stellar wind – as is probably always the case with early-type stars). It seems reasonable therefore to assume that Be/X-binaries will survive as (very-short-period) binaries consisting of a helium star and a neutron star, provided that their orbital periods were longer than about 100 days.

Systems with narrower orbits are expected to spiral-in completely and to leave Thorne-Zytkow stars, as depicted in Figure 5a.

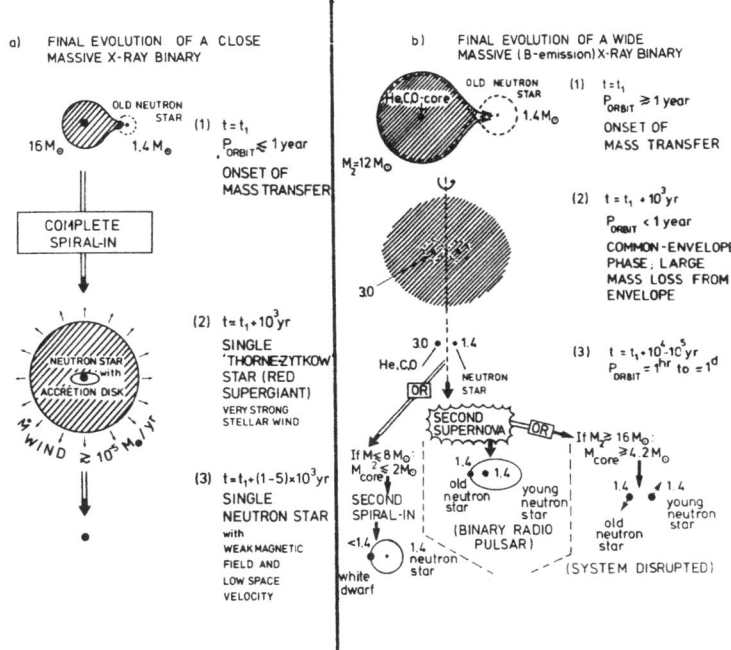

Figure 5. The various possibilities for the final evolution of a High-Mass X-ray Binary. In all cases the onset of Roche-lobe overflow leads to the formation of a common envelope and the occurrence of spiral-in. In systems with orbital periods $\lesssim$ about 1 year there is most probably not enough energy available in the orbit to eject the common-envelope, and the neutron star spirals down into the core of its companion. Subsequently, the envelope is ejected by the liberated accretion energy flux, leaving a single recycled radio pulsar. In systems with orbital periods $\gtrsim$ about a year the common envelope is ejected during spiral-in, and a close binary can be left, consisting of the neutron star and the core (consisting of helium and heavier elements) of the companion. Companions initially more massive than 8–12 $M_\odot$ leave cores that will explode as a supernova, leaving an eccentric-orbit binary pulsar or two runaway pulsars. Systems with companions less than massive than $\approx$ 8–12 $M_\odot$ leave close binaries with a circular orbit and a massive white dwarf companions, similar to PSR 0655+64.

b) The fraction of Be/X-ray binaries that survives as binaries after spiral-in

For 9 of 22 known Be/X-ray binaries the orbital periods are known. Four of these are $> 100^d$; the longest one is 580^d (corresponding to $a_1 = 3,5AU$), which for $M_2 = 15M_\odot$, $M_{2f} = 4 - 5M_\odot$ would lead to $a_2 \sim 2,5 - 3,5R_\odot$ ($P \sim 4,7^h - 7^h, 1$). If the observed fraction of Be-systems with $P \simeq 100 - 600^d$ is representative for the entire Be/X-ray binary population, one would expect some 45 percent of the Be/X-binary population to leave ultra-short period binaries consisting of a helium star and a neutron star. On the other hand, if the 4 observed systems with $P > 100^d$ are the only ones among the 22 known systems that have $P > 100^d$, one would expect some 20 percent of the Be/X-binaries to spiral-in and leave ultrashort period binaries ($a_2 = 1R_\odot - 3,5R_\odot$) consisting of a helium star and a neutron star. Thus, some 20–45% of all Be/X-binaries are expected to produce such systems.

c) Survival probability after the second supernova explosion

The survival probability after the second supernova explosion is expected to be very high for such systems, provided that the kick imparted in the explosion to the newborn neutron star does not exceed 100–200 km/s. This is because the orbital velocities in these systems are very high (> 400 km/s), and the kicks do therefore not significantly alter the survival probability. In fact, since kicks are expected to be imparted in random directions, one expects that in half of the cases (retrograde kicks with respect to the orbital motion) the systems will be more tightly bound after the explosion. We thus expect that, including kick velocities of 100–200 km/s, $\gtrsim 50\%$ of systems with $M_{He} < 4, 2M_\odot$ will remain bound (see also Dewey and Cordes 1987). Also, such kicks will make that a considerable fraction of systems with helium star masses $> 4, 2M_\odot$, which would all be disrupted in case of a spherically symmetric explosion, will now remain bound. We estimate this fraction to be of order 10–30 % for helium stars of $4, 2 - 6M_\odot$ and kick-velocities of 100–200 km/s (*cf.* de Cuyper 1985).

d) Second spiral-in phase during later stages of helium burning

The survival probability in the second SN-explosion is further enhanced by the occurrence of a second spiral-in phase in many systems with $M_{He} \lesssim 3, 5 - 4M_\odot$. Such helium stars, during later phases of core-helium burning and during helium shell burning undergo a considerable radius expansion (see Habets 1985, 1986). In narrow systems this will lead to a second spiral-in phase, in which part of the helium-rich envelope of the helium star is lost, and the orbital radius is further decreased. For example, a $4M_\odot$ helium star at carbon-ignition has a radius of $1,08R_\odot$ and a carbon core of $2,7M_\odot$. This core will certainly evolve to core collapse, regardless of the absence or presence of a helium-rich envelope.

Assuming an orbital separation of $2R_\odot$ after the first spiral-in, the Roche-lobe radius of the helium star will be $\sim 0,95R_\odot$, and the star begins to overflow its Roche-lobe just prior to carbon ignition. This leads to a second spiral-in phase and, according to Eq. (2), to a further reduction in orbital radius by a factor 15. However, since this is a rather marginal case for spiral-in (in view of the mass ratio of 0,35), one might also assume more or less "conservative" mass transfer here, in which case the orbit would shrink by only a factor 2,7.

In any case, since the mass of the exploding star now is reduced from $4, 0M_\odot$ to $2, 7M_\odot$, and the orbital velocities are even higher now, survival after the second supernova explosion becomes very likely.

We therefore conclude that the ultrashort-period binaries resulting from the common-envelope evolution of Be/X-ray binaries have a very high probability (presumably $> 50\%$) to survive the second supernova explosion and to produce a double neutron star system, like PSR1913+16.

Thus, between 10 and 20 percent of all Be/X-ray binaries are expected to produce such binary pulsar systems.

e) Expected orbital periods of double neutron-star binaries, and the timescales of orbital decay by gravitational wave emission

Assuming a post-explosion orbital eccentricity $e \sim 0,6$ (as observed in most double neutron star binaries), and pre-explosion orbital radii between $0,33R_\odot$ and $1,2R_\odot$ (after double spiral-in; resulting from the reduction by a factor 3 of orbits in the range $R_\odot$ to $3,5R_\odot$), the semi-major axes of the post-SN systems will be between $\sim R_\odot$ and $3,4R_\odot$, and their orbital periods will be between $0,06876^d$ and $0,43^d$ (1,65 hours and 10,3 hours). This will have been the case for most systems with helium star companion masses $<$ $3,5M_\odot$, and still for a fraction of the systems with M_{He} between $3,5M_\odot$ and $4,2M_\odot$. As these systems will make up the bulk of the remnants of Be/X-ray binaries with $P \gtrsim 100^d$, we expect most of the double neutron star binaries to have orbital periods in this range.

The gravitational radiation decay timescale of the orbit of a double neutron star system with component masses of $1,4M_\odot$, $e = 0,6$ and $a = R_\odot$ is $\tau_{GR} = 5,5 \times 10^6$ yrs; for $a = 3,4R_\odot$ it is $7,26 \times 10^8$ yrs. For these values of the eccentricity and the component masses, the general equation for this decay timescale is (Shapiro and Teukolsky, 1984):

$$\tau_{GR} = 3.4 \times 10^8 (P/7,75^h)^{8/3} \text{ yrs.} \tag{3}$$

If $f(P)$ is the distribution over orbital periods of newly formed double neutron-star binaries, the "observed" distribution of orbital periods of such systems can be predicted, for a steady state of star formation as:

$$F(P) = f(P) \cdot \tau_{GR}(P). \tag{4}$$

It will be clear, from the very strong dependence of $\tau_{GR}(P)$ on P, that systems with the longest orbital periods (8–10 hours) can be expected to dominate the observed distribution of orbital periods of double neutron stars.

The total number of double neutron stars in the galaxy can be obtained by integration of Eq. (4):

$$N_{BP} = \int_{1,65H}^{10,3H} f(P) \cdot \tau_{GR}(P)dP. \tag{5}$$

2.5. THE FORMATION RATE OF DOUBLE NEUTRON STAR BINARIES IN THE GALAXY AND THE RATE OF NEUTRON STAR-NEUTRON STAR MERGERS

Our above estimates indicate that some 10–20% of the observed Be/X-ray binaries are expected to produce close double neutron-star binaries. The Be-stars in these progenitor binaries are, in large majority, stars with masses in the range $8 - 15M_\odot$, which produce helium cores in the mass range $\sim 2, 2 - 4M_\odot$.

An additional number of double neutron star binaries will be produced by systems with primary stars more massive than $15M_\odot$, but in view of the steepness of the initial mass function $\psi(M) \sim M^{-2.5}$, and of the higher disruption probability of such systems in the second supernova explosion, these systems are expected to make only a minor ($< 20\%$) contribution to the double-neutron star formation rate.

We will therefore restrict our discussion to the Be/X-ray binaries.

The galactic number of such systems has been estimated on theoretical grounds as $\sim 2 \cdot 10^4$ while the observations suggest a number of at least 10^3 (Meurs and van den Heuvel 1989). Adopting a lifetime of $\sim 10^7$ yrs of the B-stars in these systems, one obtains a galactic formation rate 10^{-4} to 2×10^{-3}/yr. With a production efficiency of double neutron star binaries from these systems of 10–20% (see above), one finds that the birthrate of PSR1913+16-like systems in the galaxy is between $(1-2) \times 10^{-5}$/yr and $(2-4) \times 10^{-4}$/yr. The uncertainty in these numbers is expected to be about a factor of two (upwards as well as downwards; cf. Meurs and van den Heuvel 1989).

So, conservatively speaking, the lower limit to the formation rate of PSR1913+16-like binaries following from detailed calculations of close binary evolution and stellar population statistics is at least $0,5 \times 10^{-5}$/yr. The most likely value is, in our opinion, the one corresponding to 2.10^4 Be/X-ray binaries in the galaxy, which yields a formation rate of $(2-4) \times 10^{-4}$/yr in the galaxy.

With this formation rate, if one wishes to calculate the steady-state population of neutron star binaries in the galaxy, one has to calculate the integral of Eq. (5). To this end one has to adopt an orbital period distribution $f(P)$. As an example we adopt

$$f(P) = \text{const.}/P \qquad (6)$$

(as is found for spectroscopic binaries, cf. Kraicheva et al., 1978, Popova et al., 1983). With P between 1.65^h and 10.3^h and a total galactic formation rate $(2-4) \times 10^{-4}$/yr (see above) one obtains in Eq. (6): const. $= (1, 1 - 2, 2) \times 10^{-4}$/yr. Inserting this into Eq. (5) with τ_{GR} from Eq. (3) one finds a steady-state galactic population of neutron-star binaries of $(4, 5 - 9, 0) \times 10^4$ and a distribution of orbital periods given by

$$F(P)dP = 3,6 \times 10^4 (P/7,75H)^{5/3} d(P/7,75H) \qquad (7)$$

with P between $1,65^H$ and $10,3^H$. On the basis of Eq. (7) one would expect binary neutron star systems with P between 7^h and $10,3^h$ to be about twice as common in the galaxy as systems with periods $< 7^H$. This may explain why the two double neutron star systems found so far (PSR1913+16 and PSR1534+12) both have $P > 7^h$. Nevertheless, on the basis of Eq. (7) one would expect shorter-period systems to be found in the near future.

The above-given steady-state population predicts $\sim 100 - 200$ double neutron star systems per kpc^2 above the galactic disk. Assuming them to have a galactic scale height $z_0 \sim 3$ kpc one expects the nearest such system to be at a distance between 0,3 and 0,4 kpc. With a scale-height $z_0 = 1$ kpc, these numbers become between 0,2 and 0,3 kpc.

If one would assume the galactic number of Be/X-ray binaries to be only 1000, all the above numbers should be reduced by a factor of 30, and the nearest double neutron star system is expected at about one kpc distance. The nearest such system found so far, PSR1534+12 is at about 0.5 kpc (Wolszczan 1991). This suggests that the number of Be/X-ray binaries in the galaxy is probably larger than 10^3 but not much larger than $(0,5-1) \times 10^4$. This then would imply that $F(P)$ is about five times smaller than given by Eq. (7), leading to a birthrate of these systems, as well as a rate of neutron star-neutron star mergers of $(4-8) \times 10^{-5}$ yr^{-1} in the Galaxy. We consider this the most likely value at present. It should be noticed here that black hole-neutron star binaries may be formed at a rate one to two orders of magnitude lower than the double-neutron star formation rate (*cf.* McClintock, this volume). These binaries are expected to have relatively wide orbits as a $10M_\odot$ black hole like the one in Cygnus X-1 will never spiral-in as deeply into its companion as a neutron star would do. Therefore, black hole-neutron star binaries are – in general – not expected to merge within a Hubble time. They are therefore not expected to be of interest for producing bursts of gravitational waves. The same holds for black hole-black hole binaries.

It is comforting to notice that the above given theoretically predicted rate of formation of double neutron star systems fits well with the rate derived from the observations (Phinney 1991, Narayan et al. 1991).

2.6. THE INCIDENCE OF BARE-CORE SUPERNOVA EVENTS

The standard systems are expected to be relatively short-lived: the spin-up timescales of the X-ray pulsars in the systems SMC X-1, Cen X-3 and LMC X-4 are typically of order 3000–4000 years. The X-ray lifetimes of such systems probably do not exceed $2,5 \times 10^4$ years (Meurs and van den Heuvel 1989). With a total number of about 50 such systems in the galaxy their galactic birthrate must be of order $(1 - 2,5) \times 10^{-3}$/yr.

On the other hand, the Be/X-ray binaries are much more numerous. Extrapolating from the > 10 such systems found within 2.5 kpc distance and taking into account the relatively long recurrence timescale of their transient outbursts, one expects their total number in the galaxy to be at least 10^3. This is, however, a very conservative estimate: stellar population evolution calculations, in which binary evolution is included, predict a total galactic number of $\sim 2 \times 10^4$ (Meurs and van den Heuvel 1989). With an average lifetime of $\sim 10^7$ yrs the formation rate of Be/X-ray binaries is therefore of order $2 \cdot 10^{-3}$/yr.

The combined HMXB formation rate derived from these two types of systems is $(3 - 4,5) \times 10^{-3}$/yr which is of order 0.3 to 0.5 times the total supernova rate in the galaxy. These supernovae may be dim ones, however, as they are due to the explosion of a bare helium star. They do, however, produce a neutron star, which is bound to a massive star, and will be released into the galactic pulsar population only after the second supernova explosion in the system took place. Also this one is expected to be a dim one, as it is again expected to be the explosion of a bare helium core. Thus, going through all the population statistics, a large fraction of all supernova events from massive stars is expected to be bare core explosions. Meurs and van den Heuvel (1989) expect this fraction to be: between one half and two thirds of all massive star supernovae. This expectation seems in excellent agreement with the observational results of Filipenko, Nomoto *et al.*(see Nomoto, *et al.* 1990) who have identified the supernovae of types Ib and Ic as bare helium core explosions of massive stars, and have estimated that as much as more than half of all supernova events resulting from massive stars are explosions of bare helium cores.

3. Low-mass X-ray Binaries and their Evolution

3.1. INTRODUCTION

The origin of LMXB is the subject of Webbink's review (this volume) and will not be considered here in detail.

We just notice that there are basically two models for their formation: (i) the direct core-collapse model, in which the LMXB is the result of the evolution of a massive close binary with components that were very different in mass, for example $15M_\odot$ and $2M_\odot$. This is the model for the origin of Hercules X-1 originally proposed by Sutantyo (1975, e.g., see van den Heuvel, 1978, 1983; Verbunt et al. 1990, and Sutantyo, this volume).

When the massive component evolves to the giant phase, the low-mass star spirals down into its envelope and drives off this envelope in a common-envelope phase (see Webbink, this volume, and Taam, this volume). The resulting system consists of the helium core of the massive star plus the low-mass companion, in a tight orbit. When the helium star explodes there is a low, though finite, probability that the low-mass star remains bound. The resulting system may then, in a later phase, evolve into an LMXB.

In the alternative model the neutron star was formed from a massive white dwarf which was driven over the Chandrasekhar limit by accretion of matter from a low-mass companion star. Since in this model no mass loss from the white dwarf needs to occur during its collapse ("quiet collapse") the system needs not disrupt during the explosion.

Both models should occur only rarely in Nature as there are only ~ 100 LMXB in the galaxy, while their mass transfer timescales are thought to be $> 10^7$ yrs. Hence their birthrate needs not be larger than 10^{-5}/yr in the galaxy, whereas for the HMXB it is $\sim (3-4, 5) \times 10^{-3}$/yr (see Section 2.6).

3.2. MECHANISM DRIVING THE MASS TRANSFER IN LMXB AND LMXB EVOLUTION

3.2.1. Introduction. The observed orbital periods of LMXB range from 11 minutes for XB 1820–30 to 235 hours for Cygnus X-2 (*e.g.*, see Parmar, this volume, and Figure 1 of Bhattacharya, this volume).

We will first consider here the three "classical" mechanisms that are well understood and on which much research has been carried out in the past decades: (i) the interior nuclear evolution of the companion star; (ii) orbital angular momentum losses from the system due to the emission of gravitational radiation and (iii) orbital angular momentum losses due to "magnetic braking".

3.2.2. Mass transfer driven by the interior nuclear evolution of the companion star: origin of the wide radio pulsar binaries with circular orbits. Mass transfer driven by the interior nuclear evolution of the companion is important for orbital periods that at the onset of the mass transfer were $\gtrsim 1$ day. The evolution of these systems was first studied by Webbink et al.(1983) and Taam (1983). These authors showed that the evolution is in principle very simple. This is due to the relatively simple interior structure of evolved low-mass (solar-type) stars. In such stars the helium core is degenerate and the luminosity is produced by hydrogen burning in a shell around this core. As a result the core mass gradually increases, and the luminosity generated in the hydrogen-burning shell (L_s) gradually increases, which causes the outer radius of the star (R_s) to also gradually increase. For a given initial chemical composition (X, Y, Z), the luminosity and radius of these stars turn out to be a function only of the mass M_c of the degenerate helium core : $L_s(M_c), R_s(M_c)$, independent of the total mass present in the hydrogen-rich envelope (as long as that envelope mass is $> 0,01M_\odot$). Since the core mass M_c determines the stellar radius R_s, and since the Roche-lobe radius of this star is determined by the orbital period (assuming the compact star to have a standard mass, e.g. $1,4M_\odot$), the core mass at the moment that the star begins to fill its Roche lobe is entirely determined by the orbital period of the system at that time. Since this core mass determines the evolutionary rate of change of radius at that moment, it determines the ensuing rate of mass transfer. As a result, the initial orbital period P_o (at the onset of the mass transfer) uniquely determines the rate of mass transfer in these systems, as well as the entire further evolution of these systems. Figure 6 shows as an example the evolution of the core mass, radius, and orbital period of an evolved star in a system that started out with an orbital period of 12,5 days, a neutron-star mass of $1M_\odot$ and a companion mass of $1M_\odot$. The figure shows that the entire evolution of the system takes $8,1.10^7$ years, during which the companion mass decreases to $0,31M_\odot$, the core mass grows from $0,24M_\odot$ to $0,31M_\odot$ and the orbital period increases to 117 days. After an initial episode of slightly super-Eddington mass transfer ($\sim 10^7$ yrs) the mass-transfer rate settles at $\sim 1,0.10^{-8}M_\odot/yr$, making the system a very bright (near Eddington limit) X-ray source for a period of some 7.10^7 years.

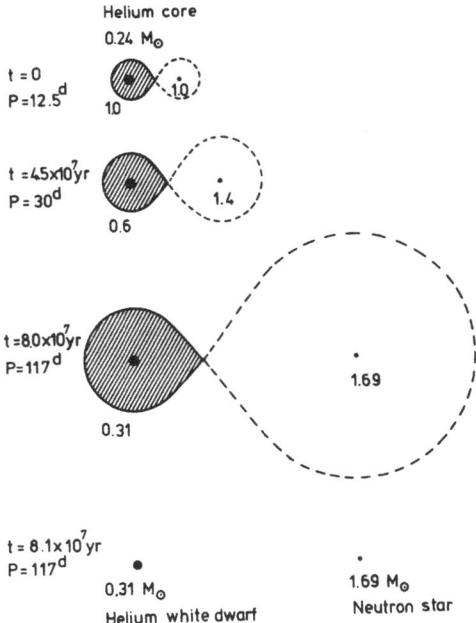

Figure 6. Evolution of a wide low-mass X-ray binary such as Cygnus X-2 into a wide radio-pulsar binary with a circular orbit and a low-mass helium white dwarf companion, such as PSR 1953+29. At the onset of the mass transfer, the low-mass companion is a (sub)giant with a degenerate helium core of 0.24 $M_\odot$. Its light is generated by hydrogen-fusion in a shell around the core. The mass-transfer from the giant to the neutron star is due to the slow expansion of the giant, driven by this hydrogen-shell burning. During the mass transfer the orbit gradually expands (due to angular momentum conservation) and after 8.10^8 yr the system terminates as a wide radio-pulsar binary (after Joss and Rappaport 1983).

As the giant companion of the system of Figure 6 fills its Roche lobe for $\sim 8.10^7$ years, the strong tidal forces exerted on its convective envelope by the neutron star are expected to completely circularize the orbit (*cf.* Savonije 1983a). Hence, the system that remains in the end will consist of a neutron star and a low-mass ($\sim 0,3M_\odot$) helium white dwarf in a wide and circular orbit. Such a system exactly resembles the second class of binary radio pulsars depicted in Figure 2 (the PSR 1953+29 class), as was noticed independently by Joss and Rappaport (1983), Savonije (1983b) and Paczynski(1983), immediately after the discovery of PSR 1953+29 by Boriakoff *et al.* (1983). Important is that, assuming a given initial chemical composition and initial neutron-star mass, the evolution and final binary period of such a system is entirely determined by the initial binary period and companion mass. (For details about the relations between initial and final binary parameters for such systems we refer to Verbunt (1989)). This implies that for the binary pulsars of the PSR 1953+29 class , it is possible to reconstruct, from the present orbital parameters, their orbital periods at the onset of the mass transfer, as well as the rate of mass transfer during the X-ray phase (which was determined by the initial orbital period). The latter can, for $P_o > 2$ days be approximated by (for a $1,0 M_\odot$ companion, $Z = 0,02$)

$$\dot{M} = -8.10^{-10} P_o [d] \tag{8}$$

Table 2 and Equation (8) show that the mass-transfer rate becomes super-Eddington for $P_o > 12$ days. This implies that for $P_o > 12$ days also mass loss from the systems will have occurred such that the "conservative" evolutionary picture that we adopted will no longer be fully correct. Therefore, for PSR 0820+02, in which the mass-transfer rate was $10^{-7} M_\odot/\text{yr}$, the relation between initial and final orbital period of Table 2 cannot be precise, and will be only approximate.

TABLE 2

Initial and final (=present) parameters of three wide binary radio pulsar systems, for an assumed initial companion mass of $0,9 M_\odot$, initial neutron star mass $1,4 M_\odot$ and initial heavy element abundance $Z = 0,02$. Assumed is conservative mass transfer (no mass and orbital angular momentum loss from the system) (*cf.* Verbunt 1989).

System Name	Binary Period (d)		Core Mass $(M_\odot)$		Mass transfer rate $(M_\odot/\text{yr})$	
	P	P_0	M_c^0	M_c^f	$\dot{M}_0$	$\dot{M}_f$
PSR 1953+29	120	6,5	0,22	0,27	$0,8 \times 10^{-8}$	$0,4 \times 10^{-8}$
PSR 1620–62	191	16	0,24	0,29	$1,6 \times 10^{-8}$	$0,6 \times 10^{-8}$
PSR 0820+02	1232	170	0,32	0,38	10^{-7}	$0,5 \times 10^{-7}$

3.2.3. Evolution driven by orbital angular momentum losses by gravitational radiation and magnetic braking.

(a) *Outline of the evolution:*

These two mechanisms are expected to be operating only in systems with very short orbital periods $(\lesssim 0,5 - 1^d)$. The timescale for orbital decay by gravitational radiation emission (assuming circular orbits) is (Faulkner 1971):

$$t_{GR} = \frac{(M_1 + M_2)^{\frac{1}{3}} (P/1,6H)^{8/3}}{2^{\frac{1}{3}} M_1 M_2} (5 \cdot 10^7 \text{ yrs}) \tag{9}$$

For $M_1 \sim M_2 \sim M_\odot$ one observes that only for $P < 12^H$, t_{GR} is shorter than a Hubble-time $(\sim 10^{10} \text{ yr})$.

Thus, in such systems, even without evolution of the companion star, continuous mass transfer from a main-sequence or helium star companion will ensue due to the shrinking of the orbit (and with it: the Roche lobe of the companion). Table 3 lists the approximate mass-radius relations for low-mass ($\leq 1, 3 M_\odot$) hydrogen-rich and helium stars in thermal equilibrium, together with the orbital periods of binaries in which such stars fill their Roche lobes (after Verbunt 1989).

The table shows that low-mass unevolved stars will fill their Roche-lobes only for orbital periods $\lesssim 12^H$. In such systems the emission of gravitational waves may, therefore be the sole cause of the mass transfer. The mass-transfer rate due to this mechanism, however, never exceeds $\sim 10^{-10} M_\odot$/yr, which would produce an X-ray luminosity of $\sim 10^{36}$ ergs/s (cf. Verbunt 1989). However, quite a number of LMXB with $P < 12^H$ show X-ray luminosities that are one to two orders of magnitude higher.

The enhancement of the mass transfer in these systems may be due to additional orbital angular momentum losses by "magnetic braking". For orbital periods between $0, 5^d$ and $1, 0^d$ angular momentum losses by magnetic braking may still help driving the mass transfer (Verbunt and Zwaan 1981; Verbunt 1989). However, for $P \gtrsim 1, 0^d$ the mass transfer can only be driven by the nuclear evolution of the companion star.

Figure 7 shows the schematic evolution of a close binary consisting of a main-sequence star and a compact star (white dwarf or neutron star) under the influence of gravitational radiation losses. The figure applies to Cataclysmic Variable (CV) binaries as well as LMXB. The system starts with a companion mass $\sim M_\odot$ at $P \sim 9^h$, and gradually evolves to shorter orbital periods, its orbital period following the orbital period vs. mass relation for main-sequence stars given in Table 3. When the companion mass falls below $\sim 0, 3 M_\odot$, the timescale for the mass transfer becomes shorter than the thermal timescale of the companion, and the mass loss causes this star to be driven farther and farther out of thermal equilibrium (it cannot cool itself as fast as it loses mass). This causes its radius to become larger than its main-sequence (thermal equilibrium) radius, such that the system deviates more and more from the orbital period vs. mass relation for main-sequence stars. At a certain point ($M < 0, 1 M_\odot$) further mass transfer causes the radius to expand: the star now gradually moves to the degenerate track in Figure 7. The minimum orbital period which it attains during this evolution is ~ 80 minutes. This is the so-called period-minimum for Cataclysmic Variables (Paczynski and Sienkiewicz, 1981). Indeed, no CVs or LMXB with hydrogen-rich companions have been found below this limiting period. All systems with shorter period appear to have hydrogen-poor companions, such that they roughly follow the relations for helium rich stars in Table 3.

(b) *The origin of the Period Gap for CV binaries*

In the case that angular momentum losses by magnetic braking are included, the mass transfer rate can become as high as several times $10^{-9} M_\odot$/yr, yielding X-ray luminosities in LMXB of $10^{37} - 10^{38}$ ergs/s, as are indeed observed in many LMXB. Such mass-transfer rates are also observed in many CV-binaries.

Now, however, in view of these much larger transfer rates, the companion will be driven farther out of thermal equilibrium than in the case of GR-losses alone. It will therefore become further inflated, and the system will follow a higher track in Figure 7, as schematically indicated in that figure. However, as soon as the companion becomes fully convective, which occurs at $M \sim 0, 3 M_\odot$, the rate of angular momentum loss by magnetic braking is expected to drop considerably (as magnetic fields allegedly need anchoring in a radiative core, cf. Spruit and Ritter, 1983)*, causing the mass-transfer rate to drop, such that the star can relax back to thermal equilibrium. This causes it to shrink and thus to

*see also Rappaport et al. 1983.

detach itself from its Roche lobe, and the mass transfer will stop completely. When its compact companion star is a white dwarf, this star will no longer have an accretion disk and, hence, will no longer appear as a CV-binary.

During this detached phase, orbital angular momentum loss by gravitational wave emission will continue and the orbit will continue to shrink. When P has been reduced to 2^H, the orbit has become so narrow that the companion (of mass $\sim 0,3 M_\odot$) again fills its Roche lobe and the mass transfer resumes. It will reappear as a CV-binary, but now with an orbital period below the "period-gap".

The above described evolution, graphically depicted in Figure 7, gives a qualitative explanation of the existence of the "period-gap" in the distribution of orbital periods of CV binaries.

TABLE 3

Approximate mass-radius relations for Roche-lobe filling low mass ($\lesssim 1, 3 M_\odot$) stars in thermal equilibrium (during the mass transfer the thermal equilibrium may gradually be disturbed such that deviations from these relations become possible) (after Verbunt 1989):

Main Sequence	$\frac{R_2}{R_\odot} = \frac{M_2}{M_\odot}$	$\mathcal{P} = 9^h \left(\frac{M_2}{M_\odot} \right)$
Helium Main Sequence	$\frac{R_2}{R_\odot} = 0,2 \frac{M_2}{M_\odot}$	$\mathcal{P} = 0^h,9 \left(\frac{M_2}{M_\odot} \right)$
Degenerate Star*	$\frac{R_2}{R_\odot} = 0,013(1 + X)^{5/3} \left(\frac{M_\odot}{M_2} \right)^{1/3}$	$\mathcal{P} = 48(1 + X)^{5/2} \frac{M_\odot}{M_2} (\text{sec})$

*here X is the fractional hydrogen abundance.

3.2.4. Mass-transfer and mass-loss mechanisms different from the three classical ones: effects of X-ray heating and evaporation of companion stars.

(a) *Reasons for the absence of a period-gap for LMXB*

The *absence* of a period gap in the distribution of orbital periods of LMXB (see Bhattacharya, this volume) and especially the lack of LMXB with $P \lesssim 3^H$ (with the exception of a few systems with hydrogen-poor companions at periods $< 1^H$, such as XB 1820-30) represents an important difference with the case of CV-binaries. A possible explanation for this difference may be the following one (proposed by van den Heuvel and van Paradijs, 1988).

A very important difference between the CV-binaries and the LMXB is that, upon entering the period gap, the LMXB contain a potentially very strong "powerhouse" in the form of a weakly magnetized neutron star, that during the preceding several tens of millions of years of mass transfer has been spun-up to a rotation period of the order of milliseconds. Once the system enters the period gap and the mass transfer stops, this neutron star will appear as a powerful millisecond radio pulsar, which starts to heavily bombard the nearby companion with its relativistic electron-positron wind, its X- and γ-radiation, and its superstrong low-frequency magnetic dipole radiation (*cf.* Krolik and Sincell 1990). This causes the onset of evaporative mass loss from the companion. If this "evaporation" takes place with a high energy efficiency (PSR 1957+20 shows that this

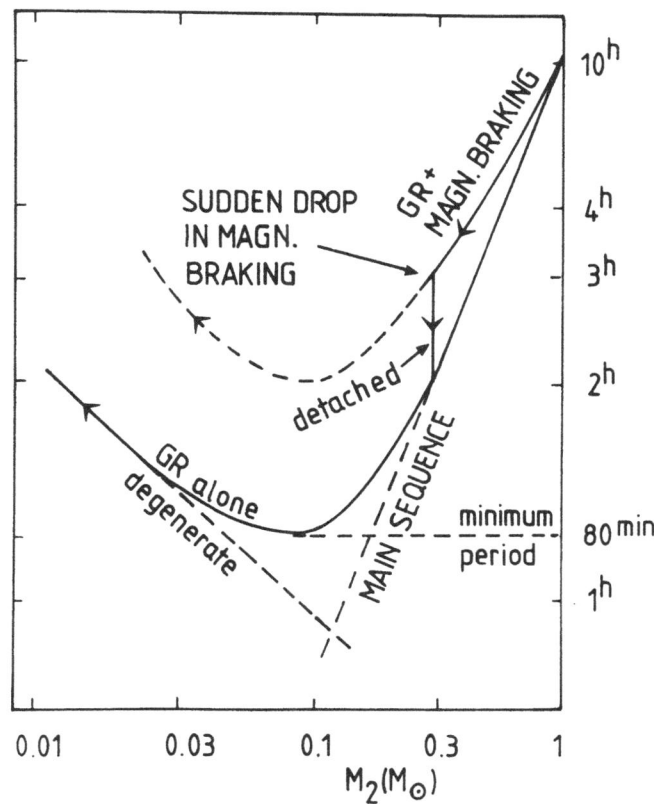

Figure 7. Schematic explanation for the origin of the "Period Gap" of Cataclysmic Variables between 2^H and 3^H (*cf.* Spruit and Ritter 1983). Indicated are evolutionary tracks of CV-binaries and LMXB in which the mass transfer is driven by angular momentum losses by Gravitational Radiation (GR) and/or Magnetic Braking. The system begins in the upper right-hand part of the diagram with $M_2 \lesssim M_\odot$, $P \leq 10^h$. If GR losses alone are operating, the system reaches a minimum orbital period of ~ 80 minutes (companion mass $\sim 0, 1 M_\odot$), after which the companion becomes degenerate and the orbit expands upon further mass transfer. If also magnetic braking operates, the mass-transfer rate may be an order of magnitude larger and the companion is driven out of thermal equilibrium, such that its radius becomes inflated. It then moves along the higher track in the diagram. However, in this case, for $M_2 \sim 0, 3 M_\odot$ the companion becomes fully convective and the magnetic braking is much reduced. The resulting reduction of the mass transfer rate causes the companion to shrink back to thermal equilibrium and to detach itself from the Roche-lobe. At this point $P_{orb} \simeq 3H$. Subsequent GR losses reduce the orbital period further during this detached phase; when P_{orb} has become $2H$ the companion fills its Roche-lobe again and mass-transfer resumes (after Spruit and Ritter 1983). Thus, between $P_{orb} \approx 2^H$ and 3^H no mass-transferring CVs or LMXB are expected to be found.

indeed appears to be the case), the powerful pulsar may possibly completely destroy the companion. Very much depends here on the efficiency of the evaporation process. For efficiencies of one per cent or more of the impinging energy flux, complete evaporation of the companion may occur, and only a single millisecond pulsar will remain (*cf.* Ergma and Federova, 1991a,b, for calculations of LMXB-evolution after entering the period gap).

(b) *Evaporation and heating during the X-ray phase*

Although the above described explanation for the virtual absence of LMXB with $P \lesssim 3^H$ seems to qualitatively fit the observations, the real-life evolution of LMXB may be more complex than sketched here, if the effects of the heating of the companion star during the preceding X-ray phase is taken into account. Such heating may possibly start the "evaporation" of the companion already during the X-ray binary phase, *i.e.*, before entering the period gap.

For a review of models including the effects of X-ray heating and evaporation during the X-ray phase we refer to Shaham (this volume) and Bhattacharya (this volume). According to the work by Shaham and collaborators, the evaporative winds during the X-ray phase might "bootstrap" themselves to produce mass-transfer rates to the neutron star that are much larger than those induced by GR and Magnetic Braking alone. Such bootstrap models predict that in short-period LMXB the mass-accretion rates should be high, close to the Eddington limit, as is indeed observed in quite a number of systems. A strong point in favor of the "evaporation" models is that these predicted the existence of binary millisecond pulsars with "evaporating" companion stars already before these systems were discovered (Ruderman, Shaham and Tavani 1989). Still there are quite a number of unsolved problems with the evaporation models, that need further investigation. The most important ones are:

(i) Why do LMXB systems with otherwise very similar characteristics (similar orbital periods, chemical composition, *etc.*) differ by sometimes as much as two orders of magnitude in mass-transfer rate?

(ii) Why does the system of PSR 1957+20 have such a long ($\sim 9^h$) orbital period, and why is its companion despite its very low mass ($0,02M_\odot$) so close to filling its Roche lobe (its radius is 2 to 3 times larger than the thermal equilibrium radius of a $0,02M_\odot$ star).

(c) *Expansion of envelopes of companion stars under the influence of X-ray heating*

In important recent papers, Podsiadlowski (1991) and Harpaz & Rappaport (1991) showed that the presence of the intense radiation field of the X-ray source changes the outer boundary conditions of the envelope of the companion star in a drastic way. Their calculations show that this change in boundary conditions forces the envelopes of low-mass stars ($M \lesssim 2M_\odot$) to expand by a large factor, to fulfill their new boundary condition. Hence, as soon as the mass transfer in a LMXB begins, and the X-ray source has turned on, the companion begins to expand, roughly on a timescale of order of the thermal timescale of its envelope, and begins to rapidly transfer mass to the neutron star. As the neutron star is the more massive component of the system, the mass transfer makes the system expand (the timescale for orbital shrinking by gravitational radiation emission is many orders of magnitude longer) until the companion's new thermal equilibrium radius (which may be two or three times larger than its radius without X-ray heating) no longer exceeds its Roche lobe radius. When this stage is reached the orbital period, which at the onset of the mass transfer for a star of, for example, $0,5M_\odot$ was some 4,5 hours will have increased by about a factor of two or more, to 9 hours of more. Since at its new orbital radius the rate of orbital angular momentum loss by gravitational radiation is much lower than at the initial orbital radius, and since the companion also does not expand further (due to heating), the mass transfer rate will now drop sharply, causing the X-ray luminosity to drop, which in turn causes the X-ray heating to drop. This will cause the companion to shrink and thus to detach itself from its Roche lobe; this causes the X-ray source to turn off completely. During the phase of rapid orbital expansion, which in the above given example lasted not longer than $\sim 10^7$ yrs, the mass transfer rate was of the order of the Eddington limit (or perhaps even larger) such that $\geq 0, 1 M_\odot$ was transferred.

Thus, during this phase the system was a very bright LMXB with a companion that had a radius that (for a main-sequence star) was much too large for its mass.

This is precisely the situation that is found in some LMXB, such as Cen X-4 and A 0620-00.

After the termination of the mass-transfer the companion, which lost over $0, 1 M_\odot$ in a short time, will have shrunk back to its unheated thermal equilibrium radius, which as a factor 2 to 3 smaller than its Roche lobe radius.

The neutron star, being spun up by accretion of $> 0, 1 M_\odot$ during $\sim 10^7$ yrs may now have become a millisecond pulsar. In this way a relatively wide ($P \sim 9 - 10^h$) detached binary with a millisecond pulsar will result.

The reader will notice the great similarity between a system resulting from the LMXB evolution sketched above according to Podsiadlowski (1991) and Harpaz & Rappaport (1991) and the eclipsing binary radio pulsar system PSR 1957+20.

It seems likely that an LMXB may go through several cycles of the type sketched above – all of relatively short duration – until the mass of the companion is exhausted. Its further ablation by the pulsar companion may then finish it off completely such that a single millisecond pulsar may remain.

Since the X-ray phases of a system together don't last more than a few times 10^7 yrs in this picture, the discrepancy between the birthrates of LMXB and millisecond pulsars would be nicely solved by this type of evolution.

The fact that the companion of PSR1957+20 is so close to filling its Roche lobe is easily explained in this model: this is where it started out – far from thermal equilibrium, with an inflated radius – when the mass transfer ended.

The pulsar now makes that the companion keeps filling its Roche lobe by causing its ablative mass loss at a rate $> 10^{-9} M_\odot$/yr (as is evidenced from the observed rate of orbital period change: $-P/\dot{P} \approx 2 \cdot 10^7$yr, cf. Ryba and Taylor, this volume). The mass-loss timescale is much shorter than the thermal timescale of the companion, causing the companion to continue its swelling up and to overflow its Roche lobe. The overflowing matter is blown out of the system by the pulsar wind. As this process keeps going the secondary – becoming a more and more bloated and fluffy star – will be completely evaporated on a timescale $\sim 2 \cdot 10^7$yr $(0, 02 M_\odot/10^{-9} M_\odot/\text{yr})$.

4. Summary and Conclusions

4.1. NEUTRON STARS IN HIGH-MASS BINARIES

a. The evolution of massive close binaries and the formation of High Mass X-ray Binaries (HMXB) seems reasonably well understood. Their formation rate in the Galaxy is expected to be of order $(3 - 4, 5) \times 10^{-3}$/yr. About half of these systems will evolve into short-lived $(2, 5 \times 10^4$ yr) "standard" High Mass X-ray Binaries. The other half will become Be/X-ray binaries, which in most cases are observable only during short-lasting transient outbursts of mass ejection from the Be-star. The number of Be/X-ray binaries in the Galaxy is estimated to be $\sim 2 \cdot 10^4$.

b. During their later evolution the standard HMXB and about half of the Be/X-ray binaries are expected to spiral-in completely, producing Thorne-Zytkow stars. The other about half of the Be/X-ray binaries, with orbital periods $\geq 100^d$ are expected to leave, after spiral-in, ultra-short period binaries consisting of helium star and a neutron star. During helium shell burning these systems will in most cases experience a second spiral-in phase, during which their orbital periods are further expected to be reduced by a factor $\gtrsim 4$ to 5, leading to orbital separations less than about a solar radius at the time of the second supernova explosion. Most of these systems remain bound after this explosion, producing eccentric double neutron star systems with orbital periods in most cases between 1,5 hour and 0,5 day.

c. Due to the short gravitational radiation decay timescale of their orbits, double neutron star systems with $P < 7^h$ are expected to be less than half as numerous as systems with $P > 7^h$, although the birthrate of the latter systems is lower. The total birthrate of close double neutron star systems in the Galaxy is estimated conservatively to be $(4 - 8) \times 10^{-5}$ yr^{-1} (most likely value) and is certainly larger than $(1 - 2) \times 10^{-5}$ yr^{-1} (lower limit). It may possibly be as high as $(2 - 4) \times 10^{-4}$/yr. As the systems are expected to have $P \lesssim 0,5^d$, this birthrate is at the same time the expected rate of merging events of double neutron star binaries in the Galaxy.

d. Although also a sizeable number of black hole-neutron star binaries is expected to be produced in the Galaxy, these systems are expected to have orbital periods of at least a few days, and probably will not merge within a Hubble time.

e. About half of all neutron-star forming events in the Galaxy are expected to be due to bare-core (helium star) explosions.

4.1. NEUTRON STARS IN LOW MASS BINARY SYSTEMS

a. The later evolution of systems that started out with orbital periods ≥ 1 day seems well understood. Here the mass transfer is driven by the interior nuclear evolution of the companion star, and the system finally evolves into a wide radio pulsar binary with a circular orbit and a white dwarf companion with mass $0, 2 - 0, 4 M_\odot$.

b. The evolution of systems with $P \lesssim 0, 5 - 1^d$ is initially driven by the "standard" mechanism of Gravitational Radiation losses and "Magnetic Braking". However, once the X-ray source has turned on, the evolution may become "self-driving" due to the X-ray heating effects on the companion.

Whether this heating will give rise to the production of a "bootstrapped" evaporative wind, which boosts the mass transfer (Shaham, this volume) or whether rapid radius expansion of the companion (Podsiadlowski 1991, Harpaz & Rappaport 1991) will be the main driver of the mass transfer here is not yet fully clear. Both mechanisms may work under certain circumstances, but both need further investigation.

In any case, both mechanisms predict a much shorter X-ray lifetime of LMXB ($\sim 10^7$ yr) than thought previously ($\sim 10^8 - 10^9$ yr). This is just what is needed to reconcile the birthrates of LMXB and millisecond radio pulsars.

c. The mass loss from very low-mass ($\lesssim 0, 1 M_\odot$) secondaries in LMXB occurs on timescales much shorter than the thermal timescale of these stars, causing them to continuously swell up and overflow their Roche lobes. Once their rapidly spinning neutron star companions have become radio pulsars, they will inhibit further mass transfer to the neutron star and will blow the overflowing matter out of the systems, causing the swelling up and Roche-lobe overflow of the secondaries to continue until this star is completely dissipated (*cf.* Kluzniak 1991).

Acknowledgements

I thank many colleagues for enlightening discussions on the topics reviewed above, in particular D. Bhattacharya, W. Kluzniak, S.R. Kulkarni, S. Phinney, M. Ruderman, J. Shaham, M. Tavani, A. Harpaz, S. Rappaport and Frank Verbunt. I especially thank S. Phinney for drawing my attention to the importance of a second spiral-in phase in most helium-star/neutron-star binaries, a fact that I should have realized myself much earlier, but did not.

This research was supported in part by the National Science Foundation under Grant No. PHY89-04035.

References

Bailes, M. (1989) *Astrophys. J.*, **342**, 917.
Biehle, G.T. (1991) *Astrophys. J.*, (in the press); Caltech preprint GRP-241.
Blaauw, A. (1961) *Bull. Astron. Inst. Netherlands*, **15**, 265.
Bodenheimer, P. and Taam, R.E. (1984) *Astrophys. J.*, **280**, 771.
Boriakoff, V., Buccheri, R., Fauci, F. (1983) *Nature*, **304**, 417.
Cannon, R.C. (1991) *Mon. Not. R.A.S.*, (in the press).
Chiosi, C. and Maeder, A. (1986) *Ann. Rev. Astron. Astrophys.*, **24**, 329.
DeCuyper, J.P. (1985) Ph.D. Thesis, Free University, Brussels.
Dewey, R.J. and Cordes, J.M. (1987) *Astrophys. J.*, **321**, 780.
Ergma, E. and Federova, A.V. (1991a) *Astron. Astrophys.*, (in the press).
Ergma, E. and Federova, A.V. (1991b) *Astron. Astrophys.*, (in the press).
Faulkner, J. (1971) *Astrophys. J.*, **170**, L99.
Fryxell, B.A. and Arnett, W.D. (1981) *Astrophys. J.*, **243**, 994.
Habets, G.M.H.J. (1985) Ph.D. Thesis, University of Amsterdam.
Habets, G.M.H.J. (1986) *Astron. Astrophys.*, **167**, 61.
Harpaz, A. and Rappaport, S.A. (1991) *Astrophys. J.*, (in press).
Joss, P.C. and Rappaport, S.A. (1983) *Nature*, **304**, 419.
Kluzniak, W. (1991) Proc. 15th Texas Symposium on Relativistic Astrophysics (in press).
Kraicheva, Z.F., Popova, E.I., Tutukov, A.V., Yungelson, L.R. (1978) in: A. Zytkow, (ed.): "Non-stationary Evolution of Close Binaries", Polish Scientific Publishers, Warsaw, p. 25.
Krolik, J.H. and Sincell, M.W. (1990) *Astrophys. J.*, **357**, 268.
Lewin, W.H.G. and Joss, P.C. (1983) in: W.H.G. Lewin and E.P.J. van den Heuvel (eds.): "Accretion-Driven Stellar X-ray Sources," Cambridge University Press, p. 41.
Maraschi, L., Treves, A. and van den Heuvel, E.P.J. (1975) *Nature*, **295**, 292.
Meurs, E.J. and van den Heuvel, E.P.J. (1989) *Astron. Astrophys.*, **226**, 88–107.
Nomoto, K.C., Filipenko, A.V. and Shigeyama, T. (1990) *Astron. Ap.*, **240**, L1.
Narayan, R., Piran, T., and Shemi, A. (1991) *Astrophys. J.*, **379**, L17.

Paczynski, B. (1971) *Ann. Rev. Astron. Astrophys.*, **9**, 183.

Paczynski, B. (1983) *Nature*, **304**, 421.

Paczynski, B. and Sienkiewicz, R. (1981) *Astrophys. J.*, **248**, L27.

Phinney, E.S. (1991) *Astrophys. J. (Letters)*, **380**, L17.

Podsiadlowski, P. (1991) *Nature*, **350**, 136.

Popova, E.I., Tutukov, A.V. and Yungelson, L.R. (1983) *Astrophys. Space Sci.*, **88**, 55.

Ruderman, M., Shaham, J., Tavani, M. (1989) *Astrophys. J.*, **336**, 507.

Savonije, G.J. (1983a) in: W.H.G Lewin and E.P.J. van den Heuvel (eds.): "Accretion Driven Stellar X-ray Sources," Cambridge Univ. Press, p. 343.

Savonije, G.J. (1983b) *Nature*, **304**, 422.

Shapiro, S.L. and Teukolsky, S.A. (1983) "Black Holes, White Dwarfs and Neutron Stars, John Wiley and Sons, New York.

Spruit, H.C. and Ritter, H. (1983) *Astron. Astrophys.*, **124**, 267.

Sutantyo, W. (1975) *Astron. Astrophys.*, **44**, 227.

Taam, R.E. (1983) *Astrophys. J.*, **270**, 694.

Taam, R.E. and Bodenheimer, P. (1989) *Astrophys. J.*, **337**, 849.

Thorne, K.S. and Zytkow, A.N. (1977) *Astrophys. J.*, **212**, 832.

van den Heuvel, E.P.J. (1974) in: "Astrophysics & Gravitation," Proc. 16th Solvay Conf. on Physics, Univ. of Brussels Press, p. 119.

van den Heuvel, E.P.J. (1976) in: P.P. Eggleton, *et al.* (eds.): "Structure and Evolution of Close Binary Systems," Reidel, Dordrecht, p. 35.

van den Heuvel, E.P.J. (1977) *Annals N.Y. Acad. of Sciences*, **302**, 14.

van den Heuvel, E.P.J. (1978) in: R. Giacconi and R. Ruffini (eds.): "Physics and Astrophysics of Neutron Stars and Black Holes," North Holland Publishing Company, Amsterdam, p. 828.

van den Heuvel, E.P.J. (1983) in: W.H.G. Lewin and E.P.J. van den Heuvel (eds.): "Accretion-Driven Stellar X-ray Sources," Cambridge University Press, p. 303.

van den Heuvel, E.P.J. (1989) in: H. Ögelman and E.P.J. van den Heuvel (eds.): "Timing Neutron Stars," Kluwer Acad. Publishers, Dordrecht, p. 523.

van den Heuvel, E.P.J. and Heise, J. (1972) *Nature Phys. Sci.*, **239**, 67.

van den Heuvel, E.P.J. and De Loore, C. (1973) *Astron. Astrophys.*, **25**, 387.

van den Heuvel, E.P.J. and Taam, R.E. (1984) *Nature*, **309**, 235.

van den Heuvel, E.P.J. and Rappaport, S.A. (1987) in: A Slettebak and T.P. Snow (eds.): "Physics of Be Stars," Cambridge Univ. Press, p. 291.

van den Heuvel, E.P.J. and van Paradijs, J.A. (1988) *Nature*, **334**, 227.

Verbunt, F. (1989) in: W. Kundt (ed.): "Neutron Stars and their Birth Events," Kluwer Acad. Publishers, Dordrecht, p. 179.

Verbunt, F. and Zwaan, C. (1981) *Astron. Astrophys.*, **100**, L7.

Verbunt, F., Wijers, R.A.M.J. and Burm, H. (1990) *Astron. Astrophys.*, **234**, 195.

Webbink, R.F. (1984) *Astrophys. J.*, **277**, 355.

Webbink, R.F., Rappaport, S.A. and Savonije, G.J. (1983) *Astrophys. J.*, **270**, 678.

Wolszczan, A. (1991) *Nature*, **350**, 688.

Rappaport, S.A., Verbunt, F. and Joss, P.C. (1983) *Astrophys. J.*, **275**, 713.

FROM LOW-MASS X-RAY BINARIES TO BINARY AND MILLISECOND PULSARS

DIPANKAR BHATTACHARYA*
Institute for Theoretical Physics
University of California
Santa Barbara, CA 93106
USA

and

Astronomical Institute
& Centre for High Energy Astrophysics
University of Amsterdam
The Netherlands

ABSTRACT. We review the origin of millisecond pulsars and low-mass binary pulsars from low-mass X-ray binaries. Observations of such pulsars suggest that the existing evolutionary picture of low-mass X-ray binaries need major modifications. In particular, careful account has to be taken of the effect of the irradiation of the secondary star by high-energy radiation during the accretion phase, and by the radiation and particle wind from the pulsar after accretion ceases. Another element that has to be incorporated into this picture is the influence of accretion onto the neutron star and its spin evolution on its magnetic field strength.

1. Introduction

Though the idea of the generation of short-period spun-up pulsars from X-ray binaries has existed in the literature for nearly two decades, the role of low-mass X-ray binaries (LMXBs) in producing very rapid pulsars became evident only after the discovery of *millisecond* pulsars (Alpar et al 1982; Joss and Rappaport 1983; Savonije 1983; Paczyński 1983).

The conventional expression for the "equilibrium spin period" (Pringle and Rees 1972; Davidson and Ostriker 1973; van den Heuvel 1977; Ghosh and Lamb 1979; see Ghosh, this volume for a discussion)

$$P_{\rm eq} \simeq 1.9 \text{ ms } \left(\frac{B}{10^9 \text{ s}}\right)^{6/7} \left(\frac{\dot{M}}{\dot{M}_{\rm Edd}}\right)^{-3/7} \tag{1}$$

implies that to be spun-up to a few millisecond period, a neutron star should have a low dipole magnetic field strength ($B \lesssim 10^9$ G), and the accretion rate $\dot{M}$ on the star must be

---

*On leave from Raman Research Institute, Bangalore, India

E. P. J. van den Heuvel and S. A. Rappaport (eds.), X-Ray Binaries and Recycled Pulsars, 257–268.
© 1992 *Kluwer Academic Publishers.*

fairly large (a good fraction of the Eddington rate $\dot{M}_{\rm Edd}$). An additional requirement is that enough matter should be accreted for the spin-up to a short period to be achieved. If the specific angular momentum deposited by the accreted matter is similar to that at the magnetospheric boundary, it would require an amount $\Delta M \sim 0.1 \, M_\odot \, (P_{\rm eq}/1.5 \, {\rm ms})^{-4/3}$ of accretion before the star can be spun up to its equilibrium period $P_{\rm eq}$.

All of the above conditions can be met in a low-mass X-ray binary. The absence of X-ray pulsations in LMXBs indicates that the accreting neutron stars in them have low magnetic field strengths (see Parmar, this volume; van der Klis, this volume). Further, since in these systems the transfer of matter takes place from a $\lesssim 1 \, M_\odot$ star to a more massive neutron star, the mass transfer is dynamically stable, contrary to that in a more massive system, and hence can last for a significant length of time. According to the standard evolutionary models of LMXBs, a good fraction of the donor's mass is transferred to the neutron star, at rates $\sim 0.01 - 1$ times the Eddington rate (see van den Heuvel, this volume, for a review), enough to spin the star up to $P \lesssim 10$ ms.

Since an LMXB has a low-mass donor, the evolutionary product at the end of mass transfer is expected to be a low-mass ($< 0.5 \, M_\odot$) degenerate dwarf in orbit around the spun-up neutron star. Binary pulsars originating from massive systems, however, have either massive white dwarfs or neutron stars as companions. This allows the classification of binary pulsars into "High-Mass" and "Low-Mass" categories, the labels indicating their present companion masses, and hence their ancestry (van den Heuvel and Taam 1984). In section 2 we discuss the expected properties of LMXB products, and compare them with those of low-mass binary pulsars (LMBPs). As we shall see, recent observations and statistical arguments (section 3) suggest major modifications in the standard evolutionary scenario of low-mass X-ray binaries. These modifications will be discussed in section 4.

One distinctive property of most low-mass binary pulsars (LMBPs) and the neutron stars in most LMXBs is their low magnetic field strength—about 3–4 orders of magnitude below that of a typical isolated young pulsar. As mentioned above, this is a necessary condition for spin-up to short periods to occur. However, if these neutron stars were born with strong magnetic fields as in ordinary pulsars, their field strengths must have been reduced in the course of evolution. Until recently it has been thought that magnetic fields of neutron stars undergo spontaneous ohmic decay with a time scale of $\sim 10^7$ yr, and the low field strengths of the neutron stars in low-mass systems therefore only reflect their large ages. Of late evidence is mounting *against* spontaneous ohmic decay of neutron star magnetic field. In section 5 we shall discuss the present status of our understanding of neutron star magnetic field evolution, and its relation to binary evolution and the origin of spun-up pulsars.

2. LMXB evolution and the origin of LMBPs: the standard model

2.1. EVOLUTION OF LMXBS AND EXPECTED END PRODUCTS

Here we shall recall the basic features of LMXB evolution, referring the reader to the review by van den Heuvel (this volume) for details. LMXBs can be divided into three broad classes, according to their initial orbital period $P_{\rm b0}$:

Wide systems: $P_{b0} \gtrsim$ 1-2 d: In these systems the secondary star fills its Roche lobe by evolving away from the main sequence. As mass is transferred from the less massive to the more massive component, the orbit expands, but nuclear evolution makes the radius of the secondary expand too, and thereby sustains mass transfer. The orbital expansion continues till only the degenerate core of the donor is left, whereupon the binary detaches and mass transfer stops. The expected end product is a 0.2-$0.4 M_\odot$ white dwarf in a wide (with period $P_b \sim$ tens of days to years), circular orbit.

Close systems: $P_{b0} \lesssim$ 12 h: These systems are brought into contact by loss of orbital angular momentum due to gravitational radiation and magnetic braking, while the secondary is still on the main sequence. The secondary shrinks on loss of mass, but the orbit also keeps shrinking due to angular momentum loss, thus keeping the mass transfer going. If the mass transfer continues uninterrupted, these systems would pass through a minimum orbital period of $\sim$ 80 min, following which the secondary becomes degenerate, its mass-radius relation reverses, and orbit begins to expand again. If this evolutionary course is strictly followed, mass transfer is unlikely to finish in a Hubble time, but as we shall discuss in section 4, the spun-up neutron star may turn on as a pulsar during the course of evolution and inhibit further accretion of matter.

Intermediate systems: 12 h $\lesssim P_{b0} \lesssim$ 1-2 d: In these systems both processes, namely, nuclear evolution of the secondary and angular momentum loss from the system would be important. The expected end product would be a low-mass (core of a subgiant) degenerate dwarf in circular orbit around the neutron star, with orbital period between several hours and a few days (see Pylyser and Savonije 1988, 1989 for the rich variety of outcomes expected from systems of this category).

2.2. Observed LMBPs and their origin

Low-mass binary pulsars have been discovered both in the galactic disk and in globular clusters. In this section we shall discuss only the disk LMBPs. In the origin of the pulsars in globular clusters a major role is played by stellar encounters (a process entirely unimportant in the galactic disk), which will be discussed in detail by Phinney (these proceedings). Table 1 lists the spin period P, surface dipole magnetic field B_s, orbital period P_b, orbital eccentricity e and the most likely companion mass M_c of the low-mass binary pulsars known in the galactic disk.

As can be seen from table 1, PSRs 1953+21 and 0820+02 have just the right characteristics expected from the evolution of wide LMXBs. From the evolutionary model by Webbink, Rappaport and Savonije (1983) the orbital periods P_{b0} at the start of mass transfer for these systems can be estimated to be $\sim$ 12 and $\sim$ 300 days respectively (see, e.g. Verbunt 1990).

According to this model, the average mass transfer rate driven by nuclear evolution from a $1 M_\odot$ star can be written as $< \dot{M} > \approx 8 \times 10^{-10} P_{b0}$ (d) $M_\odot$ yr^{-1}, which yields $\dot{M} \sim 6 \times 10^{-9}$ $M_\odot$ yr^{-1}, i.e. a little below the Eddington limit, for PSR 1953+29 and $\dot{M} \sim 10^{-7}$ $M_\odot$ yr^{-1}, a super-Eddington rate, for PSR 0820+02. In the latter case it is likely that only $\sim$ 10% of the transferred matter actually accreted onto the neutron star. If so, then the picture of "conservative" evolution adopted in the model of Webbink,

Table 1: Low-mass binary pulsars and millisecond pulsars in the galactic disk.

PSR	P ms	$\log B_s$ (G)	P_b days	e	likely M_c $M_\odot$	Ref.
0820+02	864.87	11.48	1232.5	0.012	0.2–0.4	1,2
1257+12[a]	6.22	?	-	-	-	3
1831−00	520.95	10.94	1.81	< 0.004	0.06–0.13	4,2
1855+09	5.36	8.48	12.33	2.1×10^{-5}	0.2–0.4	6,7
1937+21[a]	1.56	8.61	-	-	-	8,5
1953+29	6.13	8.63	117.35	3.3×10^{-4}	0.2–0.4	9,5
1957+20[b]	1.61	8.21	0.38	$< 2 \times 10^{-5}$	~ 0.01	10,11

Notes:
[a]No evidence of binary motion seen, most probably isolated pulsar.
[b]Companion is being vaporized by pulsar wind.

References:
1. R.N. Manchester et al, 1978 MNRAS 185, 409
2. J.H. Taylor & R.J. Dewey, 1988 ApJ 332, 770
3. A. Wolszczan, 1990 IAU Circ 5073
4. R.J. Dewey et al, 1986 Nat 322, 712
5. L.A. Rawley et al, 1988 ApJ 326, 947
6. D.J. Segelstein et al, 1986 Nat 323,714
7. M.F. Ryba & J.H. Taylor, 1991 ApJ 371, 739
8. D.C. Backer et al, 1982 Nat 300, 615
9. V. Boriakoff et al, 1983 Nat 304, 417
10. A.S. Fruchter et al, 1988 Nat 333, 237
11. A.S. Fruchter et al, 1990 ApJ 351, 642

Rappaport and Savonije (1983) breaks down, and the initial orbital period of the binary can no longer be estimated with precision. Depending on the angular momentum carried away by the escaping matter, the orbit may expand faster or slower than that expected from conservative evolution. If most of the matter was expelled from the vicinity of the neutron star, the initial orbit could have been somewhat smaller than the above estimate. Further, it is unclear whether the neutron star would have been visible as an X-ray source during the accretion phase: the large cocoon of matter surrounding the neutron star undergoing super-Eddington accretion would absorb the X-rays and reprocess them to lower energies, probably to optical/ultraviolet band. It has been suggested that the supersoft X-ray sources recently discovered by ROSAT (Trümper et al 1991; Greiner et al 1991) might be objects of a similar kind.

The predicted values of accretion rates, and the amount of accreted matter are sufficient for both these pulsars to have been spun up to periods shorter than those presently observed. The large difference in spin period between the two cases is a result of their magnetic field strengths. PSR 0820+02 has a much stronger field—as a result the corresponding "equilibrium period" is longer (see eq. 1), and the spindown timescale is shorter. The white dwarf nature of the companion has been confirmed by optical observation of PSR 0820+02 (Kulkarni 1986). PSR 1953+29 is too distant for its companion white dwarf to be optically detected.

PSRs 1855+09 and 1831−00 fall in the intermediate category of evolution, where both nuclear evolution and angular momentum loss are important. The outcome depends rather strongly on the assumed rate of magnetic braking. The relatively wide orbit of PSR 1855+09 indicates that the effect of angular momentum loss was marginal in this case. Calculations by Pylyser and Savonije (1988) suggest an initial orbital period ~ 1 d and an initial donor mass in the range 1.0–2.0 $M_\odot$ for this system. If the donor mass was high (i.e. $\sim 2\,M_\odot$), the mass transfer would have been super-Eddington, and only $\sim 0.2\,M_\odot$ would have been accreted by the neutron star.

For PSR 1831−00 the initial orbit must have been smaller, allowing the system to come into contact at a very early stage of post-main sequence evolution. Of the various models presented by Pylyser and Savonije (1988) the one that comes closest to reproducing the properties of this system has an initial donor mass $\sim 1.0\,M_\odot$ and initial orbital period ~ 0.7 d. The predicted white dwarf mass is $\sim 0.18 - 0.20\,M_\odot$.

The remaining three pulsars in table 1 do not fit into the standard evolutionary scheme of LMXBs. Two of them, PSRs 1257+12 and 1937+21, do not have binary companions, but resemble the binary millisecond pulsars in spin period and magnetic field strength (the field strength of PSR 1257+12 is yet to be measured). If they have also been spun-up due to accretion, their companions must have been lost in some way, and the standard model provides no satisfactory mechanism for achieving this.

A clue to what might have happened to the companions of the single millisecond pulsars is being provided by PSR 1957+20. In this system the intense radiation of the pulsar is vaporizing its companion away. The recent measurement of the rate of change of orbital period of this system (see Taylor, this volume), if interpreted as being due to mass loss, suggests that the $\sim 0.01\,M_\odot$ companion will completely evaporate in $\sim 10^7$ yr. The effect of pulsar radiation and of the high-energy radiation in the mass transfer phase on the secondary star adds a new dimension to the evolution of LMXBs, suggesting a major modification of the standard evolutionary scenario.

3. Some statistical issues

That the standard picture of LMXB evolution may require some modification is also indicated by several statistical arguments. Three most important of them are the following:

3.1. THE BIRTHRATE OF LMBPs

From the observed population of low-mass binary pulsars it has been estimated (Kulkarni and Narayan 1988) that the birthrate of low-mass binary pulsars (with $P_b < 25$ d) in the galactic disk exceeds that of low-mass X-ray binaries by more than an order of magnitude, assuming the lifetimes of LMXBs to be equal to the duration of mass-transfer phase predicted by the standard model (see Kulkarni, this volume for a detailed review). This means that either the progenitors of low-mass binary pulsars should be sought among objects other than low-mass X-ray binaries, or that the standard model has to be modified to incorporate shorter X-ray lifetimes of LMXBs. It is to be noted that spinning up a neutron star to a few millisecond period requires only $\lesssim 10^7$ years at Eddington accretion rate, whereas the standard model of evolution predicts an average LMXB lifetime $\gtrsim 10^8$ yr.

3.2. ABSENCE OF LOW-LUMINOSITY LMXBs

The reader will notice from the discussion in section 2 that those cases in which reasonable models for observed LMBPs could be constructed from the standard scenario all corresponded to relatively wide systems, where the secondaries fill their Roche lobe after evolving away from the main sequence. This is due to the fact that the standard model makes no definite prediction about how the mass transfer in the close systems ($P_{b0} \lesssim 12$ h) should end. If the full course of standard evolution is followed, the mass transfer rate in such a system should continue to drop and the system should become progressively fainter in X-rays as it crosses the period minimum (see van den Heuvel, this volume). Further, the evolution would also become slower and slower, and as a result the sources would "pile up" at low X-ray luminosities $\lesssim 10^{35}$ erg s^{-1}. This, however, is not observed (see Long and van Speybroeck 1983). If this is not due to observational selection, it indicates that either most LMXBs are "born wide", or mass transfer is somehow prevented from entering the low-$\dot{M}$ phase. The former seems unlikely, since many LMXBs are *known* to have $P_b < 10$ h.

If prevention of low-$\dot{M}$ can indeed be achieved, then these systems may eventually leave rapidly spinning pulsars as end products. If not, slow accretion would spin the neutron star down to too long a period to function as a radio pulsar (Jeffrey 1986).

3.3. PAUCITY OF LMXBs WITH $P_b < 2$ H

It is instructive to compare the distribution of orbital periods of low-mass X-ray binaries and Cataclysmic Variables (CVs). The latter are systems similar to LMXBs, but the accreting compact objects in them are white dwarfs, rather than neutron stars. As can be seen from fig. 1, the main difference in the period distributions of these two species is that a large number CVs have orbital periods between 1.3 h and 2h, where LMXBs are virtually absent. In fact the distribution of the orbital periods of CVs shows a "gap" between 2h and 3h. This is thought to occur due to a sudden loss of magnetic braking as the donor becomes completely convective (Spruit and Ritter 1983, Rappaport, Verbunt and Joss 1983). Above the period gap magnetic braking drives the mass transfer in a time scale shorter than the thermal time scale of the secondary, causing it to depart from thermal equilibrium and inflate. When magnetic braking disappears, the time scale for orbital decay (which is now driven by gravitational radiation alone) lengthens by an order of magnitude, during which time the secondary can cool and shrink within the Roche lobe. The system comes into contact again when the orbital period has been reduced to ~ 2 h.

LMXBs, on the other hand, become rare at periods below the gap. This indicates that most LMXBs are unable to emerge from the period gap and resume mass transfer. The low-$\dot{M}$ phase of standard evolution, which lies below the period gap, apparently does not occur in most LMXBs.

4. Beyond the standard model

The above arguments clearly suggest that the standard model needs modification. It is also clear that this modification should take into account the irradiation and ablation of the secondary. However, there is as yet no complete model of what changes this might cause in the evolutionary scenario. Ideas abound, and we shall discuss a few of them below.

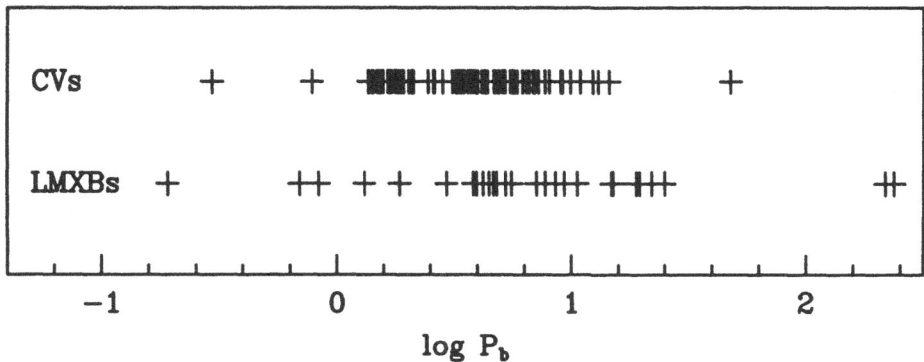

Figure 1: Distributions of orbital periods of cataclysmic variables (CVs) and low-mass X-ray binaries (LMXBs); data from Ritter (1990).

An interesting suggestion to provide a common answer to the above problems was made by van den Heuvel and van Paradijs (1988). According to this hypothesis close LMXBs continue their magnetic braking-driven large-$\dot{M}$ till they enter the period gap. At this point the binary detaches, and the neutron star, which has by this time been spun up to a few millisecond period, switches on as a pulsar. The radiation and e^+e^- wind from the pulsar impinges on the companion and erodes it away, leaving a solitary millisecond pulsar at the end. This way the absence of low luminosity systems and LMXBs below the period gap, as well as the origin of solitary millisecond pulsars can be explained. It would also reduce the active X-ray lifetime of these LMXBs, though it is not clear that this reduction would be large enough to solve the LMBP birthrate problem.

A different school of thought (Ruderman, Shaham and Tavani 1989; Ruderman et al 1989; see Shaham, this volume for a discussion) attributes the short X-ray lifetimes of LMXBs to an accelerated evolution caused by the irradiation of the secondary by high-energy radiation *during* the X-ray phase. According to this model irradiation of the secondary drives a strong wind, which results in a near-Eddington mass transfer on to the neutron star, and probably a large quantity of mass loss from the system. This would explain the absence of low-luminosity LMXBs, and may also shorten LMXB lifetimes enough to bring the derived birth rates of LMXBs and LMBPs into agreement (Tavani 1991). However, no observational evidence for strong winds from LMXBs exist except in only two cases (AC 211, Cyg X-3), and the X-ray luminosity of many LMXBs are well below the Eddington limit, contrary to the prediction of this model. In this model the orbital evolution would be strongly affected by mass loss, and it has been suggested that systems will shrink at first, but will begin to widen again at periods $\gtrsim 3$ h, explaining the paucity of systems below the so-called period gap. This explanation is somewhat doubtful, however, considering the fact that even in cataclysmic variables, for which the evolution is not expected to be significantly affected by the irradiation of the secondary, the "gap" begins at almost exactly the same period (see fig. 1). Further, this model by itself does not provide a way to form single millisecond pulsars. The evolution ends with a $\sim 0.4\,M_\odot$ main-sequence star underfilling its Roche lobe, and one has to invoke the radiation from the pulsar to eventually vaporize it away (Kluźniak et al 1988, Tavani et al 1989).

An important effect of the illumination of the secondary has recently been noted by Podsiadlowski (1991). If the illuminating flux is above a certain critical value, the radius of a low-mass secondary would increase considerably due to a change in the ionization structure of its envelope. This would cause enhanced mass transfer, expansion of the orbit, consequent reduction in $\dot{M}$ and illuminating flux, and finally a detachment of the orbit when the illuminating flux falls below the critical value causing the radius of the secondary to shrink again. If the neutron star has been spun up to a short enough spin period in the meantime, it may be in a position to ablate the companion away without further recurrence of the LMXB phase.

A similar effect can also occur if there are large variations in $\dot{M}$ of LMXBs due to other reasons (see, e.g. Hameury, King and Lasota 1986). It is well possible that at low-$\dot{M}$, LMXBs become transient X-ray sources (White, Kaluzienski and Swank 1984). These sources have bright (10^{37-38} erg s^{-1}) outburst phases separated by long, quiescent phases in which the mass accretion rate on the neutron star is very low ($\lesssim 10^{-12}\ M_{\odot}$ yr^{-1}). If the neutron star has achieved a short enough spin period, such a drop in $\dot{M}$ may move the magnetospheric boundary outside the speed-of-light cylinder. This would allow the neutron star to turn on as a pulsar, following which its wind and radiation pressure may halt any further matter inflow and also evaporate the companion away (Ruderman et al 1989, Srinivasan and Bhattacharya 1989, see Shaham, this volume).

A more radical suggestion has been that millisecond pulsars form due to accretion-induced collapse, and then become single by evaporating their companions, without going through an intermediate LMXB phase (Bailyn and Grindlay 1990). This would sever the evolutionary connection between LMXBs and millisecond pulsars (though, in some cases, the pulsar may turn into an LMXB if not enough spindown power is available to completely evaporate the companion away—see Kluźniak, this volume), and would render much of the above discussion meaningless. However, it is not clear why a neutron star born in accretion-induced collapse should necessarily have a low magnetic field as in a millisecond pulsar. Further, it appears that for an accretion-induced collapse to occur rather special conditions are necessary (see Nomoto, this volume); so the importance of this route for pulsar formation is very uncertain (see Bhattacharya and van den Heuvel 1991; Bhattacharya 1991b).

A more likely way to hide some progenitors of LMBPs is by having a phase of super-Eddington mass transfer, during which the emerging X-rays may be screened. $\dot{M} > \dot{M}_{\rm Edd}$ would result if the initial orbital period is very large, and also if the secondary mass is slightly above the stable regime (see discussion of the origin of PSRs 0820+02 and 1855+09 in section 2.2). As pointed out by Coté and Pylyser (1989), this might make an important contribution towards reconciling the birth rates of LMXBs and LMBPs.

Evaporation of the companion due to pulsar wind may play a major role even in globular clusters. Most of the pulsars so far found in globular clusters are single, and one among the binaries (PSR 1744−24A) has an evaporating companion (see Lyne, this volume). There are, of course, other ways for forming single recycled pulsars in globular clusters. Stellar encounters can release a neutron star from a binary, or a neutron star may collide with a star, disrupt it, and get spun-up by accreting some of its remains (see Phinney, this volume). It turns out that release of neutron stars from binaries via stellar encounters is not enough to explain the high abundance of singles, and it is unlikely that accretion from the remains of a disrupted star would last long enough to be able to spin a neutron star up to the very short periods seen in several globular cluster pulsars, especially those in 47 Tuc (see Bhattacharya and van den Heuvel 1991; Bhattacharya 1991b)

5. The evolution of neutron star magnetic fields

As mentioned in section 1, the low magnetic field strengths of the neutron stars in most LMXBs and LMBPs have till recently been thought to arise due to a spontaneous exponential decay of much higher fields ($\gtrsim 10^{12}$ G) at birth, in a time scale of $\lesssim 10^7$ yr, as originally hypothesised by Ostriker and Gunn (1969) and Gunn and Ostriker (1970).

Such a decay of the field strength has, however, been difficult to understand from a theoretical point of view (see, e.g. Baym, Pethick and Pines 1969; Sang and Chanmugam 1987). Observational evidence for long-lived magnetic fields of neutron stars has also emerged in several contexts:

1. A lower limit of $\sim 10^9$ yr to the age of the binary pulsar PSR 0655+64 has been obtained from the cooling age of its white dwarf companion (Kulkarni 1986). This pulsar has retained a magnetic field strength of $\sim 10^{10}$ G.

2. The upper limit to the optical brightness of the companion of PSR 1855+09 suggests a cooling age of $\gtrsim 10^9$ yr for this white dwarf, which would be a lower limit to the age of the pulsar (Callanan et al 1990; Kulkarni, Djorgovski and Klemola 1991). This pulsar has a magnetic field strength of 3×10^8 G.

3. The large inferred number of millsecond pulsars in the galaxy can only be explained if their active lifetimes are long, $\gtrsim 10^9$ yr. This is possible only if their magnetic fields are stable (Bhattacharya and Srinivasan 1986; van den Heuvel, van Paradijs and Taam 1986).

4. It has been shown from evolutionary considerations that the high-field ($\sim 10^{12}$ G) neutron star in the X-ray binary Her X-1 must be at least 10^8 yr old (Verbunt, Wijers and Burm 1990).

5. Several gamma-ray bursters, which are thought to be very old neutron stars, have shown cyclotron line features which indicate magnetic field strengths $\sim 10^{12}$ G in the emitting region (Murakami et al 1988).

6. Recent measurements of pulsar proper motions have found a number of pulsars moving towards towards the galactic plane from large heights (see Harrison et al, this volume). These are likely to be objects that have completed at least a quarter of an orbit in the galactic potential. Since this requires a good fraction of 10^8 yr, the magnetic fields of these pulsars must be long-lived too.

In view of these observations it seems very likely that the magnetic fields of neutron stars *do not* undergo significant spontaneous ohmic decay. The fact remains, however, that most of the binary pulsars have field strengths considerably lower than that of the isolated pulsar population. This can be understood if the magnetic field of a neutron star decays only if it has been a member of a binary system (see Bailes 1989; Bhattacharya 1991a; Bhattacharya and Srinivasan 1991; Bhattacharya and van den Heuvel 1991 for detailed discussions).

Evolution in binaries can affect the magnetic fields of neutron stars in at least two ways. The matter accreted onto the neutron star may screen and bury the magnetic field, reducing the exterior dipole component (Bisnovatyi-Kogan and Komberg 1974; Taam and van den Heuvel 1986; Romani 1990).

The second possibility is that the field evolution is related to the spin history of the neutron star. The magnetic field in the superconducting interior of the neutron star is carried by quantized proton fluxoids, which have to be expelled from the superconducting region for field decay to occur. As suggested by Srinivasan et al (1990), such an expulsion may be achieved due to an interaction of these fluxoids with the vortex lines in the neutron superfluid, which carry the angular momentum of the star. As the star spins down, the vortices migrate towards the surface, carrying the fluxoids with them. An isolated neutron star does not spin down very much in a Hubble time, but one in a binary system can be spun down to very long periods by the wind of the companion, before the onset of Roche-lobe overflow and spin-up, causing the expulsion of a large amount of magnetic flux.

Yet another mechanism for relating the magnetic field and spin evolution of a neutron star has recently been suggested by Ruderman (1991). In this picture the spin-down of a neutron star would result in the motion of crustal plates towards the equator, carrying the magnetic flux frozen in them. In the equatorial region the magnetic field would have opportunity to reconnect, and to be buried deep under the crust due to subduction (see Ruderman, this volume, for a discussion).

6. Summary

It has conventionally been thought that low-mass X-ray binaries are the ideal progenitors for the low-mass binary pulsars and millisecond pulsars. Several aspects of recent observations have, however, become difficult to explain based on the conventional model for LMXB evolution. It is now clear that new elements have to be incorporated into the evolutionary scheme of LMXBs to maintain the credibility of LMXBs as the progenitors of binary and millisecond pulsars. The two most important of these are:

1. The irradiation and ablation of the secondary.

2. The behaviour of the magnetic field of the neutron star as a consequence of the evolution of the binary system.

Acknowledgements

The author is grateful to Ed van den Heuvel and Frank Verbunt for many discussions. This research was supported in part by NSF Grant No. PHY89-04035.

References

Alpar M. A., Cheng A. F., Ruderman M. A., and Shaham J., 1982, Nat. 300, 728
Bailes M., 1989, ApJ 342, 917

Bisnovatyi-Kogan G. S. and Komberg B. V., 1974, Astron. Zh. 51, 373 (English translation in 1975 Sov. Astron. 18, 217)

Bailyn C. D. and Grindlay J. E., 1990, ApJ 353, 159

Baym G., Pethick C., and Pines D., 1969, Nat. 224, 673

Bhattacharya D., 1991a, in *IAU Colloquium 128: Magnetospheric Structure and Emission Mechanisms of Radio Pulsars*, ed. J. Gil, T. H. Hankins, and J. M. Rankin, Pedagogical University of Zielona Gora press, Zielona Gora, Poland, in press

Bhattacharya D., 1991b, in *Neutron Stars: Theory and Observation*, ed. J. Ventura and D. Pines, Kluwer Academic Publishers, Dordrecht, in press

Bhattacharya D. and Srinivasan G., 1986, Curr. Sci. 55, 327

Bhattacharya D. and Srinivasan G., 1991, in *Neutron Stars: Theory and Observation*, ed. J. Ventura and D. Pines, Kluwer Academic Publishers, Dordrecht, in press

Bhattacharya, D. and van den Heuvel, E. P. J., 1991, Physics Reports, in press

Callanan P. J., Charles P. A., Hassal B. J. M., Machin G., Mason K. O., Naylor T., Smale A. P. and van Paradijs J., 1990, MNRAS 238, 25P

Coté J. and Pylyser E. H. P., 1989, A&A 218, 131

Davidson K. and Ostriker J. P., 1973, ApJ 179, 583

Ghosh P. and Lamb F. K., 1979, ApJ 234, 296

Greiner, J., Hasinger, G., Kahabka, P., 1991, A&A, in press

Gunn J. E. and Ostriker J. P., 1970, ApJ 160, 979

Hameury, J. M., King, A. R. and Lasota, J. P., 1986, A&A 162, 71

Jeffrey L. C., 1986, Nat. 319, 384

Joss P. C. and Rappaport S. A., 1983, Nat. 304, 419

Kluźniak W., Ruderman M., Shaham M., and Tavani M., 1988, Nat. 334, 225

Kulkarni S. R., 1986, ApJ 306, L85

Kulkarni, S. R., Djorgovski, S., Klemola, A. R., 1991, ApJ 367, 221

Kulkarni S. R. and Narayan R., 1988, ApJ 335, 755

Long, K. S. and van Speybroeck, L. P., 1983, in *Accretion Driven Stellar X-ray Sources*, eds. W. H. G. Lewin and E. P. J. van den Heuvel, Cambridge University Press, p. 117

Murakami, T., Fujii, M., Hayashida, K., Itoh, M., Nishimura, J., Yamagami, T., Conner, J. P., Evans, W. D., Fenimore, E. E., Klebesadel, R. W., Yoshida, A., Kondo, I. and Kawai, N., 1988, Nat. 335, 235

Ostriker J. P. and Gunn J. E., 1969, ApJ 157, 1395

Paczyński B., 1983, Nat. 304, 421

Podsiadlowski, P., 1991, Nat. 350, 136

Pringle J. E. and Rees M. J., 1972, A&A 21, 1

Pylyser E. H. P. and Savonije G. J., 1988, A&A 191, 57

Pylyser E. H. P. and Savonije G. J., 1989, A&A 208, 52

Rappaport S., Verbunt, F. and Joss, P. C., 1983, ApJ 275, 713

Ritter H., 1990, A&A Suppl 85, 1179

Romani, R. W., 1990, Nat. 347, 741

Ruderman, M., 1991, ApJ 366, 261

Ruderman M., Shaham J., and Tavani M., 1989, ApJ 336, 507

Ruderman M., Shaham J., Tavani M. and Eichler D., 1989, ApJ 343, 292

Savonije G. J., 1983, Nat. 304, 422

Spruit, H. C. and Ritter, H., 1983, A&A 124, 267

Srinivasan G. and Bhattacharya D., 1989, Curr. Sci. 58, 953

Srinivasan G., Bhattacharya D., Muslimov A. G., and Tsygan A. I., 1990, Curr. Sci. 59, 31

Taam R. E. and van den Heuvel E. P. J., 1986, ApJ 305, 235

Tavani, M., 1991, ApJ 366, L27

Tavani, M., Kluźniak, W., Ruderman, M., Shaham, J., 1989, Ann. N. Y. Acad. Sci., 571, 427

Trümper, J., Hasinger, G., Aschenbach, B., Bräuninger H., Briel, U.G., Burkert, W., Fink, H. Pfeffermann, E., Pietsch, W., Predehl, P., Schmitt, J.H.M.M., Voges, W., Zimmermann, U. Beuermann, K., 1991, Nat. 349, 579

van den Heuvel E. P. J., 1977, Ann. N. Y. Acad. Sci., 302, 14

van den Heuvel E. P. J. and Taam R. E., 1984, Nat. 309, 235

van den Heuvel E. P. J. and van Paradijs J. A., 1988, Nat. 334, 227

van den Heuvel E. P. J., van Paradijs J. A. and Taam R. E., 1986, Nat. 322, 153

Verbunt F., 1990, in *Neutron Stars and Their Birth Events*, ed. W. Kundt, Kluwer Academic Publishers, Dordrecht, p. 179

Verbunt F., Wijers R. A. M. J. and Burm H., 1990, A&A, 234, 195

Webbink R. F., Rappaport S. A. and Savonije G. J., 1983, ApJ 270, 678

White, N.E., Kaluzienski, J.L., Swank, J.H., 1984, in *High Energy Transients in Astrophysics*, ed. S. Woosley, American Institute of Physics, p. 31

COMMON ENVELOPE EVOLUTION AND FORMATION
OF CATACLYSMIC VARIABLES AND LOW-MASS X-RAY BINARIES

R.F. WEBBINK
Department of Astronomy
University of Illinois
1002 West Green St.
Urbana, IL 61801
U.S.A.
and
Institute for Theoretical Physics
University of California
Santa Barbara, CA 93106

ABSTRACT. Common envelope evolution plays a critical role in the formation of both cataclysmic variables and low-mass X-ray binaries. Conditions necessary for a binary to survive this phase and produce a viable cataclysmic variable are outlined, as are their consequences for the physical properties of the system which survives. Three proposed models for low-mass X-ray binary formation – accretion-induced collapse, helium star supernovae, and triple star evolution – are then examined, with emphasis on the conditions which initial systems must satisfy to follow and survive each of these evolutionary channels.

1. Introduction

From a morphological standpoint, the existence of both cataclysmic variables (CVs) and low-mass X-ray binaries (LMXBs) poses similar evolutionary problems. Both types of binary possess a low-mass donor star together with a stellar remnant in an orbit much too small to have accommodated the progenitor of that remnant. The solutions which have been proposed for this dilemma in each case involve a phase of common envelope evolution – the engulfment of one star by its close binary companion, and the dissipation of enough orbital energy of the embedded binary in the engulfing envelope to produce a wholesale reduction of the orbital separation, and ultimately to eject that envelope.

In the following discussion I will outline the rationale for invoking this process, and the conditions under which it is likely to occur. We will examine first the origin of CVs, in whose formation common envelope evolution is now widely accepted as indispensible, and give consideration to the conditions which progenitor systems must meet, and to the consequences which follow from them. We then turn to the LMXBs, for which we will examine three formation scenarios:

  (i) accretion-induced collapse, by which LMXBs are created directly from CVs;

 (ii) helium star supernovae, whereby LMXBs are created by a separate evolutionary channel paralleling CV formation; and

E. P. J. van den Heuvel and S. A. Rappaport (eds.), X-Ray Binaries and Recycled Pulsars, 269–280.
© 1992 *Kluwer Academic Publishers.*

(iii) triple-star evolution, which combines merger of a massive LMXB with the subsequent common envelope engulfment of a more distant low-mass companion.

2. Cataclysmic Variables

It is now widely accepted that cataclysmic variables must initially have been relatively long period close binary systems. They contain white dwarfs which are on average somewhat more massive than single white dwarfs (see, for example, Webbink 1990 for a recent review). This was almost certainly the case when the CVs were created. As we shall see below, it is unlikely that individual white dwarfs in CVs have grown much in mass since their birth, and indeed most of them have probably *decreased* in mass. One is therefore obliged to look for the progenitors of CVs among binaries capable of accommodating a giant or asymptotic giant branch star, as only such stars contain degenerate cores as massive as the white dwarfs found in CVs (Ritter 1976). An initial orbital period of order $10^2 - 10^4$ days is required.

Paczyński (1976) first proposed that the transformation of a long-period progenitor into a CV occurred through common envelope evolution. In this process, tidal mass transfer, once initiated, accelerates so rapidly that synchronization of the donor star's envelope breaks down completely. The mass-losing star cannot track the contraction of its tidal limit as mass transfer proceeds, and so engulfs its binary companion within a few dynamical time scales (*i.e.*, within a few orbital periods). The companion star can scarcely have evolved off the zero-age main sequence unless its mass nearly equals the initial mass of the giant. It is therefore much denser than the common envelope in which it now orbits, as does the degenerate core of the giant.

The embedded binary core creates a gravitational wake in the common envelope, which lags behind it in rotation. Each core component experiences a drag force of order

$$F \sim \pi R_A^2 \rho v_{\rm rel}^2,$$

where R_A is the accretion radius (Bondi and Hoyle 1944; R_A characterizes the gravitational sphere of influence of a point mass), ρ the ambient envelope density, and $v_{\rm rel}$ the velocity of the core relative to the envelope. This drag extracts energy and angular momentum from the binary core and imparts it to the common envelope. Provided that the orbital separation never becomes so small that the degenerate core tidally disrupts the nondegenerate core, this inflation of the common envelope can ultimately unbind it. The orbital parameters of the binary when the common envelope is ejected can be estimated crudely by equating the initial binding energy of that envelope to the degenerate core when mass exchange first occurs,

$$E_{\rm bind} = -\frac{GM_1 M_e}{\lambda r_L A_i},$$

with some fraction $\alpha \leq 1$ (an efficiency parameter) of the change in orbital energy between initial and final states,

$$\Delta E_{\rm orb} = -\frac{GM_1 M_2}{2A_f} + \frac{G(M_1 + M_e)M_2}{2A_i}.$$

In these expressions, M_1 is the mass of the degenerate core, M_2 the mass of the nondegenerate companion star, M_e the mass of the envelope, A_i the initial orbital semimajor axis, A_f the final orbital semimajor axis, r_L the dimensionless Roche lobe radius of the initial giant (a function only of $(M_1 + M_e)/M_2$ – see Eggleton 1983), λ a measure of the degree of central concentration of the envelope of the initial giant (of order unity for convective envelopes, but smaller for radiative envelopes), and G the universal gravitational constant. All available estimates of the time scale on which common envelope evolution occurs (typically $\lesssim 10^3$ years) justify the conclusion that neither core can have assimilated a significant amount of mass from the common envelope during this process (Webbink 1988; Hjellming and Taam 1991), so M_1 and M_2 remain essentially constant. Preliminary attempts to model common envelope evolution hydrodynamically indicate efficiencies $\alpha \sim 0.3$ to 0.7 (Taam and Bodenheimer 1989; Livio and Soker 1988), but existing model codes cannot yet combine the spatial and temporal resolution in three dimensions needed for a fully satisfactory treatment of this problem.

The issue of whether or not a primordial binary will follow this evolutionary path revolves around the inability of the donor star to accommodate the contraction of its tidal limit at the onset of mass loss to its companion star. Among donors with deep surface convection zones this is invariably the case when they equal or exceed their companions in mass. Indeed, even stars with radiative envelopes may be unable to track their tidal limits if they exceed their companion's mass by a factor of 2.5 or more (Hjellming 1989). Thus common envelope evolution is the principal channel for close binary interaction among binaries with long orbital periods or extreme mass ratios.

Because the radii of giant and asymptotic branch giant stars are sensitive functions of their core masses, the white dwarf mass which appears following common envelope evolution, and with which the binary will begin a new life as a cataclysmic variable, is fixed by the orbital separation (or orbital period) of the binary at the onset of common envelope evolution. Combined with the rudimentary energy equation outlined above describing the outcome of common envelope evolution, this circumstance enables one to estimate the properties of a proto-cataclysmic binary (that is, a very close, but detached white dwarf – red dwarf pair) from the initial masses and orbital period of its progenitor. The further inclusion of angular momentum loss mechanisms which drive a proto-cataclysmic to interaction, and of a statistical distribution of primordial binaries, then provides the basis for theoretical models of the formation rates and parameter distributions of CVs (Politano 1991a,b; Webbink and Politano 1991).

The statistical simulations just cited reveal several facets of CV formation of immediate relevance to the scenarios outlined below for LMXB formation. The most important of these are the following:

Only progenitor systems with very small initial mass ratios, $M_2/M_1 < 0.28$, can contribute to the CV birthrate. This constraint arises in part because the dissipative envelope of the giant must be massive enough to absorb the orbital energy dissipated in reducing the separation of its binary core to a few solar radii, and in part because of stability requirements against rapid mass transfer in the resultant CV. For donor stars with deep connective envelopes, stability against dynamical time scale mass transfer poses the more severe constraint. Thus, main sequence donors with masses $M_2 \lesssim 0.8 M_\odot$ must satisfy the condition

$$M_2/M_1 < 0.67 \qquad \text{for} \qquad M_2 \leq 0.43~M_\odot$$
$$M_2/M_1 < 0.67 + 2.2(M_2/M_\odot - 0.43)^{4/3} \qquad \text{for} \qquad 0.43 M_\odot < M_2 \leq 0.8 M_\odot$$

(Politano 1991a, after Hjellming 1989) in order to avoid a new common envelope phase which would end in the tidal disruption. For more massive donor stars, with largely radiative envelopes, stability against thermal time scale mass transfer is the stronger constraint, and limits viable donor stars to mass ratios $M_2/M_1 \lesssim 1.25$ for main sequence donors with $M_2 > 0.8 M_\odot$. This constraint generally grows more stringent as the donor evolves away from the main sequence (Hjellming 1989).

These stability constraints also severely limit the possibility of evolved donor stars in CVs. The requirement of dynamical stability effectively excludes any putative CV having a giant branch donor more massive than two-thirds the white dwarf mass. Stars of Population I composition, age 10^{10} years, and mass less than two-thirds of the Chandrasekhar mass are still on the main sequence. The only evolved donors which can satisfy stability constraints therefore lie in the Hertzsprung gap, such as it is for stars not much more massive than the Chandrasekhar mass, as they need satisfy only the thermal stability constraint. Thus, CVs with evolved donors must have very massive white dwarfs, and do not form at all with orbital periods exceeding $\sim 2^d$.

3. Low-Mass X-Ray Binaries

Formally, the origin of low-mass X-ray binaries (LMXBs) in the galactic field poses issues very similar to those attending the formation of the cataclysmic variables. The collapsed remnant of fairly massive star is found in a binary too small to have accommodated its original precursors. (In our discussion here, the LMXBs in globular clusters are excluded, since tidal capture – a process of inconsequential frequency in the galactic disk – can create significant numbers of such systems after creation of single neutron stars: *e.g.,* Verbunt and Hut 1987; Predehl *et al.* 1991). There are some important differences in the kinds of constraints which LMXB progenitors must fulfill, however.

The formation scenario for CVs outlined above involves dissipation of most of the initial orbital angular momentum of the binary only after formation of an electron-degenerate core. The analog for LMXBs would require creation of giant stars with extended envelopes and neutron-degenerate cores – Thorne-Żytkow objects (Thorne and Żytkow 1977). However, core collapse leading to neutron star formation is generally associated with Type II supernova explosions, in which the entire envelope is ejected. Since supernova progenitors among single stars already have dimensions exceeding those of LMXBs, a binary can become an LMXB with orbital period as short as those observed only if it can dissipate most of its angular momentum, presumably in a common envelope, *before* core collapse occurs.

The other obvious difference between CV and LMXB formation channels is that CVs outnumber LMXBs in the galactic disk by a factor of order 10^4. Since the typical mass transfer rates, donor star masses, and hence presumably the lifetimes of these two classes of objects are quite comparable, it is immediately apparent that the creation of an LMXB is, relatively as well as absolutely, a very rare event. One must conclude that any scenario which is to succeed at explaining their origins must appeal to an improbable chain of events.

Let us turn now to examining three possible scenarios which have been put forward for the origin of LMXBs.

3.1 ACCRETION-INDUCED COLLAPSE

Single stars in a narrow initial mass range, $\sim 8 - 12 M_\odot$, develop degenerate ONeMg cores late in their evolution (see, *e.g.,* Nomoto 1984). If these stars are left undisturbed, their cores continue to grow in mass, and quickly reach the point of electron-capture induced collapse (Miyaji, *et al.* 1980; Sugimoto and Nomoto 1980). In a binary, this core growth can be arrested by sudden loss of the stellar envelope which feeds it, and collapse delayed until the binary resumes mass transfer at a later time.

Unlikely as it is that a binary should just reach interaction during the very brief interval between creation of an ONeMg core and its collapse, we nevertheless see clear evidence of such white dwarfs in nearly a third of classical nova outbursts (Truran and Livio 1986). This remarkable prevalence among novae of intrinsically rare white dwarfs is a direct consequence of a very strong selection mechanism favoring the most massive white dwarfs. Crudely, nova explosions occur when the hydrostatic pressure, P, at the base of an accreted envelope (mass M_e) exceeds a critical pressure, P_c ($\sim 2 \times 10^{19}$ dynes cm^{-2}), for degenerate hydrogen ignition:

$$P = \frac{GMM_e}{4\pi R^4} > P_c,$$

where M and R are the mass and radius, respectively, of the underlying white dwarf. Since R decreases rapidly as the white dwarf mass approaches the Chandrasekhar limit, the envelope mass to be supplied by the companion star needed to reach outburst also decreases rapidly. Thus, at a given accretion rate massive white dwarfs outburst far more frequently than do their more common, but less massive counterparts (Truran and Livio 1986; Ritter *et al.* 1991).

The presence of a significant contingent of ONeMg white dwarfs among classical novae does not preordain a correspondingly large accretion-induced collapse rate, however. Their presence is in the first place confirmed by the presence of large overabundances of neon-group elements in the ejecta of these novae – masses in these elements alone exceeding the mass of hydrogen which must be burned to helium to power the outburst. Thus the white dwarfs in these systems are almost certainly being eroded in mass with time. The same is true of their less massive carbon-oxygen counterparts.

The most massive white dwarfs among CVs are expected to be found not among classical novae, but among the recurrent novae. Only white dwarfs precariously close to the Chandrasekhar limit can be refueled by accretion rapidly enough to outburst as frequently as once per century or even once per decade (Fujimoto 1982; Starrfield, Sparks, and Truran 1985). While there exists some controversy over whether those recurrent novae with giant donor stars owe their outbursts to thermonuclear events (see, for example, Webbink, *et al.* 1987; Webbink 1991), those with orbital periods within the range of ordinary CVs appear to conform to the thermonuclear model. Furthermore, such abundance studies as exist (*e.g.*, Williams, *et al.* 1981) find no compelling evidence for heavy-element enhancement in their ejecta. A watershed thus appears to exist in white dwarf mass, such that extremely massive white dwarfs grow toward the Chandrasekhar limit, whereas all others retreat from it. Let us therefore consider the recurrent novae as candidates for the creation of LMXBs by accretion-induced collapse.

In Figure 1, the orbital periods of all CVs with known orbital periods, including both classical and recurrent novae, are compared with those of the LMXBs. The short-period cutoff to the CV-period distribution is produced by a competition between angular momentum loss by gravitational radiation and thermal relaxation of the donor star (Paczyński and Sienkiewicz 1981; Rappaport, Joss and Webbink 1982); its sharpness is testament to the unevolved status of the donor stars of essentially all CVs which evolve to such short periods (see, *e.g.*, Rappaport and Joss 1984). At long periods, $\log P_{\rm orb} \gtrsim -0.5$, all donor stars in CVs must be evolved to some degree, since we see none earlier than late G in spectral type. The behavior of the period distribution in both these limits is thus consistent with the evolutionary scenario outlined above for their origin. The pronounced dearth of systems near $\log P_{\mathrm{orb}} = -1$, on the other hand, must be attributed largely to secular evolution. Its cause is not yet established with any certainty, but is widely believed to be associated with an abatement of magnetic braking when the donor star becomes fully connective (at the upper edge of this gap), followed by a phase of detached evolution as the donor relaxes to thermal equilibrium before mass transfer resumes at the lower edge of the gap (Spruit and Ritter 1983).

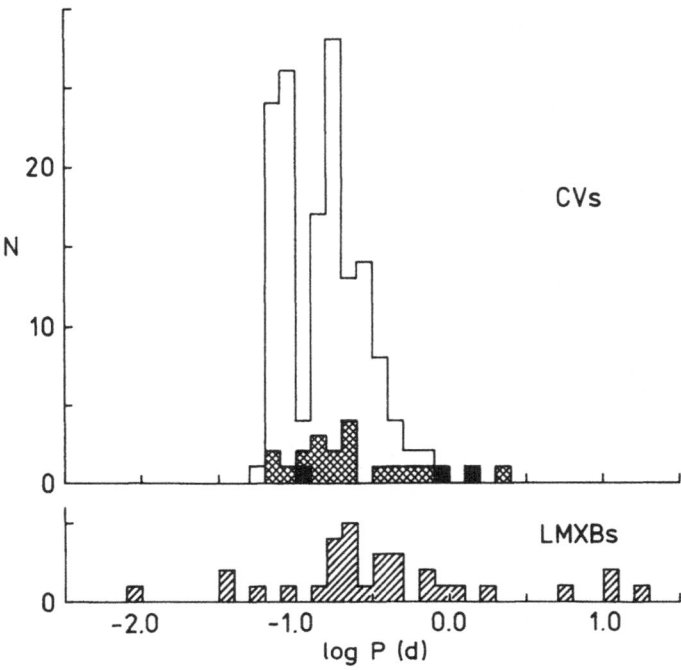

Figure 1. The distribution of cataclysmic variables (CVs) and low-mass X-ray binaries (LMXBs) of known orbital period. Classical novae and recurrent novae have been highlighted among CVs as subsets which select strongly in favor of the most massive white dwarfs, approaching the neutron star masses characteristic of LMXBs.

A comparison of the orbital period distribution of classical novae with that of all CVs suggests that they are somewhat more broadly distributed in orbital period, and relatively more frequent at long periods. These features clearly hint that the novae, which have more massive white dwarfs on average, and thus allow more massive donor stars, are more likely to be represented among systems with evolved donor stars. The trend appears to be carried further among the very few recurrent novae with known periods. It is evident in any case that the orbital period distribution of LMXBs points to a very large representation indeed from evolved donors, since fully 14 of the 31 systems with known periods plotted here have orbital periods $P_{\mathrm{orb}} > 0.^{d}32$ ($\log P_{\mathrm{orb}} > -0.5$). Another four have $P_{\mathrm{orb}} < 0.^{d}056$ (the shortest-period CV), a period range apparently inaccessible to donor stars with primordial compositions. Whether a fuller sample of recurrent novae would reveal an LMXB-like span in orbital periods remains problematic, although it should be noted that secular evolution beyond collapse of the white dwarf to a neutron star (which one might expect early in the lifetime of a recurrent nova, considering the very small core growth required) will certainly tend to drive long-period systems to even longer periods, and short-period systems to shorter periods (Taam, Flannery, and Faulkner 1980; Pylyser and Savonije 1988).

If in fact accretion-induced collapse among recurrent novae leads to the formation of LMXBs, one can obtain a very rough estimate of the LMXB birthrate by this channel as follows: The nova outburst frequency throughout the Milky Way is estimated to be $\dot{N}_N \approx 73$ (Liller and Mayer 1987). Only a small fraction f_{NR} of these outbursts, say $f_{NR} \approx 0.1$ involve recurrent novae evolving towards white dwarf collapse. A typical recurrent nova needs to accrete a mass $\Delta M_{\rm ob} \approx 10^{-7} M_\odot$ (at a rate of a few $\times 10^{-9} M_\odot$ yr^{-1} for $\sim$ 30 years) to achieve outburst, with a fraction perhaps $f_{\rm ret} \approx 0.5$ retained by the white dwarf. If, for the sake of argument, one supposes that a recurrent nova must lie within $\sim 0.04 M_\odot$ of the Chandrasekhar mass to avoid erosion and evolve towards collapse, an average system might need to add half this amount, $\Delta M_{NC} \approx 0.02 M_\odot$ to become an LMXB. The deduced birthrate is then

$$\dot{N}_{\rm LMXB} \sim \dot{N}_N f_{NR} \left(\frac{\Delta M_{\rm ob} f_{\rm ret}}{\Delta M_{NC}} \right) \sim 10^{-5} \text{ yr}^{-1}.$$

Although this estimate exceeds the LMXB birthrate inferred observationally above by two orders of magnitude, it is clear that several of the factors used to derive it (especially $f_{\rm ret}$) might easily be substantially in error. The hypothesis should not be rejected on these grounds.

3.2 HELIUM STAR SUPERNOVAE

Another possibility for the origin of LMXBs parallels the scenario outlined above for CV formation followed by accretion-induced collapse, but appeals now to somewhat more massive progenitors for the compact component. Single stars with initial masses above $\sim 12 M_\odot$ exhaust cores exceeding a Chandrasekhar mass while still on the main sequence. These massive cores pass through helium and carbon ignition nondegenerately, only later cooling to degeneracy, whereupon they immediately collapse. In a binary system, rapid mass loss from such a star tends to continue until the molecular weight gradient at the hydrogen-helium interface is reached, leaving the helium core as the mass transfer remnant. That helium core, which may itself contain a helium-exhausted carbon-oxygen core, will continue to evolve, unlike the cores exposed as white dwarfs in CVs, because it is not yet degenerate. It thus becomes possible for the core to complete its evolution and collapse *after* completion of a mass transfer or common envelope phase of binary evolution, and without having to accrete additional mass from its companion (Sutantyo 1975; van den Heuvel 1981).

As was the case among CV progenitors, a candidate binary can only produce a long-lived LMXB if the prospective donor star is of sufficiently low mass to remain dynamically and thermally stable during mass transfer. The initial mass ratios of LMXB progenitors by this channel must therefore have been quite extreme, $M_2/M_1 \lesssim 0.12$, the upper limit corresponding roughly to the ratio of the Chandrasekhar mass to the initial mass of the primary. Even stars with deep radiative envelopes are subject to developing dynamical time scale mass transfer if they are the more massive components in such disparate systems (Hjellming 1989), and so it is inevitable that LMXBs produced by helium star supernovae must have survived a phase of common envelope evolution. At the extreme initial mass

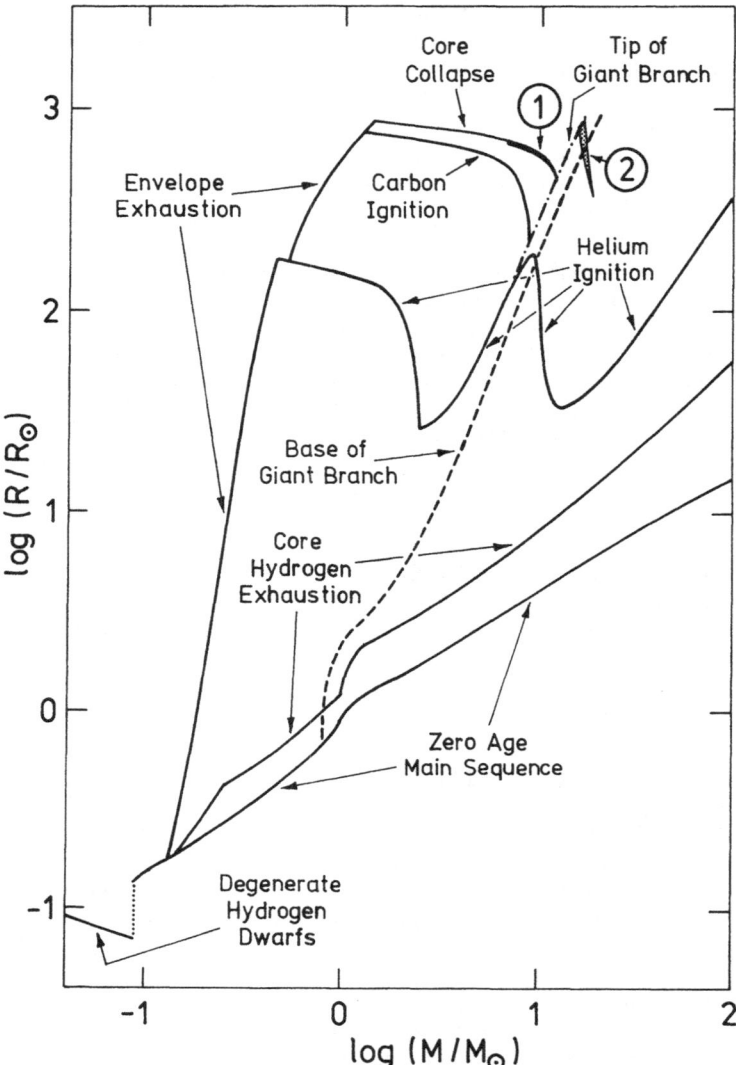

Figure 2. The radius-mass diagram for Population I stars. The radii of stars at various evolutionary states of interest are indicated. The two shaded regions labeled 1 and 2 correspond, respectively, to the progenitors (i) of ONeMg white dwarfs in recurrent novae, which may form LMXBs by accretion-induced collapse, and (ii) of helium cores capable of surviving common envelope evolution, evolving to the supernova stage and still retaining close low-mass companions to become LMXBs. Note that stars less massive than the recurrent nova progenitors here probably do not survive carbon ignition to reach core collapse. Both scenarios appeal to very narrow ranges of initial conditions.

ratio needed, this circumstance pushes prospective initial binaries to the outermost limits of binary separations still capable of tidal mass transfer (see Figure 2).

The creation of LMXBs by this evolutionary channel is thus a tenuous affair. The need for initial orbital periods near the limit for binary interaction places the survival of the binary through common envelope evolution in doubt if the efficiency of envelope ejection falls too far below unity. This limit is sufficiently severe to preclude systems surviving with main sequence stars much below a solar mass – less massive secondaries cannot provide enough orbital energy to eject the common envelope without leaving a remnant binary too small to accommodate the remaining evolution of the helium star. The helium star remnant must in fact have a mass $M_{He} \gtrsim 3M_\odot$ because less massive helium stars would expand enough during shell helium burning (Paczyński 1971; Habets 1986) to engulf the companion in a second common envelope phase. By the same token, it must be low enough in mass ($M_{He} \lesssim 4.2M_\odot$) to keep the mass expelled in the supernova below the mass of the remnant binary, lest it come unbound. This relatively narrow range in helium star masses immediately implies that the post-supernova system, if it remains bound, will have a very eccentric orbit ($e > 0.5$). And finally, both pre- and post-supernova orbits must be large enough to accommodate the main sequence star, but the post-supernova orbit, once circularized, cannot be so much larger that the main sequence star will have evolved all the way to the giant branch, rendering it unstable to dynamical time scale mass transfer for otherwise plausible mass ratios. Obviously, the number and severity of the constraints a system must satisfy to produce LMXBs in this way make it a very fragile channel, but it bears reiterating that the required birthrate is likewise very low.

3.3 TRIPLE STAR EVOLUTION

A third possibility, put forward with the apparent low-mass black hole binary A0620-00 specifically in mind, is that LMXBs are produced by triple star evolution (Eggleton and Verbunt 1986). A massive close binary system is postulated to possess a distant low-mass companion still within reach of tidal interaction. Mass exchange in the rapidly-evolving close massive system produces a Wolf-Rayet binary and thence a massive X-ray binary, upon explosion of the Wolf-Rayet component as a supernova, according to the conventional scenario for the origin of the observed massive X-ray binaries (van den Heuvel and Heise 1972; Tutukov and Yungel'son 1973). The usual massive X-ray binary scenario ends with the massive non-degenerate companion to the X-ray component overfilling its Roche lobe and engulfing the compact star. These systems typically have very extreme mass ratios, and so rapid mass exchange smothering the X-ray source, and indeed common envelope evolution appear inevitable. Calculations by Taam (1979) indicate that, at least among those systems in which the nondegenerate star has not yet developed a very dense, evolved core, the compact component will spiral to its very center, unable to eject the envelope. Accretion onto the central compact object, and residual nuclear burning in that accretion flow, lead to envelope expansion on a thermal time scale, and transformation of the remnant of the massive X-ray binary into a red supergiant of the sort described by Thorne and Żytkow (1975, 1977). Any third component lying within $\sim 2000 R_\odot$ (orbital period $\lesssim 5$ years) at this time stands to be engulfed in this expanding envelope, and the dissipation of its orbital energy within the common envelope is presumed to lead to envelope ejection. If that third component is sufficiently low in mass to satisfy the stability constraints discussed above in the contexts of CV formation and LMXB formation by helium star supernovae, an LMXB may be produced.

Although the evolution of close triple systems has received little attention, and the detection of close triple systems poses especially difficult observational problems, examples of such objects are by no means unknown. Eggleton and Verbunt (1986) cite λ Tau and VV Ori as two examples of short-period early-type close binaries with close low-mass third components. Fekel (1981) surveyed the properties of 43 known multiple systems, the great majority of them appearing in the Bright Star Catalog (Hoffleit and Jaschek 1982), with nearly half qualifying as close triple (or quadruple!) systems. The true fraction of solar-type close binary systems which have close third components has been estimated at $\gtrsim 10\%$ (Webbink 1985). It is not implausible, therefore, to suppose that triple systems of the type envisioned in this scenario occur with some frequency, although it is not yet possible to quantify that frequency.

4. Discussion and Conclusions

Which, if any, of these formation scenarios is correct remains problematic. The accretion-induced collapse scenario was revived in large measure because in the context of widely-accepted field-decay models for pulsar statistics (Lyne 1981; Lyne, Manchester, and Taylor 1985), the magnetic moments deduced for a number of the first binary millisecond pulsars to be discovered implied pulsar ages much less than the mass transfer time scale for the preceding evolutionary state (Helfand, Ruderman, and Shaham 1983; van den Heuvel and Taam 1984; van den Heuvel 1984). The high magnetic field deduced for Her X-1 in particular is difficult to reconcile with the long hiatus between neutron star creation and evolution of the donor away from the main sequence unless the neutron star were recently created by accretion-induced collapse (Sutantyo, van der Linden, and van den Heuvel 1986). By the same token, accretion-induced collapse is incapable of producing black holes in excess of $3M_{\odot}$, as apparently observed in A0620-00 (McClintock and Remillard 1986), a conundrum which prompted proposal of a triple-star origin for this system (Eggleton and Verbunt 1986).

The complexion of the evolutionary problem changes significantly if neutron star magnetic fields are long-lived. Lamb (1981) in fact argued this case on the basis of the large field deduced for Her X-1. Kulkarni's (1986) discovery of relatively cool, and therefore old, white dwarf companions to two binary millisecond pulsars has commanded a reassessment of neutron star field decay (e.g., Taam and van den Heuvel 1986), and may relieve the difficulties on this account for the helium star supernova scenario, which typically involves a long dormant period between creation of the neutron star and the onset of the next phase of mass transfer in the LMXB phase. Unlike accretion-induced collapse, moreover, the helium star supernova scenario can account for the existence of systems like A0620-00 by a relatively straightforward extension to even more massive initial primaries (de Kool, van den Heuvel, and Pylyser 1987).

A comprehensive assessment of the relative likelihood of each of the LMXB formation scenarios outlined here will require a more thorough understanding of several aspects of stellar evolution and statistics. All three scenarios depend sensitively on common envelope ejection efficiencies; models for this process have yet to take full account of its three-dimensional nature, and the possible importance of resonant pumping of the envelope (Soker 1991) deserves further investigation. Complete stellar models through core carbon exhaustion are needed to assess the window available for formation of LMXBs through either accretion-induced collapse or helium star supernovae. And for the triple star scenario, one must contend with the extremely difficult observational problem of detecting faint, solar-mass companions orbiting only a few AU from bright OB binaries.

I would like to thank E.P.J. van den Heuvel for many useful conversations in the preparation of this overview. This work was supported in part by National Science Foundation under grants AST86-16992 and PHY89-04035.

References

Bondi, H. and Hoyle, F. (1944) *Monthly Notices R. Astr. Soc.*, **112**, 195.

de Kool, M., van den Heuvel, E.P.J., and Pylyser, E. (1987) *Astr. Astrophys.*, **183**, 47.

Eggleton, P.P. (1983) *Astrophys. J.*, **268**, 368.

Eggleton, P.P., and Verbunt, F. (1986) *Monthly Notices R. Astr. Soc.*, **220**, 13P.

Fekel, F.C. Jr. (1981) *Astrophys. J.*, **246**, 879.

Fujimoto, M.Y. (1982) *Astrophys. J.*, **257**, 767.

Habets, G.M.H.J. (1986) *Astr. Astrophys.*, **167**, 61.

Helfand, D.J., Ruderman, M.A., and Shaham, J. (1983) *Nature*, **304**, 423.

Hjellming, M.S. (1989) Ph.D. thesis, University of Illinois.

Hjellming, M.S. and Taam, R.E. (1991) *Astrophys. J.*, **370**, 709.

Hoffleit, D., and Jaschek, C. (1982) *The Bright Star Catalogue* (New Haven: Yale U. Obs.).

Kulkarni, S.R. (1986) *Astrophys. J. Letters*, **306**, L85.

Lamb, F.K. (1981) in *IAU Symposium No. 95, Pulsars*, ed. W. Sieber and R. Wielebinski (Dordrecht: Reidel), p. 357.

Liller, W., Mayer, B. (1987) *Publ. Astr. Soc. Pacific*, **99**, 606.

Livio, M., and Soker, N. (1988) *Astrophys. J.*, **329**, 764.

Lyne, A.G. (1981) in *IAU Symposium No. 95, Pulsars*, ed. W. Sieber and R. Wielebinski (Dordrecht: Reidel), p. 423.

Lyne, A.G., Manchester, R.N., and Taylor, J.H. (1985) *Monthly Notices R. Astr. Soc.*, **213**, 613.

McClintock, J.E., and Remillard, R.A. (1986) *Astrophys. J.*, **308**, 110.

Miyaji, S., Nomoto, K., Yokoi, K., and Sugimoto, D. (1980) *Publ. Astr. Soc. Japan*, **32**, 303.

Nomoto, K. (1984) *Astrophys. J.*, **277**, 791.

Paczyński, B. (1971) *Acta Astr.*, **21**, 1.

Paczyński, B. (1976) in *IAU Symposium No. 73, Structure and Evolution of Close Binary Systems*, ed. P. Eggleton, S. Mitton, and J. Whelan (Dordrecht: Reidel), p. 75.

Paczyński, B., and Sienkiewicz, R. (1981) *Astrophys. J. Letters*, **248**, L27.

Politano, M. (1991a) *Astrophys. J.*, submitted.

Politano, M. (1991b), in preparation.

Predehl, P., Hasinger, G., and Verbunt, F. (1991) *Astr. Astrophys.*, submitted.

Pylyser, E., and Savonije, G.J. (1988) *Astr. Astrophys.*, **191**, 57.

Rappaport, S., and Joss, P.C. (1984) *Astrophys. J.*, **283**, 232.

Rappaport, S., Joss, P.C., and Webbink, R.F. (1982) *Astrophys. J.*, **254**, 616.

Ritter, H. (1976) *Monthly Notices R. Astr. Soc.*, **175**, 279.

Ritter, H., Politano, M., Livio, M., and Webbink, R.F. (1991) *Astrophys. J.*, **376**, in press.

Soker, N. (1991) *Astrophys. J.*, **367**, 593.

Spruit, H.C., and Ritter, H. (1983) *Astr. Astrophys.*, **124**, 267.

Starrfield, S., Sparks, W.M., and Truran, J.W. (1985) *Astrophys. J.*, **291**, 136.

Sugimoto, D., and Nomoto, K. (1980) *Space Sci. Rev.*, **25**, 155.

Sutantyo, W. (1975) *Astr. Astrophys.*, **41**, 47.

Sutantyo, W., van der Linden, T.J., and van den Heuvel, E.P.J. (1986) *Astr. Astrophys.*, **169**, 133.

Taam, R.E. (1979) *Astrophys. Letters*, **20**, 29.

Taam, R.E., and Bodenheimer, P. (1989) *Astrophys. J.*, **337**, 849.

Taam, R.E., and van den Heuvel, E.P.J. (1986) *Astrophys. J.*, **305**, 235.

Taam, R.E., Flannery, B.P., and Faulkner, J. (1980) *Astrophys. J.*, **239**, 1017.

Thorne, K.S., and Żytkow, A.N. (1975) *Astrophys. J. Letters*, **199**, L19.

Thorne, K.S., and Żytkow, A.N. (1977) *Astrophys. J.*, **212**, 832.

Truran, J.W., and Livio, M. (1986) *Astrophys. J.*, **308**, 721.

Tutukov, A.V., and Yungel'son, L.R. (1973) *Nauchn. Inf. Akad. Nauk S.S.S.R.*, **27**, 58.

van den Heuvel, E.P.J. (1981), in *IAU Symposium No. 93, Fundamental Problems in the Theory of Stellar Evolution*, ed. D. Sugimoto, D.Q. Lamb, and D.N. Schramm (Dordrecht: Reidel), p. 137.

van den Heuvel, E.P.J. (1984) *J. Astrophys. Astr.*, **5**, 209.

van den Heuvel, E.P.J., and Heise, J. (1972) *Nature, Phys. Sci.*, **239**, 67.

van den Heuvel, E.P.J., and Taam, R.E. (1984) *Nature*, **309**, 235.

Verbunt, F., and Hut, P. (1987) in *IAU Symposium No. 125, The Origin and Evolution of Neutron Stars*, ed. D.J. Helfand and J.H. Huang (Dordrecht: Reidel), p. 187.

Webbink, R.F. (1985) in *1982-1984 Eclipse of Epsilon Aurigae*, ed. R.E. Stencel (NASA Conference Publication 2384), p. 49.

Webbink, R.F. (1988) in *Critical Observations vs. Physical Models for Close Binary Systems*, ed. K.-C. Leung (New York: Gordon and Breach), p. 403.

Webbink, R.F. (1990) in *Accretion-Powered Compact Binaries*, ed. C.W. Mauche (Cambridge: Cambridge U. Press), p. 177.

Webbink, R.F. (1991) in *IAU Colloquium No. 122, The Physics of Classical Novae*, ed. A. Cassatella (Berlin: Springer), in press.

Webbink, R.F., and Politano, M. (1991) in *New Frontiers in Binary Star Research*, ed. K.-C. Leung (San Francisco: ASP Conf. Series), in press.

Webbink, R.F., Livio, M., Truran, J.W., and Orio, M. (1987) *Astrophys. J.*, **314**, 653.

Williams, R.E., Sparks, W.M., Gallagher, J.S., Ney, E.P., Starrfield, S.G., and Truran, J.W. (1981) *Astrophys. J.*, **251**, 221.

THE COMMON ENVELOPE EVOLUTION OF MASSIVE STARS

RONALD E. TAAM[1] and PETER BODENHEIMER[2]
[1]Department of Physics and Astronomy, Northwestern University
Evanston, Illinois 60208, U. S. A.
[2]University of California Observatories/Lick Observatory
University of California, Santa Cruz, California 95064, U. S. A.

ABSTRACT. The progress in our understanding of the common envelope phase of binary evolution is reviewed. The results of a series of multi-dimensional hydrodynamical simulations suggest that the interaction between the two stellar components is sufficiently rapid and energetic to eject a large fraction of the common envelope and orbital angular momentum of the binary system. These results are used as the basis for an examination of the evolution of binary systems containing massive stellar components. As possible applications of the common envelope phase we focus on the evolutionary scenarios for the formation of binary radio pulsars. We also consider, within the same framework, the origin of high velocity pulsars from disrupted massive binary systems.

1. Introduction

The common envelope phase has become standard in the phenomenology for the formation of close binary star systems. The basic premise underlying the paradigm is the requirement that an expanding red giant star not synchronously rotate with the orbital motion of its companion at the onset of or during mass transfer. It is hypothesized that the interactions between stellar components result in processes which lead to the ejection of a large fraction of the mass of the system and to the shrinkage of the binary orbit. The system is thus transformed from one consisting of a giant star with a centrally condensed core in a long period system (of the order of years) to one consisting of the remnant core and companion in a short period system (of the order of days).

Although the concept of a compact companion spiraling within a common envelope was first developed more than a decade ago by Ostriker (1975) and Paczynski (1976), progress has been hampered by the inherent difficulties of the problem. The detailed investigations of Taam, Bodenheimer, and Ostriker (1978), Meyer and Meyer-Hofmeister (1979), Delgado (1980), Bodenheimer and Taam (1984), Livio and Soker (1984, 1988), and Taam and Bodenheimer (1989) have been primarily exploratory in scope in view of the complexities involved with studying important thermodynamic and multi-dimensional hydrodynamic effects over length scales which vary from 10^9 cm to 10^{13} cm and time scales which range

E. P. J. van den Heuvel and S. A. Rappaport (eds.), X-Ray Binaries and Recycled Pulsars, 281–291.

from hours to thousands of years. As a consequence, many investigators have constructed evolutionary scenarios, in which the common envelope phase plays an integral role, making use of only rather general considerations involving energetic arguments (see, for example, Tutukov and Yungel'son 1979; Iben and Tutukov 1984) to predict the properties of the post common envelope system. The basic requirement is that sufficient energy be lost from the orbit to unbind the common envelope. However, even if the requirement is met, it is unclear whether the two stars will stop spiraling to form a short period binary system or will merge.

In this paper we outline the conditions that must be satisfied in order that the common envelope be ejected before the companion closely approaches the boundary of the giant core, and we summarize some of the general results that can be gleaned from the detailed numerical simulations. We apply the general results from these studies to the construction of evolutionary scenarios for the formation of various classes of binary radio pulsars. We also address the relevance of the common envelope stage to the origin of high velocity pulsars.

2. Evolution to the Common Envelope Stage

The formation of a common envelope is a natural outcome of two classes of binary evolution. In general, a common envelope stage is expected to be reached whenever the stellar radii of the two components in the binary are disparate. In one case, the mass transfer process from the more massive star to its companion is unstable because its Roche lobe contracts faster than its radius. This leads to a very rapid phase of mass loss from the more massive component and to the engulfment of the companion star in the binary. This circumstance is characteristic of evolutions involving red giant stars with deep convective envelopes (see Paczynski and Sienkiewicz 1972; Webbink 1979). Because the time scale of the mass transfer phase is so rapid (approaching the dynamical timescale) uniform rotation cannot be maintained throughout the common envelope and the giant star loses synchronism with the orbital motion of its companion. Based upon the work of Hjellming (1989), such a situation is expected to arise for mass ratios > 0.8.

An alternative evolution leading to the common envelope stage occurs when the mass ratio of the system is extreme. In this case the giant is not rotating synchronously with the orbital motion since there is insufficient angular momentum available in the orbit to bring it to such a state. Consequently, upon further expansion of the red giant, the companion is engulfed and plunges into the red giant interior. Here, the rotational momentum of the red giant is comparable to the total orbital rotational momentum, and tidal effects are unable to force the giant into a state of corotation. This result is quantitatively given by the condition that the sum of the moments of inertia of the two stellar components exceed one third the orbital moment of inertia (Counselman 1973; Kopal 1978) of the binary system. For red-giant structures, this condition is satisfied for mass ratios > 5.

3. Possible Outcomes of Common Envelope Evolution

The ultimate fate of the components in the common envelope scenario depends very criti-

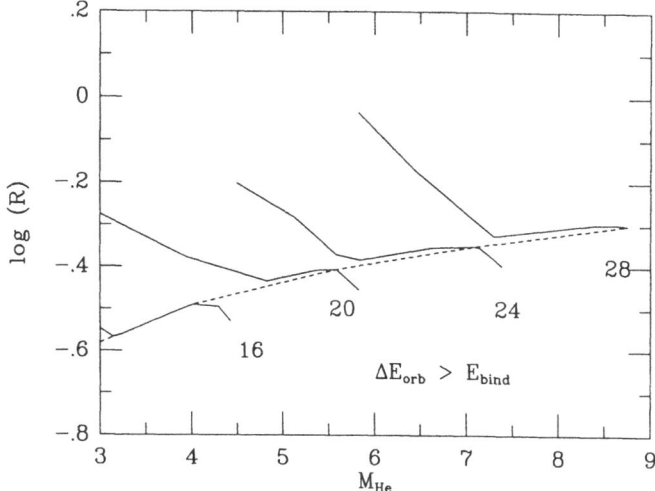

Figure 1: The radius (in $R_\odot$) of the helium core as a function of its mass for massive stars in the range $16M_\odot$ to $28M_\odot$ is denoted by the solid curve. The dashed curve corresponds to the locus for which the energy lost from the orbit is equal to the binding energy of the common envelope.

cally upon the structure of the more massive component. The companion star may eject a significant fraction of the common envelope, but it may, nevertheless, merge with the core instead of forming a short period binary system. In general, the former fate is expected for primaries not significantly evolved from the main sequence, whereas the latter fate is expected when the massive component is more evolved and centrally condensed. This tendency follows from the fact that in a more evolved system the orbital energy that can be lost is greater and the binding energy of the common envelope is much reduced. In order for the entire common envelope to be ejected before the companion merges with the core, a necessary condition is that sufficient energy be released from the orbit to unbind the envelope. The evolutionary phases where this condition is met are shown in Figure 1 for stars ranging in mass from $16M_\odot$ to $28M_\odot$. For the dashed curve it is assumed that the companion star is a neutron star with a mass equal to $1.4M_\odot$. Below this curve ejection of the entire envelope is in principle possible. It can be seen that the condition is met for helium core masses $> 3.1M_\odot$ for a $16M_\odot$ star and $> 8.7M_\odot$ for a $28M_\odot$ star.

We point out that the above estimates are lower limits to the mass of the remnant helium cores that would form in the post common envelope systems since the ejection process is not 100% efficient. The ejection of the common envelope is not spherically symmetric since matter is preferentially ejected along the equatorial plane of the binary system (Bodenheimer and Taam 1984; Livio and Soker 1988). This computational result is supported by the observational appearance of planetary nebulae with binary nuclei (i.e. low- to intermediate-mass post common envelope systems), which exhibit elliptical or butterfly shapes (Bond and Livio 1990). This nonspherical ejection is attributable to

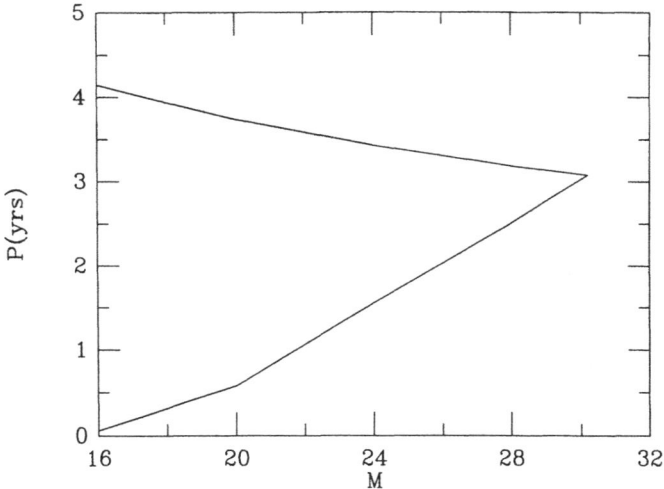

Figure 2: The orbital period (in years) of the binary system as a function of the mass of the more evolved star. Below the lower curve the system is expected to merge whereas above the upper curve the system never enters into the common envelope phase. Within the boundaries the formation of short period systems is energetically possible.

the fact that the frictional energy dissipation is distributed in the equatorial plane and not uniformly over the common envelope. Estimates for the efficiency of the ejection process range from 30% to 60% (Taam and Bodenheimer 1989). In order to successfully produce a short period system rather than a merged object, the ejection process must also be sufficiently rapid that the companion does not continue to spiral into the core. That is, the time scale for mass loss from the common envelope must be short compared to the time scale for orbital decay.

The critical orbital periods which delineate the range of evolutionary stages where the transformation of long period systems to short period systems can be accomplished is shown in Figure 2. The lower curve defines the lower bound on the orbital period for which a post common envelope binary is possible. Below this curve it is energetically impossible to eject the common envelope, and a merged system must result. On the other hand, above this curve, a binary system orbiting at a small separation is possible. For sufficiently long orbital periods (above the upper curve) the more massive star never expands to such a size as to bring the system into a common-envelope phase. For this upper bound we assume an orbital separation of 10^{14} cm. It can be seen that the formation of a short period system is possible for a progenitor system whose orbital period is in the range of 21 days to 4.2 years for a $16 M_\odot$ star, whereas for stars more massive than about $30 M_\odot$ such evolution is not found to be possible.

Provided that the system emerges from the common envelope phase as a short period binary system, the future evolution of the system will depend on its orbital separation and period. For example, for very small orbital separations the two components may still

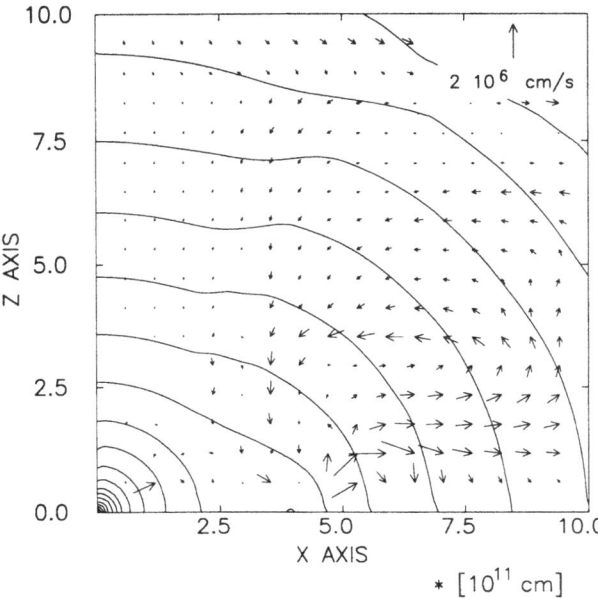

Figure 3: The velocity field and density distribution within the inner 10^{12} cm of the common envelope at a late phase. The system consisted initially of an AGB star of 2 $M_\odot$ and a main-sequence companion of 1 $M_\odot$. The maximum velocity in the grid is 2 x 10^6 cm s^{-1} and the density contours are logarithmically spaced from $\rho = 5$ x 10^{-7} g cm^{-3} to 2.2 g cm^{-3}.

merge as a result of further orbital decay or mass transfer, while for larger separations the remnant helium core may evolve as if it were in isolation. Unfortunately, little is known about how the common envelope phase actually terminates, and so previous investigators have had to rely upon estimates for the final orbital parameters based upon energetic arguments alone. Clues may be gleaned from the recent study of Taam and Bodenheimer (1991) on the common envelope evolution of low to intermediate mass binary systems. In this study, a binary system consisting of a $2M_\odot$ asymptotic giant branch star (character- ized by a C/O white dwarf core of $0.67M_\odot$) and a $1M_\odot$ main sequence star was examined. The numerical results demonstrated that the energy available from the orbital motion of the binary was sufficient to eject the mass contained within the common envelope without requiring the main sequence companion to spiral into the core. By the end of the calcula- tion, the orbital decay timescale had increased to > 200 years as a result of the removal of mass from the common envelope and the spin up of the envelope to near corotation.

The final phase of evolution of the common envelope for this case is shown in Figure 3. Note the circulating flow pattern in the vicinity of the main sequence star located at 4 x 10^{11} cm. This pattern has resulted from the low energy dissipation rates generated in the common envelope caused by the continous loss of matter and resulting reduced density near the orbit. Because of the reduced energy dissipation the remaining matter

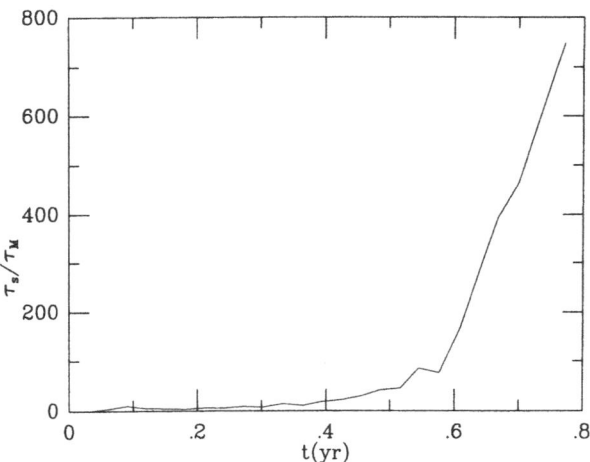

Figure 4: The ratio of the spiral timescale, τ_s, to the mass loss timescale, τ_M, for matter external to the white dwarf core of the red giant is illustrated as a function of evolution time (in years) for the common envelope configuration shown in Figure 3.

is not accelerated to escape speeds. Consequently, the angular momentum lost from the orbit is not removed very efficiently and, hence, the common envelope spins up (further accentuating the decline in the rate of energy dissipation).

Due to the nearly complete removal of mass from the common envelope the spiral time scale increases rapidly in comparison with the mass loss time scale especially after the frictional luminosity generated in the double core declines (see Figure 4). By the end of the calculation all matter exterior to the binary orbit had been ejected in an equatorial outflow, leaving only $0.01 M_\odot$ in the common envelope. Because the spiral time scale is approaching the thermal response time scale of the mass layers immediately interior to the orbit, this region is expected to contract to the white dwarf core, thereby terminating the common envelope phase.

An important feature of the structure of the red-giant progenitor which facilitates the termination of the common envelope phase is the presence of very sharp density gradients exterior to the hydrogen and helium burning shells. In the models studied, the mass contained outside the white dwarf core and within 10^{12} cm is only 0.01 - 0.02 $M_\odot$. It is a property of giant stars that when the mass above the white dwarf core approaches such a small value, the giant-like configuration can no longer be maintained and the star contracts. The occurrence of such a structure depends upon the mass of the red giant. For example, for a $5 M_\odot$ ($2 M_\odot$) star, the condition is fulfilled for white dwarf cores more massive than $\sim 0.96 M_\odot$ ($0.67 M_\odot$). In contrast, such sharp density gradients are not present in the regions exterior to the hydrogen burning shell (see Figure 5) in massive stars. On the other hand, such stars respond differently to mass loss. For example, for mass loss from a giant with a white dwarf core, the radius does not appreciably change

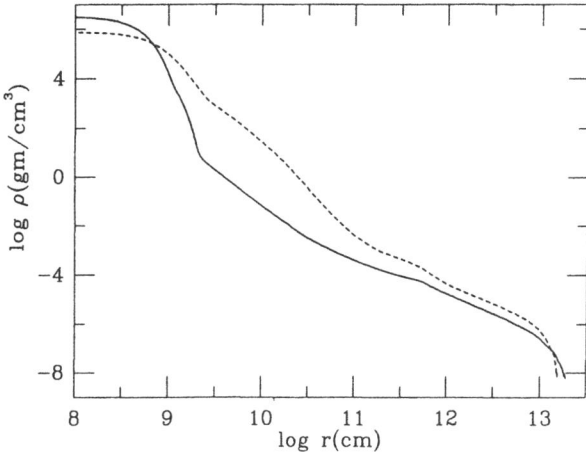

Figure 5: The density distribution as a function of radius for red giants of $2M_\odot$ (solid line) and $16M_\odot$ (dashed line). The star of $2M_\odot$ has a white dwarf core of $0.67M_\odot$, and the massive star has a helium core of $4M_\odot$.

since it depends primarily upon the mass of the core and not on the total mass. In contrast, a massive star will contract upon mass loss, but the much higher densities (and, therefore, higher energy dissipation rates) outside the burning shell characteristic of more massive stars (see Fig. 5) may make termination of the common envelope phase different from that found in the low to intermediate mass systems. A possibility that is yet to be explored in this context is the effect of material circulation induced in the inner regions by the nature of the two dimensional flow on the activity of the hydrogen burning shell (Taam and Bodenheimer 1989). If mixing is capable of significantly enhancing energy generation rates in the nuclear burning shell, then it may facilitate the ejection of the remaining matter above the helium core.

4. Applications to Binary and Single Pulsars

Assuming that the spiral-in phase terminates as outlined in the previous section, we now explore the further evolution of the post common envelope binary. In the following we assume that the binary system enters into the common envelope stage with a neutron star as one of the components. As implied in the previous sections, the common envelope phase is expected to be important in understanding the origin of short period (P < 1 day) binary systems containing very evolved components. The binary radio pulsars PSR 0655+64, PSR 1913+16, and PSR 2303+46 fit this categorization quite well, since it is known that the companion to PSR 0655+64 is a massive white dwarf whereas the companions to PSR 1913+16 and PSR 2303+46 are neutron stars (van den Heuvel 1984, 1987; Taylor and Steinebring 1986). For PSR 0655+64 the progenitor of the white dwarf was $< 9M_\odot$ whereas in both PSR 1913+16 and PSR 2303+46 the progenitor of the second-born neutron star was in the range between 9 and 16 $M_\odot$. The upper limit for the latter constraint follows from the fact that the remnant helium core mass in the post common envelope binary must have been less than $4.2M_\odot$ for, otherwise, the system would have been disrupted following the evolution of the helium core to collapse to a neutron star in the second supernova event. Because of the strong tidal interactions while the neutron star spirals through the common envelope, the orbits of the post common envelope binary are expected to be very nearly circular. This is in accordance with the low eccentricity of PSR 0655+64 (e = 0.0003, Taylor 1981). The large eccentricities of both PSR 1913+16 and PSR 2303+46 (e > 0.6), on the other hand, can be understood as due to the effect of mass loss and the possible kick given to the neutron star during the second supernova explosion. The detailed scenario for the formation of these systems has been presented in the early works of Smarr and Blandford (1976), Srinivasan and van den Heuvel (1982), van den Heuvel and Taam (1984), and reviewed by van den Heuvel (1984). From the results presented in the previous section it is expected that the orbital period at which the systems entered into the common envelope phase was long (P about a year) and, consequently, that the progenitors of these short period binary radio pulsars belong to either the class of objects known as Be X-ray binaries or VV Cephei binaries (van den Heuvel 1987a).

It has been argued that the common envelope phase of massive close binary systems plays a role in the formation of single radio pulsars (van den Heuvel 1981). In particular, the high velocities of such pulsars may reflect the orbital velocities of neutron stars in binary systems which have been disrupted. Dewey and Cordes (1987) studied the origin of the pulsar velocities within the context of disrupted binaries and found that neither the correlation between velocity and magnetic moment inferred by Anderson and Lyne (1983) nor the low incidence of binary pulsars could be reproduced. Bailes (1989), on the other hand, was able to construct a consistent scenario provided that the second supernova event occurred in a post common envelope binary with rather wide orbital periods (P about 100 days). In both studies asymmetries in the supernova event were invoked, as suggested by Shklovskii (1970), to produce the high observed velocities. From the viewpoint of binary evolution it is difficult to terminate the common envelope phase of evolution at the long orbital periods required by Bailes (1989) since the theoretical studies indicate that orbital decay will continue until the entire hydrogen-rich envelope is removed. This will occur at typical orbital separations of the order of 10^{11} - 10^{12} cm, corresponding to periods less than 4 days. This complication could be circumvented, however, if the giant star avoided the common envelope phase altogether by losing its envelope prior to the unstable mass transfer phase. However, then a large fraction of the mass would need to be removed by a super stellar wind. Furthermore, the recent work by Narayan and Ostriker (1990) seems to rule out Bailes' scenario, since they find that the slowly spinning pulsars should have higher spatial velocities than the fast spinning pulsars, while Bailes finds the opposite.

Because of the difficulties in forming most pulsars in massive binaries, Radhakrishnan and Shukre (1987) argue that the pulsar population originates from two classes of binary systems. The high velocity pulsars are produced in massive binaries which are disrupted, whereas the lower-velocity and more rapidly spinning pulsars with lower magnetic moment are produced either from the high-mass binaries in which a neutron star coalesced with its

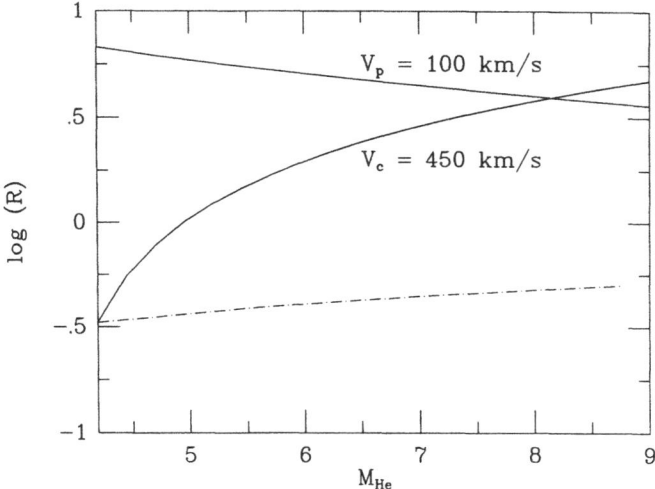

Figure 6: The orbital separation in units of $R_\odot$ between the stellar components versus the remnant helium core mass (in $M_\odot$). The solid lines correspond to the velocities of the pulsar and its companion, and the dash-dotted curve is taken from Figure 1.

companion (Radhakrishnan 1986) or from low mass binaries in which the companion to the neutron star has been either disrupted or, perhaps, evaporated as appears to be the case in PSR 1957+20 (Fruchter, Steinebring, and Taylor 1988; Phinney et al. 1988; Kluzniak et al. 1988; van den Heuvel and Van Paradijs 1988). Although the asymmetry required to give the pulsar its velocity is small (the observed velocities imply an asymmetry of about 1% based upon momentum arguments; Woosley 1991), it is possible that some systems do not require kicks (Radhakrishnan 1991). In such a case, by assuming a spherically symmetric and instantaneous loss of mass from the binary system during the second supernova event, the velocities of the single radio pulsars can be related to the velocities of the components prior to the disruption of the system (Radhakrishnan and Shukre 1987). This is shown in Figure 6. It is evident that the small orbital separations (or equivalently, the short orbital periods between 0.01 to 1 day) required by the observed pulsar velocities can be accommodated by the energetics of the common envelope scheme. Comparison of the orbital separations and masses allowed by the velocity data with the energy constraint imposed by the assumed occurrence of the common envelope evolution indicates that the range of progenitor masses is limited to a narrow range between 16 and 24 $M_\odot$ and an even narrower range of helium core masses (see above). Based on the range of parameter space allowed by the velocity data it is more likely that the relevant mass range only extends between 16 and 20 $M_\odot$. Since the system is to disrupt, the evolutionary state of the progenitor star (especially for the 16 $M_\odot$ case) must be more advanced than that indicated by energy arguments alone for the successful formation of a short period binary (see Figure 2). Accordingly, the period for which common envelope evolution might form a short period system which is then disrupted after the second supernova event would be on the order of a year.

5. Future Work

The existence of binary radio pulsars in short period systems has provided important insight into the late stages of binary evolution of massive stars. The common envelope phase has become an important link in understanding their manner of formation and their progenitor systems. Much of the work has been qualitative in nature and is very suggestive. The recent quantitative work has confirmed that the physical processes acting during this phase can, indeed, lead to a very nonconservative evolution in which substantial mass and angular momentum are lost.

Future studies will need to be directed toward determining the fraction of binaries containing compact stars which form short period systems and, at the same time, establishing the relationship between the orbital parameters of the post common envelope binary and those prior to the common envelope phase. Detailed studies are required on the manner in which the common envelope phase terminates. These studies will be important for understanding the orbital evolution of the systems themselves and will be especially relevant for placing on a firmer footing the role of the common envelope phase of evolution with regard to the origin of pulsar velocities. Another area ripe for further study is the investigation of the structure and evolution of merged systems containing a neutron star core (Thorne and Zytkow 1977). Particular attention should be paid to investigations of their stability.

The outlook for further understanding of stellar evolution in massive binaries is bright, and future work in these areas will help clarify our understanding of this very important phase of binary evolution.

A portion of the results reported in this paper were obtained from calculations performed at the National Center for Supercomputer Applications in Urbana, Illinois. This research has been supported in part by the NSF under grant AST-8608291.

References

Anderson, B., and Lyne, A. G. 1983, Nature, **303**, 597.
Bailes, M. 1989, Ap. J., **342**, 917.
Bodenheimer, P., and Taam, R. E. 1984, Ap. J., **280**, 771.
Bond, H. E., and Livio, M. 1990, Ap.J., **355**, 568.
Counselman, C. C. 1973, Ap. J., **180**, 307.
Delgado, A. J. 1980, Astr. Ap., **87**, 343.
Dewey, R. J., and Cordes, J. M. 1987, Ap. J., **321**, 780.
Fruchter, A. S., Steinebring, D. R., and Taylor, J. H. 1988, Nature, **333**, 237.
Hjellming, M. S. 1989, unpublished Ph. D. Thesis, Univ. of Illinois, Urbana-Champaign.
Iben, I. Jr., and Tutukov, A. V. 1984, Ap. J. Suppl., **54**, 335.

Kluzniak, W., Ruderman, M., Shaham, J., and Tavani, M. 1988, Nature, **334**, 225.

Kopal, Z. 1978, *Dynamics of Close Binary Systems*, (Dordrecht: Reidel).

Livio, M., and Soker, N. 1984, M.N.R.A.S., **208**, 763.

Livio, M., and Soker, N. 1988, Ap. J., **329**, 764.

Meyer, F., and Meyer-Hofmeister, E. 1979, Astr. Ap., **78**, 167.

Narayan, R., and Ostriker, J. P. 1990, Ap. J., **352**, 222.

Ostriker, J. P. 1975, paper presented at IAU Symposium No. 73, Cambridge, England.

Paczynski, B. 1976, in *IAU Symposium No. 73, Structure and Evolution of Close Binary Systems*, ed. P. Eggleton, S. Mitton, and J. Whelan (Dordrecht: Reidel), p. 75.

Paczynski, B., and Sienkiewicz, R. 1972, Acta Astr. **22**, 73.

Phinney, E. S., Evans, C. R., Blandford, R. D., and Kulkarni, S. R. 1988, Nature, **333**, 832.

Radhakrishnan, V. 1986, *Highlights of Astronomy*, **7**, 3.

Radhakrishnan, V. 1991, this volume.

Radhakrishnan, V., and Shukre, C. S. 1987, *High Energy Phenomena Around Collapsed Stars*, ed. F. Pacini (Dordrect; Reidel), p. 271.

Shklovskii, I. S. 1970, Sov. Astr., **13**, 562.

Smarr, L. L., and Blandford, R. 1976, Ap. J., **207**, 574.

Srinivasan, G., and van den Heuvel, E. P. J. 1982, Astr. Ap., **108**, 143.

Taam, R. E., and Bodenheimer, P. 1989, Ap. J., **337**, 849.

Taam, R. E., and Bodenheimer, P. 1991, Ap. J., in press.

Taam, R. E., Bodenheimer, P., and Ostriker, J. P. 1978, Ap. J., **222**, 269.

Taylor, J. H. 1981, in *IAU Symp. No. 95, Pulsars*, ed. W. Sieber and R. Wielebinski (Dordrecht: Reidel), p. 361.

Taylor, J. H., and Steinebring, D. R. 1986 Ann. Rev. Astr. Ap., **24**, 285.

Thorne, K. S., and Zytkow, A. N. 1977, Ap. J., **211**, 832.

Tutukov, A. V., and Yungel'son, L. R. 1979, in *IAU Symposium No. 83, Mass Loss and Evolution of O-Type Stars*, ed. P. S. Conti and C. W. H. de Loore (Dordrecht: Reidel), p. 415.

van den Heuvel, E. P. J. 1981, in *IAU Symp. No. 95, Pulsars*, ed. W. Sieber and R. Wielebinski (Dordrecht: Reidel), p. 379.

van den Heuvel, E. P. J. 1984, J. Astr. Ap., **5**, 209.

van den Heuvel, E. P. J. 1987a, in *High Energy Phenomena Around Collapsed Stars*, ed. F. Pacini (Dordrecht: Reidel), p. 1.

van den Heuvel, E. P. J. 1987b, in *The Origin and Evolution of Neutron Stars*, ed. D. J. Helfand and J. H. Huang (Dordrecht: Reidel), p. 393.

van den Heuvel, E. P. J., and Taam, R. E. 1984, Nature, **309**, 235.

van den Heuvel, E. P. J., and Van Paradijs, J. 1988, Nature, **334**, 227.

Webbink, R. F. 1979, in *IAU Coll. No. 46 Changing Trends in Variable Star Research*, ed. F. M. Bateson, J. Smak, and I. H. Urch (Hamilton, NZ: U. Waikato), p. 102.

Woosley, S. E. 1991, private communication.

THE EVOLUTION OF HER X-1 AND WHY HER X-1 LIKE SYSTEMS ARE VERY RARE

W. SUTANTYO

Institute for Theoretical Physics
University of California
Santa Barbara, CA 93106

and

Department of Astronomy & Bosscha Observatory
Institute of Technology Bandung, Indonesia

ABSTRACT. We desribe the evolutionary history of Her X-1 and present several combinations of orbital parameters of the progenitor system which will lead to the formation of a system like Her X-1 through a supernova (SN) explosion and a tidal evolution. We consider both, the symmetric and asymmetric SN explosion. We also investigate why Her X-1 like systems are very rare. Two mechanisms which may prevent the formation of such systems, i.e., the evaporation of the companion star by the high energy radiation from the pulsar and the disruption of the binaries by a supernova explosion, are discussed. We find that in all considered cases the pulsar energy is far from sufficient to evaporate its companion. We also study the evaporation process in a particular model of a system consisting of a 2.25 $M_\odot$ main-sequence star and a 1.44 $M_\odot$ neutron star with an initial orbital period of 1 d. We find that a 6 and 10 ms pulsar can only evaporate ~ 0.036 and ~ 0.016 $M_\odot$, respectively, from its companion. The evaporation process only happens in 10^3–10^4 yr as the pulsar energy diminishes very rapidly. Assuming the mass of the exploding star is 3.2 $M_\odot$ and the kick velocity due to the asymmetric SN explosion is 100–200 km s^{-1}, most systems will be disrupted provided that the pre-supernova orbital period is ≤ 1 d and the ejection velocity of the SN shell is in excess of 1–1.5 $\times 10^9$ cm s^{-1}.

1. Introduction

The X-ray binary Her X-1 is unique among the known X-ray binaries. The system consists of a neutron star of ~ 1 $M_\odot$ and a nondegenerate star of ~ 2 $M_\odot$. The mass of the nondegenerate star is intermediate between those of high mass X-ray binaries (≥ 8 $M_\odot$) and low mass X-ray binaries (≤ 1 $M_\odot$). So far, Her X-1 is the only such a system known.

The system is located at high galactic latitude ($b^{II} = 35°$). The distance estimate of the system yield a value of ~ 4.5 kpc (Thomas et al. 1986). These parameters imply a distance of ~ 2500 pc above the galactic plane. According to its total mass the system must belong to disk population stars. This indicates that the system must have moved out from the

293

E. P. J. van den Heuvel and S. A. Rappaport (eds.), X-Ray Binaries and Recycled Pulsars, 293–309.
© 1992 *Kluwer Academic Publishers*.

galactic plane where it was born, less than $\sim 10^9$ yr ago (the maximum age of a star of $>$ 1.8 $M_\odot$ which is the smallest possible companion mass; Rappaport and Joss, 1983).

To reach its present distance above the galactic plane and to overcome the galactic gravitational field, the system must have acquired a runaway velocity of > 100 km s^{-1}. This velocity can be derived from the sling shot effect due to the supernova explosion which has formed the neutron star (Sutantyo, 1975). To reach z $\sim$ 2500 pc, the initial velocity must be $\sim$ 100–150 km s^{-1} and the system must be at least a few times 10^7 yr old (Lyne et al., 1982).

We will discuss the possible evolutionary scenario of the system and offer an explanation of why systems like Her X-1 (i.e. X-ray binaries with a nondegenerate component of 2–4 $M_\odot$) are very rare.

2. The Evolutionary Scenario

Following Sutantyo (1975; see also reviews of van den Heuvel, 1976, 1983) the evolution of Her X-1 can be divided into four phases; i.e., spiral-in or common-envelope phase, supernova phase, sleeping or quite phase and X-ray phase.

2.1. SPIRAL-IN OR COMMON-ENVELOPE PHASE

The progenitor of Her X-1 is very likely a binary system consisting of a massive ($\sim$ 20 $M_\odot$) and a low mass ($\sim$ 2–4 $M_\odot$) star. Sutantyo (1975) and Van den Heuvel (1976, 1983) suggests that in a system with such a low mass-ratio the mass transfer which happens when the more massive component is filling its Roche lobe is unstable. The mass transfer, in combination with the tidal instability, causes the orbit shrinks rapidly. The less massive star will enter the envelope of the mass loser and move deeper and deeper approaching the core in a spiral-in orbit. Both stars will be embedded in a common envelope. The friction between the less massive component and the envelope will generate enough energy to blow off the envelope. The mass loss will also carry angular momentum from the system. The outcome of such an evolution is a very short period (< 1 d) binary consisting of a helium star (which is originally the helium core of the primary) and a low mass main-sequence star.

2.2. SUPERNOVA (SN) PHASE

The helium star will eventually explode as a supernova and leave behind a neutron star as a remnant. As the exploding star is expected to be more massive than its companion, the system will, in most cases, be disrupted by the explosion. This may explain why systems like Her X-1 are very rare. We will discuss this in more detail in Section 5.

In much rarer cases, when the system is not disrupted by the SN explosion, the system will consist of a neutron star and a main-sequence star. As a result of the explosion the system has an eccentric orbit. The explosion will in most cases impart a runaway velocity of ≥ 100 km s^{-1} to the system (see Section 3).

2.3. Sleeping phase

After the explosion the nondegenerate star is still on the main-sequence and still underfills its Roche lobe. If the mass of the main-sequence star is $\leq 2.35\ M_\odot$, it needs at least $\sim 4 \times 10^8$ yr (the main-sequence lifetime) before it evolves to fill up its Roche lobe and the X-ray phase begins. This gives the lower limit for the age of the system. Within this time, if the runaway velocity is ≥ 100–150 km s^{-1}, the system may have undergone a few cycles of oscillation with respect to the galactic plane with an amplitude of a few kpc (Oort, 1965; Lyne et al., 1982). Hence, the present high latitude of the system can easily be explained. During this phase the tidal interaction between both stars may circularize the orbit.

The neutron star can be very active at the moment of its birth. However, its energy will diminish rapidly and within 10^5–10^6 yr it becomes a quiescent neutron star. Therefore this phase is called a "sleeping" or "quiet" phase. In Section 5 we will investigate whether during this phase the newly born neutron star is capable to evaporate its companion.

2.4. X-ray phase

When the nondegenerate star evolves to fill up its Roche lobe, the mass transfer begins and the system becomes an X-ray source. The X-ray lifetime is $\sim 10^4$–10^6 yr (Savonije, 1978). Her X-1 is in this phase now.

Verbunt et al. (1990) show that the above described scenario seems to be the most reasonable one, as another scenario (which involves the accretion induced collapsed of a white dwarf; Sutantyo et al., 1986) gives an age and runaway velocity for the system which are too small to reach its present distance above the galactic plane.

3. The Orbital Evolution

We will derive some combinations of the orbital parameters at the pre- and post-SN-phase which will result to a system like Her X-1. We denote the orbital parameters before and after the SN explosion and after the tidal circularization as follows,

	Before SN	After SN	After tidal-evolution
Mass of the exploding star (star 1)	M_1	M_n	M_n
Mass of the companion (star 2)	M_2	M_2	M_2
Eccentricity	0	e	0
Orbital period	P_0	P_f	P_n
Semi-major axis	a_0	a	a_n

Here various parameters used in the following discussions are defined. We assume that the orbital parameters after the tidal evolution is equal to the present observed parameters. Unless it is indicated otherwise, throughout the following discussions we adopt $M_2 = 2.25\ M_\odot$, $M_n = 1.44\ M_\odot$ and $P_n = 1.7$ d.

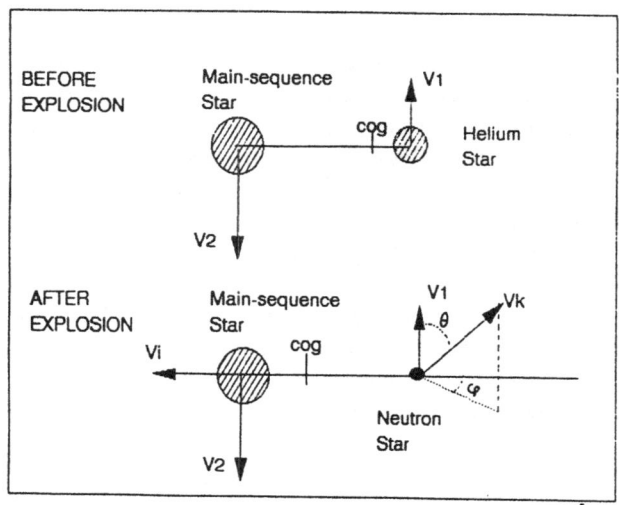

Figure 1: Velocity components of both stars before and after the SN explosion as seen from the center gravity of the system (cog). See text for the definition for the velocity components.

3.1. THE GEOMETRY OF THE EXPLOSION

We assume the explosion happens instantaneously. The velocity components of both stars before and after the explosion *with respect to the center of gravity of the system* are shown in Figure 1. V_1 and V_2 are, respectively, the orbital velocity of the exploding (star 1) and the unexploding star (star 2) before the explosion.

$$V_2 = \sqrt{\frac{GM_1^2}{a_0(M_1 + M_2)}} \tag{1}$$
$$V_1 = \frac{M_2 V_2}{M_1}$$

V_k is the kick velocity acquired by the exploding star due to asymmetries in the explosion. V_i is the velocity imparted by the supernova shell to the unexploding star due to the impact and the ablation (Colgate, 1970; Fryxell and Arnett, 1981).

3.2. THE CHANGE OF THE ORBITAL PARAMETERS DUE TO THE EXPLOSION

We will consider now the *relative orbit* of star 1 with respect to star 2. The semi-major axis and the eccentricity after the explosion can be derived from,

$$\frac{a_0}{a} = 2 - Y F_2 \tag{2}$$

$$(1 - e^2) \ = \ \frac{a_0}{a} Y F_1 \tag{3}$$

where Y is the ratio of the pre- to the post-explosion total mass,

$$Y = \frac{M_2 + M_1}{M_2 + M_n} \tag{4}$$

and,

$$F_1 \ = \ [1 + 2v_k cos\theta + v_k^2(cos^2\theta + sin^2\theta sin^2\phi)] \tag{5}$$
$$F_2 \ = \ (1 + 2v_k cos\theta + v_k^2 + 2v_i v_k sin\theta cos\phi + v_i^2) \tag{6}$$

v_k and v_i are, respectively, the ratio of the kick and the impact velocity to the initial orbital velocity V_0,

$$v_k \ = \ \frac{V_k}{V_0}, \quad v_i = \frac{V_i}{V_0} \tag{7}$$
$$V_0 \ = \ V_1 + V_2 = \sqrt{\frac{G(M_1 + M_2)}{a_0}}$$

V_i is calculated from the formula given by Colgate (1970) multiplied by a factor α.

$$V_i = \alpha(\frac{R}{2a_0})^2 \frac{M_{shell}}{M_2} V_{ej}(1 + ln\frac{V_{ej}}{V_{es}}) \tag{8}$$

where $0 \le \alpha \le 1$, R is the radius of the unexploded star, V_{ej} is the ejection velocity of the SN shell, M_{shell} is the mass of the SN shell, V_{esc} is the escape velocity from the surface of the unexploded star. The factor α is included as in the original formula the dissipative and the stripping effects [see Wheeler et al. (1975) and Dewey and Cordes (1987)] are neglected and plane parallel layers are assumed so that the formula may overestimate the effect of impact. In the following discussions we assume that $\alpha = 0.5$. [Two dimensional hydrodynamical analyses by Fryxell and Arnett (1981) and Taam and Fryxell (1983) indicate that the ratio of the imparted momentum to the incident momentum is ~ 0.3–0.8].

3.3. THE RUNAWAY VELOCITY

We will now consider the orbital motion of both components with respect to the original center of gravity of the system (see Figure 1). Let the line connecting the two stars be the x-axis and the orbital plane be the (x, y) plane. Let the velocity components of both stars after the explosion be (u_1, v_1, w_1) and (u_2, v_2, w_2), respectively. The velocity of the center of gravity of the system with respect to the original center of gravity can then be calculated from a general formula,

$$V_g = \sqrt{u_g^2 + v_g^2 + w_g^2} \tag{9}$$

where,

$$u_g = \frac{M_2 u_1 + M_n u_1}{M_2 + M_n} \tag{10}$$

$$v_g = \frac{M_2 v_1 + M_n v_1}{M_2 + M_n}$$

$$w_g = \frac{M_2 w_1 + M_n w_1}{M_2 + M_n}$$

From Figure 1 one can derive that,

$$u_1 = V_k sin\theta cos\phi \quad v_1 = V_1 + V_k cos\theta \quad w_1 = V_k sin\theta sin\phi \tag{11}$$
$$u_2 = -V_1 \qquad v_2 = -V_2 \qquad w_2 = 0$$

We call V_g the runaway velocity of the system.

3.4. THE TIDAL CIRCULARIZATION

As described above, after the SN explosion the system will enter the sleeping phase and stay in this phase for $\sim 4 \times 10^8$ yr. During this phase the tidal interaction between the two components will circularize the orbit. The tidal effect will dissipate energy from the system but will conserve the total angular momentum. Therefore we expect that the total angular momentum of the system at the birth of the neutron star and at the X-ray (present) stage is the same. Neglecting the rotation angular momentum, as this is in most cases only a few percent of the total angular momentum, the present total angular momentum is,

$$H = A P_n^{\frac{1}{3}} \tag{12}$$

where,

$$A = \left(\frac{G^2}{2\pi(M_2 + M_n)}\right)^{\frac{1}{3}} M_n M_2 \tag{13}$$

As mentioned above, H must be equal to the total angular momentum immediately after the explosion,

$$H = A P_f^{\frac{1}{3}} (1 - e^2)^{\frac{1}{2}} \tag{14}$$

Using Equations (2) and (3), Equation (14) can be expressed as,

$$P_n^{\frac{1}{3}} = P_0^{\frac{1}{3}} Y^{\frac{2}{3}} F_1^{\frac{1}{2}} \tag{15}$$

Note that this equation is independent of the post-explosion parameters a and e as well as of the effect of impact of the SN shell. P_n, M_2 and M_n are the present (observed) parameters. So assuming a value for the mass of the exploding star (M_1) and a certain kick velocity vector, we can calculate P_0. Then Equations (2) and (3) give the post explosion orbital parameters a and e.

3.5. Symmetric SN explosion

First, we will consider spherically symmetric SN explosion. In this case $V_k = 0$. The exploding star is very likely a helium star with a mass between 2.3 to 5 $M_\odot$ (see van den Heuvel, 1983; for a larger mass the system will be disrupted). We assume various values of M_1 ranging from 2.5 to 5 $M_\odot$. We calculate P_0, P_f and e which will yield the final orbital parameters, $M_2 = 2.25 M_\odot$, $M_n = 1.44 M_\odot$ and $P_n = 1.7$d, in the way described above.

The typical velocities of supernova shells (V_{ej}) is of the order of 10^8–10^9 cm s^{-1} (see Kundt, 1989). We assume models with supernova shell velocities of 0 (no impact), 0.5, 1 and 1.5×10^9 cm s^{-1}. We also assume that 0.18 $M_\odot$ is lost in the form of the binding energy of the neutron star (Verbunt et. al. 1990). The results are given in Table 1. In this table we also calculate the runaway velocity V_g acquired by the system after the SN explosion.

As we can see from Table 1, even if we neglect the impact of the SN shell, the system will acquire a runaway velocity of > 100 km s^{-1} if the mass of the exploding star > $3.5 M_\odot$. If we assume $V_{ej} = 0.5 \times 10^9$ cm s^{-1} no bound orbit solution is found if $M_1 > 4.3 M_\odot$, while for $V_{ej} = 1 \times 10^9$ cm s^{-1} and 1.5×10^9 cm s^{-1}, the upper limit for M_1, if the binary to remain bound, is 3.5 and 3.0 $M_\odot$, respectively.

3.6. Asymmetric SN explosion

We will now consider asymmetric SN explosion. Assuming the kick velocity to the exploding star (V_k) is 100 km s^{-1} we can calculate the runaway velocity as a function of the direction of the kick. Here we assume that the mass of the exploding star is 3.2 $M_\odot$ and the final system is the same as above. The ejection velocity of the SN shell (V_{ej}) is assumed to be 1×10^9 cm s^{-1}.

The results are shown in Figure 2. Each line gives the runaway velocity as a function of θ for a given value of ϕ indicated on the line. Beyond the tip of each line the binary is disrupted. The horizontal line indicates the runaway velocity in case of symmeric explosion. Note that the system will get a runaway velocity > 85 km s^{-1} in all cases.

We also calculate the post-SN eccentricity as a function of the direction of the kick (Figure 3). For all values of ϕ, the binaries are disrupted if $\theta \leq 60°$, and remain bound if $\theta \geq 95°$.

4. Is the Old Age of Her X-1 in Conflict with the Strong Observed Magnetic Field?

Truemper et al. (1978) reported evidence for a strong and narrow line feature at ~ 58 keV in the X-ray spectrum of Her X-1. If this feature is interpreted as due to cyclotron emission, it implies a total magnetic field strength of 5.3×10^{12} G. Another way to infer the magnetic field strength is that, if we assume Her X-1 is rotating near its equilibrium period one can derive that the magnetic field strength should be $\sim 2.5 \times 10^{11}$ G. These facts suggest that a surface magnetic field strength of 10^{11}–10^{12} G seems to be a reasonable estimate for Her X-1.

It has been proposed that the magnetic field strength of neutron stars may decay on timescale of 10^6–10^7 yr (Taylor and Manchester, 1977; Radhakrishnan and Srinivasan, 1981; Lyne et al. , 1982). This field decay hypothesis has been used to explain the existence

Table 1: Orbital parameters before and after a symmetric SN explosion which will result in a system like Her X-1. It is assumed that $M_2 = 2.25\ M_\odot$, $M_n = 1.44\ M_\odot$ and $P_n = 1.7$ d. Four models are presented here: a) the impact of the SN shell is neglected (no impact), b) the velocity of the SN shell (V_{ej}) is 0.5×10^9 cm s^{-1}, c) $V_{ej} = 1 \times 10^9$ cm s^{-1} and d) $V_{ej} = 1.5 \times 10^9$ cm s^{-1}. We also present the value of g (the product of the evaporation efficiency and the geometric factor; see Section 5) and gE_{rot}/E_b (the ratio of the pulsar energy which can be used for evaporation to the binding energy of the companion star) for pulsars with an initial spin period of 6 and 10 ms.

M_1 $(M_\odot)$	P_0 (d)	P_f (d)	e	V_g (km/s)	g	gE_{rot}/E_b $P = 6$ ms	gE_{rot}/E_b $P = 10$ ms
No impact							
2.50	1.03	1.93	0.29	48.3	0.0005	0.0131	0.0047
3.00	0.84	2.28	0.42	71.1	0.0004	0.0105	0.0038
3.50	0.70	2.98	0.56	93.8	0.0003	0.0074	0.0027
4.00	0.59	4.55	0.69	116.6	0.0002	0.0042	0.0015
4.50	0.51	9.74	0.83	139.4	0.0001	0.0015	0.0005
5.00	0.44	93.35	0.96	162.1	0.0000	0.0001	0.0000
$V_{ej} = 0.5 \times 10^9$ cm s^{-1}							
2.50	1.03	1.97	0.31	51.4	0.0005	0.0128	0.0046
3.00	0.84	2.46	0.47	78.6	0.0004	0.0095	0.0034
3.50	0.70	3.83	0.65	108.7	0.0002	0.0053	0.0019
4.00	0.59	11.50	0.85	142.7	0.0000	0.0012	0.0004
4.50	No bound orbit solution						
$V_{ej} = 1 \times 10^9$ cm s^{-1}							
2.50	1.03	2.17	0.39	65.0	0.0004	0.0113	0.0041
3.00	0.84	3.90	0.65	109.6	0.0002	0.0052	0.0019
3.50	0.70	334.52	0.99	165.6	0.0000	0.0000	0.0000
4.00	No bound orbit solution						
$V_{ej} = 1.5 \times 10^9$ cm s^{-1}							
2.50	1.03	2.73	0.52	87.3	0.0003	0.0083	0.0030
3.00	0.84	35.10	0.93	156.5	0.0000	0.0003	0.0001
3.50	No bound orbit solution						

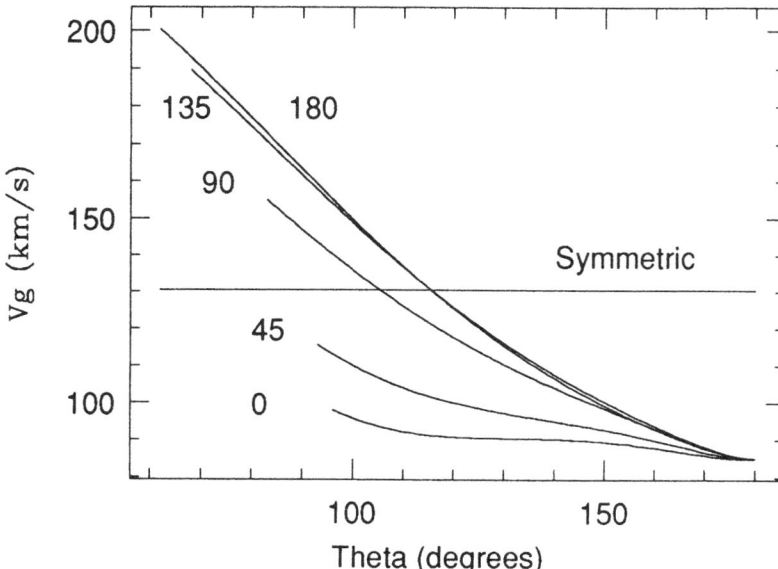

Figure 2: Runaway velocity as a function of the direction of the kick velocity. We assume $M_1 = 3.2 M_\odot$, $V_k = 100$ km s^{-1}, and $V_{ej} = 1 \times 10^9$ cm s^{-1}. The final system after the SN explosion and the tidal circularization is a neutron star binary with $M_2 = 2.25 M_\odot$, $M_n = 1.44 M_\odot$ and $P_n = 1.7$ d. The absisca shows the value of θ. Number on each line indicates the value of ϕ. Beyond the tip of each line the binary is disrupted. The horizontal line gives the value for symmetric explosion.

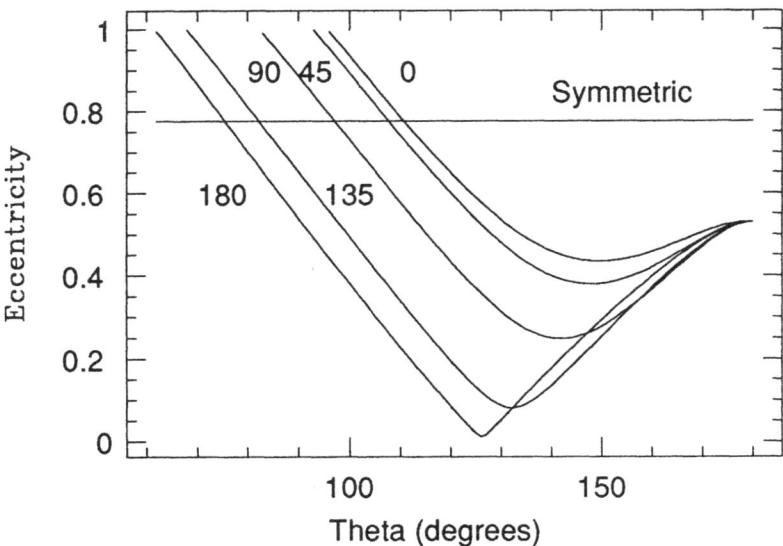

Figure 3: Post-SN eccentricity as a function of the direction of the kick velocity. We use the same parameters as in Figure 2.

of low magnetic field pulsars as well as the distribution of pulsars in the magnetic field-pulse period ($B–P$) diagram.

If we adopt the above described evolutionary scenario for Her X-1, the age of the neutron star in the system should be $\geq 4 \times 10^8$ yr. If the magnetic field does decay on a 10^6 to 10^7 year timescale, it is difficult to understand why Her X-1 still possesses strong magnetic field at such an old age.

Recent observations indicate, however, that beside Her X-1, there are some other old binaries with a neutron star component which shows a long term stability of its magnetic field. One of the best examples is the binary pulsar PSR 0655+64. The companion of the pulsar is a cold white dwarf with a cooling time of 10^9 yr (Kulkarni, 1986). As the neutron star was formed before the white dwarf (van den Heuvel, 1989), the age of the neutron star should be at least $\sim 10^9$ yr. On the other hand, the magnetic field as derived from P and dP/dt is fairly strong, i.e., $\sim 10^{10}$ G, which, based on field decay hypothesis, is much too high for its age. Further, some of gamma-ray bursters, which are probably very old neutron stars, also show evidence of strong surface magnetic field (Murakami et al., 1988; Fenimore et al., 1988). These facts indicate that magnetic field decay is not a general phenomena.

All observed pulsars which have significantly low magnetic field are, or show indication that they have been, members of accreting binary systems. This leads to an explanation that the magnetic field of a neutron star decays appreciably only if it accretes matter from a binary companion (Taam and van den Heuvel, 1986; Bailes, 1989; Bhattacharya and van den Heuvel, 1991). As statistical studies of pulsars seem compatible with the field decay hypothesis, possibly many pulsars, even those with high surface magnetic field, have been members of acrreting binaries as well.

After the formation of the neutron star, Her X-1 system spent most of its time in the sleeping phase where accretion to the neutron star did not take place. The accretion started only recently ($< 10^6$ yr) when the system entered the X-ray phase. This may explain why its magnetic field did not decay significantly despite its old age.

5. Why Her X-1 Like Systems are very rare?

Based on the distribution of mass ratio derived by Halbwachs (1987) and Hogeveen (1990), Bhattacharya and van den Heuvel (1990) indicate that 65 to 85 % of binaries with a B-type component and an orbital period less than a few years have small mass ratio ($q < 0.4$). Such systems are expected to evolve to Her X-1 like systems. However, systems like Her X-1 are extremely rare in nature. There should, therefore, be some mechanisms which in most cases prevent the formation of Her X-1 like systems. One can envisage two mechanisms, i.e., the evaporation of the nondegenarate star by the high energy radiation from the pulsar, and the disruption of the binary system by the SN explosion. We will discuss each of those mechanisms.

5.1. IS THE NEUTRON STAR CAPABLE TO EVAPORATE ITS COMPANION?

Bhattacharya and van den Heuvel (1991) suggested an attractive though speculative hypothesis, that the very active newborn pulsars in these systems may completely evaporate their companions on a relatively short timescale ($\leq 10^5$ yr). Besides advancing the idea, the authors also indicated some difficulties which this hypothesis may face, i.e. the low values for the product of the evaporation eficiency and the geometric factor (will be discussed below) and the absence of many supernova remnants with a central source activity in the form of an evaporating binary system. We will explore this hypothesis in detail.

The condition under which the pulsar is able to completely evaporate its companion is (Bhattacharya and van den Heuvel, 1991),

$$\frac{gE_{rot}}{E_b} > 1 \tag{16}$$

where E_{rot} is the rotational energy of the pulsar,

$$E_{rot} = \frac{1}{2}I\omega^2 \tag{17}$$

(ω is the spin angular velocity of the pulsar), E_b is the binding energy of the companion, and g is the product of the evaporation efficiency f (the fraction of the pulsar energy which can be used for the evaporation process) and the geometric factor $(R/2a)^2$.

We calculate the value of g and the ratio gE_{rot}/E_b for all combinations of parameters given in Table 1. The radius and the binding energy of the companion star are derived from the model of a 2.25 $M_\odot$ main-sequence star computed by Iben (1967). This model gives $R = 1.46R_\odot$ and $E_b = 2.17 \times 10^{49}$ erg.

M önchmeyer and Müller (1989) indicate that a newly formed neutron stars can be a fast rotator with a spin period between 6 to 13 ms. For this reason we assume two values

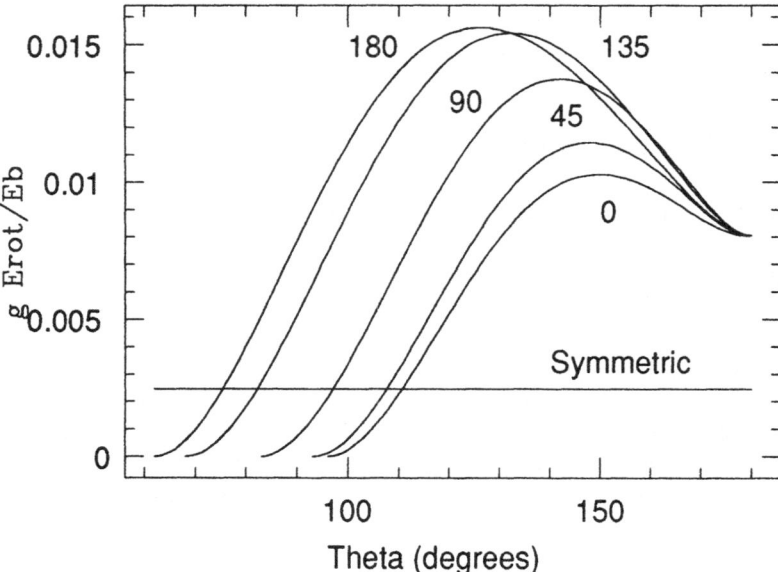

Figure 4: Ratio of the pulsar energy available for the evaporation process, to the binding energy of the companion as a function of the direction of the kick velocity. We use the same parameters as in Figure 2. The pulsar is a standard neutron star ($I = 10^{45}$ g cm^2, $B_s = 4 \times 10^{12}G$, $R_n = 10^6$ cm) with a spin period of 6 ms. The evaporation efficiency is $f = 0.1$.

for the initial pulsar spin period, i.e., 6 and 10 ms. Ruderman et al. (1989) and Krolik & Sincall (1990) show that the evaporation efficiency f might be as high as 0.1 and may possibly approach unity. We assume here that $f = 0.1$.

The results, in case we assume symmetric SN explosion, are shown in Table 1. We find that for all cases the value of g, also the ratio gE_{rot}/E_b is $\ll 1$; which means that the total energy of pulsar is far from enough to evaporate completely its companion. This is exactly what Bhattacharya and van den Heuvel indicated, that it is not easy to meet the rather stringent condition for the evaporation process to work; i.e. the requirement of a fairly large value of g.

We also look to the case of an asymmetric SN explosion. Figure 4 shows the value of gE_{rot}/E_b as a function the direction of the kick velocity. Here we assume the same parameters as used in Figure 2. Also here we find that this ratio is $\ll 1$ in all cases.

It is interesting to know to what extent the pulsar can evaporate its companion. We do the calculation for a system consisting of a 2.25 $M_\odot$ main-sequence star and a 1.44 $M_\odot$ neutron star with an initial orbital period of 1 d. Using the main- sequence model of a 2.25 $M_\odot$ star computed by Iben (1967; cf. Novotny, 1973) we calculate the binding energy as a function of M_r (the mass inside the radius r) from,

$$E_b(M_r) = \int_0^{M_r} \frac{GM_r}{r} dM_r \qquad (18)$$

The pulsar energy flux used for evaporating the companion is (van den Heuvel and van Paradijs, 1988; Bhattacharya and van den Heuvel, 1991),

$$F_{evap} = f(\frac{R}{2a})^2 10^{41} (\frac{B_s}{4.10^{12}})^2 (\frac{10\ ms}{P_{rot}})^4 \tag{19}$$

where R is the companion radius, a is the orbital separation, B_s is the surface magnetic field strength, P_{rot} is the rotation period of the pulsar and f is the evaporation efficiency. We assume $f = 0.1$ and $B_s = 4 \times 10^{12}$ G

As pulsars are emitting energy at the cost of their rotational energy, the pulsar will spin down during its evolution. Assuming the pulsar magnetic field strength to be constant one can derive that the pulsar rotation period at a time t is,

$$P_{rot}(t) = \sqrt{(P_{rot})_0^2 + 2Kt} \tag{20}$$

where $(P_{rot})_0$ is the pulsar initial period and,

$$K = \frac{8\pi^2 B_s^2 R_n^6}{3Ic^3} \tag{21}$$

(I is the moment of inertia and R_n is the radius of the neutron star; for a standard neutron star $I = 10^{45}$ g cm^2 and $R_n = 10^6$ cm).

In this calculation we peel off the nondegenerate star layer by layer and each time we calculate the energy required to remove this layer, i.e. the binding energy of this layer. We can then calculate the amount of pulsar energy which has been radiated away to provide this energy,

$$\Delta E_{rot} = \frac{\Delta E_b}{g} \tag{22}$$

Equations (17) and (20) give the time t required by the pulsar to radiate this energy.

We assume that, due to the mass loss, the orbital angular momentum is decreased by the amount of,

$$\Delta J_{orb} = (\frac{J_{orb}}{M_2(t) + M_n}) \Delta M_2 \tag{23}$$

where ΔM_2 is the amount of mass loss and $M_2(t)$ is the mass of the companion at time t. As we know the orbital angular momentum we can calculate the orbital separation a which is required to calculate g. We also calculate the evaporation timescale as a function of t from,

$$\tau_{evap} = \frac{E_b(t)}{F_{evap}(t)} \tag{24}$$

It is assumed that the evaporation process does not affect the interior structure of the evaporating star, as (see below) the evaporation only takes place in 10^3–10^4 yr, which is very short in comparison to the evolutionary and thermal timescales of the star, and only a very small amount of mass is blown off by the process. We stop the calculation when the remaining pulsar energy has become so low that it is not sufficient to remove $10^{-5} M_\odot$ from its companion.

The results are depicted in Figure 5 for a pulsar with an initial period of 6 ms. This figure shows the companion mass and the evaporation timescale as a function of t. One

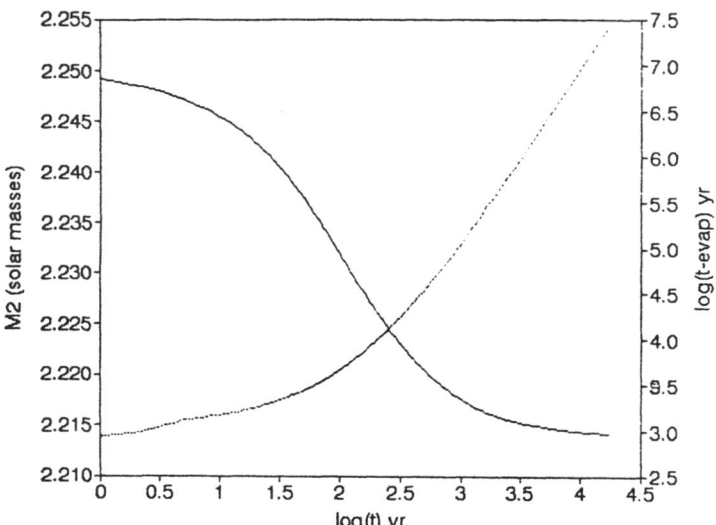

Figure 5: Mass of the evaporating star (solid line) and evaporation timescale (dotted line) as a function of t. The initial pulsar spin period is 6 ms, the initial mass of the evaporating star is 2.25 $M_\odot$ and the initial orbital period is 1 d. The mass of the pulsar is 1.44 $M_\odot$.

can see from the figure that a 6 ms pulsar is only capable of evaporating $\sim 0.036 M_\odot$ from its companion. Using the same method we find that a 10 ms pulsar can only evaporate $\sim 0.016 M_\odot$.

Note that although the evaporation timescale is very short at the beginning ($\sim 10^3$–10^4 yr), it increases to 10^6–10^7 yr in a few thousands years. This is due to the fact that the pulsar spin period grows very rapidly in the course of time. A standard neutron star with an initial period of 10 ms will spin down by a factor of 10 in 10^4 yr [see Equation (20)]. Hence, within the same time the pulsar energy flux will decrease by a factor 10^4 [see Equation (19)] and the evaporation timescale will increase by the same factor. We conclude therefore that the pulsar, despite being very energetic at its birth, is not able to evaporate its companion by a considerable amount.

5.2. THE DISRUPTION OF THE BINARY BY AN SN EXPLOSION

From Equations (2) and (3) one can derive that the binary will be disrupted by the SN explosion if,

$$2 - Y F_2 \leq 0 \qquad (25)$$

It can be shown from this inequality that the disruption will occur if the kick velocity vector is inside a solid angle with a critical semi-angle $\theta_{cr}(\phi)$, where $\theta_{cr}(\phi)$ is the solution of $2 - Y F_2 = 0$. Assuming the kick velocity to have a random direction, one can derive

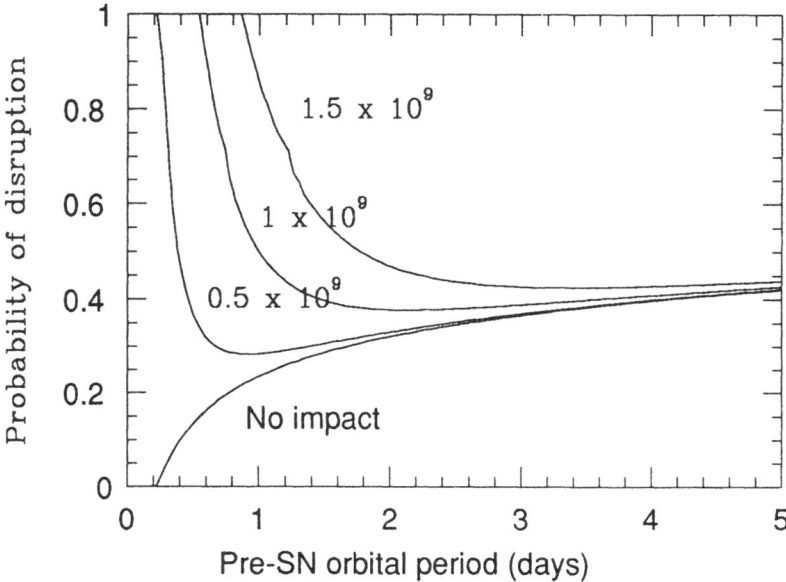

Figure 6: Probability of disruption as a function of the pre-SN orbital period for a system with $M_1 = 3.2M_\odot$, $M_2 = 2.25M_\odot$, $M_n = 1.44M_\odot$. The kick velocity is 100 km s^{-1}. Number on each line indicates the ejection velocity of the SN shell, i.e. 0 (no impact), 0.5, 1 and 1.5×10^9 cm s^{-1}.

that the probability of disruption is given by,

$$f = \frac{\int_0^{2\pi} \int_0^{\theta_{cr}(\phi)} \sin\theta \; d\theta \; d\phi}{4\pi} \qquad (26)$$

We calculate f by assuming that the mass of the exploding star before and after the explosion is, respectively, 3.2 and 1.44 $M_\odot$, and the mass of the unexploding star is 2.25 $M_\odot$. We present the result in P_0–f diagram as shown in Figure 6. Each line represents the disruption probability for a given value of V_{ej} (the velocity of SN shell) which is indicated on each line, i.e., 0 (no impact), 0.5, 1 and 1.5×10^9 cm s^{-1}. The kick velocity V_k is assumed to be 100 km s^{-1}.

Notice that the binary will in most cases be disrupted if $V_{ej} \geq 1$–1.5×10^9 cm s^{-1} and $P_0 < 1$ d. This conclusion is still valid for $V_k = 200$ km s^{-1}. We wish to stress the importance of the effect of the impact of the SN shell for the occurrence of disruption since without the impact most short period binaries will remain bound.

This gives a promising explanation of why a bound system like Her X-1 is exceptional provided that V_{ej} and P_0 are bounded by those above values. Taking into account the uncertainty in the ratio of the imparted to the incident momentum of the SN shell [which is incorporated in the value of α—see Equation (8)—which we assume to be 0.5], it does not seem difficult to meet the lower limit of V_{ej} as it is of the order of the observed ejection velocities of SN shell.

The pre-SN system is the outcome of a spiral-in process. Though our understanding of the spiral-in process is still far from complete, one expects that such a process will result in a very tight binary [the binding energy of the final binary must exceed the binding energy of the original envelope; see reviews by de Kool (1987) and van den Heuvel (1989)]. The requirement of a pre-SN orbital period of ≤ 1 d agrees with this view.

6. Conclusions

We have considered several combinations of orbital parameters of binaries in the pre- and post-SN phase which will lead to the formation of a system like Her X-1. Even a moderate mass (~ 3–$4\ M_\odot$) of the exploding star is enough to bring the system to its present distance above the galatic plane within its lifetime as too large mass will disrupt the system. The evaporation process cannot explain why Her X-1 like systems are very rare. For all considered cases the pulsar energy is far too weak to completely evaporate the pulsar's companion. A more detailed study for a model of system consisting of a 2.25 $M_\odot$ main-sequence star and a 1.44 $M_\odot$ neutron star with an orbital period of 1 d indicates that only very small amount of mass (< 2 %) can be evaporated by the pulsar. We conclude, therefore, that the only mechanism which may prevent the formation of many Her X-1 like systems seems to be the disruption of the binary by a SN explosion. Assuming that $M_1 = 3.2 M_\odot$, $M_2 = 2.25 M_\odot$, $M_n = 1.44 M_\odot$, and $V_k = 100$–200 km s^{-1}, we find that most systems will be disrupted if the ejection velocity of the SN shell is ≥ 1–1.5×10^9 cm s^{-1} and the pre-SN orbital period is ≤ 1 d. The impact of the SN shell plays an important role in disrupting the systems. Without the impact, most systems will remain bound. The requirement for such a short period pre-SN system is not unexpected as the system is the outcome of a spiral-in process where a lot of energy and angular momentum has been lost from the original system.

Acknowledgements
The author wishes to thank E. van den Heuvel, D. Bhattacharya and F. Verbunt for their helpful advice. Travel support from the Leids-Kerkhoven Bosscha Fonds is gratefully acknowledged. This research was supported in part by the National Science Foundation under Grant No. PHY89-04035.

References

Bailes, M. (1989), *Astrophys. J.* **342**, 917

Bhattacharya, D. and van den Heuvel, E.P.J. (1991), *Physics Report*, (in press)

Colgate, S.A. (1970), *Nature* **225**, 247

de Kool, M. (1987), PhD Thesis Univ. Amsterdam, Chapter V

Dewey, R.J. and Cordes, J.M. (1987), *Astrophys. J.* **321**, 780

Fenimore, E.E., Conner, J.P., Epstein, R.I., Klebesadel, R.W., Laros, J.G., Yoshida, A., Fujii, M., Hayashida, K., Itoh, M., Murakami, T., Nishimura, J., Yamagami, T., Kondo, I. and Kawai, N. (1988), *Astrophys. J. Letters* **335**, L71

Fryxell, B.A. and Arnett, W.D. (1981), *Astrophys. J.* **243**, 994

Halbwachs, J.L. (1987), *Astron. Astrophys.* **183**, 234

Hogeveen, S. (1991), *Astrophys. Space Sci.* **173**, 315

Iben, I. (1967), *Astrophys. J.* **147**, 62

Krolik, J.H. and Sincell, M.W. (1990), *Astrophys. J.* **357**, 208

Kulkarni, S. (1986), *Astrophys. J. Letters* **306**, L85

Kundt, W. (1989), in *Neutron Stars and Their Birth Events*, ed. W. Kundt, Advanced Science Institutes, Kluwer Academic Publ., Dordrecht, p. 1

Lyne, A.G., Anderson, B. and Saltere, M.J. (1982), *Monthly Notices Roy. Astron. Soc.* **201**, 503

M önchmeyer, R. and Mü ller, E. (1989), in *Timing Neutron Stars*, eds. H. Ö gelman and E.P.J. van den Heuvel, Advanced Science Institutes, Kluwer Academic Publishers, Dordrecht, p. 549

Murakami, T., Fujii, M., Hayashida, K., Itoh, M., Nishimura, J., Yamagami, T., Conner, J.P., Evans, W.D., Fenimore, E.E., Klebesadel, R.W., Yoshida, A., Kondo, I. and Kawai, N. (1988), *Nature* **335**, 235

Novotny, E. (1973), *Introduction to Stellar Structure and Interior*, Oxford University Press, New York

Oort, J.H. (1965), in *Galactic Structure*, Eds. A. Blaauw and M. Schmidt, Univ. Chicago Press, Chicago, p. 445

Radhakrishnan, V. and Srinivasan, G. (1981), *Proc. IAU Second Regional Meeting*, eds B. Hidayat and M.W. Feast, Tira Pustaka, Jakarta, p. 423

Rappaport, S. and Joss, P.C. (1983), in *Accretion Driven Stellar X-Ray Sources*, Eds. W.H.G. Lewin and E.P.J. van den Heuvel, Cambridge University Press, p. 1

Savonije, G.J. (1978), *Astron. Astrophys.* **62**, 317

Sutantyo, W. (1975), *Astron. Astrophys.* **41**, 47

Sutantyo, W., van der Linden, T.J. and van den Heuvel, E.P.J. (1986), *Astron. Astrophys.* **169**, 133

Taam, R.E. and Fryxell, B.A. (1984), *Astrophys. J.* **279**, 166

Taam, R.E. and van den Heuvel, E.P.J. (1986), *Astrophys. J.* **305**, 235

Taylor, J.H. and Manchester, R.N. (1977), *Astrophys. J.* **215**, 885

Thomas, H., Schmidt and H.U., Schoembs, R. (1986), in *The Evolution of Galactic X-ray Binaries*, eds. J. Truemper, W.H.G. Lewin, and W. Brinkmann, NATO Advanced Science Institutes, Reidel, Dordrecht, p. 221

Truemper, J., Pietsch, W., Reppin, C., Voges, W., Staubert, R. and Kendizorra, E. (1978), *Astrophys. J. Letters* **219**, L105

van den Heuvel, E.P.J. (1976), *IAU Symp.* No. 73, p. 35

van den Heuvel, E.P.J. (1983), in *Accretion Driven Stellar X-ray Sources*, Eds. W.H.G. Lewin and E.P.J. van den Heuvel, Cambridge University Press, p. 303

van den Heuvel, E.P.J. (1989), in *Timing Neutron Stars*, eds. H. Ögelman and E.P.J. van den Heuvel, Advanced Science Institutes, Kluwer Academic Publishers, Dordrecht, p. 523

van den Heuvel, E.P.J. and van Paradijs, J. (1988), *Nature* **334**, 227

Verbunt, F., Wijers, R.A.M.J. and Burm, H.M.G. (1990), *Astron. Astrophys.* **234**, 195

Wheeler, J.C., Lecar, M. and McKee, C.F. (1975), *Astrophys J.* **200**, 145

Collisions Between a White Dwarf and a Main-Sequence Star

Maximilian Ruffert
Max-Planck-Institut für Astrophysik
Karl-Schwarzschild-Str. 1
D-8046 Garching bei München
Germany

ABSTRACT. We model collisions between a white dwarf and a main-sequence star using a three-dimensional hydrodynamical code based on the piecewise parabolic method. Initially the main-sequence star has the structure of a polytrope with index $n = 1.5$ and a mass of 0.5 $M_\odot$. The white dwarf is treated as a gravitating point of equal mass. We only consider initially parabolic orbits, because in globular clusters the velocity dispersion is much smaller than the escape velocity at the surface of a star. We compare the results of two sets of calculations: The first uses only one grid, the second has five nested and refined grids.

Initial Conditions

The main-sequence star is modeled as a polytrope with index $n = 1.5$. This is appropriate for lower mass stars present in globular clusters. Its mass is 0.5 $M_\odot$ and its radius $R_* = 6 \cdot 10^{10}$ cm. The white dwarf is treated as a point mass of 0.5 $M_\odot$. The initial distance between the two objects at which the calculations are begun is two stellar radii ($2\ R_*$). Only initially parabolic orbits are considered, since the velocity dispersion in globular clusters is much smaller than the escape velocity at the surface of a 0.5 $M_\odot$ star. So at the distance of $2\ R_*$ the white dwarf is sent on a collision course with a velocity of approximately 500 km/s. Only one parameter is left to be varied: the impact parameter. It is fixed by the angle (between the velocity vector and the axis of symmetry) at which the white dwarf is set on course. The minimal distances aimed for are 0 (i.e. a central collision), 0.25 R_*, 0.5 R_*, 1 R_* and 1.5 R_*. The initial conditions are kept as close as possible to those used in Różyczka *et al.* (1989).

Computational Procedure

The hydrodynamic evolution is done by a finite difference code based on the piecewise parabolic method of Colella and Woodward (1984). The equation of state contains the pressure of an ideal gas plus radiation pressure. The self-gravity of the gas is calculated by a fast Fourier-transform. This implies that we use an equidistant Cartesian grid. To

E. P. J. van den Heuvel and S. A. Rappaport (eds.), X-Ray Binaries and Recycled Pulsars, 311–315.
© 1992 *Kluwer Academic Publishers*.

the gas potential we add the point mass potential of the white dwarf. Both are softened for computational convenience. The softening of the white dwarf potential is also used to imitate its finite size. The motion of the white dwarf in the potential of the gas is integrated by a leap-frog scheme. The only symmetry implied in the problem is the mirror-symmetry about the orbital plane, which allows us to halven the amount of computations. Because the temperatures never exceed 10^8 K and are high only in the closest vicinity of the white dwarf, the influence of nuclear burning is negligible. Thus it is not included in the calculations.

Monogrid Collisons

All calculations are done with $32^3/2$ and $64^3/2$ zones to check the influence of the numerics. Two interesting sets of parameters are calculated with $128^3/2$ cells. Only the latter grid is sufficiently fine to resolve the white dwarf (cell size $\approx 10^9$ cm).

At small impact parameters (nearly central collisions) the structure of the main-sequence star is completely disrupted within an hour. Between 10 and 30 % of the matter becomes unbound and leaves the system. The remaining matter eventually settles around the white dwarf and forms a thick disk. The long term evolution towards a red giant has not been considered in detail yet. A detailed discussion of the results can be found in Ruffert and Müller (1990).

Larger impact parameters are more difficult to model, because both the timescales become longer and mass leaves the grid too quickly.

Multigrid Collisions

Following and simplifying the descriptions of Berger and Colella (1989) we recalculated some of the monogrid collisions using the same initial conditions and focussing on grazing collisions. The problems of the monogrid calculations—insufficient resolution around the white dwarf; the spatial extent of the grid being too small— are solved by using 5 refined and nested grids. While the coarsest grid has a size of 10^{12} cm, the finest grid has cells of only $2 \cdot 10^9$ cm, so $32^3/2$ calculations are largely sufficient. An additional run with $64^4/2$ zones was done for comparison. The evolution of the latter model is followed for 20 hours.

The almost central collisions evolve to a steady state on timescales of hours. The large grids now allow us to determine the amount of unbound mass much more precisely than was possible before. The monogrid calculations are confirmed.

At larger impact parameters (the white dwarf only grazes the main-sequence star) the star is left elongated but intact. However, energy and angular momentum are transferred from the orbit to the main-sequence star. The orbit of the white dwarf is changed from parabolic to elliptic, i.e. the white dwarf will eventually fall back. During the second impact the main-sequence star is destroyed. The evolution resembles the one obtained from more head-on collisions.

Presentation of Results

The global values like total energies and angular momentum show that they are transferred between the orbit and the gas. Contour plots of the density and temperature distribution

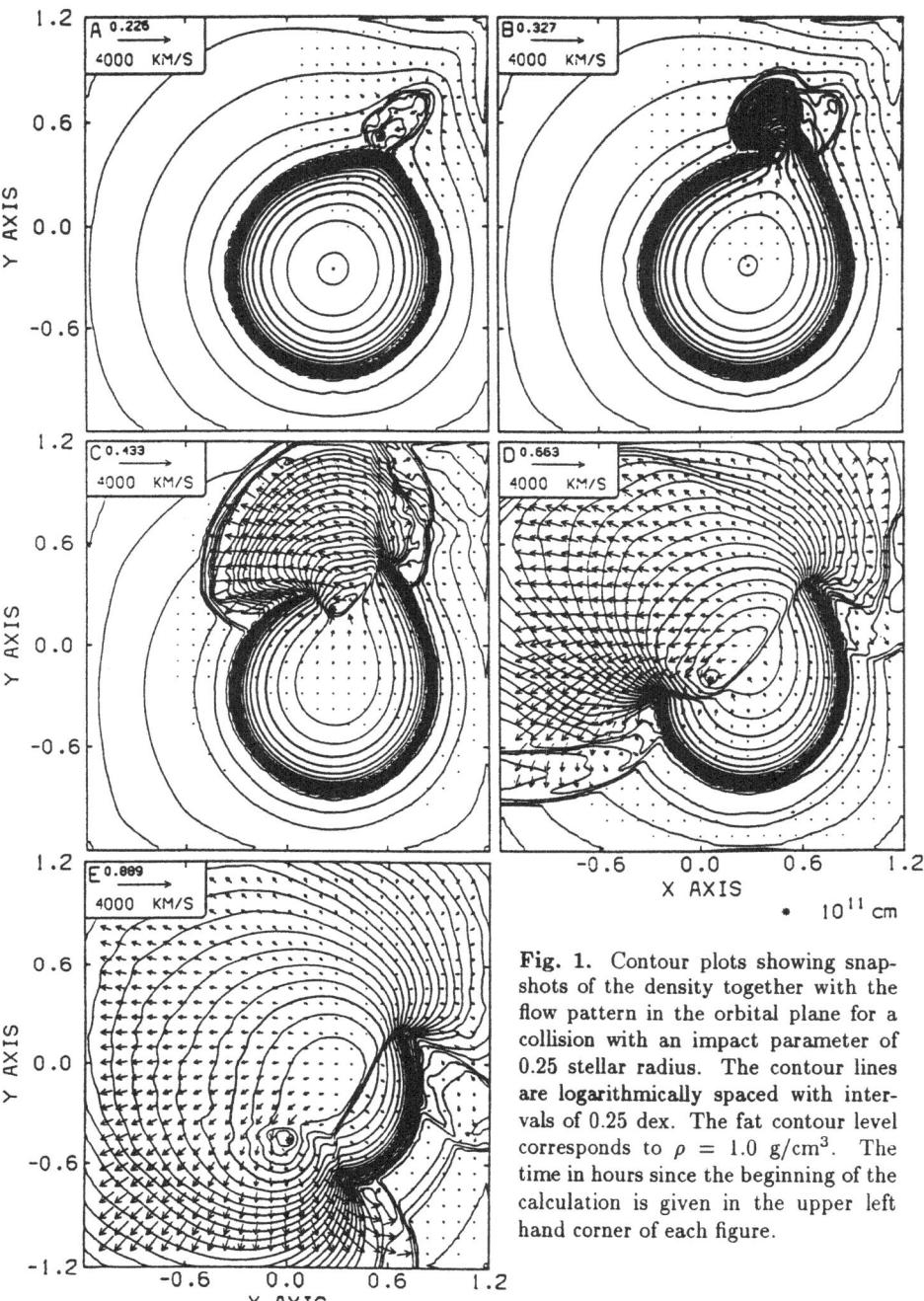

Fig. 1. Contour plots showing snapshots of the density together with the flow pattern in the orbital plane for a collision with an impact parameter of 0.25 stellar radius. The contour lines are logarithmically spaced with intervals of 0.25 dex. The fat contour level corresponds to $\rho = 1.0$ g/cm^3. The time in hours since the beginning of the calculation is given in the upper left hand corner of each figure.

314

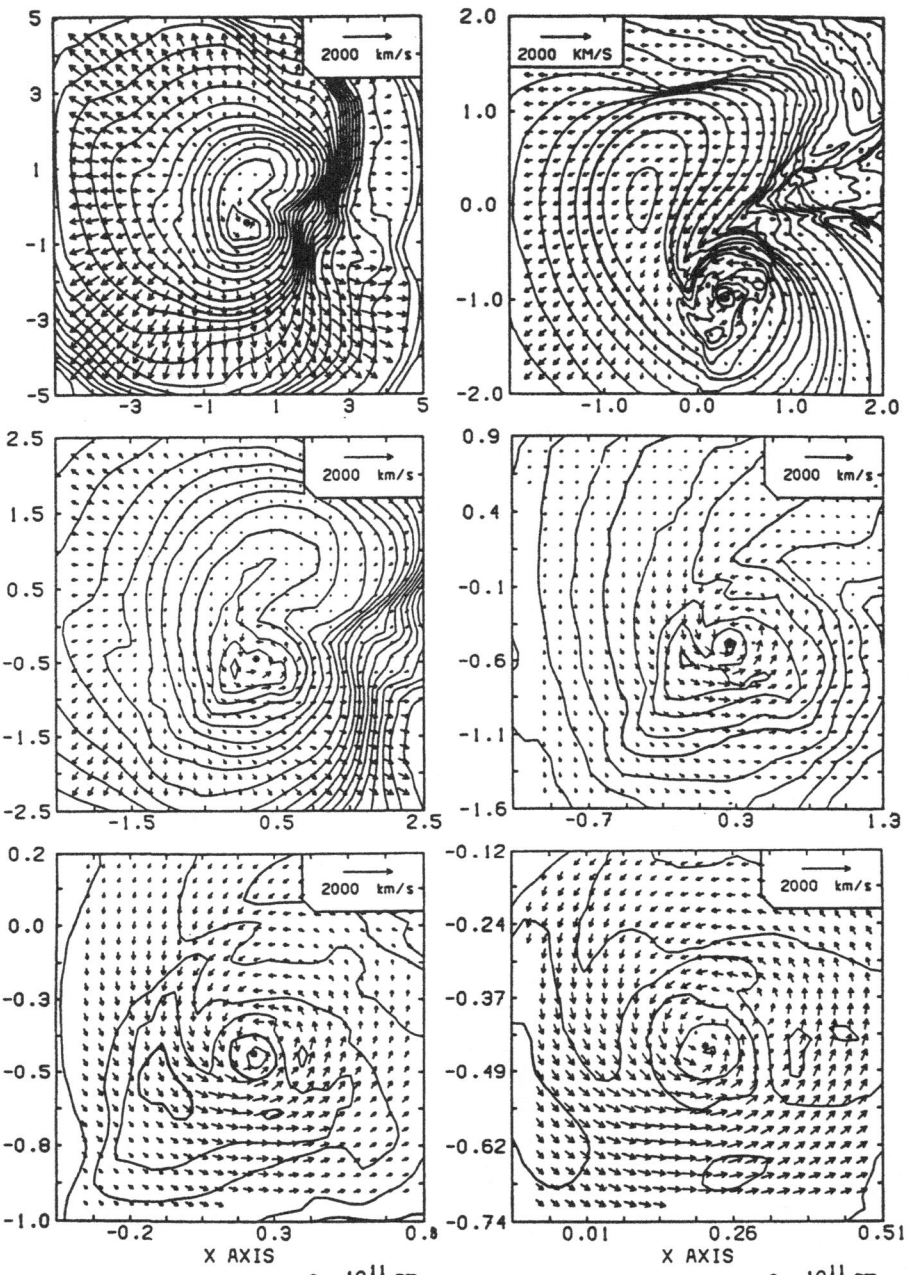

Fig. 2. Contour plots showing snapshots of the density together with the flow pattern. The top right figure shows the distribution for a model with an impact parameter of 0.5 stellar radius at $t = 2.50$ h, calculated on a grid with $128^3/2$ cells. The other five figures show the five nested grids for the same impact parameter at nearly the same time $t = 2.48$ h. Each of the five grids has $32^3/2$ cells. The two fat contours correspond to 0.1 and 1.0 g/cm³.

along with the velocity vectors show the flow in the orbital plane. However the temporal evolution and the three-dimensional distribution can best be seen in movies. Two movies have been produced, one for each of the calculations described above. The flow in the orbital plane is shown by false-color representations, the three-dimensional distribution by ray traced views.

Two snapshots for the evolution can be found in Figs. 1 and 2. Fig. 1 shows the evolution as the white dwarf plunges into the main-sequence star with an impact parameter of 0.25 R_*. A bow shock develops and crosses the extended star thereby heating up its matter. The matter accelerates and part of it eventually leaves the system. Fig. 2 is a comparison between the monogrid calculation (top right) and the model with five nested grids. In both calculations the impact parameter was 0.5 R_* and the snapshots are taken at approximately the same time (2.5 h).

References

Berger, M.J., Colella, P.: 1989, *J. Comput. Phys.* **82**, 64.

Colella, P., Woodward, P.R.: 1984, *J. Comput. Phys.* **54**, 174.

Różyczka, M., Yorke, H.W., Bodenheimer, P., Müller, E., Hashimoto, M.: 1989, *Astron. Astrophys.* **208**, 69.

Ruffert, M., Müller, E.: 1990, *Astron. Astrophys.* , **238**, 116.

NEW DIRECTIONS IN GLOBULAR CLUSTER MODELING

Piet Hut
Institute for Advanced Study
Princeton, NJ 08540, USA

ABSTRACT. Recent observations show a fascinating and abundant presence of many different types of binaries and binary remnants in globular clusters. What was a perceived dearth of binaries only a few years ago has now turned into a veritable zoo of interesting objects. Besides the, now classic, X-ray binaries, dozens of millisecond pulsars have been detected, scores of blue stragglers, significant numbers of spectroscopic as well as eclipsing variables, and clear evidence for main-sequence broadening caused by the presence of a significant fraction of binaries.

   This diverse collection of objects forms a gold mine for dynamicists modeling the evolution of globular clusters. While all these binaries and binary products provide useful diagnostics, some of them are even directly involved in the physical processes of energy generation, driving the core expansion after core collapse. In this sense we are beginning to get a direct look at the central engines which power the later phases in the evolution of globular clusters.

   In this paper I first give an overview of globular cluster evolution, during which I discuss two new pieces of research: 1) a general linear stability analysis of the isothermal sphere; and 2) a detailed investigation of the rate of error growth in N-body systems. I then discuss separately six more new directions in globular cluster modeling: 3) the inclusion of a mass spectrum in post-collapse simulations; 4) the addition of primordial binaries in these calculations; 5) new results in binary-binary scattering experiments; 6) the study of hydrodynamic effects during close stellar encounters; 7) approximate migration models for recycled binaries; 8) connections with the dynamics of nuclei of galaxies.

1. Introduction

 The last two years have seen an explosion of new activities in the modeling of globular cluster evolution. This has been partly triggered by a wealth of new observational data of various types. These exciting developments appear after a much slower rate of growth of our understanding of star cluster evolution in general, during the last few decades.

   In a nutshell, the study of globular cluster evolution began by the late thirties and early forties with the prediction that star clusters would eventually evaporate. Estimates of evaporation rates were worked out further in the fifties. During the nineteen sixties, the mechanism of gravothermal collapse was discovered, and studied in detail in numerical calculations in the seventies. For a long time, the fate of a globular cluster after the initial collapse phase was nearly totally unknown. The eighties finally shed light on this fundamental question, in the form of a number of

E. P. J. van den Heuvel and S. A. Rappaport (eds.), X-Ray Binaries and Recycled Pulsars, 317–348.

simulations, leading to the discovery of a new physical phenomenon: gravothermal oscillations. So far, these models have been relatively crude, and the main task of the nineteen nineties will be the development of detailed models of post-collapse evolution, detailed enough to be reliably compared with observations.

The organization of the present paper is as follows. In §2 I give an overview of the general area of globular cluster evolution. Our understanding of core collapse is summarized in §2.1, with the addition of some new results concerning the qualitative form of core contraction (§2.1.2), and the rate of error growth in collapse simulations(§2.1.3). The evolution past core collapse is discussed in §2.2. In §3 the first results of post-collapse multi-mass simulations are discused. §4 addresses the implications of primordial binaries for globular cluster evolution. After a qualitative discussion of the main physical effects, the results of detailed simulations are reviewed. Primordial binary observations and their relations with the simulations are briefly mentioned. §5 summarizes the latest developments in binary-binary scattering experiments. §6 reviews some of the recent attempts to begin to understand some of the intricacies of the microscopic processes in star cluster evolution: hydrodynamic interactions during close encounters between stars. §7 addresses the observationally important question of the distribution of binaries through a globular cluster. Recent approximate models are discussed, and compared with observations. A connection between the dynamics of globular clusters and modest galactic nuclei is discussed in §8. §9 sums up.

2. Overview of Globular Cluster Evolution

Globular clusters undergo core collapse on a time scale that is only a few times larger than the half-mass relaxation time. During and after collapse, infinite central density can be avoided only when some form of central energy source turns on. Several energy sources are possible, and they can be divided into three categories. Binaries can increase in binding energy, stars can undergo mass loss, or a black hole may form and subsequently swallow stars. All three processes result in a direct or indirect heating of the cluster.

One of the most interesting new developments is the investigations of the role played by primordial binaries, which can provide 'fossil fuel' to power the central energy source, for a considerable time after core collapse. In the present paper I will only address the theoretical aspects of globular cluster evolution with primordial binaries. For the many exciting new observational results, I refer to an extensive review which will appear separately (Hut et al. 1992). A short review of the field of globular cluster evolution was given by Elson et al. (1987). A general background for the dynamical evolution of globular clusters can be found in the excellent monograph by Spitzer (1987).

2.1. Core Collapse

The two hundred or so globular star clusters that orbit our galaxy present us with a tantalizing problem. With $\sim 10^5 - 10^6$ member stars each, their dynamics is still too complex for detailed star-by-star simulation on present-day computers. However, much insight has been gained from approximate simulations, based on Fokker-Planck or conducting gas-sphere models, as well as from direct N-body calculations with up to a few thousand particles.

2.1.1. History of Simulations

Half a century ago, Ambartsumian (1938) and Spitzer (1940) independently predicted the inevitable decay of star clusters by evaporation. Two decades later, Antonov (1962) and Lynden-Bell and Wood (1968) showed how a star cluster undergoes internal core collapse, well before its final evaporation, due to the instabilities caused by the negative heat capacity of all self-gravitating systems. Also around this time, Hénon (1961, 1965) constructed the first cluster models exhibiting core collapse.

In the nineteen-seventies, star cluster research received an enormous boost from the unexpected discovery of globular cluster X-ray sources, which were explained as neutron stars that had been tidally captured into binary stars (Fabian *et al.* 1975, Press and Teukolsky 1977). On the theoretical frontier, a variety of numerical simulations began to sketch how star clusters evolve towards core collapse. These simulations were based on the Fokker-Planck approximation for the slow drift of stars in energy and angular momentum space (*cf.* Spitzer 1975). Again a decade ahead of his time, Hénon (1975) was the first to extend a cluster evolution model past core collapse. Soon afterwards, the first observational evidence for a collapsed core, which happened to be in M15, was presented (Newell and O'Neill 1978). It took an additional thirteen years, and the launch of the Hubble Space Telescope, to resolve the core of M15 (Lauer *et al.* 1991).

After a few years of quiet work by theoreticians trying to figure out how to model a cluster after core collapse, the results of a number of different simulations were published around 1984, which were discussed and summarized in papers presented at the I.A.U. symposium 113 (Goodman & Hut 1984). Until 1990, nearly all simulations of star cluster evolution were started with initial conditions in which all stars were single. Stars were allowed to form binaries either dynamically, through near-simultaneous close encounters of three stars leaving two of them bound, or tidally, through energy dissipation in the tidal bulges of one or both of the stars involved in a close encounter. As mentioned above, more information can be found in Elson *et al.* (1987) and Spitzer (1987).

2.1.2. The Qualitative Behavior of the Simulations

The contraction of the inner parts of a star cluster, on a two-body relaxation time scale, is a direct consequence of the negative heat capacity of any self-gravitating system, as follows from the virial theorem. A clear and very physical discussion of the resulting gravothermal instability was given by Lynden-Bell and Wood (1968), who discussed the thermodynamic behavior of an isothermal gas sphere, enclosed in a spherical adiabatic boundary. Although their arguments predicted the overall contraction of the inner parts of a star cluster, they did not address the evolution of the core mass.

Hachisu and Sugimoto (1978) first investigated the form of the run-away solutions, by computing the second order variation in the entropy of the bounded isothermal gas system. A more detailed treatment of the self-similar contraction of the core of a star cluster during the later stages of core collapse was given by Lynden-Bell & Eggleton(1980). They also gave a condition which the heat conduction had to obey in order to give rise to a self-similar contraction of the inner parts of a cluster. They showed how the stellar dynamical process of two-body relaxation can be approximated in the context of a gas sphere, even though the former has a mean-free path much larger than the size of the system, while in the latter this inequality is reversed.

The conditions for core collapse to occur have been further elucidated recently by Makino & Hut (1991). They present a linear stability analysis of the isothermal sphere, for different choices of heat conductivity. Two extreme cases they treat are given by stellar dynamics and by radiative heat transfer. The latter case corresponds to a star starting out with a (nearly) constant temperature distribution throughout its interior. Such a star will start to contract on a thermal time scale, and as a result its core radius, too, shrinks. However, during the contraction the heat conductivity will be more effective in the outer layers than in the inner parts. The reason is that the mean free path length for photons is much larger in the less dense outer layers. Therefore, the innermost layers are slow to respond, and material will rain onto the core from those layers somewhat further out. As a result, the core mass will increase, even though the core radius decreases.

The other extreme case is that of stellar dynamics. Here the heat conductivity has the opposite behavior. At higher densities the frequency of two-body interactions becomes higher, and consequently heat conductivity increases. As a result, the initial contraction will take place more easily and quickly in the core. Indeed, the linear stability analysis of Makino and Hut show how both the radius and the mass of such a system decrease in value. In the non-linear regime, the smaller heat conduction time scale will quickly lead to a core contraction which decouples from the outer layers, leading to the self-similar collapse described by Lynden-Bell and Eggleton.

2.1.3. Limitations of the Simulations: Error Growth

Our understanding of the later stages of core contraction is largely based on Fokker-Planck simulations (Spitzer 1987). This name stems from the fact they are based on a description in terms of statistical diffusion of particle orbits in energy space, described by a Fokker-Planck approximation (Cohn 1979). These methods are quite fast, and have been very fruitful for furthering our understanding of pre-collapse evolution, but they have severe shortcomings when applied to the later phases, as will be discussed below in more detail. Therefore, recent simulations have been increasingly based on direct N-body integration of a system of point masses, under the influence of their mutual gravitational potential.

Even though the N-body problem is easy to formulate, and in fact forms the oldest problem in mathematical physics which could not be solved analytically, it is not easy to simulate numerically either. Not easy, that is, if we insist on getting the correct answer in a strict, mathematical sense of the word. Heggie *et al.* (1988) have shown that a correct calculation which follows an N-body system all the way to core collapse requires a computer calculation with an enormous precision: a word length of roughly N digits! Strictly speaking, this would exclude the simulation of any system with more than 25 or so particles, even when using quadruple precision (128 bit word length).

Fortunately, physicists, and especially astrophysicists, don't always insist on getting correct answers. Good enough answers are good enough for them. The problem is to determine what is good enough. At present, nobody has any formal proof that N-body models give reliable answers, but neither is there any reason to believe that they have any intrinsic bias built in. The bottom line is that there does not seem to be any reason to get worried, although it is interesting to be aware of the articles of faith underlying the simulations. Let me give the main arguments pro and con the reliability of N-body simulations.

The first warning concerning the rate of growth of errors in N-body calculations came from the pioneering investigations by Miller (1964). He considered systems with $N < 32$, and found that the e-folding time scale became shorter with increasing

N, when compared with a crossing time. The dramatic effects of the exponential instability were demonstrated forcefully by Lecar (1968), who coordinated a study, using different computers and algorithms, of the solution generated by one particular choice of initial conditions. In his case, the effect of the instability is intertwined with the effects of different truncation errors and other sources of numerical error. A much later investigation (Heggie *et al.* 1988) showed that the trend found by Miller changes for larger N, to a rate of growth approximately independent of N for $N \gtrsim 30$. They found that in this limit, errors grow with an e-folding time which is nearly one-tenth of a crossing time. More details of their work can be found in Goodman *et al.* (1991). Similar results have been obtained by Kandrup & Smith (1990).

Faced with the impossibility to trace the unique evolution of a star cluster model, starting with a particular set of initial conditions, we can ask whether the (strictly speaking) wrong result of a computer calculation can still be correct in some statistical sense. This question can be asked in two stages. First, we can ask ourselves whether the outcome of an N-body simulation may be the true result of different initial conditions, close to those actually used. This is actually a reasonable guess, since the exponential fanning out of neighboring orbits would imply an convergence upon time reversal of the inspection of a set of forwardly computed orbits [Note that this last point is important: if we would compute the orbits independently backwards in time, we would see an exponential growth towards the past as well, since the equations of motion are invariant under time-reversal].

In practice, errors in conserved quantities such as total energy and angular momentum of the system thwart such a backward shadowing approach, but to within the accuracy of those quantities, it seems likely that a true orbit exists, starting off very near to the actual initial conditions, and leading to the obtained final positions. However, even if such a shadowing theorem could be proved, we would still have no guarantee that the true orbit, although it may be everywhere close to the calculated orbit, would be a generic orbit. In the worst case, if the shadowing orbit would be drawn in a devious way from a pathologic subset of orbits with measure zero, the calculated orbit may have no relevance to the behavior of a typical orbit. However, we have no reason to believe that (computational) nature would be so devious. On the contrary, it is very difficult to imagine how and why such selection effects could take place. While this type of reasoning obviously does not constitute a proof, it will probably be sufficiently reassuring for most physicists.

2.2. Post-Collapse Evolution of Globular Clusters

It was only after the early eighties that simulations became sophisticated enough to be able to penetrate past the near-singular state of core collapse into the asymptotic regime of post-collapse evolution and eventual cluster evaporation (see Goodman & Hut 1985, and references therein). That decade saw a number of exciting theoretical discoveries. Gravothermal oscillations, first predicted on the basis of numerical simulations (Sugimoto & Bettwieser 1983; Bettwieser & Sugimoto 1984), were later also found in semi-analytical studies (Goodman 1987). At about the same time, it was realized that close encounters and physical collisions between stars, previously invoked as possible explanations of cluster X-ray sources and blue stragglers (Fabian *et al.* 1975, Krolik 1983), could also have far-reaching dynamical consequences for the cluster as a whole (Goodman 1984, Ostriker 1985, Lee & Ostriker 1986, Lee 1987ab).

Another important development was the observational discovery of many seeing-limited cores (*cf.* Djorgovski and King 1986), suggestive of remnants of core collapse, and quite well explained in terms of the results of multi-mass cluster simulations, with a judicious choice of stellar mass function (Murphy and Cohn 1988, Murphy *et al.* 1990). Around the same time, evidence was accumulating for the presence of a substantial population of primordial binaries in globular clusters (Latham *et al.* 1985; Pryor *et al.* 1987, 1989; for a review, see Hut *et al.* 1992).

2.2.1. Cluster Expansion

The evolution of a globular cluster after core collapse has only recently been studied intensively, and many aspects of our understanding of it remain uncertain and may change dramatically in the coming years. The *mean* behavior of the cluster after core collapse *is*, however, quite firmly established: the half-mass radius expands according to $r_h(t) \propto t^{2/3}$, where t is the time since core bounce, while the velocity dispersion drops according to $v \propto t^{-1/3}$. This relation may be derived from general principles, without any knowledge of the mechanism of energy generation in the core (*cf.* Hénon 1965, 1975), in a manner analogous to Eddington's (1926) prediction of the mass-luminosity relation for stars, which requires no precise knowledge of the nature of their internal energy generation.

The derivation goes as follows: (1) the half-mass relaxation time t_{hr} in a self-similar solution scales as $t_{hr} \propto t$, the time since core bounce; (2) $t_{hr} \propto N t_{hc}$, where N is the number of stars in the cluster, t_{hc} is the crossing time at the half-mass radius, and we have neglected a factor $\log N$; (3) if we neglect the slow change in mass and particle number due to escape, the virial theorem gives $t_{hc} \propto r_h^{3/2}$; (4) combining these gives $t \propto r_h^{3/2}$ which leads to the results quoted above. In contrast, the rate of expansion of the core *does* depend on the details of the central engine (Cohn 1985; Ostriker 1985). Recently, semi-analytical models for a variety of physical processes and approximations have been developed by Stodółkiewics & Giersz (1990) and Giersz (1990ab).

Thus, regardless of the precise state of the core, and the physical processes going on there, after core collapse, the cluster half-mass radius expands steadily. As it does so, the Galactic tidal field steadily removes the outermost stars, with the result that, eventually, the entire cluster is disrupted. The timescale for this process is longer than the core collapse time, but only by a factor of a few.

What is the character of the central engine in a post-collapse cluster? Several energy sources are possible: one is binding energy extracted from binaries, another is mass loss from the system by stellar evolution. Binaries can be formed by three-body dynamical capture or by two-body tidal capture. Enhanced mass loss can occur when stars collide and merge, forming heavier remnants with a much shorter lifetime under stellar evolution than original cluster stars. Finally, a black hole can form through repeated merging (for a recent review of these three mechanism, see Goodman 1989).

The heating caused by each of these mechanisms takes on quite different forms. Let us first look at binaries. A hard binaries is defined as having a binding energy $>> 1kT$, a measure for the average thermal energy of a single field star. In other words, a hard binary has an orbital velocity clearly exceeding the velocity dispersion of the system (in case of equal mass stars; for unequal masses we have to compare the kinetic energies instead). When a hard binary encounters a single star, it tends to achieve equipartition with the single star, possibly after a temporary capture and/or exchange. In the process it will give off energy to the escaping single star, and harden in the process. In this way, on average the single star comes out of the scattering process with more energy than it went in with. In this way, hard binaries tend to heat their environment.

The second mechanism, mass loss, is a more indirect way of heating a star cluster. Let us take the situation that a cluster loses a fraction ϵ of its mass, either through the escape of one or more stars, or through mass loss through stellar evolution (through a wind, or a Helium flash, or a supernova explosion of a multiple merger product). In all these cases, the mass loss will take place nearly instantaneously with respect to the evolution time scale of the cluster. As a result, the kinetic energy will on average decrease by a factor ϵ (in case of winds or explosive mass loss), or less (in case of a slow diffusion of stars toward unbounded orbits). However, the potential energy of a star cluster is quadratic in its mass, and a decrease by mass of a fraction ϵ will lead to a decrease in potential energy of a fraction 2ϵ. Therefore, the initial virial equilibrium, in which the potential energy was twice the kinetic energy of the cluster, cannot be maintained. After the mass loss, the potential energy will have decreased much more than the kinetic energy. Therefore, effectively, the cluster will have been heated with respect to the new equilibrium situation.

The third mechanism, heating by a black hole, is also an indirect process. The central hole will most likely capture stars which have orbits which are confined predominantly to the central regions of the cluster. Therefore, they carry a relatively small fraction of the kinetic energy of the stellar population. Capturing such stars again tends to increase the relative temperature of the remaining stellar population.

2.2.2. Core Oscillations

In some of the earliest post-core collapse simulations, Sugimoto and Bettwieser (1983; Bettwieser and Sugimoto 1984) found chaotic fluctuations in the size of the core radius. They explained these as a consequence of the gravothermal instability, and therefore introduced the term 'gravothermal oscillations' to describe them. In essence, the underlying physical mechanism can be simply described as follows. For a large number of stars in the system, the inner relaxation time scale is much larger than the half-mass relaxation time scale, which determines the overall rate of expansion. Therefore, the inner regions have the tendency to evolve on a time scale much smaller than the overall expansion time scale. As a result, the inner regions tend to get impatient, and a small fluctuation can trigger a local re-collapse, followed by a local re-expansion. The larger the number of stars, the more the central and outer time scales are decoupled, and the more chaotic the oscillations become. A formal demonstration that the dynamical behavior of these oscillations is characterized by a low-dimensional chaotic attractor is given by Breeden et al. (1991; for a summary, see Cohn et al. 1991).

The gravothermal character of the core oscillations was confirmed explicitly by Goodman (1987), who performed a linear stability analysis of a new regular self-similar model for post-collapse evolution, and classified the different modes of behavior according to the type of linear instability they exhibit. He found that for a total number of stars $N < N_1$ his self-similarly expanding solution is linearly stable, while for $N_1 < N < N_2$ the solution is overstable, and for $N > N_2$ it is unstable. He estimates $N_1 \approx 7000$ and $N_2 \approx 40,000$. Although the instability for very large N has indeed the expected character of a gravothermal instability, we do not yet understand the nature of the overstability. It is presently somewhat unclear how this behavior is modified by the presence of complicating physical effects, such as a stellar mass spectrum or a substantial binary population.

Like gravothermal collapse, gravothermal oscillations now appear to be a ubiquitous phenomenon, at least in the models which treat the stars and all physical processes as continuous quantities. Inagaki (1986) and McMillan (1986, 1989) have expressed doubts as to whether the oscillations persist in real clusters, where the stars and the physical processes are discrete, and statistical fluctuations may be very large. This issue can probably only be resolved by direct N-body simulations of systems containing $> 10^4$ stars.

2.2.3. Limitations of the Simulations: Time scales and Hydrodynamics

A central problem in globular cluster simulations is the occurrence of widely disparate timescales. Binary stars, with orbital periods of days or less, and separations of fractions of an astronomical unit, play an essential role in the dynamics and the overall energy budget of a cluster. However, the cluster itself evolves on a timescale of 10^9 years, 10^{12} times larger than the orbital period of a tight binary. Comparing the total lifetime of a cluster to the time step needed during periastron passage in an eccentric tight binary, containing two white dwarfs, say, we have a disparity which can easily reach a factor of 10^{18}. It is clear that special measures are needed in order to ensure an accurate treatment of star cluster evolution.

A separate, but equally important, challenge is the extension of our programs to include crude models of physical effects relating to the finite size and internal physics of stars. We wish to incorporate, at least in a rudimentary fashion, the ability to "mix and match" stellar dynamical, stellar evolution, and hydrodynamical codes. One of the most severe requirements here will be that the individual program modules co-exist peacefully during an extended simulation, without the need for intervention by a human supervisor, to guide the codes through the many unforseen bottleneck situations that can reasonably be expected to arise.

For example, allowing arbitrary configurations of binary systems to form, including the possibility of stable (or unstable) mass transfer between any pair of normal, evolved or degenerate stars will surely open a Pandora's box of possible interactions. Add to this the occasional close passage of a third star (or even another binary!) while all this is in progress, and it is clear that we will have to start with quite crude heuristic models, in order to have any hope of success. For example, common-envelope evolution might initially be treated with a rather simple recipe for when, and to what extent, two stars will spiral together, given a particular set of initial conditions. Once the proper overall framework is established, however, more realistic refinements can be added in a relatively straightforward manner.

3. Multi-Mass Models

Until recently, most simulations of the evolution of globular clusters beyond core collapse started off with a population of equal-mass stars. This is not a bad approximation for a first attempt to model an old population of stars, with a turn-off mass around $0.8 M_\odot$. However, any detailed comparison with observations requires the inclusion of the effects of the presence of a mass spectrum, especially for the central regions of a cluster where mass segregation plays an important role. The first post-collapse simulations containing a mass spectrum, as well as a large number of other physical effects, were performed in a remarkable investigation by Stodółkiewics (1982, 1985), based on a Monte Carlo Fokker-Planck code. Recently, more detailed multi-mass simulations based on a direct-integration Fokker-Planck approach, have been performed by Murphy et al. (1990), Chernoff & Weinberg (1990), and Lee et al. (1991).

The simulations by Murphy et al. and by Lee et al. were based on the Fokker-

Planck approach, in which the central energy generation was modeled on the dynamical formation of binaries in three-body encounters, based on a generalization of the equal-mass expression given by Hut (1985). Murphy *et al.* investigated how gravothermal oscillations are affected by the presence of a mass spectrum. They found that instability occurred for clusters containing more than about $10^5 M_\odot$, with a limiting mass value which had a weak dependence on the steepness of the slope of the initial mass function (note: after publication we realized that we had not used the correct heating rate. As mentioned in more detail by Grabhorn *et al.* 1991, the line in fig. 6 of Murphy *et al.* 1990 should be raised by 30 ~ 40%. The predictions for the core radii, however, do not change significantly).

As a consequence, most galactic globular clusters are expected to exhibit instability against core oscillations after core collapse. This has the important observational consequence that the core of a collapsed cluster is likely to be substantially larger than what would be expected from estimates which neglect core oscillations. The reason is that during an oscillation most of the time is spent in the expanded stages, and it is therefore extremely unlikely that we could catch a collapsed cluster in the contracted stage of an oscillation. Instead, we are likely to witness an expanded stage with a core radius typically in the range of 0.05 ~ 0.1 pc. As a specific example, Murphy *et al.* give a value of 0.07 pc for a cluster with a mass of $4 \times 10^{-4} M_\odot$ and a mass slope of 1.5, close to the Salpeter value of 1.35.

Such relatively large values for the predicted core radii of collapsed clusters are near the edge of observability for typical clusters. It is interesting that the recent HST observations of M15 report a core radius of 0.13 pc, only slightly larger than typical values found by Murphy *et al.* Alternatively, core radii around 0.1 pc can be caused by the presence of primordial binaries, as will be discussed in the next section.

The simulations by Lee *et al.* (1991) include a galactic tidal field, while averaging over gravothermal oscillations. They are thus complementary to the simulations reported by Murphy *et al.* (1990). Lee *et al.* make a number of interesting predictions for the evolution of the mass functions in a globular cluster as a function of radius in the cluster. In addition, they show how during most of the post-collapse evolution the half-mass relaxation time is of order $t_{rh} \sim 0.1t$, with t the age of the cluster. Only during the final evaporation phase does t_{rh} drop significantly, to $t_{rh} < 0.01t$. They compare this with the near absence of galactic globular clusters with $t_{rh} < 10^8$ yr.

Lee *et al.* also studied the effects of the presence of degenerate stars, neutron stars and white dwarfs, many of which are more massive than the average stars and therefore tend to concentrate in the core. This effect became more pronounced in the later stages of their calculations, because the loss of stars over the tidal boundary is more pronounced for the lighter stars which increasingly dominate the outer halo population. An observational effect of a core dominated by dark degenerate remnants would be an increase in the observed core radius, since the light would be less concentrated than the mass.

Chernoff & Weinberg's simulations were also based on a Fokker-Planck approach. They concentrated on the pre-collapse phase, and made a detailed study of the effects of mass loss by stellar evolution as well as by escaping stars, through the dynamical response of the tidal radius of a globular cluster. In addition, they presented detailed information about observational aspects of their models, in terms of colors, surface brightness, mass-to-light ratios, and mass functions. They pointed out that these realistic detailed modeling studies complement earlier, more schematic studies of the galactic family of globular clusters as a whole (*cf.* Chernoff *et al.* 1986, Chernoff and Shapiro 1987, Aguilar *et al.* 1988, Chernoff & Djorgovski 1989).

Intriguing as the above Fokker-Planck results are, it is important to realize how many effects have not yet been modeled accurately. One limitation lies in their statistical nature, which is of questionable validity after core collapse, when individual binaries can release large amounts of energy. The resulting recoil will introduce large changes in the orbital motions of single stars and centers-of-mass of binaries, in direct contradiction of the underlying Fokker-Planck assumptions. A second limitation for realistic applications lies in the relatively small number of physical parameters that can be modeled statistically. For example, providing statistical bins simultaneously for a collection of mass, energy and angular momentum choices will result in many bins containing only a fraction of one star, obviously jeopardizing the statistical assumptions of a Fokker-Planck code. These problems can be overcome by direct N-body calculations, but so far the necessary computer power, in the range of Teraflop-days (Hut *et al.* 1988).

4. Primordial Binaries

Globular clusters contain relatively fewer binaries than the galactic disk, and until only a few years ago it was not clear whether any stars had been formed in binaries at all, during the birth of globular clusters. Recently, however, observations have indicated that the number of primordial binaries is large enough to have an important influence on the dynamical evolution of globulars. This has been the motivation for several detailed simulations modeling such an evolution in the presence of primordial binaries.

4.1. Observations

Only a decade ago, there was no positive evidence that globular clusters contained significant number of binaries other than those made dynamically through two-body and three-body capture. This was rather remarkable, since in the solar neighborhood binarism is the rule rather than the exception. The first systematic search for radial velocity variables (Gunn & Griffin 1979) did not result in any positive detection, and for a number of years it was widely believed that globular clusters were severely binary-poor.

A few years later, two dwarf novae were detected in a globular cluster (Margon *et al.* 1981; Margon & Downes 1983). However, these cataclysmic binaries contain white dwarfs and may well have been formed through tidal capture in a way similar to the formation of their neutron star containing counterparts, the low-mass X-ray binaries. Thus, still no observational handle was available on the presence or absence of primordial binaries in globular clusters. This situation began to change soon after the first discovery of a red giant star showing a variable radial velocity with a strong indication for a binary orbit (Latham *et al.* 1985). A few years later, the first non-zero estimate of a primordial binary population in globular clusters was given by Pryor *et al.* (1989): they estimated that $\sim 10\%$ of all the stars in the then surveyed clusters were the primary of a binary.

Meanwhile, another exciting development was the discovery of a large number of millisecond pulsars, starting with the discovery of 3 ms pulsar in M 28 (Lyne *et al.* 1987; for a recent update, see his contribution to the present proceedings). Again, these pulsars do not provide a direct handle on the abundance of primordial binaries, since they clearly have been recycled. Indirectly, however, they do suggest that primordial binaries do play an important role in their dynamics (*cf.* the contribution by Phinney in the present proceedings, and Hut *et al.* 1991a).

Finally, the last few years have witnessed an explosion of new observational results. Besides X-ray binaries, cataclysmic variables, millisecond pulsars and radial velocity variables, we now have firm evidence for many newly discovered eclipsing variables, as well as for a significant broadening of the main sequence in the color-magnitude diagram of several globular clusters, indicating the presence of significant numbers of primordial binaries. An extensive review of all these new observational developments, together with references to the original literature, is given by Hut *et al.* (1992).

4.2. N-Body Simulations

Until recently, most N-body calculations started off from a cluster model which contained only single stars. At the late stages of core collapse, one or more binaries were formed dynamically in a simultaneous close encounter of three stars. The energy released by these stars reversed the collapse to a slow expansion of the whole cluster. For information about single-star runs, see the review by Aarseth (1985), and earlier references therein.

The earliest N-body simulations which began with a significant number of primordial binaries, are those by Aarseth (1980) and Giannone & Molteni (1985). Aarseth started his 250-body calculations with eight primordial binaries. Giannone and Molteni used a similar number of particles, while increasing the number of primordial binaries to sixty. A few years later, Leonard and Duncan (1988, 1990) published a number of runs aimed at modeling young galactic clusters rather than old globular clusters. Therefore, they contained a large fraction of stars in binaries (2/3 by mass). Because of the large computational cost implied, the total number of stars they used was relatively modest ($N = 45$), but appropriate for their main interest, which was a study of the properties of the escapers.

The most extensive N-body calculations containing primordial binaries have been published recently by McMillan *et al.* (1990, 1991). They modeled the evolution of equal-mass star clusters containing a mass fraction of approximately 20% binaries, containing more than $N = 1,000$ stars. Some of the runs extended to very late times, past the point were all the primordial binary 'fuel' was burned up, at which time the cluster had to manufacture new binaries in three-body encounters. For comparison, they also ran simulations with the same binary percentage but a smaller overall number of stars, as well as simulations without primordial binaries.

McMillan *et al.* found the following global response of a cluster to the presence of primordial binaries. The pre-core-collapse evolution was driven by mass segregation between the equal-mass single stars and the binaries, which were twice as heavy. After core collapse, the cluster showed, on average, a smooth reexpansion driven by a steady rate of burning (*i. e.* hardening) of primordial binaries. With so much primordial fuel present, the post-collapse cluster core was significantly larger than was the case in comparison runs without primordial binaries. On a time scale more than an order of magnitude larger than the original collapse time, most binaries were destroyed as a consequence of binary-binary collisions. The surviving binaries showed a gradual hardening. Typically, core collapse occurred, in a 1000 particle system with 100 mildly hard binaries, after $50t_{cr}$ (half-mass crossing times). By this time about one third of the binaries had been destroyed, but without giving off much energy. By $t = 150t_{cr}$, another one third of the binaries had been destroyed, this time accompanied by a significant production of energy. Of the remaining one third of the original binaries, half again disappeared, through destruction and escape, by $t = 500t_{cr}$ (the half-mass relaxation time is $t_{hr} \approx 10t_{cr}$).

An interesting aspect of the simulations of McMillan *et al.* was the smooth statistical nature of the process of energy production by the binaries. Previous runs had always shown strong perturbations, of order unity, of the core due to individual strong scattering events involving a single hard binary in the core. These perturbations will become smaller with larger N values, but only very slowly, since N_c, the number of core particles, shows only a weak dependence on the total number of particles: $N_c \propto N^{1/3}$ (Goodman, 1984). The reason is that the three-body mechanism for producing new binaries is highly density-sensitive. Even though the required relative energy production rate per half-mass crossing time drops linearly with N, the core cannot grow much without falling behind its obligatory energy generation rate.

This situation is very different with the presence of primordial binaries, where binary–single-star encounters and binary-binary encounters can generate heat, without any need for encounters of three independent bodies. As a result, even a modest amount of primordial binaries (a few percent will do) will give rise to a core population growing linearly with cluster mass: $N_c \propto N$ (Goodman & Hut 1989). As a result of this larger core size, fluctuations due to strong interactions involving individual binaries do not cause such large perturbations for the core as a whole. This is the reason that the simulations by McMillan *et al.* are the first ones to enter the regime in which the core shows a smooth evolution, with fluctuations significantly less than order unity. Consequently, many theoretical notions for large-N clusters can now be tested directly, rather than through Fokker-Planck simulations.

An example of such detailed analysis is the ratio of the core radius to the half-mass radius, about 0.1 immediately after core collapse in the simulations by McMillan *et al.* , dropping to about 0.05 by the time most of the binaries were destroyed. The core mass similarly dropped by a factor of two during this time. The mass fraction of binaries in the core showed a steady, near-linear decrease from 0.5 to 0.1 between $\tau \sim 20$ and $\tau \sim 250$, where the time increment $d\tau$ is measured in units of the instantaneous half-mass crossing time.

In addition, McMillan *et al.* obtained a wealth of microscopic information concerning individual interactions between single stars, binaries, triples and occasional more complex multiple star systems. They presented a study of binary interactions in the context of a real cluster environment, and compared their results with earlier "laboratory" simulations, in which similar interactions were studied in isolation (see §4.4). In addition, they studied the formation and evolution of hierarchical triple systems, and the generation of energy within short-lived overdense "clumps" of stars, rather than via binary interactions in an otherwise smooth background.

The main shortcoming of N-body calculations so far has been the relatively small N values, some two orders lower than typical globular clusters. Fortunately, computer power is growing rapidly, and as a consequence, the nineties may provide us with the first opportunity to model a globular cluster on a star-by-star basis. The hardware requirements of computing speeds in the Teraflop domain (Hut *et al.* 1988) may begin to become accessible in the coming years. An interesting approach to reaching such high speed has been pioneered in the GRAPE project (from "GRAvity PipE") at Tokyo University (Sugimoto *et al.* 1990), through the development of special-purpose hardware in the form of parallel Newtonian-force-accelerators, in analogy to the idea of using floating-point accelerators to speed up workstations. For more information about the GRAPE hardware design, see Ito *et al.* (1990, 1991). For software aspects, and choice of algorithms, see Makino *et al.* (1990) and Makino (1991).

4.3. Fokker-Planck Simulations

As long as N-body simulations are limited to at most a few thousand particles, Fokker-Planck simulations provide the most accurate way to model realistic N values for globular clusters, in the range $N = 10^5 \sim 10^6$. Earlier Fokker-Planck simulations containing primordial binaries were reported by Spitzer & Mathieu (1980). They did not find a reversal of core collapse, probably because their calculations did not extend far enough (*cf.* Spitzer 1987). Hills (1975) described an analytical model for the evolution of a cluster core containing initial binaries, but his choice of scaling for core quantities, $R_{core} \propto N_{core}^2$, turned out to be unphysical (R_{core} scales roughly proportional to N_{core} for an almost isothermal collapse). Summaries of these investigations, as well as of the early N-body explorations using primordial binaries mentioned above, can be found in the introductory section of McMillan *et al.* (1990) and Gao *et al.* (1991).

The first, and so far only, Fokker-Planck simulations with primordial binaries which extended past core collapse are the ones reported by Gao *et al.* (1991). Previous Fokker-Planck simulations (see Murphy *et al.* 1990 and references therein) had modeled energy generation only in a rather indirect way. The trick had been to estimate the net rate of energy release from binary–single-star scattering through an estimate of the bottleneck factor: the formation of new binaries in simultaneous encounters of three single stars. The latter rate had been estimated by Goodman and Hut (Hut 1985).

The problem with the presence of primordial binaries is that this trick does not work any more, since no such bottleneck is present anymore. Instead, a more accurate treatment would suggest the introduction of a two-dimensional distribution function for the binaries, with the binding energy and the energy of the center-of-mass motion as the two independent variables. However, to avoid the expense and complexity of a two-dimensional computational grid, Gao *et al.* took a short-cut by introducing a factorization of the distribution function in terms of two one-dimensional distribution functions, one for the internal and one for the external energy of the binaries. They justified this approach by noting that the two-body relaxation time is typically much shorter than the time between successive strong interactions of binaries.

They also noted the shortcomings of such an approach, in that it could not treat the ejection of binaries on elongated orbits far into the halo. As we will see in §7, such orbits do play an important role, in the simulations themselves, as well as in providing important observational handles on the evolution of globular clusters. But, as they mentioned in their discussion, an accurate description of these 'halo parking orbits' would require a three-dimensional distribution function, including not only internal and external energy but also the external angular momentum of the center-of-mass orbit. Clearly, such a treatment is not practical within the Fokker-Planck approach – not in the least because a typical cell in a three-dimensional computational grid would contain much less than one binary at any given time, making the statistical interpretation completely unrealistic. In this sense, it is clear that the simulations by Gao *et al.* , ingenious and interesting as they are, have pushed the range of applicability of Fokker-Planck treatments to the edge of what is compatible with the underlying assumptions.

One advantage of the Fokker-Planck approach of Gao *et al.* is that the large effective-N values allowed them to study the occurrence of gravothermal oscillations, which could not be studied in the relatively small-N simulations by direct integration of McMillan *et al.* They found that the primordial binary population was able to suppress these oscillations during a period equal to several initial core collapse times.

After this time, with most primordial binaries having been burned up, the core would shrink enough to allow the gravothermal oscillation instability to occur.

5. Binary-Binary Scattering

In this section I summarize the results of some of the binary-binary scattering experiments which I performed a couple years ago, and which have not been published earlier. Originally, this summary was scheduled to appear in the proceedings of the workshop on Self-Gravitating Systems in Astrophysics and Nonequilibrium Processes in Physics, which was held in Kyoto in 1989 (McMillan *et al.* 1990, 1991 refer to this reference). However, since I understand that no proceedings will be published for this meeting, I include my intended contribution here in the form of the present section.

Gravitational scattering experiments have been performed and analyzed with a new code which is capable of treating any type of self-gravitating target or projectile composed of point masses, with no restrictions on the number of particles or the hierarchical structure of their internal orbits. This computer code forms an important ingredient in a computational laboratory for stellar dynamics. The most complex part of the code is the module which recognizes the emerging of stable final states signaling the end of a scattering experiment. The analysis needed for the recognition is based on a mixture of theoretical, heuristic and empirical reasoning.

A first application of the code is discussed, in the form of a study of binary-binary encounters. Such encounters will be important in the dynamics of globular clusters if primordial binaries contain a few percent of the stars of a typical cluster, as has been suggested by recent observations. In this case, mass segregation will cause the heavier binaries to concentrate in the core and many binary encounters will involve other binaries rather than single stars. Some binary-binary scattering cross sections are presented which have been recently obtained. These can be used to estimate the relative importance of different types of scattering processes.

Another application of the new code will be a study of the formation and destruction of hierarchical triples which cannot be formed in three-body scattering events, but only in those involving four or more bodies. Additional subjects for study are binary binaries (two binaries in a stable orbit around each other), and even more complex systems, which are likely to be produced in a cluster core which is saturated with primordial binaries.

Finally, the software which handles the recognition and classification of hierarchical structure will have several applications in the context of larger N-body codes. It will provide a general way to handle the regularization and other special treatments which are needed to retain accuracy in the integration of the particle orbits in dense subsystems. It also will enable detailed analysis and diagnostics of the precise nature of the multiple encounters which fuel the post-collapse expansion in models of globular clusters.

5.1. Introduction

Scattering experiments play an important role in many areas of microscopic physics. In most systems the interactions of individual constituents display a bewildering variety of behavior when investigated in detail. In general, such information is not relevant for a global, thermodynamic description of the system. What we are mainly interested in can often be expressed by a few numbers, such as the heat capacity and the heat conductivity. A convenient way to arrive at such final bulk numbers is a three-step approach: (1) one starts by measuring a few relevant numbers at the end of each individual scattering experiment, where these numbers are the values of some quantities which are judiciously chosen so as to shed maximum light on the property of the system one is after; (2) the next step involves a statistical averaging (ensuring a proper weighting procedure) of all individual results in similar runs in order to express all these results in the form of a few cross sections and reaction rates; (3) the final step is to start with the statistical nature of the individual constituents, as characterized by the cross sections and reaction rates, and to average their behavior over the system as a whole, by using the known statistical distributions of properties of the constituents.

An astrophysical example of an application of microscopic scattering experiments is given by the low energy nuclear physics experiments which are performed to determine the cross sections and reaction rates which are used as input in stellar evolution calculations. In this case, steps 1) and 2) of the classification given above are performed in the laboratory while step 3) is realized on a computer in a stellar evolution code.

In other areas of astrophysics, scattering experiments themselves can be of macroscopic scales. Such experiments are purely computational by necessity, whenever the individual experiments are impractical. The subject or gravitational scattering falls in this category, since encounters between single stars and double stars, or for that matter between whole galaxies, are not easy to orchestrate in the real world. In the remainder of this paper, we will first give an overview of the software implementation of steps 1)-3), followed by a discussion of a specific application, namely globular clusters with primordial binaries.

5.2. Gravitational Scattering

Gravitational scattering studies the interaction and final states of encounters between two or more objects, each of which can be a single star, a double star or a hierarchical multiple such as a triple or a binary binary. Although the study of gravitational bound states lies at the very beginning of mathematical physics, gravitational scattering is a very young field. Only for the simplest case, binary - single star encounters in the limit of all three stars having equal mass, a complete theoretical and experimental treatment is available. For some other cases, such as more general binary - single star scattering and binary - binary scattering, some quantitative results have appeared, but even here detailed knowledge of cross sections and reaction rates is still lacking. For all other cases, not only is quantitative knowledge nonexistent, even a qualitative description, notation and classification is lacking. Fig. 1 summarizes the basic processes which can occur in gravitational few-body scattering.

A specific example of an application of gravitational scattering in astrophysics is based on the gravitational three-body problem. After measuring the outcome of orbit calculations for individual encounters between single stars and binary stars, it is possible to produce statistical expressions, in a given environment, for the average rate of increase of binary binding energy and the resultant rate of heating of the stellar environment. In the three-step classification given above, step 1) is relatively simple when we limit ourselves to gravitationally interacting point masses. Step 3) consists of writing a computer code which can follow the evolution of a star cluster in a statistical way, for example in the Fokker-Planck approximation. After core collapse, a star cluster can begin to re-expand when binaries are formed dynamically and begin to give off excess heat to the environment when their binding energy increases on average. A major complication, however, arises at step 2): how do we insure a proper sampling of initial conditions, as well as a proper statistical interpretation of the final results of the individual experiments?

Some of the questions involved in step 2) have been addressed by Hut and Bahcall (1983), in an extensive project of numerical orbit integrations to determine cross sections of scattering processes between single stars and binaries. In this paper an outline and background of the three-body scattering project can be found, together with technical details concerning the numerical scattering experiments of single star incident on binary stars. Also in this paper are presented the main results of the first 1.7 million scattering experiments, for equal masses and both soft and intermediate binaries. More detailed results can be found in Hut (1984), which includes an atlas of differential cross sections for a wide variety of different processes. For a modern mathematical discussion of the general three-body problem, see Marchal (1990). See also the reviews by Valtonen (1988) and Anosova (1990).

5.3. $N-$Body Scattering for $N > 3$

In general N-body scattering experiments, progress has been much more modest, compared to binary-single-star scattering. Some binary-binary scattering experiments have been reported in the literature, notably by Mikkola (1983ab, 1984ab), Hoffer (1983, 1986) and recently Leonard and Duncan (1990). For a more complete list of references, cf. the review articles by Hut (1985); Elson et al. (1987); Hut et al (1988); Valtonen and Mikkola (1990). However, the results of the published scattering experiments were not presented in such a way as to allow computation of scattering cross sections and reaction rates, as is customarily done in other branches of physics. Fig. 2 presents the first result along these lines, and will be discussed below.

One reason that N-body scattering experiments for $N > 4$ have not been reported earlier is the enormous complexity of the internal dynamics of the scattering process itself, as well as the large number of possible final states. Writing software intelligent enough to know when to terminate such a scattering process is a nontrivial exercise in qualitative reasoning. When an interacting group of particles breaks up in separate clumps, under which conditions is each clump stable against further disintegration? And under which conditions can we guarantee that none of the clumps will undergo a further encounter with any of the other clumps? Several minimal criteria will come to mind, such as the requirement that the system of clumps is unbound, and that the same holds for each pair of clumps. In addition, it seems safest to require that all clumps move radially outwards, and that each clump moves away from each other clump. However, further thought (and experimentation!) shows that some of these requirements are too rigid, and also that there still are some loopholes left open, allowing further interactions between clumps to take place.

To give an example of a too rigid restriction, consider clumps moving away in opposite directions at great speed, and a third clump moving away slowly from the previous interaction area, but then reversing its direction under the influence of the slight attraction of the receding clumps. In this case, the inter-clump interaction is clearly over, but it can take a very long time before the third clump has crossed the interaction region, and is finally moving out again.

An example of a loophole is the following. Consider two clumps moving away in nearly parallel orbits, slightly diverging so as to give them a relative motion which is receding. Assume them to be unbound, which guarantees that, by themselves, they will forever continue receding. However, there is still the possibility of a third

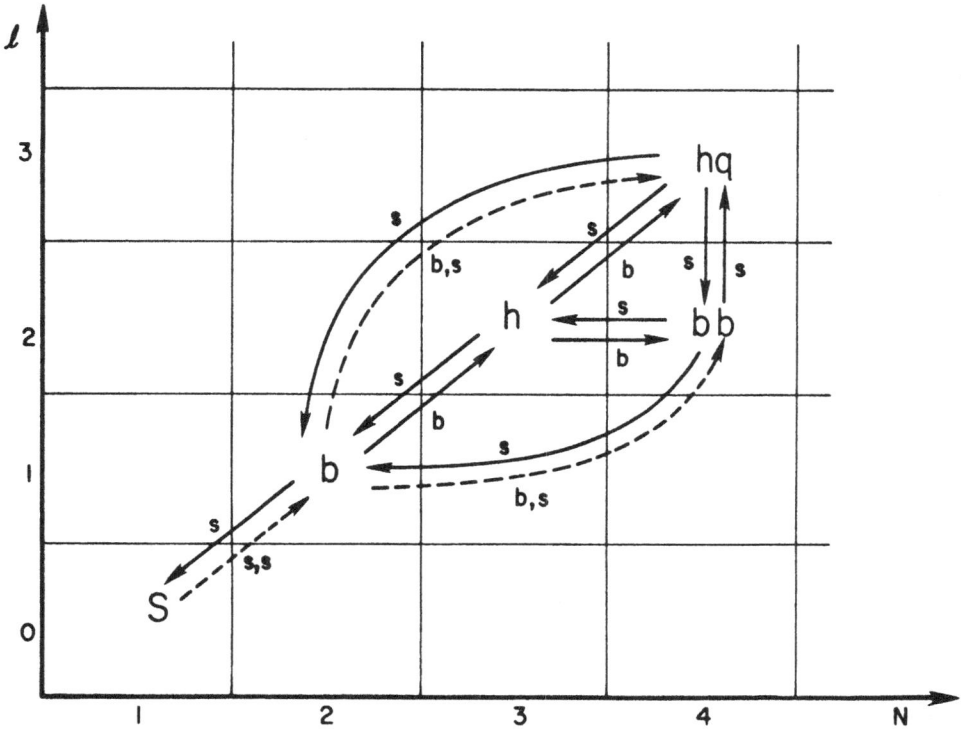

Fig.1 A chart of gravitational isotopes: stably bound systems of N point masses, in a hierarchy of l levels deep. s denotes a single star; b a binary, h a hierarchical triple, *i.e.* a single star in orbit around a binary; bb a binary binary, *i.e.* a second-order binary each of its members being a binary itself; hq a hierarchical quadruple, *i.e.* a second order hierarchical system, composed of a single star orbiting a simple hierarchical system of type h. Encounters between two objects can result in transitions indicated by the full lines; encounters which require the simultaneous interaction of three objects are indicated by dashed lines. In all transitions starting at a particular object, the additional objects involved in the encounter are listed alongside the arrow which describes the transition. Note the existence of a bottle neck in the transitions between single stars and binaries. This process is a "forbidden transition" from the point of view of the dominant two-object encounters, and requires the near-simultaneous presence of three unrelated objects (single stars in this case).

clump (or combination of several other clumps) exerting just enough force to focus the two escaping clump orbits onto each other, even though these residual forces are guaranteed not to hinder escape of the two clumps from the system as a whole. In this rare case, subsequent interaction between the clumps cannot be excluded.

These examples are mentioned here just to give an impression of the complexity of orchestrating individual scattering experiments in an automatized way. In addition, there is the problem of orchestrating whole series of scattering runs. These problems are very similar to the question of how to organize laboratory experiments in general.

5.4. A Computational Laboratory

Gravitational scattering experiments can be performed on a computer in a variety of ways. The simplest approach is to write a computer code which is restricted to solving the equations of motion and propagating the particles, while giving periodic output of the intermediate results. A human controller can then study the successive output results and determine at which point the calculation has run its course and should be terminated. Let us first look at scattering experiments involving whole galaxy models, rather than only a few particles such as in three-body scattering experiments.

In the galactic case, the criterion for halting the calculations can be relatively simple. One possibility would be simply to take the separation of two galaxies beyond a certain distance, in the case of a high-speed encounter between two galaxies where it is *a priori* known that the two galaxies will not stick together. A more complicated criterion is necessary in the case that the encounter speed is low enough to allow the possibility (but not the certainty!) of a merging of the two galaxies. Not only is a decision needed as to which of the two outcomes has been established (merging or escape), also there is a lot of freedom in determining a halting criterion in the case of a merger. It is sensible to continue the run as long as the merger remnant still shows clear signs of evolution, of not yet having settled down in its new form. However, the translation of this qualitative remark into hard objective criteria is not so simple, and certainly not unique.

This brings us to a second type of approach to performing gravitational scattering experiments on a computer. We can extend the task of the computer, by writing a second code which controls the action of the code used for getting the computational crunch work done. Again, depending on the subtleties involved in deciding when to end a calculation, this second-level code can be short and simple, or it can be longer and more complicated than the first-level code, for example if something like pattern recognition is involved in classifying the status of the system at a given time (*cf.* the discussion in the previous section).

The task of the human researcher is obviously greatly simplified with the halting decisions for each experiment being taken care of by the computer. What remains to be done by hand, though, is to choose a set of initial conditions for each experiment which will be carried out. This third-level type of activity can be automated as well, as will be discussed briefly below.

5.5. A Laboratory Assistant

The central problem in automatizing the setup of experiments is formed by the much more extended amount of knowledge which needs to be built into the computer program. On the first level, the program should have a working understanding only of the equations of motion. On the second level, not only the mechanics but also the aim of each experiment should be coded into the program. For example, the question of what exactly we want to measure to what (quantitative or qualitative) accuracy at the end of a calculation can cause the structure of the second-level program to be vastly different for different types of experiments, even when the first level program remains completely unchanged.

At the third level, we try to automatize not only the halting but also the starting-up and the choice of start-up parameters of each experiment. This requires a much wider type of knowledge than that used on the second level. In laboratory terms, a machine to perform the experiments is often modified in such a way that it can also signal the end of an experiment, and in addition that it can do some of the early data reduction and interpretation as well. However, the choice of a complicated suite of experiments is often left to a human laboratory assistant, rather than a computer (although this situation is slowly changing).

One important distinction to make at this level is that between pilot studies and production runs. When we start a new series of experiments, it may well be that we do not know the precise part of parameter space which we will want to search systematically. Instead, we may want to do a few "shots in the dark", to get a preliminary feeling for the relationship between the choice of initial conditions and the type of outcome of the experiments. Once we have a clear enough picture, we can start to carry out production runs, in which we systematically explore the parameter range which seemed appropriate as a result of the pilot studies. In general it is a good idea to use these productions runs also to keep checking the appropriateness of the calculations, as was done first in the pilot studies.

Of these two types of third-level activities, perhaps the pilot studies are the most complex from a computer science point of view (as well as from a coding point of view). The reason is that at this point most is demanded in terms of imagination and creativity, because the outcome of experiments is completely new and different parameter regimes are often unknown until the pilot experiments are actually performed. What is needed goes beyond the more familiar rule-based expert systems, in that some of the rules of the game have to be determined empirically, while performing pilot experiments.

5.6. Applications: Primordial Binaries

The software tools described above have been under development over the last several years, and have recently reached a stage in which they can be applied to real problems. Specifically, the computational scattering laboratory for general N-body scattering saw its "first light" recently, with the successful completion of initial runs of binary-binary and binary-triple scattering. Here we will give some preliminary results of the binary-binary scattering experiments.

In the dense cores of globular clusters gravitational encounters occur in a bewildering variety when mass segregation has produced a stellar population which is locally dominated by binaries. Fig. 2 presents some preliminary results which indicate the relative importance of different type of scattering processes. In Fig. 2b, I confirm in quantitative detail the result reported first by Mikkola (1983ab), that binary-binary encounters produce significant numbers of hierarchical triples. In a binary-saturated cluster core, these triples themselves undergo subsequent encounters, leading to an enormous variety of formation scenario's for observable objects such as millisecond pulsars and low-mass X-ray binaries, some of which may well have a third companion star.

Fig. 2 Preliminary results for scattering cross sections in binary-binary scattering. All masses are equal and the two binaries have identical semimajor axes a. Their eccentricities differ, and are drawn independently from a thermal distribution, flat in e^2. The cross sections are normalized by multiplying them with the square of the relative velocity between the two binaries, as expressed in the system of units used by Hut and Bahcall (1983; $v_{cr} = 1$ in equal-mass binary-binary scattering).

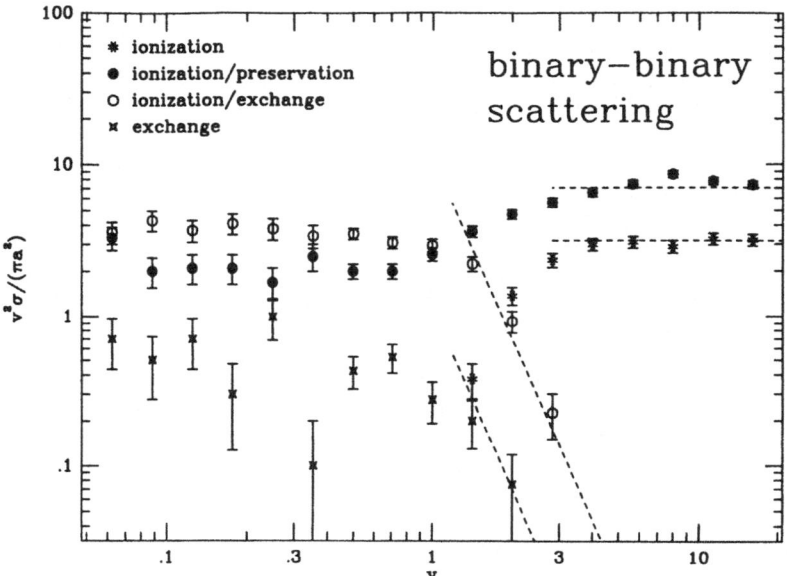

Fig. 2a. Four types of scattering cross sections are presented, as a function of relative velocity of binaries: 1) ionization, in which both binaries are broken up into single stars; 2) ionization/preservation, in which one of the binaries is broken up while the other retains its original composition; 3) ionization/exchange, in which two single stars emerge and one binary, formed by members drawn from both original binaries; 4) exchange, in which two new binaries emerge, each containing members from both original binaries. The dashed lines indicate theoretical scaling laws. These will be refined and presented in a detailed paper, to be submitted to *Astrophys. J.*

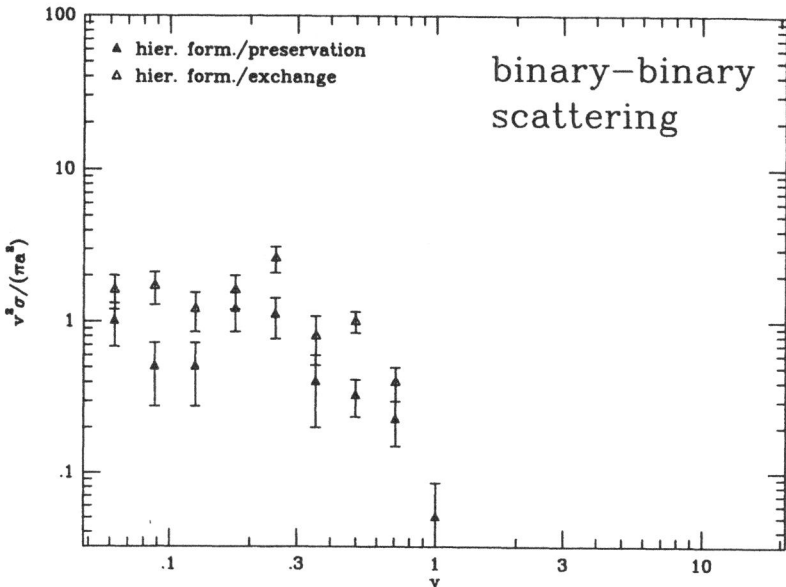

Fig. 2b. The remaining two types of scattering cross sections indicate the occurrence of formation of hierarchical triples. Preservation denotes the survival of one of the two binaries in the form of the inner binary of the triple. The alternative in which the inner binary has received members from each original binary is called exchange. Note the frequent occurrence of hierarchical triple formation, compared to the other processes depicted in fig. 2a, in the left-hand part of the figure, which describes collisions between hard binaries.

5.7. Future developments

So far our discussion has been based on interactions between point particles. In many astrophysical applications, however, we cannot neglect the physical size of the stars, because the probability of collisions is appreciable. In addition, an immediate consequence of introducing more realistic star models is the possibility of energy dissipation, even without physical collisions. For example, in a typical globular cluster theoretical models predict that hundreds, or perhaps even thousands, of binaries are formed by tidal capture even before core collapse, and a comparable number of stars will actually collide and merge. Such processes are not only interesting in themselves, but they will also have an important influence on the overall dynamics of the whole star cluster. Therefore, if we are to make a realistic globular cluster evolution model, it will be essential to take such processes into account.

A detailed realistic description of a collision between two stars would involve a three-dimensional hydrodynamic code which includes an accurate treatment of shocks, radiation transport, nuclear reactions, convection, etc. Even a collision between two stars will in general result in a merger remnant with a mass less than the combined mass of the progenitors, even if we wait sufficiently long for the gas clouds in bound orbits to fall back on the merged stars. The remaining matter will leave the interaction area, and may leave the star cluster under consideration (in case of a globular cluster), or may be reprocessed in some form (as could occur in a galactic nucleus). In a globular cluster simulation, we could decide to follow only what happens in a limited volume of space time centered on a scattering event, while simply neglecting the gas clouds and radiation emitted in the process and concentrating on the remaining merger remnants.

Until recently, available computer power was too limited to attempt three-dimensional hydrodynamical calculations of stellar interactions. However, the recent development of particle-based schemes such as Smooth Particle Hydrodynamics (*cf.* Monaghan 1985, Hernquist and Katz 1988) together with the increase in speed (as well as availability!) of supercomputers, may make a systematic study of fully 3-D stellar encounters possible.

6. Microscopic Physics

Most globular cluster model calculations so far have used the point-mass approximation to represent individual stars. At first sight, this is reasonable for a typical globular cluster star, since the chance for a physical collision or close encounter with another star is small. However, in the central regions finite-star-size effects become important.

6.1. Stellar Finite-Size Effects

Let us make a simple estimate of the importance of non-point-mass effects. For comparison, take a star at the half-mass radius of a globular cluster, with a velocity dispersion of 10 km/s. With interstellar separations of ~ 0.2 pc, it takes a star 20,000 years to cross an interstellar distance. But the radius of the target area is small, about 0.3 A.U. (the geometric mean between the sum of the diameters of the stars and the ninety degree turnaround distance, because of gravitational focusing). With an interstellar distance of 40,000 A.U., the chance to hit the next star is less than 10^{-10}. The chance to have an encounter after 10^{10} yr is thus less than 0.005 %.

These arguments indicate that the early stages of globular cluster evolution, up to core collapse, can be well approximated by point-mass dynamics. The main mechanisms, evaporation and the onset of the gravothermal instability, are not affected much by the occasional collision of a couple stars. This picture changes dramatically, however, as the gravothermal collapse progresses. When the core has shrunk down to a size of ~ 0.1 pc, the local density has gone up to a whopping $\sim 10^7 \mathrm{pc}^{-3}$. This changes the mean time between physical encounters from the half-mass radius value of more than 10^{14} yr to a threatening 10^9 yr, significantly less than the age of a globular cluster.

There are other arguments pointing to the importantce of physical encounters. Even though the average star does not have much of a chance to have a physical collision before core collapse, the number of stars which have undergone an encounter close enough to lead to a tidal capture steadily increasing to a significant fraction of a percent. We do not know whether such near-misses, with a pericenter distance of less than about three stellar radii, can lead to the formation of close binaries, or whether the stars will spiral in to form a merger remnant (see below). But if they avoid merging, the encounters could result in the presence of ~ 1000 close binaries in or near the core, just before core collapse (Statler *et al.* 1987). The presence of a number of binaries this large will certainly affect the evolution of the cluster, and models based on the point-mass approximation will not any longer give a reasonable approximation to the dynamics of the central area.

Finally, the outcome of three-body scattering events are not well represented by point-mass dynamics, even apart from the question of the feasibility of tidal capture. If we start with a moderately hard binary, say with an energy of $10kT$, in which both stars are on the main sequence, we have a semimajor axis for the orbit of order 1 A.U. – about a hundred times larger than the sum of the radii of the two stars. When a third star gets involved in a strong scattering encounter with such a binary, there is a large chance that this star will be temporarily captured. In such a resonance scattering event, many close encounters may take place, and in half of all cases the closest encounter will be at less than 3% of the original semimajor axis (Hut & Inagaki 1985). Therefore, our $10kT$ star already has a nonnegligible chance for a physical encounter, about 15% as it turns out. A somewhat harder binary of $30kT$ has a probability of 50% to suffer a collision during a resonance scattering, and the probability tends to unity for binaries which are much harder.

The situation is even worse for binary-binary scattering. Even a small abundance of primordial binaries will result in a much larger concentration in the core, due to mass segregation, and a proportionally enhanced probability for binary-binary encounters. In a typical resonance scattering event between two identical binaries, the typical closest encounter distance is only 1% of the original semimajor axis (Leonard 1991). Thus even an encounter between two $10kT$ binaries has already a fifty-fifty chance to result in a physical collision, with the chance become larger for a greater degree of hardness.

From all this it is clear that we have to take finite-size effects seriously before we can trust any detailed simulation enough to compare it reliably to observations of globular clusters. The next subsections give an overview of recent work in modeling finite-star-size effects.

6.2. Tidal Capture: Semi-Analytic Results

After the discovery of X-ray sources in globular clusters, the mechanism of tidal capture was suggested by Fabian *et al.* (1975) to explain the origin of these sources as the capture of a neutron star by a main-sequence star through the dissipation of the excess orbital energy in the tidal bulges raised on the normal star by the neutron star. They estimated the amount of energy dissipation by assuming that the energy needed to raise the tidal bulges would be dissipated efficiently (*cf.* Verbunt & Hut 1987).

More detailed calculations, for a variety of stellar models, were performed by Press & Teukolsky (1977), Lee & Ostriker (1986), Giersz (1986), McMillan *et al.* (1987), and Ray *et al.* (1987). These calculations mostly concentrated on the processes taking place during the first encounter between the two stars. They left open the question of what will happen during subsequent encounters.

In order to sketch the problem, let us start with an incoming star which brings in an excess amount of kinetic energy equivalent to that contained in its motion at infinity at, say, 10 km/sec. The amount of orbital energy transfer is highly dependent on the distance of closest approach. Therefore, most encounters will either not transfer enough orbital energy, or transfer far more than the relative kinetic energy at infinity. For definiteness, however, let us assume that the orbital energy lost is only twice the energy at infinity. As a result, the incoming star will be captured on an orbit with a binding energy corresponding to a circular orbital speed of order 10 km/sec, which implies a semimajor axis of order 10 A.U. Of course, the orbit will not be circular but highly elongated, and after some tens of years the two stars will meet again for a second close encounter.

If we can assume that the energy which was transferred in the first encounter to the tidal bulges has been dissipated, and if the star or stars in which tidal bulges were raised are not swollen significantly by the process of dissipation, then it is reasonable to assume that a similar amount of orbital energy is lost again during the second pass. If we can continue these assumptions for subsequent passages as well, we have the following scenario. If the Sun would be one of the two stars, the first capture orbit will stretch out to somewhere near Saturn's orbit, the second to Jupiter's orbit, the fourth to Mars' orbit, the sixth to that of the Earth, the eighth to Venus' and the tenth to Mercury's orbit. However, the last nine orbits will take less time to traverse than the first one, and within a hundred years the orbit is already enclosed within that of Mercury (meanwhile, the poor planets may have made off to infinity; good thing the Sun does not live in the core of a globular cluster).

There is a problem, though, with the two 'if's' involved in the previous paragraph, as was pointed out by Kochanek (1991). Perhaps the energy in the tidal bulges can be dissipated during the first passage, if the first captured orbit is wide enough. But it is not clear how the star(s) can get rid of the excess heat on the subsequent orbital timescales, of order a few years or less. To shed more light on the nature of these problems, Kochanek developed a very interesting technique based on the affine star model introduced by Carter & Luminet (1985). Although his model has some limitations (radial modes and $l = 2$ f-modes only), it has two great advantages, stemming from the fact that the model is not based on a linear mode analysis approximation. First, it can be applied to any amount of distortion, and is therefore accurate for arbitrarily close encounters (until the star flies apart, literally). Second, the dissipation is based on an actual dynamical model of a star, which can be followed from one close passage to the next.

Kochanek's study reached several interesting conclusions. Most importantly, he showed explicitly that our intuition of interacting binary evolution cannot be applied to that of tidal capture, simply because the latter takes place on an orbital time scale while the former generally happens on a thermal time scale (apart from episodes of unstable mass transfer or catastrophic mass loss). As a result, the evolution of the capture orbit will be highly stochastic, depending on the exact phases of the main modes of oscillation of the star(s) during each subsequent pericenter passage. While the orbital energy and angular momentum are most likely to decrease on average, individual encounters can lead to either a gain or a loss. The simplified picture sketched above thus turns out to be completely inaccurate.

Kochanek concludes that the large majority of close encounters probably cannot lead to tidal capture, due to a variety of problems, such as the possibilities of a stellar disruption or extensive mass loss, of a physical collision, or escape due to a third passing star. Surprisingly, those binaries that do survive may well have an orbit significantly wider than had been estimated by earlier authors. He estimates that a few percent of the encounters lead to orbits with a pericenter distance of more than 50 stellar radii, due to perturbations of a third star, which can transfer extra angular momentum to the binary. If the large majority of all other cases lead to some catastrophic development or other, these unusually wide orbits may make up a large fractions of those orbits which survive.

6.3. Hydrodynamical Effects: Numerical Results

Detailed numerical simulations of the hydrodynamical effects occurring in close encounters between stars have only recently become feasible. The main problem is that the combination of the stellar rotation and the orbital revolution makes a fully three-dimensional treatment absolutely necessary. Fortunately, computer speed and availability have dramatically increased over the last few years, to the point that the initial encounter of two or more stars can now be modeled with at least some confidence that the major features of the calculations correspond to reality. The question raised by Kochanek (1991), of what happens after successive encounters, is still completely beyond present computational reach, unfortunately. Waiting for orders of magnitude of increase in raw computer speed would be frustrating. Perhaps a clever and careful development of new algorithms will bring such calculations within the realm of possibilities. A combination of fully 3-D hydro calculations during close encounters, together with semi-analytical models which can follow the approximate damping of the oscillations during the intervening times, may be one approach.

After the first explorations by Benz & Hills (1987) a number of 3-D hydro treatments of close stellar encounters have appeared during the last couple years, by Benz et $al.$ (1989), Cleary & Monaghan (1990), Goodman & Hernquist (1991), Davies et $al.$ (1991) and Rasio & Shapiro (1991). A discussion of each of these papers is beyond the scope of the present review. A general pattern emerging from these calculations has been the very efficient merging of stars involved in scattering processes which lead to close encounters in which the outer layers of the stars began to overlap. In the treatment of binary-binary encounters by Goodman & Hernquist for example, multiple mergers were common in which three or even all four stars fused. Obviously, outcomes like this will have a very significant effect upon the outcome of detailed N-body simulations of globular clusters.

How will we be able to incorporate these hydrodynamical effects into the star-by-star simulations which will become available, several years from now? There is one piece of good news, together with the bad news that point-mass simulations of star clusters already will require Teraflop speeds (Hut et $al.$ 1988). The good news is that the addition of a modest amount of hydrodynamic modeling of close encounters will not significantly affect the total computational cost! This follows from the analysis by Makino & Hut (1990), who showed that the relative cost of computations involving binary encounters decreases $\propto N^{-25/12} \log N$, or roughly $\propto N^{-2}$, as compared to the overall computational cost of following the orbits of the single stars. Thus the binary cost drops steeply for increasing N, even though the ratio of the half-mass crossing time to the typical binary orbital time increases $\propto N$.

There are two reasons for this counter-intuitive result: on a local level, very tight binaries can be modeled as isolated point masses for an increasing fraction of the time for increasing N; while on a global level, the rate of formation of binaries per half-mass crossing time is ~ 0.1, independent of N. The detailed analysis by Makino & Hut showed that a typical binary requires a total of $\sim 10^7$ time steps to integrate the relative motion of its components, during its whole life. In contrast, the computational cost of the system as a whole increases strongly, $\propto N^2$, even per half-mass crossing time. They conclude that there are three regimes in N-body simulations of star clusters. For $N < 1000$, binaries form the computational bottleneck, requiring more computer time than the single stars. For $1000 < N < 10,000$, the fraction of computer time needed for binary integration drops steeply, from ~ 1 to $\sim 10^{-4}$. For $N > 10,000$, it becomes feasible to model stellar collisions and close encounters using Smooth Particle Hydrodynamics, using at least a few hundred particles per star, or even $> 10^4$ particles per star for $N > 10^5$. The implementation of a SPH code inside a N-body code will be far from trivial, though (see also §5.7).

A realistic simulation of globular cluster evolution should include stellar evolution effects as well, both for single stars as well as binaries. Including stellar evolution into an N-body code does not increase the computational cost in any appreciable way. The integration of a typical star, including the giant stage, requires a few hours on a VAX 11/780, which corresponds to $\sim 10^9$ floating point operations. Even if all 10^5 stars needed such treatment, the total cost would be $\sim 10^{14}$ floating point operations, which is much smaller than that required for the whole star cluster in the point-mass approximation (Hut *et al.* 1988). Thus from a hardware point of view, inclusion of stellar evolution is trivial. Of course, to develop the software needed, by integrating some form of stellar evolution code into an N-body code, will be a major undertaking, already for single-star evolution, and much more so for binary star evolution, where problems such as spiral-in in a common-envelope phase will occur.

7. Binary Migration

Giants and subgiants in globular clusters all have a mass comparable to that of the turn-off mass, around $0.8M_\odot$, as a consequence of sharing the age of the clusters, on the order of 10^{10}yr. Main-sequence stars, which make up the bulk of the mass of a cluster, are less massive than that. Some of the white dwarfs are probably heavier than $0.8M_\odot$, up to $1.2M_\odot$, but the total mass in these stars is unlikely to exceed a few percent of the total cluster mass, and may be much smaller (*cf.* Murphy *et al.* 1990). Neutron stars, around $1.4M_\odot$, are even rarer. Merger remnants are likely to be present too, but also in relatively small numbers (see §6.1).

Primordial binaries tend to be more massive than single stars, on average, simply because each member star lives under the same constraints as a single star, its mass being bounded above by the turn-off mass. As a result, primordial binaries will drift to the central regions of the cluster. This mass segregation is analogous to the effects in the Earth's atmosphere, where heavier gasses, such as CO_2, have a smaller scale height than the lighter gasses. A factor two difference in mass, on average, is enough to concentrate most binaries within the inner few core radii, in the final equilibrium distribution. This distribution can be described conveniently using a multi-mass generalization of King models. Indeed, this approach has been widely taken in the analysis of cluster observations.

The main question is then: will this thermodynamic equilibrium ever be reached? McMillan *et al.* cautioned against the assumption of equilibrium, by pointing out how the recoil of binary–binary and binary–single-star scattering events tend to place many binaries on wide 'halo parking orbits' for a considerable length of time. They observed a bimodal distribution of recycled binaries (as opposed to the pristine primordial binaries still drifting in from the far halo, even at late times): at any given time, a snapshot of their simulations revealed that most of the binaries which had passed at least once through the core could be found in or near the core, but some would be positioned far out in the halo, with relatively few in the intermediate area.

Similar conclusions were reached by Phinney & Sigurdsson (1991), who applied this reasoning to the distribution of the pulsars in M15. Specifically, they argued that the binary pulsar 2127+11C is an example of a binary currently parked in the halo, since the projected distance from the center is ~ 20 core radii (*cf.* the contribution by Phinney in the present proceedings).

Since the binary distribution through the cluster has obvious observational interest, it is really frustrating that state-of-the-art cluster simulations are not able to offer clear predictions. The problem is that Fokker-Planck models cannot handle recoil (*cf.* §4.3), and that N-body models cannot handle a large enough particle number (*cf.* §4.2). While the final solution will be to wait for N-body modeling to catch up, Hut *et al.* (1991b) decided to take a short-cut, by modeling the evolution of individual binaries in a fixed background potential determined by the single star distribution (a similar approach is being developed by Phinney and Sigurdsson; *cf.* the contribution by Phinney in the present proceedings).

In the model by Hut *et al.* binaries are characterized by a two-dimensional distribution function in binding energy and distance to the cluster center (or, equivalently, internal and external orbital energy). The model forms a middle ground between realistic N-body calculations and analytic estimates based on average amounts of binary hardening per scattering event. They present a series of Monte Carlo simulations for an initial population of 5×10^4 binaries against a fixed background population of 5×10^5 single stars in a tidally truncated cluster model. They followed the individual histories of all binaries as they experience a variety of different physical mechanisms: mass segregation, scattering recoil, escape from the cluster, and, optionally, coalescence through gravitational radiation losses and collisional mergers.

The main observational consequences of their simulations are: 1) most binaries are destroyed by binary–binary interactions. In the point-mass approximation, the rest escape. In a more realistic model, the majority of the rest merge. 2) At any instant, most binaries are drifting in toward the center, before their first strong encounter. 3) A typical binary spends most of its active life (after their first strong scattering event) in or near the cluster core. 4) The few binaries which receive a recoil sufficient to place them in the halo past the half-mass radius remain there long enough to make a significant contribution to the radial binary distribution. 5) This latter effect is strongly suppressed by collisions and spiral-in, both of which tend to lower the average distance of a binary from the cluster center.

It was an exciting exercise, to see how a population of 50,000 binaries evolved from scratch. It certainly whets one's appetite for a self-consistent N-body simulation, in which similar numbers of stars can be followed with a dynamically responding core. But even the relatively crude estimates following from the fixed-background simulations are already providing a lot of incentive to test our qualitative predictions. Fortunately, rapid progress is being made in the observational techniques for finding binaries in globular clusters, as summarized by Hut *et al.* (1992).

8. Galactic Nuclei

Much of what has been discussed so far can be applied not only to globular clusters, but to at least some galactic nuclei as well. For many galaxies, the velocity dispersion in the central regions is an order of magnitude larger. As a consequence, processes involving binaries are affected significantly. For example, physical collisions become more likely for hard binaries, since the member stars have to be closer. Also, gravitational radiation plays an important role in binary destruction at such small separations. Nonetheless, there are many interesting areas of overlap between the evolution of globular clusters and of galactic nuclei. The first meeting specifically dedicated to a comparison between these two areas was held a few years ago at CITA. Pointers to the literature can be found in the proceedings of this meeting (Merritt 1989), especially in the contributions by Murphy *et al.* (1989), Lee (1989), Quinlan & Shapiro (1989), and Rasio *et al.* (1989). More recent work was published by Quinlan

& Shapiro (1990) and Murphy *et al.* (1991).

Not all galactic nuclei have a velocity dispersion which is much larger than that typical for globular clusters. An interesting exception is the spiral galaxy M33, a member of the local group. Recent observations by Kormendy & McClure (1991) give a central line-of-sight velocity dispersion of 22 km s^{-1}, only twice that in a typical globular cluster. Their observed limit on the core radius is $r_c \simeq 0.3$ pc. From these observations, Hernquist *et al.* (1991) argue that the true core radius is likely to be $r_c \lesssim 0.1$ pc.

In short, their arguments run as follows. Using the virial theorem, or more specifically, the model of an isothermal sphere (Binney and Tremaine 1987, eq. (4-123)), the local density $\rho(0.3\,\mathrm{pc}) = 2.0 \times 10^5$ M$_\odot$ pc^{-3}. The corresponding local two-body relaxation time (Binney and Tremaine 1987, eq. (8-71), and Spitzer 1987, eq. (2.62) with $m = M_\odot$ and $\ln \Lambda = 10$) is $t_{\mathrm{relax}}(0.3\,\mathrm{pc}) = 9.4 \times 10^7$ y. Since 0.3 pc is an upper limit to the core radius and since the central density is about three times that at the core radius, the central relaxation time is $t_{\mathrm{relax}}(0) \lesssim 3 \times 10^7$ yr. Such a short relaxation time implies that the system probably has already undergone core collapse.

With the observed velocity dispersion of 22 km s^{-1}, Hernquist *et al.* suggest that a collapsed core may have a true core radius $r_c \simeq 0.1$ pc. They offer three independent arguments for this value. First, simulations without primordial binaries show that cores oscillate after core collapse and that they spend most of their time near maximum expansion. They then typically contain 0.5 % to 1% of the mass of a globular cluster. The corresponding core radius is $r_c \sim 0.1$ pc (Murphy *et al.* 1990). Second, a significant population of primordial binaries will cause core collapse to halt and reverse at relatively large core radii. Again the value is $r_c \sim 0.1$ pc (Goodman & Hut 1989, McMillan *et al.* 1990, 1991, Gao *et al.* 1991). Third, HST observations show that M15 has a core radius of $r_c \sim 0.1$ pc (Lauer *et al.* 1991).

Hernquist *et al.* conclude that the nucleus of M33 has an efficiency of low-mass X-ray binary formation equal to that of all Galactic globular cluster cores combined. This suggests that about a dozen such binaries should be present. Their combined emission may explain the enigmatic, unresolved X-ray source in M33's nucleus, with its high X-ray luminosity of 10^{39} erg s^{-1} (Long *et al.* 1981). Besides X-ray binaries, they also predict a large population of ~ 100 ms pulsars.

9. Discussion and Outlook

This review has focused on the many new directions which have been taken in the modeling of globular clusters over the last few years. There are a number of exciting developments, both in the simulations of the evolution of globular clusters as complete systems, as well as in the treatment of the microphysical processes: close encounters and collisions between individual stars.

The most exciting development lies in the possibility to compare the simulations with observations, both before and after core collapse. This is a sign that the simulations have finally reached a degree of maturity which had been absent in the first couple decades of modeling. A milestone in this respect has been the ability to model the post-collapse evolution of a globular cluster using a realistic mass spectrum (Murphy *et al.* 1990, Lee *et al.* 1991). The first examples of an application of these simulations to the real world have been given by Grabhorn *et al.* (1991), who model the evolution of the globular clusters M15 and NGC 6624, and fit the observed surface-brightness and projected velocity dispersion profiles. They find good agreement with simulations in which the core size is near the maximum value attained during oscillations.

Interesting as these developments are, it is also important to realize the many shortcomings and unwarranted approximations which are still present in the present state-of-the-art simulations, both in Fokker-Planck models which have difficulty handling binary fluctuations as well as in N-body models which cannot yet handle large-N values (see §§4.2-3). Whenever Teraflop speeds will become available, these limitations will disappear. A promising development in this direction is the GRAPE project at Tokyo University (see §4.2), aimed at developing gravitational-force-accelerators in the form of chips which can be added to workstations to reach multi-Gigaflop speeds.

Even when Teraflop speeds will be routinely available, there will be additional bottlenecks in our understanding of globular cluster evolution. The main remaining problem will then be the modeling of the micro-physical processes which occur during scattering events involving close encounters between single stars and binaries. Three-dimensional hydrodynamical modeling of such events is now beginning to be feasible on the dynamical timescale of a single close encounters. Unfortunately, we do not yet know of any good approach toward modeling the longer-term behavior of the oscillations, mass loss and break-up of stars involved in the three-body and four-body dances triggered by these scattering events, which may last many thousands of years (see §6.2). Nonetheless, it would be a very useful start to be able to run a star-by-star modeling of a globular cluster with a simple hydrodynamical treatment of close encounters, using for example a 1000-particle SPH model for each star involved in such an encounter, for the duration of the encounter. Such calculations should become possible before the end of the decade.

Acknowledgements

I thank the Institute for Theoretical Physics at the University of California at Santa Barbara for their hospitality during my stay, which was partly funded by NSF grant PHY 89-04035.

References

Aarseth, S.J. 1980, in 'Star Clusters', IAU Symposium No. 85, ed. J.E. Hesser (Dordrecht; Reidel), p. 325.

Aarseth, S.J. 1985, in Multiple Time Scales, eds. J.U. Brackbill & B.I. Cohen (New York: Academic), p. 377.

Aguilar, L., Hut, P. & Ostriker, J.P. 1988, Astrophys. J., 335, 720.

Ambartsumian, V.A. 1938, Ann. Leningr. State U., No. 22., p. 19 [A translation appeared in Goodman, J. & Hut, P. (eds.) 1985, Dynamics of Star Clusters, IAU Symp. No. 113 (Dordrecht: Reidel), p. 521].

Anosova, J.P. 1990, Celest. Mech. 48, 357.

Antonov, V.A. 1962, Vestnik Leningrad Univ., 7, 135. [A translation appeared in Goodman, J. & Hut, P. (eds.) 1985, Dynamics of Star Clusters, IAU Symp. No. 113 (Dordrecht: Reidel), p. 525].

Benz, W. & Hills, J.G. 1987, Astrophys. J., 323, 614.

Benz, W., Hills, J.G. & Thielemann, 1989, Astrophys. J., 342, 986.

Bettwieser, E. & Sugimoto, D. 1984, Mon. Not. R. astr. Soc., 208, 439.

346

Binney, J.J. and Tremaine, S.D. 1987, *Galactic Dynamics*, Princeton Univ. Press.

Breeden, J.L., Packard, N.H. & Cohn, H.N. 1991, to be submitted to *Astrophys. J.*.

Carter, B. & Luminet, J.P. 1985, *Mon. Not. R. astr. Soc.*, **212**, 23.

Chernoff, D.F., Kochanek, C. & Shapiro, S.L. 1986, *Astrophys. J.*, **309**, 183.

Chernoff, D.F. & Shapiro, S.L. 1987, *Astrophys. J.*, **322**, 113.

Chernoff, D.F. & Djorgovski, S. 1989, *Astrophys. J.*, **339**, 904.

Chernoff, D.F. & Weinberg, M.D. 1990, *Astrophys. J.*, **351**, 121.

Cleary, P.W. & Monaghan, J.J. 1990, *Astrophys. J.*, **349**, 150.

Cohn, H. 1979, *Astrophys. J.*, **234**, 1036.

Cohn, H. 1985, in *'Dynamics of Star Clusters', IAU Symposium No. 113*, eds. J. Goodman & P. Hut (Dordrecht;Reidel), p. 161.

Cohn, H.N., Lugger, P.M., Grabhorn, R.P., Breeden, J.L., Packard, N.H., Murphy, B.W. & Hut, P. 1991, in *The Formation and Evolution of Star Clusters*, A.S.P. Conference Series, 13, ed. K. Janes (ASP, San Francisco), p. 381.

Davies, M.B., Benz, W. & Hills, J.G. 1991, to appear in the Nov. 10 issue of *Astrophys. J.*.

Djorgovski, S. & King, I.R. 1986, *Ap. J. (Lett.)*, **305**, L61.

Eddington, A.S. 1926 *The Internal Constitution of the Stars* (Cambridge Univ. Press).

Elson, R., Hut, P. & Inagaki, S., 1987, *Ann. Rev. Astron. Astrophys.* **25**, 565.

Fabian, A.C., Pringle, J.E. & Rees, M.J. 1975, *Monthly Notices Roy. astron. Soc.*, **172**, 15p.

Gao, B., Goodman, J., Cohn, H.N. & Murphy, B. 1991, *Astron. J.*, **370**, 567.

Giannone, P. & Molteni, D. 1985, *Astron. Astrophys.* **143**, 321.

Giersz, M., 1986, *Acta Astron.* **36**, 181.

Giersz, M., 1990ab, Nicolaus Copernicus Astronomical Center preprints 220, 221.

Goodman, J. 1984, *Astrophys. J.*, **280**, 298.

Goodman, J. 1987, *Astrophys. J.*, **313**, 576.

Goodman, J. 1989, in *Dynamics of Dense Stellar Systems*, ed. D. Merritt, (Cambridge University Press), p. 183.

Goodman, J., Heggie, D. & Hut, P. 1991, submitted to *Astrophys. J.*

Goodman, J. & Hernquist, L. 1991, *Astrophys. J.*, in press.

Goodman, J. & Hut, P. (eds.) 1985, *Dynamics of Star Clusters*, IAU Symp. No. 113 (Dordrecht: Reidel).

Goodman, J. & Hut, P. 1989, *Nature*, **339**, 40.

Grabhorn, R.P., Cohn, H.N., Lugger, P.M. & Murphy, B.W. 1991, preprint, submitted to *Astrophys. J.*

Gunn, J.E. & Griffen, R.F. 1979, *Astron. J.*, **84**, 752.

Hachisu, I. & Sugimoto, D., 1978, *Prog. Theor. Phys.*, **60**, 13.

Heggie, D.C., Goodman, J. & Hut, P., 1988. "On the Exponential Divergence of N-Body Systems", Poster Paper, IAU Commission 37, Baltimore, August 5-6.

Hénon, M. 1961, *Annal. d'Astrophys.*, **24**, 369.

Hénon, M. 1965, *Annal. d'Astrophys.*, **28**, 62.

Hénon, M. 1975, in *'Dynamics of Stellar systems', IAU Symposium No. 69*, ed. A. Haili (Dordrecht;Reidel), p.133.

Hernquist, L. & Katz, N., 1989, *Astrophys. J. Suppl.* **70**, 419.

Hernquist, L., Hut, P. & Kormendy, J. 1991, in preparation.

Hills, J. 1975, *Astron. J.*, **80**, 1075.

Hoffer, J.B., 1983, *Astron. J.* **88**, 1420.

Hoffer, J.B., 1986, *Astrophys. Lett.* **25**, 127.

Hut, P., 1984, *Astrophys. J. Suppl.* **55**, 301.

Hut, P., 1985, in *Dynamic of Star Clusters, I.A.U. Symposium 113*, eds J. Goodman & P. Hut (Dordrecht: Reidel), p. 231.

Hut, P., & Bahcall, J.N., 1983, *Astrophys. J.* **268**, 319.

Hut, P. & Inagaki, S. 1985, *Astrophys. J.*, **298**, 502.

Hut, P., Makino, J. & McMillan, S.L.W. 1988, *Nature* **336**, 31.

Hut, P., Murphy, B.W. & Verbunt, F. 1991a, *Astron. Astrophys.* **241**, 137.

Hut, P., McMillan, S. & Romani, R. 1991b, submitted to *Astrophys. J.*.

Hut, P., McMillan, S., Goodman, J., Mateo, M., Phinney, S., Pryor, T., Richer, H. & Weinberg, M. 1992, to appear as a review paper in *P.A.S.P.*

Inagaki, S. 1986, *Publ. Astron. Soc. Japan*, **38**, 853.

Ito, T., Makino, J., Ebisuzaki, T., & Sugimoto, D. 1990 *Comp. Phys. Comm.* **60**, 187.

Ito, T., Ebisuzaki, T., Makino, J., & Sugimoto, D. 1991 *Publ. Astron. Soc. J.*, in press.

Kandrup, H.E. & Smith, H., Jr., 1990, preprint, submitted to *Ap.J.*

Kochanek, C.S. 1991, preprint, submitted to *Ap.J.*

Kormendy, J. & McClure, R.D. 1991, to be submitted to *Astrophys. J.*.

Krolik, J.H. 1983, *Nature*, **305**, 506.

Latham, D.W., Hazen-Liller, M.L. & Pryor, C.P. 1985, in *I.A.U. Coll 88, Stellar Radial Velocities*, ed. A.G.D. Phillip & D.W. Latham (Schenectady, NY: L. Davis Press), p. 269.

Lauer, T. R., *et al.* 1991, *Astrophys. J.* **369**, L45.

Lecar, M. 1968, *Bull. Astron.* **3**, 91.

Lee, H.M. 1987a, *Astrophys. J.* **319**, 772.

Lee, H.M. 1987b, *Astrophys. J.* **319**, 801.

Lee, H.M. 1989, in *Dynamics of Dense Stellar Systems*, ed. D. Merritt, (Cambridge University Press), p. 105.

Lee, H.M. & Ostriker, J.P. 1986, *Astrophys. J.* **310**, 176.

Lee, H.M., Fahlman, G.G., Richer, H.B. 1991, *Astrophys. J.* **366**, 455.

Leonard, P.J.T. 1991, *Astron. J.*, **101**, 562.

Leonard, P.J.T. & Duncan, M.J. 1988, *Astron. J.*, **96**, 222.

Leonard, P.J.T. & Duncan, M.J. 1990, *Astron. J.*, **99**, 608.

Long, K. S., D'Odorico, S., Charles, P. A., & Dopita, M. A. 1981, *Astrophys. J. Lett* **246**, L61.

Lynden-Bell, D. & Eggleton, P.P., 1980, *M.N.R.A.S*, **191**, 483.

Lynden-Bell, D. & Wood, R, 1968, *M.N.R.A.S*, **138**, 495.

Lyne, A.G., Brinklow, A., Middleditch, J., Kulkarni, S.R., Backer, D.C. & Trevor, T.R. 1987, *Nature* **328**, 399.

Makino, J. 1991, *Publ. Astron. Soc. J.*, in press.

Makino, J., & Hut, P. 1990, *Astrophys. J.*, **365**, 208.

Makino, J., & Hut, P. 1991, *Astrophys. J.*, in press.

Makino, J., Ito, T. & Ebisuzaki, T. 1990, *Publ. Astron. Soc. J.* **42**, 717.

Marchal, C. 1990, *The Three-Body Problem* (Elsevier, Amsterdam).

Margon, B. & Downes, R. A. 1983, *Astrophys. J. Lett.* **274**, L31.

Margon, B., Downes, R. A., & Gunn, J. E. 1981, *Astrophys. J. Lett.* **249**, L1.

Mateo, M., Harris, H.C., Nemec, J. & Olszewski, E.W. 1990, *Astron. J.*, **100**, 469.

McMillan, S.L.W. 1986, *Astrophys. J.*, **307**, 126.

McMillan, S.L.W. 1989, in *Dynamics of Dense Stellar Systems*, ed. D. Merritt (Cambridge University Press), p. 207.

McMillan, S.L.W., McDermott, P.N. & Taam, R.E. 1987, *Astrophys. J.*, **318**, 261.

McMillan, S.L.W., Hut, P. & Makino, J. 1990, *Astrophys. J.*, **362**, 522.

McMillan, S.L.W., Hut, P. & Makino, J. 1991, *Astrophys. J.*, **372**, 111.

Merritt, D. (ed.) 1979 *Dynamics of Dense Stellar Systems* (Cambridge University Press).

Mikkola, S., 1983a, *M.N.R.A.S.*, **203**, 1107.

Mikkola, S., 1983b, *M.N.R.A.S.*, **205**, 733.

Mikkola, S., 1984a, *M.N.R.A.S.*, **207**, 115.

Mikkola, S., 1984b, *M.N.R.A.S.*, **208**, 75.

Miller, R.H., 1964. *Ap.J.* **140**, 250.

348

Monaghan, J.J., 1985, *Comp. Phys. Reports*, **3**, 71.

Murphy, B.W. & Cohn, H.N. 1988, *Mon. Not. R. astr. Soc.*, **232**, 835.

Murphy, B.W., Cohn, H.N. & Durison, R.H. 1989, in *Dynamics of Dense Stellar Systems*, ed. D. Merritt, (Cambridge University Press), p. 97.

Murphy, B.W., Cohn, H., & Hut, P. 1990, *Mon. Not. R. astr. Soc.*, **245**, 335.

Murphy, B.W., Cohn, H.N. & Durisen, R.H. 1991, *Astrophys. J.*, **370**, 60.

Newell, B. & O'Neil, E.J. 1978, *Ap. J. Suppl.* **37**, 27.

Ostriker, J.P. 1985, in *'Dynamics of Star Clusters', IAU Symposium No. 113*, eds. J. Goodman & P. Hut (Dordecht;Reidel), p.347.

Phinney, E.S. & Sigurdsson, S. 1991, *Nature*, **349**, 220.

Press, W.H. & Teukolsky, S.A. 1977, *Astrophys. J.*, **213**, p. 183

Pryor, C., McClure, R.D., Hesser, J.E., & Fletcher, J.M. 1987, *Bull. Amer. Astron. Soc.* **19**, 676.

Pryor, C., McClure, R.D., Hesser, J.E. & Fletcher, J.M. 1989, in *Dynamics of Dense Stellar Systems*, ed. D. Merritt (Cambridge Univ. Pr.), p. 175.

Quinlan, G.D. & Shapiro, S.L. 1989, in *Dynamics of Dense Stellar Systems*, ed. D. Merritt, (Cambridge University Press), p. 113.

Quinlan, G.D. & Shapiro, S.L. 1990, *Astrophys. J.*, **356**, 483.

Rasio, F.A., Shapiro, S.L. & Teukolsky, S.A. 1989, in *Dynamics of Dense Stellar Systems*, ed. D. Merritt, (Cambridge University Press), p. 121.

Rasio, F.A. & Shapiro, S.L. 1991, Cornell preprint CRSR 968.

Ray, A., Kembhavi, A.K. & Antia, H.M. 1987, *Astron. Astrophys.*, **184**, 164.

Spitzer, L. 1940, *Mon. Not. R. astr. Soc.*, **100**, 396.

Spitzer, L. 1975, in *'Dynamics of Stellar Systems', IAU Symp. No. 69*, Ed A. Haili (Dordrecht;Reidel), p. 3.

Spitzer, L. 1987, *Dynamical Evolution of Globular Clusters*, Princeton University Press.

Spitzer, L. & Mathieu, R.D. 1980, *Astrophys. J.*, **241**, 618.

Statler, T.S., Ostriker, J.P., & Cohn, H. 1987, *Astrophys. J.*, 316, 626.

Stodólkiewics, J.S. 1982, *Acta Astron.*, **32**, 63.

Stodólkiewics, J.S. & Giersz, M., 1990, Nicolaus Copernicus Astronomical Center preprint 218.

Sugimoto, D. & Bettwieser, E. 1983, *Mon. Not. R. astr. Soc.*, **204**, 19p.

Sugimoto, D., Chikada, Y., Makino, J., Ito, T., Ebisuzaki, T. & Umemura, M. 1990, *Nature* **345**, 33.

Valtonen, M. 1988, *Vistas in Astronomy*, **32**, 23.

Valtonen, M. & Mikkola, S., 1990, *The Few Body Problem in Astrophysics*, Turku Univ. preprint.

Verbunt, F. & Hut, P. 1987, in *I.A.U. Symposium 125, The Origin and Evolution of Neutron Stars*, eds. D.J. Helfand & J.-H. Huang (Dordrecht: Reidel), p.187.

STATISTICS OF PULSARS IN GLOBULAR CLUSTERS

H. M. JOHNSTON, S. R. KULKARNI and E. S. PHINNEY
California Institute of Technology, 105-24
Pasadena, CA 91125
USA

ABSTRACT. More than eighty globular clusters have now been surveyed for pulsars; nearly thirty cluster pulsars have so far been discovered. We review the current status of cluster pulsar searches and their selection effects. We discuss various methods which have been used to extrapolate the observed pulsar population to the underlying distribution. We show that the dependence on assumptions about the luminosity function, and in particular the minimum pulsar luminosity L_{min}, can lead to the large discrepancies between various authors in the estimates of the total cluster pulsar population. We investigate the way in which number of pulsars in a cluster, N_p, scales with cluster parameters such as central density ρ_c or core mass M_c. Our best estimate is that N_p scales as $M_c \rho_c^{0.5}$, significantly less steep than predicted by the two-body tidal capture model. This yields $2500/\bar{f}$ as the total number of pulsars produced in the globular cluster system, where $\bar{f}$ is the mean beaming factor. Such a high number is also supported by the high birthrates inferred from the young ages of the known cluster pulsars. The discrepancy between these high birthrates and the low birthrates of LMXBs in clusters suggests that there may be two channels of pulsar formation, one of which produces most pulsars, and the other most of the LMXBs. However, the eleven fast pulsars in 47 Tuc are a problem for this scenario. We suggest that there may be another parameter involved in determining the number of pulsars in clusters. The most likely candidate is a time-dependence of the rate of pulsar formation, which some authors attribute to the onset of core collapse.

1. Introduction

It is now clear that globular clusters contain a rather large number of pulsars (see Lyne, this volume). While the current count stands at 28, estimates of the total numbers range from 2×10^2 to 2×10^3 and more if beaming is taken into account. Cluster pulsars, like cluster Low Mass X-ray Binaries (LMXBs; see contributions by van den Heuvel and Bhattacharya), their presumed progenitors, offer us new insights into the formation and evolution of binaries in globular clusters. However, more uniquely, the statistics of cluster pulsars allow us to address observationally several global issues: the Initial Mass Function (IMF) of clusters, the stability of proto-clusters and the important role of primordial binaries.

A complete list of clusters in which pulsars and/or LMXBs have been discovered is presented in Table 1. We first identify the key issues which motivate the study of pulsar and LMXB statistics, followed by presentation and discussion of the various estimates. We conclude with a discussion of the birthrates and future observational directions. Related

E. P. J. van den Heuvel and S. A. Rappaport (eds.), X-Ray Binaries and Recycled Pulsars, 349–364.

articles are those by Phinney and by Hut (this volume) and a recent review by Phinney and Kulkarni (1991; hereafter PK).

Abundance of Cluster Neutron Stars. The simplest hypothesis is that cluster pulsars are primordial[†] neutron stars spun up by mass transfer. A significant fraction of the primordial neutron stars may escape from the cluster. In the disk, pulsars are observed to have large space velocities with a mean value of 200 km s^{-1} (see Harrison and Lyne, this volume). Following Paczyński (1990), we describe the velocity distribution as

$$p(u)\,du = \frac{4}{\pi}\frac{du}{(1+u^2)^2}, \quad u = \frac{v}{v_*}, \quad v_* = 270\,\text{kms}^{-1}. \tag{1}$$

Using the escape velocities listed in Table 1 we find that the fraction of neutron stars retained, f_v, to range from 0.11 to 0.32 with a mean value of 0.19. Note that Hut et al. 1991 show that the escape velocities remain essentially constant with time, despite significant dynamical evolution of clusters. (We exclude the core-collapsed clusters since they have undergone significant dynamical evolution, and their present parameters bear no relation to their initial parameters).

Recently, Narayan and Ostriker (1990) claimed a bimodal distribution with two populations, containing approximately equal numbers of pulsars, with mean space motions of 60 km s^{-1} and 200 km s^{-1}, in which case f_v ranges from 0.05 to 0.25 with a mean value of 0.12, not significantly different from the previous estimates.

Assume a power-law differential IMF, $N(m) \propto m^{-(1+x)}\,dm$, valid for $m > m_l$, the minimum mass. The number of neutron stars per unit initial mass of the cluster, the specific incidence, is then

$$\nu = \frac{x-1}{8x}\left(\frac{m_l}{8\,M_\odot}\right)^{x-1} M_\odot^{-1}$$

where we assume that stars with $m > 8\,M_\odot$ evolve to neutron stars. For a proto-cluster with Salpeter IMF, $x = 1.35$, total mass $10^6\,M_\odot$ and $m_l = 0.08\,M_\odot$, we obtain $\nu = 0.6\%$ and 6×10^3 primordial neutron stars, of which $\sim 10^3$ can be expected to be retained.

Mass loss due to stellar evolution can potentially unbind clusters. (The collective effect of the supernovae does not necessarily disrupt the cluster; see PK). Assuming that stars $0.8\,M_\odot < m < 8\,M_\odot$ become 0.6 $M_\odot$ white dwarfs, the mass fraction lost by a Salpeter IMF is 0.36, i.e. the cluster will not disrupt due to mass loss. The fractional mass loss and ν as functions of x and m_l is shown in Figure 1 (after Goodman 1991).

Chernoff and Weinberg (1990) found that mass loss is accentuated by tidal stripping of the outer parts of the cluster by the Galactic gravitational field, and concluded that only clusters with IMFs steeper than Salpeter survive. Bailyn and Grindlay (1990) present this as strong objection to the primordial neutron star hypothesis and suggest that cluster pulsars are created by accretion induced collapse (AIC) of massive white dwarfs (see Nomoto, this volume). Apart from the uncertain physics of the AIC process, this reasoning is quite insecure. Using $m_l = 0.4\,M_\odot$ (the value used by Chernoff and Weinberg), *guarantees* that the cluster will not survive, as can be seen from Figure 1[‡]. Clusters with

---

[†] The term 'primordial', here and in related literature, refers to neutron stars descended from the early massive stars, as opposed to any recently transmuted white dwarfs.

[‡] Chernoff and Weinberg, in their cluster models, actually used several white dwarf masses, as well as a range of progenitor stars that disrupt completely; however, the point remains that the cluster survival is sensitive to m_l.

Table 1. Cluster Pulsars and LMXBs

Cluster names		d (kpc)	R_{gc} (kpc)	$\log \rho_c$ $M_\odot/pc^3$	v_{esc} km/s	[m/H]	Sources	Survey	Ref.
1821−249	M28	5.8	3.2	5.5	29	−1.8	P	JB	9
1639+365	M13	7.1	8.9	4.0	44	−1.6	2P	AO	7
1516+022	M5	7.6	6.6	4.7	61	−1.6	2P	AO	16
2127+11	M15	9.7	10.4	c	50	−2.1	X,5P	AO	2,6,15,16
1310+186	M53	18.5	19.2	3.5	35	−2.0	P	AO	7
1620−264	M4	2.1	6.8	4.4	27	−1.1	P	JB	14
1802−07	NGC 6539	3.1	6.0	4.4	23	−1.1	P	PK	5
1908+00	NGC 6760	4.1	6.0	4.2	26	−0.8	P	AO	1
0021−772	47 Tuc	4.6	8.1	5.0	62	−0.8	11P	PK	12,13
1850−08	NGC 6712	6.2	4.2	3.9	29	−1.3	X		
1745−24	Terzan 5	7.1	1.8	5.9	—	−0.7	X,P	PK	10
1745−20	NGC 6440	7.1	2.1	5.9	72	−0.5	X,P	PK	11
1730−335	Liller 1	7.9	1.2	c?	—	−0.3	X		
1820−30	NGC 6624	8.0	1.5	c	—	−0.8	X,2P	PK	4
1724−30	Terzan 2	10.0	1.4	c	—	−0.5	X		
1732−30	Terzan 1	10.6	1.9	c	—	+0.1	X		
1746−37	NGC 6441	11.7	3.3	5.8	71?	−0.1	X		
0512−40	NGC 1851	12.0	17.2	5.7	32?	−1.2	X		
1747−31	Terzan 6	12.8	4.0	c	—		X		
1832−33	NGC 6652	14.3	6.0	5.1	23	−0.9	X		

Table 1. Clusters containing pulsars and low-mass X-ray binaries. The clusters are divided into two groups, of low and high metallicity. Within each group, they are ordered by distance from the Sun. The first seven columns show: cluster name (IAU designation and common name); distance from the Sun; distance from the Galactic centre; central density (from Chernoff and Djorgovski 1989; Djorgovski, *pers. comm.*); escape velocity from the core for non-core collapsed clusters (from the literature, or calculated from data provided by Djorgovski, *pers. comm.*); metallicity (from Webbink 1985). The last two columns show whether the cluster contains an LMXB (X) and/or a pulsar (P) (the numeral indicates the number of objects), and the survey which discovered the pulsar: JB=Jodrell Bank, AO = Arecibo Observatory, PK = Parkes. References to pulsar surveys: 1. Anderson et al. 1990a; 2. Anderson et al. 1990b; 3. Biggs et al. 1989; 4. Biggs et al. 1990; 5. D'Amico et al. 1990; 6. Kulkarni et al. 1990; 7. Kulkarni et al. 1991; 8. Lyne & Biggs 1989; 9. Lyne et al. 1987; 10. Lyne et al. 1990; 11. Manchester et al. 1989; 12. Manchester et al. 1990; 13. Manchester et al. 1991; 14. McKenna & Lyne 1988; 15. Prince et al. 1991; 16. Wolszczan et al. 1989. The LMXB list is from Verbunt and Hut (1987). 1728−34/Grindlay 1 has been excluded since Grindlay 1 is no longer believed to be a cluster. Two new X-ray sources (Terzan 6 and NGC 6652) discovered by ROSAT are from Predehl et al. (1991).

$x = 1.35$ but with reasonable values of $m_l \sim 0.1$ M$_\odot$ can probably survive. Indeed, Richer and Fahlman (1989) find a global $x \sim 1$ for M71, and Phinney (1991) has convincingly argued, on dynamical grounds, for a global $x \sim 1.3$ for M15.

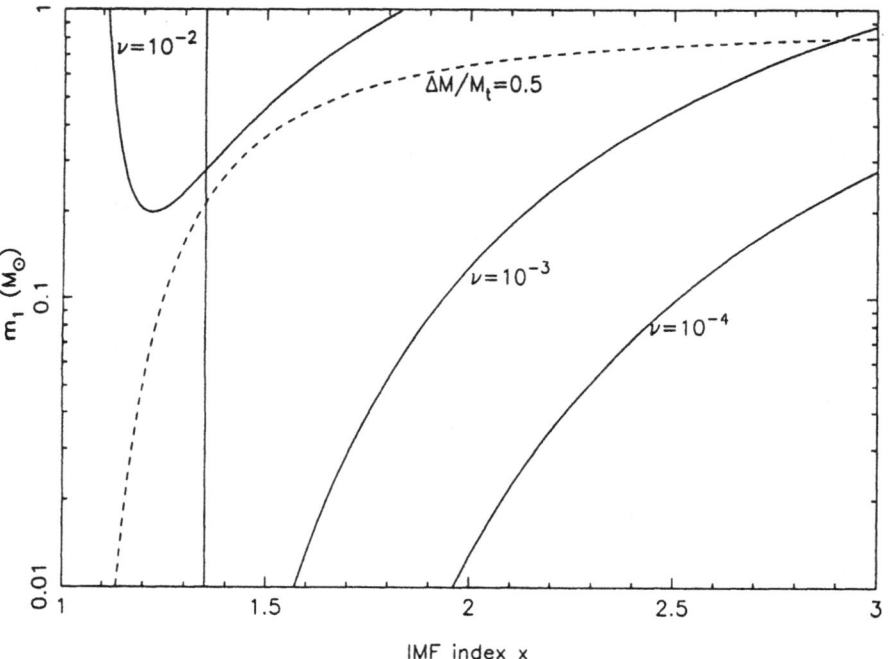

Figure 1: Specific incidence of neutron stars $\nu = N_{ns}/M_t$ as a function of minimum mass m_1 and IMF power law index x. The curve denoting fractional mass loss $\Delta M/M_t = 0.5$, where the cluster becomes unbound, is also shown.

Weighting Factor. Procedures to determine the weighting factor or the specific incidence of systems evolving to the observed radio pulsars or LMXBs are an integral part of statistical studies. Immediately after the discovery of LMXBs, the very high specific incidence of LMXBs in clusters relative to the field (see §6) was attributed to the formation of binary systems by tidal capture (Katz 1975).

The volumetric rate of tidal encounters is proportional to the product of the volume densities of the two species, and since neutron stars are heavier than field stars the encounters take place in the cluster core. Thus the number of two-body tidal products can be expressed as

$$N_p \propto n_{ns} n r_c^3 \sigma_c^{-1} \propto \nu M_t \rho_c^\alpha \sigma_c^{-1} T \tag{2}$$

where n_{ns} and n are the number densities of neutron stars and other stars, M_t, r_c, σ_c and ρ_c are the mass, core radius, core velocity dispersion and core density of the cluster respectively and T is the interval over which the process lasts. Note that only core parameters are relevent, since the neutron stars are mostly confined to the core. If the fraction of neutron stars ν is independent of the structural parameters, i.e. if the IMF is fixed, then $\alpha = 1$. T is usually assumed to be constant and set to the age of Galaxy. The assumption of constant T is certainly questionable since clusters evolve considerable over the age of the Galaxy.

Several lines of reasoning — the discovery of primordial binaries in optical spectroscopic surveys (Pryor et al. 1989), the discovery of pulsars (Kulkarni et al. 1991) or pulsar-like emission (Fruchter and Goss 1990) in low density clusters, etc. — have highlighted the dominant role of three-body (primordial binary and neutron star) collisions (**T3**) in all but the densest clusters (PK). While the cross-section for two-body encounters ('**T2**' collisions, in the notation of PK) goes like $\sigma_2 \sim \pi R_*^2 a_c$, the cross-section for **T3** interactions is $\sigma_3 \sim \pi a a_c$, where a_c is the characteristic radius, $a_c \sim GM/\sigma_c^2$. Since a, the orbital radius of a binary, is much greater than R_*, the radius of a typical star, we can see that for a reasonable fraction of binaries, **T3** dominates over **T2**. In dense clusters, stellar encounters destroy or harden (i.e. reduce a) wide binaries, reducing the efficiency of **T3** encounters. This can be effectively parametrized by making α less than unity.

The IMF, through ν (Figure 1), can also affect the weighting function. However, there are not much observational data in this regard. Also, it is not even clear, on observational or theoretical grounds whether the IMF can be characterized by a simple power law. Variations in the IMF have been reported, some of which are probably genuine, e.g. in the LMC, clusters with x of ~ 0.3 to 2.5 have been found (Elson et al. 1989; Mateo 1988[†]). However, here one must be cautious of the many observational studies which suffer from improper accounting of the distortions to the observed MF introduced by mass segregation, e.g. McClure et al. (1986) found a trend of x with metallicity, but their conclusions were based on luminosity functions obtained at distances of several core radii from the cluster centers.

Given the current observational situation, we use the bare minimum number of parameters to derive the weighting function: ρ_c, M_t and α. For lack of detailed data and understanding, we assume f_v, ν and T are the same for all clusters.

2. Searches for Cluster Pulsars and Selection Effects

Table 1 shows the current status of cluster pulsar searches. It is immediately obvious that pulsars are only being found in the most nearby clusters. LMXBs, however, can be discovered anywhere in the Galaxy, and thus the surveys are complete. On the other hand, with sufficiently deep searches, using Arecibo, pulsars are being found in even the most unprepossessing clusters, while the LMXBs are in general confined to the metal-rich clusters near the Galactic centre.

This complicates comparison of the two classes of objects. The large distances to globular clusters ensure that only the brightest pulsars are detected, and so a careful analysis of selection effects is necessary. The principal effects that limit the ability to detect pulsars are: the finite sampling time and interstellar scattering which discriminates against finding fast pulsars; dispersion measure limitations, which select against distant pulsars or pulsars in regions of high electron density; and the Galactic background radiation, which reduce sensitivity.

In addition, pulsars in tight binary systems are selected against, because the changing velocity of the pulsar leads to a changing pulse period, and hence blurring of the peak in the power spectrum. Reducing the integration time reduces the blurring, as the pulsar's velocity is more nearly constant, but also reduces the sensitivity of the observations. Some groups (Anderson et al. 1990) are now incorporating partial compensation for this effect into their search codes, by assuming the acceleration of the pulsar is constant over the observation (Middleditch and Priedhorsky 1986; Hertz et al. 1990). Johnston and Kulkarni (1991) have estimated the parameter space covered by such algorithms and show that the classical **T2** binaries ($a \sim 3R_*$) are accessible to such searches. Tighter binaries are missed,

---

[†] Richtler et al. (1991) recently attributed these earlier results to problems with crowding and incompleteness, and claim that all IMFs are consistent with a Salpeter slope.

though. Following Kulkarni et al. (1990) the fraction missed is usually parametrized by the factor β, which is the ratio of the tight binaries to pulsars in wide binaries (including single pulsars).

Most of these selection effects, such as scattering, dispersion, and the binarity effect, do not apply to imaging surveys. The absence of sources in a VLA map puts strict limits on the presence of pulsars, binary or otherwise, down to the flux limit of the image.

The following is a brief description of the major cluster pulsar surveys.

Arecibo. All clusters observable with the Arecibo dish ($0° \leq \delta \leq 37°$) were searched at 1400 MHz with a sampling time of 507 μs. This resulted in the discovery of PSR2127+11A, and PSR1908+00 in NGC 6760. A second search at 430 MHz resulted in the discovery of seven more pulsars (Table 1). Partial details of the two surveys may be found in Wolszczan et al. (1989) and Anderson et al. (1990).

Jodrell Bank. A total of 65 clusters were observed using the 76 m Lowell telescope, at two different frequencies: 610 MHz and 1420 MHz, using a sampling interval of 300 μs (Lyne et al. 1988; Lyne and Biggs 1989). This resulted in the discovery of PSR1821–24 in M28, and PSR1620–26 in M4.

Parkes. Southern globular clusters are being searched by the Parkes telescope at 1400 MHz and 600 MHz, using 300 μs sampling. Partial details of the survey may be found in Manchester et al. 1991 and Robinson et al., this volume. The survey has resulted in the discovery of many pulsars, including eleven in 47 Tuc (Manchester et al. 1989; Biggs et al. 1990; D'Amico et al. 1990; Manchester et al. 1990; Lyne et al. 1990).

VLA surveys. There have been four surveys of clusters using the VLA. *(i)* Hamilton et al. (1985) imaged twelve clusters at 1400 MHz, to a flux limit of 0.7 mJy; they detected one point source in M28, which led to the discovery of the first cluster pulsar. *(ii)* Fruchter and Goss (1990) imaged seventeen clusters at 1400 MHz to a similar flux level, and found five point sources, four of which have since been shown to be pulsars. The fifth one is likely to be a pulsar as well. *(iii)* Kulkarni et al. (1990) imaged four clusters at 1400 MHz to levels between 0.1 mJy and 0.3 mJy. They identified already known pulsars in M15, M4 and M28, and located an intriguing point source in M3 which has yet to be identified. *(iv)* Johnston et al. (1991) imaged four clusters at 1400 MHz to levels between 0.1 mJy and 0.2 mJy. They found several sources in M13 and M22, not yet identified. The VLA surveys show that there cannot be a large number of tight binaries and limit $\beta < 2$ (Johnston et al. 1991).

3. Methodologies

That the observed cluster pulsar population is perhaps the tip of a much larger underlying distribution can be graphically appreciated by comparing the histogram of the observed pulsar luminosities with the minimum luminosities of the observed sample of disk millisecond pulsars, ~ 10 mJy kpc^2, which itself is suspected to be biased to the brighter pulsars. An estimation of the total number of cluster pulsars thus requires large extrapolation, a process necessarily fraught with large uncertainties.

The Birthrate Method. (Introduced by PK). If a group of clusters has been producing pulsars at a constant rate for time T, then the expected number of pulsars of age less than t is $\sim N(t/T)$, where N is the total number of pulsars. The characteristic age, $\tau_c = P/2\dot{P}$ is a reliable measure of the age of a pulsar. Magnetic field decay only makes τ_c an upper limit. However, in some dense clusters such as M15, $\dot{P}$ could be corrupted by dynamical effects

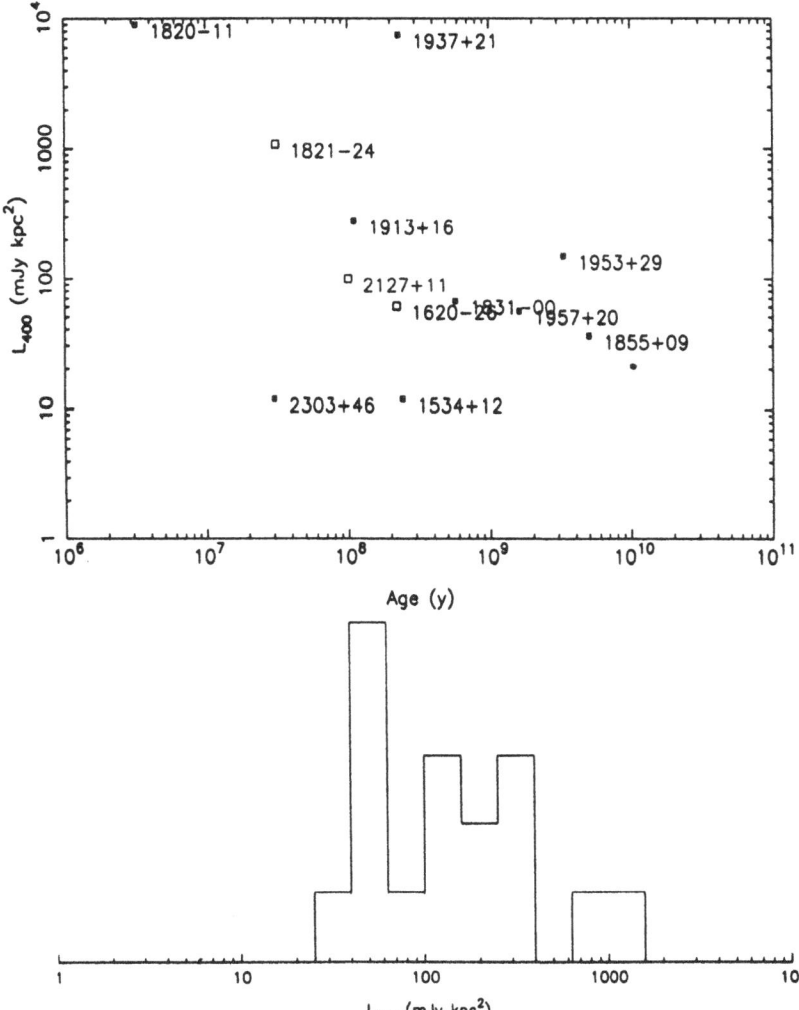

Figure 2: (a) Luminosity as a function of spin down age for disk and cluster recycled pulsars: cluster pulsars are shown as open boxes. (b) Histogram of luminosities of all cluster pulsars.

(Wolszczan et al. 1989, Phinney 1991). If younger pulsars are brighter than older pulsars, then the birthrate can be estimated with minimal correction for sample incompleteness. From Figure 3, we see that such a trend, is at best, rather modest. Thus, most likely we need to apply an *upward* statistical correction to the observed birthrate to obtain the true birthrate. Despite this potential complication, the few measured $\dot{P}$s indicate high birthrates in M28 ($\tau_c = 3 \times 10^7$ y; Foster et al. 1990), M4 ($\tau_c = 3 \times 10^8$ y; McKenna and Lyne 1988) and M15 ($\tau_c \sim 3 \times 10^8$ y; Prince et al. 1991) and M13 ($\tau_c \sim 10^9$ y; Kulkarni et al. 1991). This has to be balanced by the non-detections and the choice of T. PK obtain a rough estimate $N_p \sim 2 \times 10^3 / \bar{f}$ pulsars over the whole cluster system, over Hubble lifetime; here $\bar{f}$ is the mean beaming factor and is expected to be between 0.2 and 1.

The Luminosity Function Method I. Kulkarni et al. (1990, henceforth KNR) used a semi-empirical luminosity law (see below) to model the true underlying pulsar population in clusters. The selection effects of five major pulsar surveys (§2) were modelled. KNR assumed that the pulsars are born uniformly in $\log B$ in the range $8 < \log B < 10$ and evolved the population under assumption of constant B. An empirical relation, based on the properties of the observed disk millisecond pulsars, was used to derive the mean 400 MHz luminosity of the pulsars L_m: $\log L_m = (\dot{P}_{-15}/3P^3) + 1.1$, where P is the pulsar period in seconds, and $\dot{P}_{-15}$ is the period derivative in units of 10^{-15} s s^{-1}. The spread around this distribution was adapted from Narayan and Ostriker (1990). The resulting population was then to found to have a luminosity distribution,

$$dN/dL \propto L^{-\gamma}, \qquad \gamma = 2,$$

with a turnover at $L_{400} \sim 3$ mJy kpc^2. This is remarkably similar to the luminosity function of the single, slow pulsars. Here we note that a luminosity decay law $L \propto t^{1/(1-\gamma)}$ would give such a power law luminosity function. Thus for $\gamma = 2$, the implied luminosity decay law is t^{-1}.

By comparing the estimates of the number of detectable pulsars with the actual detections, KNR estimated $N_p \sim 1000(1 + \beta)/\bar{f}$ (this value interpolated to $\alpha = 0.5$ to facilitate consistent comparison with other estimates.) It is interesting to note that KNR could *not* reproduce the young ages of the observed sample. This suggests that cluster pulsars fade more rapidly than t^{-1}, which implies an even higher birthrate and hence a substantial cluster pulsar population.

The Luminosity Function Method II. Fruchter and Goss (1990; hereafter FG) imaged 17 globular clusters with the VLA and detected pulsar-like emission (i.e. steep spectrum emission close to the core) in five clusters. Assuming the standard luminosity function, $\gamma = 2$, $L_{min} = 3$ and $\alpha = 0.5$, they estimate $N_p \sim 500/\bar{f}(3/L_{min})$ to $2000/\bar{f}(3/L_{min})$.

The Luminosity Function Method III. Wijers and van Paradijs (1991; hereafter WvP) used the integrated flux from three globular clusters containing at least one detected pulsar to constrain the underlying population. The basis of this powerful method is the statistic, $f_1 = L_1/L_t$, the ratio of the luminosity of the brightest (detected) pulsar to the integrated luminosity. The underlying pulsar population was assumed to be restricted to the same region of the $B - P$ diagram as in KNR; here B is the dipole field strength of pulsars. The use of the model luminosity law $L_m(P, \dot{P})$ (see above) then specified L_{max}. The value of L_{min} was assumed to be 0.076 mJy kpc^2 at 1400 MHz, and 1 mJy kpc^2 (by a ν^{-2} extrapolation) at 400 MHz.

The method was applied to three clusters for which VLA data, at 1400 MHz were available (Terzan 5, NGC 6440 and NGC 6539). f_1 for two of these clusters is between 0.6 and 0.5 (after accounting for measurement errors). Recent observations (see by Fruchter and Goss) reveal two pulsar-like source in Terzan 5, one in the core and the other outside the core (the famous eclipsing pulsar, see contribution by Lyne) and diffuse emission from the core region with steep spectrum and hence attributable to a sea of faint pulsars. The f_1 for the whole cluster is $\lesssim 0.3$ and the f_1 for the core region alone is $\lesssim 0.4$.

For each cluster, the total luminosity of the cluster was fixed to the observed luminosity and pulsars drawn from a variety of luminosity functions, $dN/dL \propto L^{-\gamma}$. The luminosity function with $\gamma = 1.5$ gave the maximum number of pulsars satisfying the observed f_1 values. Assuming $\alpha = 0.5$, N_p (based on data excluding Terzan 5) was constrained to be less than $300/f_b$ to $700/f_b$, significantly less than the estimates of KNR and FG. Note that this estimate, as well as that of FG, is independent of β because it is based on imaging observations.

Shortfalls of the Methods. The principal weakness of all these methods, except the birthrate method, is their dependence upon the assumptions of luminosity functions and in particular their extreme sensitivity to L_{min}. This topic deserves a full discussion and is treated in section 5. The birthrate method is the least model dependent but is hampered by the absence of a significant number of $\dot{P}$ measurements. ·

All the methods suffer from the drawback of needing to assume a weighting scheme for the cluster. It is better to derive this dependence from the data itself since in that case we would also learn something about the tidal capture process (see Phinney's contribution). This is attempted in the next section.

4. The Fully Self-Consistent Approach

The luminosity function and the choice of weighting function can be obtained from the data using a generalization of a method first pioneered by Verbunt and Hut (1987) and introduced by PK. Verbunt and Hut showed that $\alpha = 1$ is an acceptable choice for the weighting function for LMXBs. The difference here is that unlike X-ray surveys, the radio pulsar searches are quite incomplete and the extent to which they are incomplete depends upon the underlying luminosity function. Thus a two-dimensional diagram is called for: one for α and the other for the luminosity function, parametrized by γ (Figure 4, from PK; see also Johnston et al. *in prep.*, henceforth JKP).

In Figure 4, each box represents a globular cluster which has been searched by one of the surveys described in section 2. The clusters are ordered by density, with densest clusters at the top of the figure. The height of each box is proportional the weight assigned to the cluster, $M_c \rho_c^{\alpha \dagger}$. The length of the box is proportional to the fraction of all pulsars that would have been discovered by the survey. In case of multiple surveys, the most sensitive survey is considered. This is calculated by using the published flux limits S, the distance to the cluster, and a luminosity function, $dN/dL \propto L^{-\gamma}$. Limits are always scaled to 400 MHz assuming an intensity spectrum, $\propto \nu^{-2.5}$; we use a spectral index of 2.5 instead of 2.0 since recent work (Foster et al. 1991) has shown that most millisecond pulsars have exceedingly steep spectra. The integral form of the luminosity function is shown at the top of the figure (reversed from the usual orientation). The known pulsars are indicated by crosses.

The basis of the method is that if both α and γ have been chosen correctly, then the crosses will be distributed uniformly through the area defined by the boxes. Since we do not have many pulsars, we resort to 1-d collapses of the boxes, which can be found at the bottom and to the right of the main diagram.

Focussing on the bottom strips, we see that that the $\gamma = 1.5$ and logarithmic luminosity functions have too many pulsars clustered towards the faint end of the surveys (Kolmogorov-Smirnov probabilities 0.001 and 10^{-13} respectively). Both the $\gamma = 2$ and the KNR luminosity function (essentially $\gamma = 2$ with a turnover at faint luminosities) are acceptable (KS probabilities 0.26 and 0.27).

---

† Note that we have omitted the σ_c^{-3} dependence (eq. [2]) from our definition of the cluster weight, in order to minimize the number of parameters. This could make a difference, as low density clusters tend to have low core velocity dispersions as well, which would thus tend to reduce the dependence on ρ_c.

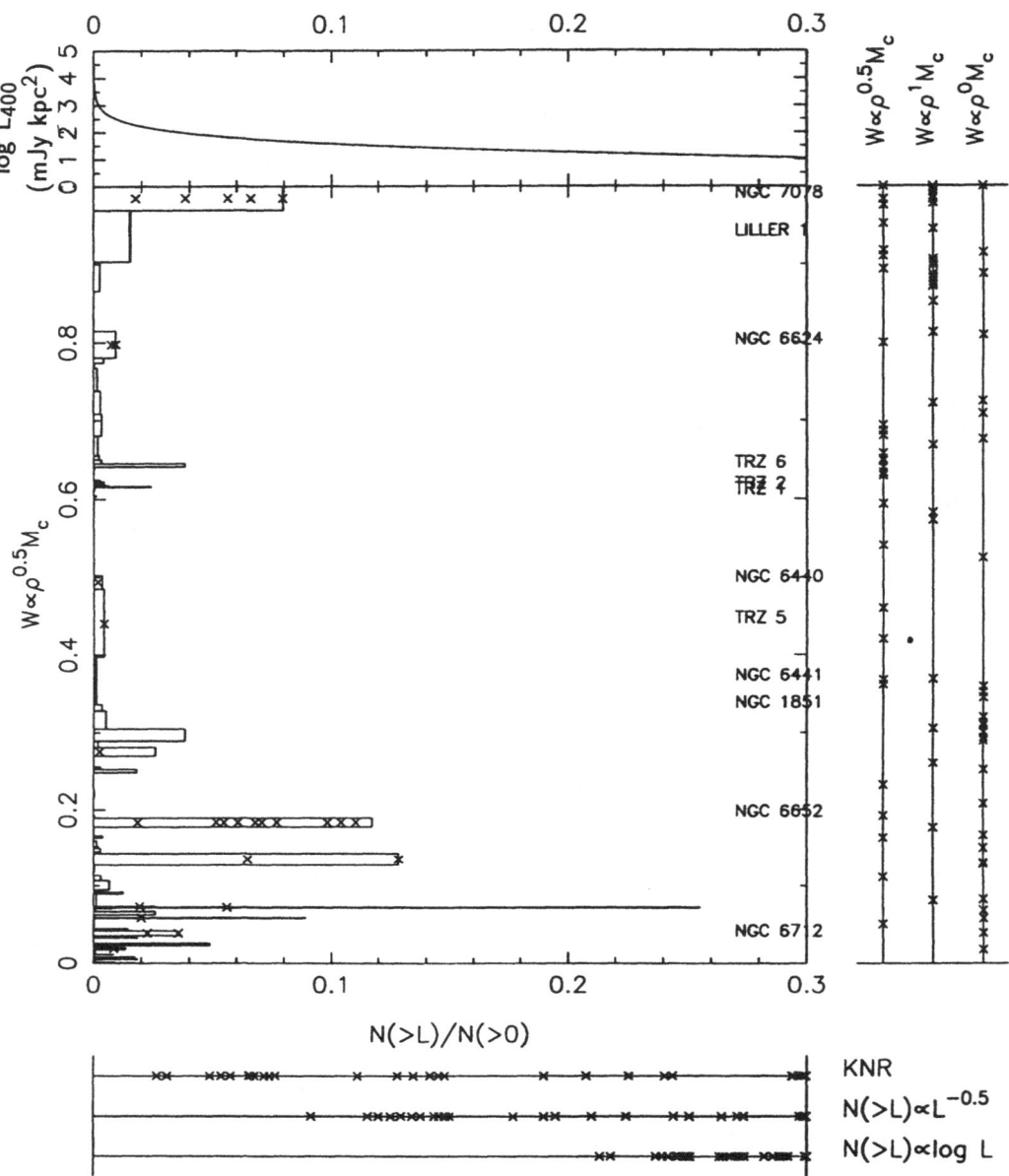

Figure 4: Two-dimensional version of the Verbunt and Hut (1987) diagram, for determining the luminosity function and density weighting of cluster pulsars, from PK.

Next we construct a cumulative distribution of the area of the boxes. If the area of each box is x_i, then we lay segments of length x_i end to end to form a line of length $X = \Sigma_{i=1}^{n} x_i$, the fraction of all pulsars so far discovered. If α and γ have been chosen correctly, the pulsars should be uniformly distributed along this segment as well. The $\alpha = 1$ weighting function has too many pulsars in clusters with low weight (KS probability 7×10^{-6}). The $\alpha = 0.5$ is our preferred choice (KS probability 0.21).

In the figure, we also plot down the right hand side of the figure the positions of cluster LMXBs (see Table 1). Their distribution is consistent with either $\alpha = 1$ or $\alpha = 0.5$ (KS probability 0.52 and 0.90 respectively).

Note that 47 Tuc is clearly anomalous, producing a large bunching of pulsars at one place. While the distribution is still consistent with being uniform, this may be a suggestion that our model is too simplistic. In §6, we argue that the distribution of pulsars may be a function of more than just current cluster parameters.

From the fraction of the entire space available, using our preferred density weighting scheme and luminosity function, we predict there are ~ 2500 pulsars in the entire cluster system.

5. The Total Number of Cluster Pulsars

From the discussion in sections 3 and 4, we find that estimates for N_p, the total number of pulsars in the cluster system vary widely: from the low estimate of $200/\bar{f}$ (WvP) to $2500/\bar{f}$ (PK and JKP) with KNR and FG in between at $\sim 1000/\bar{f}$. Some of this difference arises because not all methods are attempting to infer the same quantity. There are two questions one can ask:

- What is the number of active pulsars, N_p^a (pulsars above a minimum luminosity threshold)?
- What is the total number of pulsars produced in the cluster system so far, N_p^t?

Some general comments are in order. WvP and FG are both attempts to obtain N_p^a, whereas the birthrate method yields N_p^t. KNR, by explicitly assuming a model for the birth of pulsars, attempt to estimate N_p^t. For this reason, any method which utilizes the KNR luminosity function essentially estimates N_p^t; this comment applies in particular to the JKP scheme. The KNR luminosity function results in effectively a $\gamma = 2$ luminosity function and thus this comment is applicable to results derived using this luminosity function. Finally, $N_p^t > N_p^a$ which explains why the birthrate, KNR and JKP estimates are systematically larger than WvP.

The principal weakness of all luminosity methods is their sensitivity to the choice of L_{min}. Let us consider the case where the observed luminosity function arises due to luminosity decay in time. Then *a cutoff of L_{min} implies a cutoff in time.* For example, let the pulsar's luminosity decay as $L(t) = L_0 t_d (t + t_d)^{-\delta}$; t_d is the decay timescale, which we (arbitrarily) choose to be 10^7 y. This corresponds to a differential luminosity function $dN/dL \propto L^{-(1+\delta)/\delta} = L^{-\gamma}$. Then with $\delta = 1$ (or $\gamma = 2$), after 10^{10} y, the pulsar has decayed to 10^{-3} of its original luminosity. With $\delta = 1.5$ ($\gamma = 5/3$) the same pulsar would have decayed to a similar luminosity in only 10^9 y; with $\delta = 2$ ($\gamma = 3/2$), the pulsar is only 3×10^8 y old.

This means that the figure of 200 pulsars that WvP estimate for the cluster population, using $\gamma = 3/2$ ($\delta = 2$), refers only to pulsars born in the last $\sim 3 \times 10^8$ y. This is illustrated in Table 2 where we show the results of a simulation in which pulsars were generated every τ_x y, subject to Poisson statistics between now ($t = 0$) and a Hubble time

Table 2.
Luminosity function simulations

Decay law	dN/dL	τ_x (10⁷ y)	$L_{min} = 10^{-3}L_0$ f_1 10%	50%	90%	f_{12} 10%	50%	90%	n_{psr}	L_{min} free f_1 10%	50%	90%	f_{12} 10%	50%	90%	n_{psr}
t^{-1} L^{-2}		3.3	0.1	0.2	0.3	1.0	1.5	3.9	300	0.1	0.2	0.3	1.0	1.8	4.6	300
		10	0.1	0.2	0.5	1.1	2.0	5.8	100	0.1	0.2	0.5	1.1	1.9	6.4	100
		33	0.1	0.3	0.7	1.1	2.0	9.4	30	0.1	0.3	0.7	1.1	1.9	11.8	30
		—	0.2	0.4	0.7	1.1	2.0	10.1	10	0.2	0.4	0.7	1.1	2.1	8.0	10
$t^{-1.5}$ $L^{-5/3}$		3.3	0.2	0.5	0.7	1.0	1.0	1.0	30	0.2	0.4	0.7	1.0	2.3	10.7	300
		10	0.3	0.5	0.9	1.0	1.0	1.0	10	0.2	0.5	0.8	1.2	2.7	17.7	100
		33	0.4	0.7	1.0	1.0	1.1	1.3	3	0.2	0.5	0.9	1.2	2.7	48.9	30
		—	*	0.7	1.0	1.1	—	—	1	0.3	0.5	0.9	1.1	3.0	24.0	10
t^{-2} $L^{-3/2}$		3.3	0.3	0.6	0.9	1.0	1.0	1.0	10	0.3	0.6	0.9	1.0	3.3	26.7	300
		10	0.4	0.8	1.0	1.0	1.0	—	3	0.3	0.6	1.0	1.2	3.9	50.9	100
		33	*	0.7	1.0	1.0	—	—	1	0.3	0.6	1.0	1.2	3.8	190.	30
		—	*	*	1.0	—	—	—	0.3	0.3	0.6	1.0	1.2	4.3	70.1	10
e^{-t/t_d} L^{-1}		3.3	*	0.9	1.0	—	—	—	2	0.5	0.9	1.0	1.0	7.4	—	300
		10	*	0.4	1.0	—	—	—	0.7	*	1.0	1.0	2.7	—	—	100
		33	*	*	1.0	—	—	—	0.2	*	*	1.0	403	—	—	30

* Total integrated flux from cluster too small
— $f_{12} > 100$

($t = -10^{10}$ y). Each pulsar is born with a luminosity L_0, and decays according to the decay law indicated in the first two columns. We have tabulated the values of f_1 and f_{12}, the ratio of the brightest pulsar to the total cluster flux, and of the brightest pulsar to the second brightest, for the 10th, 50th and 90th percentiles. The first section shows the result when we assume $L_{min} = 10^{-3}L_0$; the second is where we do not constrain L_{min}.

In order to satisfy the observed $f_1 \sim 0.4$ to 0.7, it can be seen that for $\gamma < 2$ luminosity functions, large values of N_p are indicated when the luminosity is allowed to decay with no lower threshold. Only for the $\gamma = 2$ luminosity law do we find the inferred N_p is the same in either case. If the luminosity function is steeper than $\gamma = 2$ then the f_1 test shows that typically clusters cannot have more than 10–30 pulsars whereas for steeper luminosity laws the number of pulsars per cluster is large and somewhat unconstrained (30 to 300, depending on choice of L_{min}).

From Table 2, we can also conclude that the $\gamma = 2$ luminosity function is the only one luminosity function for which $N_p^a \sim N_p^t$. For other choices ($\gamma < 2$), we find $N_p^a \ll N_p^t$. This explains the agreement between FG, KNR and JKP.

It still remains to be understood why WvP and the other estimates do not agree on N_p for the choice of $\gamma = 2$, the preferred luminosity function of KNR, FG and JKP. This could be small number statistics or due to subtle selection effects. This is especially important since the WvP data were a subset of the FG data, an already biased sample. For $\gamma < 2$

luminosity functions we find from our simulations a strong correlation between f_1 and L_{tot}, the total luminosity, which implies that clusters selected on the basis of their luminosity will not be a fair sample. Clearly, these issues only highlight the importance of expanding the application of the powerful f_1 test from the current sample of 2 to a larger number.

To summarize: The currently favored luminosity function, from a variety of evidence, is $dN/dL \propto L^{-2}$ and the choice of the weighting function is $\propto M_c \rho_c^{0.5}$. With this choice of luminosity function and weighting function, the inferred number of pulsars produced in the cluster system (active or otherwise) range from $N_p^t \sim 500/\bar{f}$ to $2500/\bar{f}$. The high observed birthrates favor the larger numbers. This appears to be confirmed in the case of 47 Tuc and Terzan 5 where either a large number of pulsars have been discovered or a large number is indicated.

6. Birthrates

Having estimated the total number of pulsars, we now estimate their birthrate, and compare this to the birthrate of the LMXBs. If the latter are the progenitors of recycled pulsars, then the birthrates should be equal: $B_x = N_x/\tau_x = B_p = N_p/\tau_p$, where N and τ are the number of objects and their lifetimes.

For pulsars, we need to be careful whether we are using N_p^t or N_p^a. If the former, then the appropriate age to use is half a Hubble time; if the latter, we need to use the mean age of the observed pulsars. For the KNR luminosity function, which explicitly assumes pulsars are visible for a Hubble time (since pulsars never reach the death line), we use $\tau_p \sim 5 \times 10^9$ y, the mean age.

We assume $\bar{f} = 1$, and adopt a value $N_p^t \sim 2000$, consistent with both KNR and JKP, and favored by the birthrates. Then we obtain a birthrate of $B_p \sim 2000/\tau_p \sim 4 \times 10^{-7}$ y^{-1}. Since there are 12 LMXBs known in globular clusters (see Table 1), then $B_x \sim 12/\tau_x$. We can interpret this in two ways. If we assume the canonical lifetime for LMXBs of $\tau_x \sim 10^9$ y (Verbunt 1988), then $B_x \sim 10^{-8}$ y^{-1}, which implies $N_p \sim 60$, a value much lower than any of the estimates, which would lead us to conclude that LMXBs are not the progenitors of the recycled pulsars. On the other hand, letting τ_x be a free parameter, we find the mean duration of the X-ray phase of LMXBs to be $\tau_x \sim 12/4 \times 10^{-7} \sim 3 \times 10^7$ y. Such a value is not implausible: in this time, a neutron star accreting at the Eddington limit can accrete $0.3 - 0.4$ M$_\odot$ of material, sufficient to spin it up to a period of a few ms. This requires, however, that most cluster LMXBs be accreting near the Eddington luminosity. In fact, cluster LMXBs have luminosities between 10^{36} and 10^{37} erg s^{-1}, a factor of $10 - 100$ below Eddington (Verbunt et al. 1984). This limits the resulting pulsar period to several tens of milliseconds, or implies that not all LMXBs result in spun-up pulsars.

Table 1 shows that the birthrate problem is even worse than this. If we divide clusters by their metallicity, considering clusters with $[m/H] > -1.5$ and $[m/H] < -1.5$ separately, the latter group contains only one LMXB compared to eleven, but an approximately equal number of pulsars. In these clusters, we must have $\langle \tau_p^{-1} \rangle \sim 10^3 \langle \tau_x^{-1} \rangle$. This requires that at least some LMXBs have lifetimes much shorter than 10^9 y. PK suggest that there might be two populations of LMXBs in globular clusters, one with a lifetime $\tau_x \sim 10^5$ y, and one with $\tau_x \sim 10^9$ y. They identify the former as products of interactions involving catastrophic destruction of the field star, resulting in a short-lived accretion phase (the "fast" channel). The outcome is a moderately spun up *single*, pulsar ($P \gtrsim 3$ ms). It is possible that this is the more likely outcome when a neutron star encounters a field star (Kochanek 1991). A small number of tidal captures lead to a long lived accretion phase and are identified with the observed LMXBs (the "slow" channel). Highly spun-up ($P \lesssim 3$ ms) pulsars in a *binary* system emerge from this channel. If the birthrate of the fast-channel products is 10^{-6} y^{-1} while the birthrate of the slow-channel products is 10^{-8} y^{-1}, then we would not expect to see any of the former, even though they are responsible for the formation of 99% of the pulsars.

This scenario agrees well with the current state of observations, where most pulsars are single and relatively long period, and can therefore be identified with products of the fast channel whose progenitors are too short-lived to be visible today. However, the eleven pulsars in 47 Tuc all have periods less than 6 ms, and more than half are in binary systems (Manchester et al. 1991), suggesting they are products of the slow channel. This raises the birthrate problem again.

These results suggest there is another parameter involved in determining the number of pulsars in clusters. The most likely candidate is a time dependence of the rate of pulsar formation (corresponding to relaxing the assumption that T is constant in §1). Sigurdsson and Phinney (1991) suggest that during the onset of core-collapse, the pulsar formation rate is greatly accelerated. As the core starts to collapse, binaries are "burnt" in encounters, gradually hardening the binaries until the time-scale for interaction becomes very large or until an encounter ejects the binary from the core completely. At this stage core-collapse recommences, and further pulsar formation is unlikely. This model predicts that the number of pulsars in a cluster is strongly dependent on the past history of the cluster, in particular the time since core collapse. Further evidence is suggested by clusters like Liller 1, a dense, core-collapsed cluster, whose integrated flux is very small (Johnston et al. 1991). In this model, Liller 1 would represent a cluster long past core collapse, whose pulsars have faded to lower luminosities.

References

Anderson, S., Gorham, P., Kulkarni, S.R., Prince, T. & Wolszczan, A. 1990. Nature 346 42

Anderson, S., Kulkarni, S.R., Prince, T. & Wolszczan, A. 1990, *I.A.U. Circ.* #5013.

Bailyn, C.D. & Grindlay, J.E. 1990, *Ap. J.*, **353**, 159.

Biggs, J.D., Lyne, A.G. & Brinklow, A. 1989, in *Timing Neutron Stars*, eds. H. Ögelman, E.J.P. van den Heuvel, Kluwel, p. 157-162.

Biggs, J.D., Lyne, A.G., Manchester, R.N. & Ashworth, M. 1990, *I.A.U. Circ.* #4988.

Chernoff, D.F. & Djorgovski, S.G. 1989, *Ap. J.*, **339**, 904.

Chernoff, D.F. & Weinberg, M.D. 1990, *Ap.J.*, **351**, 121.

D'Amico, N., Lyne, A.G., Bailes, M., Johnston, S., Manchester, R.N., Staveley-Smith, L., Lim, J., Fruchter, A.S. & Goss, W.M. 1990, *I.A.U. Circ.* #5013.

Elson, R.A.W., Fall, S.M. & Freeman, K.C. 1989, *Ap. J.*, **336**, 734.

Foster, R.S., Fairhead, L. & Backer, D.C. 1991, *Ap. J.*, submitted.

Fruchter, A.S. & Goss, W.M. 1990, *Ap. J. Letters*, **365**, L63 (FG).

Goodman, J., talk presented at AAS Meeting, Philadelphia, Jan. 1991.

Hamilton, T T., Helfand, D.J. & Becker, R.H. 1985, *A. J.*, **90**, 606.

Hertz, P., Norris, J.P., Wood, K.S., Vaughan, B.A. & Michelson, P.F. 1990, *Ap.J.*, **354**, 267.

Hut, P., Murphy, B.W. & Verbunt, F. 1991, *Astron. Astrophys.*, **241**, 137.

Johnston, H.M. & Kulkarni, S.R. 1991, *Ap. J.*, **368**, 504.

Johnston, H.M., Kulkarni, S.R. & Goss, W.M. 1991, *Ap. J.*, in press.

Johnston, H.M., Kulkarni, S.R. & Phinney, E.S. 1991, *in prep* (JKP).

Katz, J.I. 1975, *Nature*, **253**, 698.

Kochanek, C.S. 1991, *Ap. J.*, submitted.

Kulkarni, S.R., Anderson, S.B., Prince, T.A. & Wolszczan, A. 1991, *Nature*, **349**, 47.

Kulkarni, S.R., Goss, W.M., Wolszczan, A. & Middleditch, J. 1990, *Ap. J. Letters*, **363**, L5.

Kulkarni, S.R., Narayan, R. & Romani, R.W. 1990, *Ap. J.*, **356**, 174 (KNR).

Lyne, A.G. & Biggs, J.D. 1989, *pers. comm.*

Lyne, A.G., Biggs, J.D., Brinklow, A., Ashworth, M. & McKenna, J. 1987, *Nature*, **332**, 45.

Lyne, A.G., Manchester, R.N., D'Amico, N., Staveley-Smith, L., Johnston, S., Lim, J., Fruchter, A.S., Goss, W.M. & Frail, D. 1990, *Nature*, **347**, 650.

Manchester, R.N., Lyne, A.G., Johnston, S., D'Amico, N., Lim, J., Kniffen, D.A., Fruchter, A.S. & Goss, W.M. 1989, *I.A.U. Circ.* #4905.

Manchester, R.N., Lyne, A.G., Johnston, S., D'Amico, N., Lim, J. & Kniffen, D.A. 1990, *Nature*, **345**, 598.

Manchester, R.N., Lyne, A.G., Robinson, C., D'Amico, N., Bailes, M. & Lim, J. 1991, *Nature*, **352**, 219.

Mateo, M. 1988, *Ap. J.*, **331**, 261.

McClure, R.D., Vandenburg, D.A., Smith, G.H., Fahlman, G.G., Richer, H.B., Hesser, J.E., Harris, W.E., Stetson, P.B. & Bell, R.A. 1986, *Ap. J. Letters*, **307**, L49.

McKenna, J. & Lyne, A.G. 1988, *Nature*, **336**, 226.

Middleditch, J. & Priedhorsky, W.C. 1986, *Ap. J.*, **306**, 230.

Narayan, R. & Ostriker, J.P. 1990, *Ap. J.*, **352**, 222.

Paczyński, B. 1990, *Ap. J.*, **348**, 485.

Phinney, E.S. 1991, *Mon. Not. R. astr. Soc.*, in press.

Phinney, E.S. & Kulkarni, S.R. 1991, *Nature*, submitted (PK).

Predehl, P., Hasinger, G. & Verbunt, F. 1991, *Astron. Astrophys.*, **246**, L21.

Prince, T., Anderson, S.B., Kulkarni, S.R. & Wolszczan, A. 1991, *Ap. J.*, submitted.

Pryor, C., McClure, R.D., Fletcher, J.M. & Hesser, J.E. 1989, *A. J.*, **98**, 596.

Richer, H.B. & Fahlman, G.G. 1989, *Ap. J.*, **339**, 178.

Richtler, T., de Boer, K.S. & Sagar, R. 1991, *ESO Messenger*, **64**, 50.

Sigurdsson, S. & Phinney, E.S. 1991, *in prep.*

Verbunt, F. 1988, in *The Physics of Compact Objects: Theory vs. Observation, COSPAR/IAU* Symposium, Sofia July 1987, eds. N.E. White, L. Filipov, Pergamon Press, p. 529.

Verbunt, F. & Hut, P. 1987, in *The Origin and Evolution of Neutron Stars, IAU Symp. No. 125*, eds. D.J. Helfand, J.-H. Huang, Reidel, Dordrecht, p. 187.

Verbunt, F., van Paradijs, J. & Elson, E. 1984, *Mon. Not. R. astr. Soc.*, **210**, 899.

Webbink, R.F. 1985, in *Dynamics of Star Clusters*, IAU Symp. **113**, eds. J. Goodman, P. Hut, Reidel, p. 541.

Wijers, R.A.M.J. & van Paradijs, J. 1991, *Astron. Astrophys.*, **241**, L37 (WvP).

Wolszczan, A., Kulkarni, S.R., Middleditch, J., Backer, D.C., Fruchter, A.S. & Dewey, R.J. 1989, *Nature*, **337**, 531.

CVs in Globular Clusters: Clues to Compact Binary Production

JONATHAN E. GRINDLAY
Harvard Observatory
60 Garden St.
Cambridge, MA 02138, USA

ABSTRACT. Cataclysmic variables (CVs) ought to be the most numerous compact binaries in globular clusters, given standard theories for their formation by capture or exchange and the large excess of white dwarfs over neutron stars expected in globulars. Yet their numbers thus far detected are vastly surpassed by the numbers of either Low Mass X-ray Binaries (LMXBs) or their possible successors, millisecond pulsars (MSPs). Where are the cluster CVs ? Here we briefly describe both the previous detections and limits, the ongoing searches and several observations of LMXBs which may relate to the cluster CV enigma.

1. Introduction

For the past 15 years it has been clear that globular cluster cores are remarkably efficient in producing compact low mass x-ray binary sources (LMXBs), essentially all of which are also x-ray burst sources and are thus neutron stars in binary systems. For the past three years it has become clear that globular clusters are at least equally prolific factories for producing millisecond pulsars (MSPs), and that the number and distributions of MSPs vs. LMXBs, which have been supposed to be their forebearers, can constrain the origin and evolution of each (e.g. Grindlay and Bailyn 1988, Kulkarni, Narayan and Romani 1990). For the past one year it has become clear that globular clusters are not so devoid of primordial binaries as previously supposed (e.g. Clark 1975), but rather that they may contain originally perhaps 10% of their stars in primordial binaries (Pryor et al 1991).

In this paper we review some of the observations and constraints on the numbers and formation of compact binaries and CVs in globular clusters. Our observations and analysis are still in progress, so that interim results are presented.

2. Previous Observations and New Considerations

Until recently, we have thought the total number of LMXBs in all globulars was "fixed" at about 10 and the number of possible CVs (e.g. low luminosity x-ray sources) was probably one to two orders of magnitude larger per cluster (cf. review by Grindlay 1988). However, much lower CV populations have been derived including effects of mass segregation (Verbunt and Meylan 1988) and stable mass transfer vs. cluster IMF (Bailyn, Grindlay and Garcia 1990=BGG). Recent considerations of "previous" observations may also change the conclusions for the total LMXB vs. CV populations.

365

E. P. J. van den Heuvel and S. A. Rappaport (eds.), X-Ray Binaries and Recycled Pulsars, 365–373.
© 1992 *Kluwer Academic Publishers.*

2.1 MORE LMXBS: M15 TRANSIENT SOURCE ?

Although Lightman and Grindlay (1982) concluded the probability of finding two active LMXBs within a single cluster was small, it is not negligible. The recent discovery that the bright LMXB identified with the optical star AC211 apparently underwent an x-ray burst (Dotani et al 1990) motivated us to reconsider this question. What if the burst were actually from a second source, fainter than AC211, in the cluster core ? This would remove the need for the surprising conclusion of Dotani et al that the persistent source is *not* in fact a high luminosity source obscured through an accretion disk corona (ADC). Rather, if the burst is actually from a second (weaker) source in the cluster, the ADC picture could be preserved. In this case the puzzling reduced persistent flux between the burst precursor and main peak would be due to AC211 alone; the actual burster persistent flux would have 'disappeared' (i.e. become too soft) as in other luminous bursts with precursors. An ADC model (e.g. Naylor et al 1988) is suggested by both the *presence* of smooth x-ray modulation (Hertz 1986) and not eclipses and the optical modulation as well as line spectrum.

The Einstein images of M15 which accurately located the LMXB (Grindlay et al 1984) do not show, when stacked, any obvious second source. However, due to the extremely compact core, a second source could be expected within just $\sim$3 arcsec of AC211. The FWHM of the Einstein HRI and mirror is comparable, so that a second source would only be detected if its flux were comparable to that from AC211. We estimate that a second source fainter than $\sim$0.3 of AC211 would be lost.

There is a possible second source in M15. On July 20, 1979, an apparent transient source was detected with the SSI instrument on the Ariel V satellite (Pye et al 1983). This source had duration of $\gtrsim$ 2 days in its partially observed decay and a maximum brightness of $\gtrsim$ 150 mCrab (it was detected only after the peak). Although this outburst could be due to AC211, its persistent flux is usually only about 5-10 mCrab and no such sustained outbursts by a factor of $\gtrsim$ 15 have been reported from this frequently observed LMXB before or since (to our knowledge). Thus it is possible that M15 contains (at least) one other LMXB, a transient source similar to the transient source previously found in the globular NGC 6440 or the field transient Cen X-4, which also undergoes x-ray bursts during its long-term outburst phases. Although the Ariel V "transients" of Pye et al (1983) contain a number of RS CVn stars, and these chromospherically active stars may now be indicated in M15 given the detection of Ca II H-K emission in the core of the cluster (Murphy et al 1991) and RS CVns have also been suggested as possible counterparts for some of the *low* luminosity cluster sources by BGG, this transient is much too luminous at maximum ($\gtrsim 10^{37}$ erg/s) to be an RS CVn in M15; it must be an LMXB. There is even yet another possible Ariel V transient: Pye et al (1983) also point out that NGC 6809 = M55 is in the error box of another of their (short duration) transients:

A1933-312, a 1.6 hour transient from 31 August 1977.

If future high resolution imaging observations (possibly ROSAT but probably AXAF required) show a second source in M15, it will not be so surprising given the high incidence of millisecond pulsars (MSPs) already found and the extremely high central density now inferred (Phinney 1991) for the central cusp. It would be significant in that, along with the other two new cluster LMXB "transients" apparently discovered with ROSAT (cf. section 3), the total number of LMXBs would increase (by 30 % at least !) and the possible discrepancy with MSP statistics reduced.

2.2 FEWER CVs: M5, M30, and ω-Cen RESULTS ?

Whereas the number of LMXBs may be going up, the number of confirmed CVs in clusters seems to be going down. Of the two dwarf novae (DN) proposed as cluster members, V101 in M5 (Margon et al 1981) and V4 in M30 (Margon and Downes 1983), the second has now been shown to be almost certainly a foreground object (Machin et al 1991). This is based on both photometry, which shows the object is at least two magnitudes brighter than typical DN in quiescence (or in outburst), its colors, and – most significantly – its radial velocity, which is inconsistent with the cluster velocity. If it is a cluster member, it is anomalously bright and is apparently being ejected from the cluster at a velocity $\sim+320$ km/s (though mass outflow, as likely indicated by the blueshifted He I lines in AC211/M15, cannot be excluded it would seem less likely since V4/M30 is redshifted with respect to the cluster).

Similarly, the searches for the optical counterparts for the low luminosity x-ray sources (Hertz and Grindlay 1983) in ω-Cen (Margon and Bolte 1987, Shara et al 1988) and 47 Tuc (Shara et al 1988) have been negative. These authors have concluded that the lack of detection of either variable (Shara et al) or sufficiently blue objects (Margon and Bolte) in the x-ray error circles limits the space density of cluster CVs to be no greater than the solar neighborhood value. If CVs were distributed like LMXBs, they might be expected to show instead a factor 100 enhancement over their galactic plane/bulge counterparts, so these limits are extremely important. However they should be treated with caution: the Margon and Bolte photometry concentrated on the EXOSAT error circles for the source positions, yet we now note these are in significant disagreement ($\sim$30 arcsec) with the archived Einstein HRI positions for the sources 'A' and 'D' studied. (Revised EXOSAT positions, and new optical searches, have also been reported by Auriere and Koch-Miramond 1991). We are now examining our own CCD photometry to search for possible counterparts in the $\sim$5 arcsec HRI circles. The variability limits must also be treated with caution since the CCD limits of Shara et al are restricted in both time coverage (4 hours) and sensitivity ($M_B \lesssim 7.5$). Furthermore, while it is relatively easy to pick out dwarf novae such as V4/M30 at $\sim$50 core radii from the cluster center (or V101/M4 at $\sim$10 core radii), these objects are usually much fainter at maximum (e.g. $M_V \gtrsim 4$) than RR Lyrae stars ($M_V \sim 1$) and thus will be more difficult to find on historical plates in the central cores where even RR Lyrae counts are incomplete and where CVs are predominately expected. Thus Margon and Bolte's argument that CVs are limited by the lack of historical discoveries of dwarf novae on plates is overstated.

2.3 M31 NOVAE

On the other hand, novae are difficult to miss in globulars. Ciardullo et al (1989) have reported the negative results of a search for novae in M31 globulars based on 6 years of $H\alpha$ CCD monitoring of a sample of 52 globulars. They conclude that ~ 6 novae would have been expected if the novae were as over-abundant in globulars as LMXBs appear to be (in M31). This result is marginally significant but could be extended to a strong limit (with additional M31 cluster monitoring) on the number of cluster CVs producing novae. At face value, and with large uncertainties, it might suggest that the number of cluster CVs is $\lesssim 15$ times the galactic plane value, and that the early estimates of Hertz and Grindlay (1983) should be decreased by a factor of $\gtrsim 6$.

3. Current Observations and Limits

3.1 MAGNETIC CVs in 47 TUC and NGC 5824 ?

The central x-ray source in 47 Tuc (Grindlay et al 1984) is one of the two brightest low luminosity sources in the survey of Hertz and Grindlay (1983). The other is at a projected offset of 15 core radii from the center of the highly condensed, but distant, globular NGC 5824. We are accumulating optical data on both fields that suggests both objects must be magnetic CVs (i.e. either DQ Her or AM Her systems) if indeed they are cluster CVs. Such objects have been recently considered by Chanmugam et al (1991). Foreground sources (e.g. dMe stars) can probably be ruled out in both cases, and a background QSO is also unlikely (though a QSO was found near the NGC 5824 source IPC position but well outside the HRI position). Thus if both sources are indeed cluster members, they are most likely magnetic CVs or perhaps highly attenuated ADC/LMXB sources. Here we briefly review the arguments for each.

3.1.1. *47 Tuc.* A possible optical counterpart for this source has been found by Auriere et al (1989), whose analysis of the Einstein IPC data on the source also suggested a possible 120 sec quasi-periodicity and thus possibly a DQ Her type CV system. The optical object was found to be variable (it dimmed by $\gtrsim 1.5$ mag) by Bailyn and Cool (1990), further strengthening its probable identification with the x-ray source and the derived (Auriere et al 1989) constraint that $L_x/L_{opt} \gtrsim 1$. This value is consistent with that for a CV but not for a LMXB unless the LMXB is strongly attenuated by an accretion disk corona. Although such an ADC model is argued above (still) for the LMXB AC211 in M15, it seems less likely for the central 47 Tuc source since its optical absolute magnitude ($M_V \sim$ 3-4) would be faint for an LMXB (cf. van Paradijs 1983). Precisely the opposite argument applies if the system is a non-magnetic CV: its x-ray luminosity is so high that its L_x/L_{opt} value should be $\lesssim$ 0.01, not 1, given the general correlations for non-magnetic CVs found in Patterson and Raymond (1985). Thus the optical counterpart should be much brighter (by 5 magnitudes !) unless it differs greatly from the general trend. Since these L_x/L_{opt} correlations do not generally apply for magnetic CVs, which undergo accretion through a polar cap rather than through an optically thick boundary layer, a magnetic CV is consistent with the observations.

3.1.2. *NGC 5824*. A similar argument can be made for the HRI source in NGC 5824, for which no optical candidate is within the HRI error circle and is sufficiently bright that $L_x/L_{opt} \lesssim$ 0.1. We have noted (Grindlay 1988) the presence of an interesting uv-bright 21st mag candidate near the HRI position. Low resolution spectra obtained by us at CTIO reveal it to be an SdB star, without emission lines and thus similar to the several we have found in ω-Cen source error circles. However another interesting candidate has been found in our spectroscopy of other blue objects in the 10 arcsec radius circle: a probable barium star, or CH star, with Ba II (4554 Å) in absorption as well as strong CH absorption and $m_v \sim$ 18. Since McClure and Woodsworth (1990) have recently shown these stars in the field to be binary systems (with $\gtrsim$ 80 day periods for those measured) containing a subgiant and evolved component (white dwarf, whose envelope was transferred to the giant thereby polluting the latter with, e.g., barium), this system contains at least the fundamental ingredients for the source to be a high luminosity CV: a WD in a binary with a giant companion (with mass transfer now back onto the WD if indeed it is the x-ray counterpart). A full discussion of these observations is given in Grindlay et al 1991a.

The optical flux from the sub-giant CH star companion ($M_V \sim$ 0.5) will overwhelm the light expected from the accretion disk, since applying the expression given by Webbink et al (1987) for the absolute magnitude of a disk with an accretion rate of 6×10^{-10} $M_\odot$/yr onto a WD of mass 1.2 $M_\odot$ yields an expected absolute magnitude of $M_V \sim$ 5.7. This is fainter than the CH star by at least 4-5 magnitudes so that it would be lost against the giant continuum. However if the accretion rate were higher (as required for a less massive WD primary), then the disk luminosity would increase. In any event, to get the observed $\gtrsim 1.2 \times 10^{34}$ erg/s luminosity in the Einstein x-ray band (0.5 - 3.5 keV), the accretion must again probably be magnetic. A pure disk accretor (e.g. dwarf nova) would likely convert the entire accretion luminosity to XUV flux given the large optical depths expected in the boundary layer. Thus the lack of a luminous uv-bright optical counterpart suggests, like 47 Tuc, it is likely to be a DQ Her or possibly AM Her system. Our spectrum, which shows the Ba II absorption feature, shows no emission lines such as He II, however. This may be due to the overwhelmingly bright continuum (and absorption) spectrum of the giant companion. However, higher sensitivity spectra (with less contamination from other stars in the slit) shall be obtained to set more stringent limits on faint emission lines.

3.2 CV SEARCH BY Hα vs. R AND U IMAGING

As reported in Grindlay et al 1991b (GCB), we are carrying out a narrow band imaging study of several globular clusters to search for stars that are both bright in Hα and in the U band. Narrow band images have been obtained in Hα (30-50 Å) and in adjacent narrow ($\sim$70 Å) continuum bands, together with broad band U images, on 47 Tuc, NGC 6397 and NGC 6752 with sufficient total exposure times to reach $M_V \sim 8$ in each cluster. The goal is to find CV candidates, to be then confirmed spectroscopically, and to measure or limit the production of CVs in globulars. The 47 Tuc fields (4 fields, each 4 x 4 arcmin) cover a roughly 30 degree wedge from the cluster center out to 10 core radii and are approximately centered on regions of diffuse or multiple point source low luminosity x-ray emission as found by Hartwick et al (1982). The other two clusters, with smaller apparent core radii, were observed centered on the cluster centers again in a 4 x 4 arcmin field.

The initial analysis of these data has been done using the ratio technique, or color maps analysis, used by us (Bailyn et al 1988) in our study of high luminosity LMXB clusters. Our ratio analysis been able to reach the Hα vs. M_V sensitivity limits for NGC 6397 and NGC 6752 shown in GCB; the limits for 47 Tuc are very similar to those shown for NGC 6752. We have verified these ratio technique sensitivities for all three clusters by injecting artificial stars with a range of Hα equivalent widths (EW) and absolute magnitudes (Cool et al 1991). No objects with obvious Hα *and* U excesses show up in visual examination of the three clusters whereas about half of the control population of field CVs (cf. Figure 1 of GCB) would have been detected for *average* cluster surface brightness values. However, as Cool et al also show, this ratio or color map technique is primarily useful only for relatively less crowded regions of the frame(s), since an object with extreme colors can and will be lost if it is in the wings of superposed (brighter) stars. Thus full DAOPHOT photometry of the frames is required and is still in progress.

Initial DAOPHOT analysis of our largest core radius ($\sim$10 r_c) field in 47 Tuc shows that for stars bright enough ($m_v \lesssim 21$) that their Hα - R effective colors can be measured to $\lesssim 0.2$ mag uncertainty, there are no CV candidates in the field with EW(Hα) $\gtrsim 15$ Å and $M_V \lesssim 8$ together with large ($\gtrsim 0.5$ mag) U excess. (However there are two possible candidates with EW(Hα) ~ 15 Åand $M_V \sim 7$ but only $\lesssim 0.2$ mag uv excesses—enough to rule out dMe stars such as the Hα excess object shown in the image reproduced in Figure 2 of GCB and certainly our first candidates for spectroscopy.) Thus a conservative preliminary limit would be that in this field at $\sim$8-12 r_c there are no CVs brighter than $M_V \sim 7$ with EW(Hα) $\gtrsim 15$

Å. These boundaries would encompass more than half of the field CVs plotted in GCB. Thus our initial search is sensitive to most dwarf novae in outburst and to perhaps half "high" accretion rate DN in quiescence, for which Patterson (1984) gives a mean $M_V \sim 7.5$. Our final results should be $\sim$1-2 mag more sensitive for all three clusters, especially NGC 6397 with distance modulus 1.5 mag closer, and thus sensitive to even low luminosity DN in quiescence.

3.3 HST OBSERVATIONS

The Hα searches just described will be severely limited by cluster crowding in the central cluster cores ($\lesssim$ 1-2 r_c). For this, we require HST and shall in fact soon obtain Hα and R images for the core of NGC 6752. These observations should reach $M_V \sim 9$. We shall propose similar HST observations of both NGC 6397 and 47 Tuc to complete this study.

3.4 ROSAT OBSERVATIONS

Confirmation of the CV nature of Hα and U-bright objects requires optical spectroscopy, which will be difficult for the faintest and most crowded candidates. Detection of these objects with ROSAT soft x-ray images provides a complementary technique, which is of course also motivated by seeking to identify the low luminosity x-ray source class(es) in globulars. We eagerly await our HRI images obtained on NGC 6397 (and several other clusters) for comparison with the optical candidates and general study.

Preliminary results from the ROSAT survey already hint at the surprises to come: two new high luminosity LMXBs were reported in the globulars Terzan 6 and NGC 6652 (Predehl et al 1991). These are ostensibly "transient" sources in that their apparent luminosities are above the HEAO-1 survey limits of Hertz and Wood (1985). Their detections reinforce the notion introduced in section 2.1 above: that the total LMXB population in globulars may be significantly larger due to transient sources. Both of these two new suggested cluster sources should be confirmed (i.e. accurately positioned) with the ROSAT HRI along with the possible Ariel V transients in M15 and M55 suggested here.

4. Conclusions

It appears that the number of cluster LMXBs may be significantly larger ($\gtrsim$ 30%) than usually assumed due to the presence of even more x-ray transients. This would reduce the discrepancy with the MSP population in globulars (Grindlay and Bailyn 1988) and thus the need for alternative MSP production channels such as AIC. Reconsideration of earlier observations for counterparts of the low luminosity x-ray sources somewhat reduces the constraints they have placed on cluster CVs, but new observations will surpass these limits. Ground-based optical (e.g. Hα and U, as well as variability studies) observations for radii $\gtrsim$ 2 core radii already suggest that CVs are reduced in their relative enhancement in globulars over their numbers in the galactic plane by factors of $\gtrsim$ 2-6 below the relative enhancement ($\sim$100) shown by both LMXBs and MSPs. HST observations will extend these limits into the inner core. ROSAT observations will provide not only a sensitive follow-up for any CV candidates, but also independently constrain the nature of the low luminosity cluster x-ray sources. The emerging CV deficit, if measured, would suggest that the mass segregation (Verbunt and Meylan 1988) and mass transfer considerations (BGG) are operative or that additional considerations are needed. One possibility, if the deficit is found to be extreme, is that CVs are preferentially destroyed upon capture or exchange formation vs. the same processes for the much less numerous neutron stars. This might occur if all such formation processes are inherently "dirty" and a brief period of Eddington-limited accretion luminosity, without formation of a giant envelope, is required to expel gas from the system. On the other hand, if luminous magnetic CVs such as possibly identified with the 47 Tuc and NGC 5824 sources are confirmed, other end-products may be possible. One such possibility is discussed by us (Grindlay 1991) for the possible interpretation of the super-soft ROSAT sources in the LMC/SMC reported by Trumper at this meeting.

I thank A. Cool for the Hα analysis and discussions, along with P. Callanan and C. Bailyn. The staff at CTIO provided their usual superb support for our optical search for CVs. This work was partially supported in part by NASA grants NAGW-624 and NAS8-30751.

5. References

1. Auriere, M. et al 1989, *Astr. Ap.*, **214**, 113.

2. Auriere, M. and Koch-Miramond, L. 1991, in *ASP Conf. Series*, Vol 13, 360.

3. Bailyn, C. et al 1988, *Ap. J.*, **331**, 301.

4. Bailyn, C., Grindlay, J. and Garcia, M. 1990, *Ap. J. Letters*, **357**, L35. (paper BGG).

5. Chanmugam, G. et al 1991, preprint.

6. Ciardullo, R. et al 1989, *BAAS*, **21**, No. 4, 1127.

7. Clark, G. 1975, *Ap. J. Letters*, **199**, L43.

8. Cool, A., Grindlay, J. and Bailyn, C. 1991, in preparation.

9. Dotani, T. et al 1990, *Nature*, **347**, 534.

10. Grindlay, J. 1988, in *Proc. IAU Symp. No. 126: Globular Cluster Systems in Galaxies*, J. Grindlay and A. G. D. Philip, eds., Dordrecht:Reidel, p. 347.

11. Grindlay, J. et al 1984, *Ap. J. Letters*, **282**, L13.

12. Grindlay, J. and Bailyn, C. 1988, *Nature*, **336**, 48.

13. Grindlay, J., Cool, A. and Bailyn, C. 1991a, in preparation.

14. Grindlay, J., Cool, A. and Bailyn, C. 1991b, in *ASP Conf. Series*, Vol 13, 396. (paper GCB)

15. Grindlay, J. 1991, in preparation.

16. Hartwick, F., Cowley, A. and Grindlay, J. 1982, *Ap. J. Letters*, **254**, L11.

17. Hertz, P. and Grindlay, J. 1983, *Ap. J. Letters*, **267**, L85.

18. Hertz, P. and Wood, K. 1985, *Ap. J.*, **298**, 95.

19. Hertz, P. 1986, *IAU Circular No. 4272*.

20. Kulkarni, S, Narayan, R. and Romani, R. 1990, *Ap. J.*, **356**, 174.

21. Lightman, A. and Grindlay, J. 1982, *Ap. J.*, **262**, 145.

22. Margon, B. and Bolte, M. 1987, *Ap. J. Letters*, **321**, L61.

23. Machin, G. et al 1991, *MNRAS*, in press.

24. McClure, R. and Woodsworth, A. 1990, *Ap. J.*, **352**, 709.

25. Murphy, B. et al 1991, *Nature*, **351**, 130.

26. Naylor, T. et al 1988, *MNRAS*, **233**, 285.

27. Patterson, J. 1984, *Ap. J. Suppl.*, **54**, 443.

28. Patteson, J. and Raymond, J. 1985, *Ap. J.*, **292**, 535.

29. Phinney, S. 1991, these proceedings.

30. Predehl, P., Hasinger, G. and Verbunt, F. 1991, *Astron. Ap.*, in press.

31. Pryor, C. et al 1991, in *ASP Conf. Series*, Vol 13, 439.

32. Shara, M. et al 1988, *Ap. J.*, **328**, 594.

33. Van Paradijs, J. 1983, in *Accretion Driven Stellar X-ray Sources* , (W.H.G. Lewin and E.P.J. Van den Heuvel, eds.), Cambridge Press, p. 189.

34. Verbunt, F. and Meylan, G. 1988, *Astr. Ap.*, **203**, 297.

EVAPORATION OF COMPANIONS IN VLMXBS
AND IN BINARY MILLISECOND PULSARS

J. SHAHAM
Physics Department and Columbia Astrophysics Laboratory
Columbia University
538 West 120th Street
New York, NY 10027
and
Institute for Theoretical Physics
University of California
Santa Barbara, CA 93106

I. Abstract. The principles underlying the process of formation of a wind from a stellar atmosphere by external heating are applied to binary companions of neutron stars (NSs) which are being heated by radiation from the NS in Very-Low-Mass X-ray Binaries (VLMXBs) and in binary millisecond pulsar (BMP) systems. Among others, the possibilities of companion evaporation and of self-excited X-ray systems is discussed. The fast changes in the binary period of the "windy" BMP 1957+20 and the nature of the newly discovered "windy" BMP 1744-24A are discussed as well.

II. Introduction

Several problems associated with our understanding of LMXBs and millisecond pulsars have pointed in the direction of companion winds even before the first such wind was discovered in PSR1957+20:

One problem was that of the overabundance of bright LMXBs shining at, or close to, Eddington luminosities. Theoretical models suggest that the lifetime τ of a LMXB at mass transfer rate $\dot{m}$ is a steep inverse power law in the companion mass m, $\tau \equiv m/|\dot{m}| \propto m^{-\alpha}$ (α is 11/3 for Gravitational Radiation Driven mass transfer from a degenerate companion). Therefore, most observed binary systems in which mass accretion onto a compact object occurs should have low mass-transfer rates. While this seems to be true for Cataclysmic Variables it is manifestly not so for LMXBs, many of which are very bright.

Another problem had to do with the spin-up scenario for millisecond pulsars: if these are indeed spun-up in LMXBs then their periods should slow down as $|\dot{m}|$ drops so that when they emerge as radio pulsars their period p will only be

$$p \sim (6 \text{ msec}) B_9^{6/7} \left(\frac{m}{.1 M_\odot} \right)^{-3(\alpha+1)/7}$$

E. P. J. van den Heuvel and S. A. Rappaport (eds.), X-Ray Binaries and Recycled Pulsars, 375–386.
© 1992 *Kluwer Academic Publishers.*

with B_9 their sufrace dipolar magnetic field in units of $10^9 G$. For $\alpha = 11/3, p \propto m^{-2}$, hence a slowdown to p values of order seconds could be expected. How do millisecond pulsars ever emerge from LMXBs?

A related problem was the solitary nature of PSR1937+214: a millisecond pulsar without a companion. If the spin-up scenario was right – how did the companion disappear?

Indications for the very low mass of the companion of Cyg X-3 put it deep inside its Roche Lobe (RL). Can one understand mass transfer without RL overflow? And in 1957+20 itself the companion seemed again to be sunk in its RL. If the only mechanism for mass transfer was RL overflow and 1957+20 was a spun-up pulsar, how was RL overflow terminated and how did the companion sink in its Lobe?

All of these problems can, in principle, be solved if radiation from the NS can induce a wind from the companion. When accretion from the wind onto the NS suffices to supply the energy of its radiation, bootstrapping can occur and luminosities can be driven up to the Eddington maximum until such time that the companion gets too far away and mass transfer rates drop by several orders of magnitude. The companion need not fill its RL for mass transfer to take place and may, occasionally, fully evaporate to leave behind a solitary millisecond pulsar.

Winds in close binaries with NSs seem, therefore, to play an important role in determining the behavior and evolution of these systems.

III. General Concepts

When mass is strictly conserved and when viscosity is neglected, one may write down three equations that govern steady wind flow in radial symmetry [1]: the continuity equation

$$\frac{d}{dr}(\rho u r^2) = 0, \tag{3.1}$$

Newton's second law

$$\rho u \frac{du}{dr} = -\frac{dP}{dr} - \rho \frac{d\Phi}{dr} \tag{3.2}$$

and the energy flow equation

$$\frac{1}{r^2} \frac{d}{dr} \left[\rho u r^2 \left(\frac{1}{2} u^2 + \omega + \Phi \right) \right] = H + \frac{1}{r^2} \frac{d}{dr} \left(r^2 \kappa \frac{dT}{dr} \right). \tag{3.3}$$

The symbols in (3.1) – (3.3) are as follows:

r radial distance from the stellar center
ρ wind density
u wind radial velocity
P pressure
T temperature
Φ gravitational potential
ω enthalpy per unit mass
H energy production rate per unit volume
κ thermal conductivity, assumed isotropic.

For an ideal gas the enthalpy function ω is

$$\omega = \frac{\gamma}{\gamma - 1}\frac{P}{\rho} = \frac{\gamma}{\gamma - 1}\frac{kT}{\mu} \tag{3.4}$$

where μ is the mass per pressure particle and γ is the polytropic exponent

$$\gamma = \left(\frac{\partial \log P}{\partial \log \rho}\right)_s .$$

The limit $\gamma \to 1$ would correspond to an isothermal flow, as only $\frac{d\omega}{dr}$ enters the equations of motion, hence $\gamma \to 1$ should signal $\frac{dT}{dr} = 0$ by (3.4). In general,

$$d\omega = Tds + \frac{1}{\rho}dP$$

where s is the entropy per unit mass.

Equation (3.2) can be rewritten as

$$\frac{d}{dr}\left(\frac{1}{2}u^2 + \omega + \Phi\right) = T\frac{ds}{dr} \tag{3.2a}$$

so that

$$T\frac{Ds}{Dt} = uT\frac{ds}{dr} = \Lambda + \frac{u}{F}\frac{d}{dr}\left(r^2\kappa\frac{dT}{dr}\right) \tag{3.5}$$

where

$$F \equiv \rho u r^2 \qquad (conserved\ by\ (3.1)), \tag{3.6}$$

$\frac{D}{Dt}$ is the comoving derivative and where $\Lambda \equiv \frac{H}{\rho}$ is the energy production rate per unit mass. Equation (3.5) shows the simple fact of life, that as a given chunk of matter moves downstream in the wind, its entropy increases (decreases) by the heating (or cooling).

The basic point to understand here is why would a wind form in the first place, and at what flux value, F, will it form when it does. We have discussed the $H = \kappa = 0$ case previously [2]; here we turn directly to the case $H > 0$. Equation (3.3) readily integrates to a generalized conservation law,

$$\mathcal{E}(r) = \mathcal{E}_0 + \int_{r_0}^{r} dr \frac{r^2 H(r)}{F} + \left[\frac{1}{\rho u}\kappa\frac{dT}{dr}\right]_{r_0}^{r}$$

$$= \mathcal{E}_0 + \int_{r_0}^{r} TdS \tag{3.7}$$

Consider then, first, the case of $\kappa = 0$. It is clear that for a heating function with a given profile $H = \tilde{H}g(r/r_0)$, a scaling law

$$F \propto \tilde{H}$$

holds, so that the wind flux will be proportional to the amount of net heating. A transonic solution readily exists, and satisfies the condition

$$u_0^2 = \frac{1}{2}r_0\Phi_0' + \frac{1}{2}(\gamma - 1)\frac{r_0\Lambda_0}{u_0} \tag{3.8}$$

where the index 0 denotes quantities at the sonic point and where $u_0 = c_0 \equiv \sqrt{\gamma(p_0/\rho_0)}$ is the flow speed there. With $\Phi = -\frac{Gm}{r}$, (3.8) can be rewritten as [2]

$$\eta_0^2 = 1 + \frac{x}{\eta_0} \tag{3.9}$$

where

$$\eta_0 = \frac{2u_0}{v_e} \qquad x = \frac{4r_0\Lambda_0(\gamma - 1)}{v_e^3} \qquad v_e = \left(\frac{2Gm}{r_0}\right)^{1/2} \tag{3.10}$$

(v_e is the escape velocity from r_0).

(3.9) restricts $\eta_0 \geq 1$; that, in turn, requires the temperature T_0 at the sonic point to satisfy

$$T_0 \geq \frac{\mu v_e^2}{4\gamma k}, \tag{3.11}$$

which is a powerful condition on T_0 and r_0.

x measures the ratio between the flow time scale and the heating time scale. We can distinguish between the strong heating (SH, $x \gg 1$) and weak heating (WH, $x \ll 1$) cases:

$$\eta_0 \simeq \begin{matrix} 1 & \qquad x \ll 1, & \text{WH} \\ \\ x^{1/3} & \qquad x \gg 1, & \text{SH} \end{matrix} \tag{3.12}$$

From Eqs (3.2a) and (3.5) we see that when the flow speed is still low, $\frac{dT}{dr} \sim \frac{\gamma-1}{\gamma}\frac{\mu}{k}\frac{\Lambda}{u}$ so the first thing that happens is a quick rise in temperature and subsequent decrease in density, while the pressure does not change by much. For SH, the distance δr traveled to reach $T_0 \sim \frac{\mu v_e^2}{4\gamma k}$ satisfies $\frac{\delta r}{R} \sim \frac{1}{x}$, where R is the radius of the star emitting the wind (i.e., the companion).

As one can see from (3.7), a non-negligible thermal conductivity will make the local effective heating drop due to its redistribution. This, in turn, could diminish the wind [3] for small Peclet numbers (i.e. for cases when the flow time is much longer than the heat conduction time). However, the effective heat conductivity in stellar atmospheres is quite sensitive to local magnetic fields [1] and it is therefore hard to assess its importance for VLMXBs [4].

IV. Cooling and the Concept of P_{min}

To model the role played by radiative cooling in the formation of a wind in a stellar atmosphere heated externally at a rate Λ_H, McCray and his collaborators used the expression [5], [6]

$$\Lambda_c = 3.2 \times 10^{12} PT^{-3/2}(T_R - T) \qquad \text{erg/gsec} \tag{4.1}$$

which approximates free-free (FF) processes in ionized H, excited via photospheric photons at temperature T_R. For fixed P, $-\Lambda_c$ maximizes at $T = 3T_R$, with

$$-\Lambda_{c,max} \cong 1.2 \times 10^{12} P T_R^{-1/2} \qquad \text{erg/gsec} \tag{4.2}$$

Thermal equilibrium can be maintained as long as there is at least one temperature T for which

$$\Lambda \equiv \Lambda_H + \Lambda_c = 0 \tag{4.3}$$

If Λ_H is constant, (4.3) has generally two solutions, unless

$$P < P_{min} \simeq 10^5 \Lambda_{H,15} T_{R,4}^{1/2} \left(\frac{\alpha_{FF}}{\alpha}\right) \qquad \text{dyne/cm}^2 \tag{4.4}$$

where the α's are the cooling coefficients, ($\alpha_{FF} \equiv 3.2 \times 10^{12} \text{cm}^3 (^\circ\text{K})^{1/2}/\text{g}$ for FF alone). Then, there is no solution. As one climbs up in the atmosphere at $u = 0$, (3.2) requires P to drop; but, once P_{min} is reached, $\Lambda = 0$ is impossible to maintain. Around that point is when a flow speed begins to build up and a wind begins to form. When Λ_H arises not from acoustic energy, say (like in the sun) but from external radiative heating, heating may become saturated (due to ionization) once a temperature T_s is reached, with $T_s \gg T_R$ (this condition is not really necessary – but it describes the reality of X-ray heating and simplifies the algebra). Thus [5], [6]

$$\Lambda_H = \begin{cases} \Lambda_{H,0}\left(1 - \frac{T}{T_s}\right) & T \leq T_s \\ 0 & T \geq T_s \end{cases} \tag{4.5}$$

In this case, one could maintain $\Lambda = 0$ below P_{min} only with a discontinuous jump of the temperature to values very close to T_s. Subsonic flow begins in this region, which becomes an *isothermal* flow at T_s. This simplifies the analysis considerably: The (isothermal) sound speed at the sonic point equals about half the escape velocity v_e from it and one obtains [5], [2]

$$\rho u \sim \frac{P_{min}}{v_e}. \tag{4.6}$$

This is why P_{min} is so important a concept, and its detailed calculation through the calculations of Λ_H and Λ_c is of utmost importance for wind strength determinations.

For a star of cross sectional area $\mathcal{A}$ which intercepts a total external luminosity $\hat{L}$,

$$\Lambda_H = \frac{\hat{L}}{\rho \mathcal{A} \ell} \tag{4.7}$$

where $\rho \ell$ is the column density absorbing the radiation,

$$\rho \ell \sim \frac{m_p}{\sigma}$$

with m_p the proton mass and σ the effective heating cross section in the stellar atmosphere. Thus,

$$\Lambda_{H,15} = 4 \times 10^{-16} \frac{\hat{L}}{\mathcal{A}} \left(\frac{\sigma}{\sigma_T}\right) \tag{4.8}$$

with σ_T the Thomson cross section. Since $\rho u \sim \frac{\dot{m}_w}{\mathcal{A}}$ with $\dot{m}_w$ the total wind mass outflow per unit time,

$$\dot{m}_w \simeq \rho u \mathcal{A} \sim 10^5 \Lambda_{H,15} T_{R,4}^{1/2} \times 2.3 \times 10^{-8} \left(\frac{R}{R_\odot}\right)^{1/2} \left(\frac{M_\odot}{m}\right)^{1/2} \left(\frac{\alpha_{FF}}{\alpha}\right)$$

or

$$\dot{m}_w \simeq 1.3 \times 10^{-8} \hat{L}_{36} \left(\frac{\sigma}{\sigma_T}\right) \left(\frac{R}{R_\odot}\right)^{1/2} \left(\frac{M_\odot}{m}\right)^{1/2} \left(\frac{\alpha_{FF}}{\alpha}\right) T_{R,4}^{1/2} \qquad M_\odot/\text{yr.} \qquad (4.9)$$

McCray, Sunyaev and their collaborators calculated Λ for Her X-1 in great detail [7],[8],[9],[10]; they found $\frac{\sigma}{\sigma_T} \simeq 15$, $\frac{\alpha_{FF}}{\alpha} \sim .1$, but various authors obtained values in a two-orders-of-magnitude range for $\left(\frac{\sigma}{\sigma_T}\right)\left(\frac{\alpha_{FF}}{\alpha}\right)$! In the end, the issue of having self-excited accretion in Her X-1 was regarded unsettled. However, chances for bootstrapping are much greater in VLMXBs [2].

V. The Slightly Underfilled Roche Lobe

One key feature distinguishes this configuration from that in which the companion is buried inside its lobe: when $z > 0$ (where the z coordinate is measured from the L1 point, along the line of centers and towards the NS), the wind is being helped by the gravity of the compact star, hence heating need only expand the atmosphere to the RL boundary to form a wind. Indeed, the sonic point condition (3.8) becomes

$$\Lambda_0 = \frac{f u_0 z_0}{\gamma - 1} \qquad (5.1)$$

showing that the sonic point has a positive z_0, hence it is located beyond L1 (right at L1 if $\gamma = 1$).

Again, two time-scales are of importance here [6]: that of getting to L1 (τ_F) and that of heating up to T_s (τ_H). High values for $\dot{m}$ will again be obtained when a situation similar to that of (3.11) exists, where here v_e is the velocity needed only to climb from the stellar surface to L1. When $\tau_F \gg \tau_H$, the flow will arrive at L1 essentially with Mach number $M = 1$ and speed $\left(\frac{kT_s}{\mu}\right)^{1/2}$. But the more interesting case is that of $\tau_F \ll \tau_H$. In this case the flow will arrive at L1 with a *low* Mach number and a *low* temperature, possibly close to $3T_R$.

If the pressure at the sonic point is (P_{min}/q) and the temperature $3\theta T_R$ $(q, \theta > 1)$, then

$$\rho u = \frac{\mu P_{min}}{3k\theta T_R q} \cdot \left(\frac{5 \cdot 3k\theta T_R}{3\mu}\right)^{1/2}$$

hence, by (4.4) and (4.8), we find

$$\dot{m}_w \simeq 3 \times 10^{-7} \hat{L}_{36} \left(\frac{\sigma}{\sigma_T}\right) \left(\frac{\alpha_{FF}}{\alpha}\right) \left(\frac{1}{q\sqrt{\theta}}\right) \qquad M_\odot/\text{yr} \qquad (5.2)$$

A comparison with (4.9) shows a possibility for higher $\dot{m}_\omega$ values here. We shall later adopt a representative value of

$$\dot{m}_\omega \simeq 4.5 \times 10^{-7} \hat{L}_{36} \qquad M_\odot/\text{yr} \tag{5.3}$$

based on $q\sqrt{\theta} \sim 10$ and $\left(\frac{\sigma}{\sigma_T}\right)\left(\frac{\alpha_{FF}}{\alpha}\right) \sim 15$, a value that is likely to hold for VLMXBs [2].

VI. VLMXB Bootstrapped Evolution

This being the case, VLMXBs ($m < 0.08M_\odot$) have a good chance of emitting self excited X-rays in a bootstrapping process, in which the X-rays are emitted due to the accretion of some of the wind that they themselves generate at the companion [11]. The self-excited luminosity will build itself up to the largest value permitted by the surroundings, possibly to close to the Eddington luminosity. The fraction of the wind leaving the system will determine the evolution of the companion inside its lobe. When the companion illumination becomes such that $\dot{m}_\omega$ decreases below the self-sustained value, $\dot{m}_\omega$ will drop abruptly – either to the $\dot{m}_\omega$ expected from the radiation of the neutron-star-turned-Radio-pulsar if the companion is inside its Lobe, or to the $\dot{m}_\omega$ value of RL overflow. In either case, a large luminosity gap will form in the LMXB statistics and millisecond pulsars will not lose their fast spins in a dwindling $\dot{m}$. Occasionally, a binary millisecond radio-pulsar with a "windy" companion could emerge [12], [13]; or a VLMXB will have the companion deep in its Lobe [4]; or the companion may be fully consumed, leaving behind a solitary millisecond pulsar.

To lowest order in $\frac{m}{M}$ (where M is the mass of the neutron star), the binary evolution is controlled by the following equations:

$$\dot{\eta}/\eta = \frac{2\dot{m}}{m}\left(1 - h + \frac{h\dot{m}_s}{|\dot{m}|} - \beta\right) \tag{6.1}$$

$$\dot{a}/a = \frac{2\dot{m}}{m}\left(\frac{h\dot{m}_s}{|\dot{m}|} - \beta\right) \tag{6.2}$$

where in (6.1) and (6.2), $\dot{m}\,(\equiv -|\dot{m}|)$ is the net mass loss rate from the companion, a is the binary separation and

$$\eta = \frac{R_L}{R_s} \tag{6.3}$$

with R_L the Lobe radius and R_s the stellar one. For R_s one usually adopts the relation

$$\log\left(\frac{R_s}{R_\odot}\right) = -n\log\left(\frac{m}{M_\odot}\right). \tag{6.4}$$

Also in (6.1) and (6.2), β is the fraction of $|\dot{m}|$ that ends up being accreted and

$$h = \frac{5}{6} - \frac{1}{2}n \tag{6.5}$$

where n is defined in Eq. (6.4). For $n = -0.82$, appropriate, e.g., to moderately massive MS stars in thermal equilibrium, $h = 1.24$. The quantity h is always > 0 but for convective stars, where the adiabatic index n_{ad} is $1/3$, h can have values between 1.24 and 0.67 depending on whether or not the mass transfer time is longer than thermal times [14]. Under standard Gravitational Radiation (GR) or Magnetic Breaking (MB) evolution or when the star loses mass by external heating, h will be close to 1.24 until nuclear burning shuts off.

Finally in (6.1) and (6.2),

$$\dot{m}_s = \dot{m}_{so} + (1 - \beta)(Q - 1)|\dot{m}|h^{-1}, \tag{6.6}$$

where $\dot{m}_{so}$ is the mass loss rate expected solely from GR or MB if the evolution were conservative ($\beta \equiv 1$) and if the companion were to fill its Lobe ($\eta \equiv 1$) at the present configuration, and where Q is the effective angular momentum per unit mass carried by the escaping particles, in units of the angular momentum per unit mass of the companion.

While it seems to be possible to maintain steady-state RL-overflow bootstrapping under a variety of relevant conditions when the companion is degenerate and h = 2/3, this is not the case when the companion is a Main Sequence (MS) star and when $h > 1$. If, in Eqs. (6.1) and (6.2), we assume that $|\dot{m}| \gg \dot{m}_{so}$, we have $\dot{\eta}/\dot{m}(\propto ((1 - \beta)Q - h)) < 0$ whenever $Q < h/(1 - \beta)$. This inequality can hold, in the bootstrapped state, for any suitable value of β, because values of Q that exceed unity by any significant amount are very hard to come by when simple flows are involved (the situation could be different during the *millisecond-radio-pulsar* phase, when shocks develop close to the companion, see below). Hence, even if at the beginning of the bootstrapped phase $\eta = 1$, we see that as m decreases, η *increases*, meaning $R_L > R_s$: the companion begins to sink inside its Lobe. We find that during a bootstrapping episode with $|\dot{m}| \gg \dot{m}_{so}$,

$$a \propto m^{2[Q(1-\beta)-1]}, \tag{6.7}$$

$$\eta \propto m^{2[Q(1-\beta)-h]} \quad (\equiv 1 \text{ for RL overflow }) \tag{6.8}$$

and

$$\Gamma \propto m^{4h-4Q(1-\beta)+2/3} \quad (\propto m^{2/3} \text{ for RL overflow }), \tag{6.9}$$

with $4\pi\Gamma$ the solid angle subtended by the companion star. Bootstrapping will cease when, at a mass $m = m_c$, Γ becomes too small to trap enough radiation to produce the necessary wind. The value of m_c depends on the value of the *initial* mass m_0, i.e. the mass of the companion when bootstrapping began and it was still filling its Lobe. We find

$$\nu_c^{4h-4Q(1-\beta)+2/3} = \frac{0.2}{\beta} \left(\frac{f\chi\xi}{0.06}\right)^{-1} M_{1.4}^{2/3} \nu_0^{4(h-Q(1-\beta))} \tag{6.10}$$

where $\nu_c = \frac{m_c}{0.1 M_\odot}$, $\nu_0 = \frac{m_0}{0.1 M_\odot}$, $M_{1.4}$ is the neutron stellar mass in $1.4 M_\odot$ units, f measures the efficiency of the wind formation process, χ is the fraction of the neutron-stellar radiation in the direction of the companion that is not obstructed and re-directed by matter, and ξ is the fraction of the rest mass energy of accreted matter that is converted to energy at the neutron stellar surface. Clearly, we must have $\frac{0.2}{\beta} \left(\frac{f\chi\xi}{0.06}\right)^{-1} \nu_0^{-2/3} < 1$ or else m_0 itself would already not satisfy the bootstrapping conditions (i.e. the value of Γ would be too small even when the companion is still filling its RL). As the radius of capture of matter moving at the escape velocity from the companion is so much in excess of the system's dimension, $\beta \sim 1$ seems to be a fair approximation during the bootstrapping episode. With $\beta = 1$, (6.10) reads

$$\nu_c = 0.75\nu_0^{0.88}\zeta^{-0.18} \tag{6.11}$$

VII. The "Windy" BMP 1957+20: Why does the Orbit Decay so Fast?

Not unlike the situation in several LMXBs, it now appears [15] that the orbital period-change time-scale in 1957+20 is also quite short: $|P_{orb}/\dot{P}_{orb}| \sim 3 \times 10^7$ yrs and, in this case, the binary period *decreases*; hence (unless center-of-mass accelerations are involved), the orbit *decays*. To understand the meaning of such behavior let us write, this time to *first* order in $\frac{m}{M} \equiv q$,

$$\frac{\dot{P}_{orb}/P_{orb}}{\dot{m}/m} = 3 \times \left[(1-\beta)(1-q)(Q-1) - \beta(1-q) - \frac{2}{3}(1-\beta)q + \frac{\dot{m}_{so}(h-q)}{|\dot{m}|}\right], \tag{7.1}$$

whence the condition for $\dot{P}_{orb} < 0$ becomes

$$Q > \frac{1}{1-\beta}\left[1 - \frac{\dot{m}_{so}\{h(1+q)-q\}}{|\dot{m}|}\right] + \frac{2}{3}q \tag{7.2}$$

when $\beta < 1$, and

$$|\dot{m}| < \dot{m}_{so}[h(1+q)-q] \sim h\dot{m}_{so} \tag{7.3}$$

when $\beta \equiv 1$.

In the latter case, (7.3) clearly sets the spin-up time scale to be no shorter than the one that would have obtained with RL- filling GR- or MB- driven mass loss. The inequality could be satisfied when the companion is buried within its Lobe and a wind is excited off it (that remains bound to the binary) or, if it RL overflows, when $h > 1$ (i.e., when it is not totally degenerate or convective). But when some mass loss from the system does occur, what does (7.2) mean?

For a wind that is leaving the companion through the part of the surface facing the NS and is carrying to infinity solely the angular momentum its particles had while on the companion, $Q < 1$ if the companion rotates synchronously, because these particles had the same orbital frequency as the companion but were moving at a *smaller* average distance from the center-of-mass. A retrograde companion spin may push Q higher, if the spin is fast enough. Naturally, the particles in the wind would acquire a higher (or lower) Q *after* they've left the companion, in a process similar to the one that must be taking place in accretion disks, where most particles lose their angular momentum to tidal torques *and* to a few non-accreting particles that escape to infinity or return to the companion with a high Q (but probably still with $(1-\beta)Q < 1$). The jury is still out on the details of such a process. One could also make, quite generally, the following argument: for a particle to escape from the binary, it should, by energy conservation, start out with a velocity that is at least

$\sqrt{2}v_o\left[1 + \frac{1}{2}\left(\frac{v_*}{v_o}\right)^2\right]^{\frac{1}{2}}$, with v_o the orbital velocity of the companion. If escaping particles would only start out moving also in the right *directions*, a suitable Q would certainly be obtained. That can actually happen in VLMXBs: escape velocities from companions there are of order of the escape velocity from the binary, so wind particles emitted in a general direction which is *opposite* to the motion of the companion may actually not escape at all but end up being recaptured. Thus, the wind would indeed be effectively emitted in the "right" directions, with $Q > 1$ [16].

For 1957+20, Eq.(7.2) actually translates to $Q \geq 1.01$. As we shall now see, however, the actual $Q - 1$ value required here is likely to be much higher, not unlike the values needed for $\beta < 1$ VLMXBs. The difference is, that in 1957+20 pulsar radiation pressure may actually drive *all* wind particles to infinity (the pulsar can certainly supply the needed escape energy). From the spin-down rate of the pulsar [15] we can estimate its luminosity and for a companion radius of $10^{10} cm$, Eq. (4.9) gives $\dot{m}_w \sim 2 \cdot 10^{15} g/sec$. This seems to be a rather optimistic value for the wind-flow rate, unless the effective area involved, $\mathcal{A}$, is larger than the companion's projected surface area (or else the companion is out of thermal equilibrium and happens to be close to filling its Roche lobe; then both $\mathcal{A}$ will increase by a factor of 4 and Eq. (5.3) will give another factor of ~ 3). The observed $\dot{P}_{orb}$ implies, by (7.2),

$$\left(\frac{\dot{m}}{2 \cdot 10^{15} g/sec} \right) (Q - 1.01) \simeq 8, \tag{7.4}$$

and unless $\dot{m}_w$ is 10-100 times higher than what (4.9) predicts, Q needs to be very large indeed.

Can observations point us towards a better estimate of $\dot{m}_w$? Given enough time they could actually give a *definite* answer [17], but that may take too long... Unless matter distribution in the system is very anisotropic and our line of sight is sufficiently above the orbital plane, the emergence of high frequency radio signals during eclipse time [18] would permit $\dot{m}_w$ to be ~ 10 times the above value but not much more, just by plasma frequency considerations. This leaves Q to be not much below 1.8.

We are presently carrying out calculations [16] to determine possible Q values in 1957+20. A clue can be found in observations of the structure of the bow-shock and the density discontinuities close to it: A region in the *back* of the companion, that is sufficiently close to it and carries enough mass, can contribute substantially to a high Q. Such a region will, on the one hand, exert a constant negative torque on the companion. On the other hand, its relatively higher density would mean a longer residence time for wind particles there, hence a chance for them to be torqued for a longer time by the reaction torque from the close companion, thus acquiring the higher Q. To produce the necessary asymmetry in the matter distribution around the companion, the latter should probably be *rotating*, in a prograde sense, a possible left-over from the very short bootstrapping era that sank the companion in its Lobe to begin with.

With the above Q values the companion will hit its Lobe and totally disrupt in less than $\sim 4 \cdot 10^7 yrs$ (see Eq. (6.8)).

VIII. The "Windy" BMP 1744–24A: A Test Lab for Accretion onto Millisecond Pulsars?

Eclipse durations of 1744–24A vary considerably [19]: eclipses covering as much as $\sim 180°$ in orbital phase and even disappearance of pulses for complete orbital periods have been observed. Given the characteristics of the eclipses Tavani and I suggested [2], [20] that the mass loss in this BMP is due again to pulsar radiation, which drives a wind from the illuminated atmosphere of the companion, possibly this time a $\sim 0.1 M_\odot$ cool MS star which slightly underfills its Roche Lobe. Radio signals may be both reflected/refracted and free-free absorbed in the cool streams. Unlike in 1957+20, however, the NS might attempt to accrete some of the $\sim 2 \times 10^{-12} M_\odot \mathrm{yr}^{-1}$ mass flow, since the pulsar pressure in this system is only marginally strong enough to expel the flow. In this case the NS could occasionally produce X-ray bursts, as the propeller effect [21] is likely to quench steady accretion; thus, an identification of PSR 1744–24A with the X-ray burster seen in the direction of Ter 5 [22] is possible. The system could have evolved from a NS $+0.14 M_\odot$ MS companion binary in a bootstrapped accretion process which spun the pulsar up to a ~ 6 msec period while pushing the companion out. We are now at the end of the evolutionary phase during which the system has been coming back into contact to become, again, a VLMXB. It is possible, in fact, that this system has already alternated more than once between a VLMXB and a BMP phase. Such evolutionary scenarios could make the system some $(5 - 10) \times 10^9$ years old. If our interpretation is correct, 1744–24A may turn out to be the best laboratory found to date to study the accretion process on a millisecond period neutron star.

A major ingredient of a scenario like this is the assumption that, during the present evolutionary phase, accretion has gone on at a rate $|\dot{m}|_s$ until some neutron star spin period p_c was reached (which was, of course, no faster than the equilibrium period for the rate $|\dot{m}|_s$), at which bootstrapping stopped and mass accretion dropped to a rate $|\dot{m}|_e$, which was sufficiently low that pulsar radiation pushed the wind out of the system. From then on, no steady accretion has yet begun. Also, p has slowed down to ~ 11msec in a time $\lesssim 10^{10}$yrs. These conditions produce the following constraints on the scenario:

$$0.33 \leq \mu_{27} \leq 10^{-3} \left(\frac{|\dot{m}|_s}{|\dot{m}|_e} \right) \left(\frac{|\dot{m}|_s}{2.3 \times 10^{-11} M_\odot \mathrm{yr}^{-1}} \right)^{1/2} \tag{8.1}$$

and

$$1 \text{ msec} \leq p_c \leq 0.08 \left(\frac{|\dot{m}|_s}{|\dot{m}|_e} \right)^{6/7} \text{ msec}, \tag{8.2}$$

where μ_{27} is the magnetic moment of the NS in units of $10^{27} G cm^3$ and where the l.h.s. of (8.2) comes from rotational stability requirements. The present upper limit on μ_{27} from spin-down measurements is 0.7, quite consistent with (8.1).

IX. Acknowledgements

It is a pleasure to thank the directors of this ASI for their kind hospitality and for partial support. NASA support for travel, preparing the lecture and manuscript and for carrying out some of the research described here through grant NAGW-1618, is gratefully acknowledged. I thank Menashe Banit and Marco Tavani for many helpful discussions. This is contribution number 457 of the Columbia Astrophysics Laboratory.

X. References

[1] Holzer, T.E. and Axford, W.I., Ann. Rev. Astr. Ap., 8, 31-60 (1970).
[2] Shaham, J., in Proceedings of "Neutron Stars - an Interdisciplinary Field", a NATO ASI, Crete 1990 (in press).
[3] Eichler, D. and Levinson, A., Ap.J. (Letters), 335, L67-L70 (1988).

[4] Tavani, M., Ruderman, M., and Shaham, J., *Ap.J. (Letters)*, **342**, L31-L34 (1989).

[5] London, R., McCray, R. and Auer, L.H., *Ap.J.*, **243**, 970-982 (1981).

[6] London, R. and Flannery, B.P., *Ap.J.*, **258**, 260-269 (1982).

[7] Basko, M.M., Hatchett, S., McCray, R., and Sunyaev, R.A., *Ap.J.*, **215**, 276-284 (1977).

[8] Basko, M.M., and Sunyaev, R.A., *Astr. Sp. Sci.*, **23**, 117-158 (1973).

[9] Buff, J. and McCray, R., *Ap.J.*, **189**, 147-155 (1974).

[10] McCray, R. and Hatchett, S., *Ap.J.*, **199**, 19s6-205 (1975).

[11] Ruderman, M., Shaham, J., Tavani, M., and Eichler, D., *Ap.J.*, **343**, 292-312 (1989).

[12] Ruderman, M., Shaham, J., and Tavani, M., *Ap.J.*, **336**, 507-518 (1989).

[13] Kluzniak, W., Ruderman, M., Shaham, J., and Tavani, M., *Nature*, **334**, 225-227 (1988).

[14] Rappaport, S., Verbunt, F., and Joss, P.C., *Ap.J.*, **275**, 713-731 (1983).

[15] Taylor, J.H., this volume.

[16] Banit, M. and Shaham, J., in preparation; Banit, M., Shaham, J. and Tavani, M., in preparation.

[17] Chen, K., *Nature*, **341**, 576 (1989).

[18] Fruchter, A.S., this volume.

[19] Lyne, A.G., Manchester, R.N., D'Amico, N., Staveley-Smith, L., Johnston, S., Lim, J., Fruchter, A.S., Goss, W.M., and Frail, D., *Nature*, **347**, 650-652 (1990).

[20] Shaham, J. and Tavani, M., *Ap.J.*, (in press, 1991).

[21] Illarionov, A.F. and Sunyaev, R.A., *Astr. Ap.*, **39**, 185-195 (1975).

[22] Makishima, K. *et al.*, *Ap.J. (Letters)*, **247**, L23-L25 (1981).

RADIATION-DRIVEN EVOLUTION OF LOW-MASS X-RAY BINARIES AND THE FORMATION OF MILLISECOND PULSARS

MARCO TAVANI[1]
University of California
Institute of Geophysics and Planetary Physics, LLNL, CA 94550
and Department of Astronomy, UC Berkeley, CA 94720.

ABSTRACT. Recent data on low-mass X-ray binaries (LMXBs) and millisecond pulsars (MSPs) pose a challenge to evolutionary theories which neglect the effects of disk and companion irradiation. Here we discuss the main features of a radiation-driven (RD) evolutionary model that may be applicable to several LMXBs. According to this model, radiation from the accreting compact star in LMXBs 'vaporizes' the accretion disk and the companion star by driving a self-sustained mass loss until a sudden accretion-turn off occurs. The main characteristics of the RD-evolution are: (1) the lifetime of RD-LMXB's is of order 10^7 years or less; (2) both the orbital period gap and the X-ray luminosity may be consequences of RD-evolution of LMXB's containing lower main sequence and degenerate companion stars; (3) the companion star may transfer mass to the primary even if it underfills its Roche lobe; (4) a class of recycled MSPs can continue to vaporize the low-mass companions by a strong pulsar wind even after the accretion turn-off; (5) the RD-evolutionary model resolves the apparent statistical discrepancy between the number of MSPs and their LMXB progenitors in the Galaxy. We discuss the implications of the discovery of single MSPs in low-density globular clusters and the recent measurements of short orbital timescales of four LMXBs.

1. Introduction

Although there are approximately 100 known low-mass X-ray binaries (LMXB's) in the Galaxy (and ~ 10 in globular clusters), in many cases the nature of the primary and companion stars, the mass transfer mechanism, and the lifetimes remain elusive. Evolutionary scenarios neglecting the effect of radiation from the primary star have been guiding the study of LMXB's, and have accounted for the properties of several LMXB's studied *individually* (e.g., [1]). Depending on the nature of the Roche lobe-filling companion star, mass transfer may be driven by angular momentum loss caused by gravitational radiation (GR), magnetic braking (MB), or core expansion for sub-giant companions. In the 'standard model' the mechanism of mass transfer is independent of the nature of the primary star, and works in the same way for binaries containing black holes, neutron stars and white dwarfs (cataclysmic variables, CVs). However, several questions arise both from the *statistical* features of LMXB's, and from new data on orbital timescales: (1) *Why are most of the LMXB luminosities close to the Eddington limit of one solar mass primary ?* The luminos-

---
[1]Affiliated with the International Center of Relativistic Astrophysics (ICRA)

E. P. J. van den Heuvel and S. A. Rappaport (eds.), X-Ray Binaries and Recycled Pulsars, 387–399.
© 1992 *Kluwer Academic Publishers.*

ity distribution of LMXBs with steady emission is peaked near the maximum possible value rather than toward the lower end of the luminosity range. The observed distribution is the opposite of what expected in the standard model which predicts that the X-ray sources spend most of their time at low values of the luminosity [1]. (2) *Why have there been no observed steady LMXB's with persistent emission in the range* $L_X \sim 10^{36} - 10^{34.5}$ *erg* s^{-1} *?* In what follows we will refer to this luminosity range as the 'luminosity gap'. (3) *Why is the orbital period distribution of LMXB's different from the corresponding distribution of CV's?* Fig. 1 gives the orbital period distributions for LMXBs and CVs. The CV 'period gap' is between 2 hrs and 3 hrs with a significant peak near $P_{orb} \simeq 2$ hrs. The LMXB period gap is wider than the CV gap with no galactic binary between 1 hr and ~ 3 hrs. (4) *Is the birthrate of millisecond pulsars in the Galaxy and globular clusters the same as the birthrate of their LMXB progenitors ?* A statistical analysis of galactic MSPs and LMXBs gives an apparent discrepancy by a factor ~ 100 if the standard model is applied to the LMXB progenitors [2]. The same apparent discrepancy may exist in globular clusters ([3]; see, however, [4]).

Here we address these questions by showing that radiation from the primary star (operating effectively only for a primary with mass-energy conversion efficiency of order 10 %) can influence binary evolution so as to produce the observed luminosity and orbital period 'gaps' of the LMXB distributions, as well as to reconcile the LMXB/MSP birthrates. We discuss in Sect. 2 several issues of the RD-evolution such as lifetimes, orbital evolution and LMXB period gap, and recycled pulsars in binaries with main sequence companions. In Sect. 3 we briefly address some remarkable features of the binaries containing MSPs in the process of vaporizing their companion stars. In Sect. 4 we discuss the apparent statistical discrepancy between the estimated number of galactic MSPs and their LMXB progenitors evolved according to the standard model. It is shown that the RD-model can possibly resolve the statistical discrepancy of the MSP/LMXB birthrates. Finally, we present in Sect. 5 data concerning the orbital evolution of a very interesting class of short-lived LMXBs. Recent observations of $\dot{P}_{orb}/P_{orb}$ for the sources Cyg X-3, X 1822-371, EXO 0748-676 and 4U 1820-30 are interpreted in terms of the RD-evolution model.

2. Radiation-Driven Evolution of LMXB's

Irradiating photons in the energy band between 0.3 - 10 keV as well as soft γ-rays of energy $\sim 1/2$ MeV are the most efficient in driving an evaporative wind from an illuminated atmosphere [5,6,7]. In the presence of an external irradiating flux the ionization-recombination and heating-cooling balance conditions are satisfied at a relatively large depth (photosphere) but they cannot be maintained for low values of both density and temperature in the outer and more dilute atmospheric layers. As the density in the outer layers decreases, only a rapid rise in temperature can keep the gas medium in balance with the radiation field. A RD-wind is a consequence of a thermal instability of the irradiated atmosphere which develops a high temperature corona. Complex radiative, hydrodynamic and atomic physics effects make it difficult to compute the many features of the RD-wind mechanism. However, the progress obtained in twenty years of investigations of this subject (e.g. [8] and references therein) together with more recent results concerning the case of LMXBs [5, 6, 7, 9, 10] provide a firm basis of discussion of RD-winds [2].

---

[2] The reader is referred to Refs. [7, 9, 11] for further details of the RD-wind formation mechanism. A detailed and comprehensive calculation of atomic physics processes affecting the formation of RD-winds

The RD-*mass loss* rate can be written as $|\dot{m}_{rad}| = 10^{-17} f \hat{L}$ g s^{-1} [5,6], where the dimensionless quantity f depends on the details of energy deposition, radiation transport and hydrodynamics at the base of the corona [7, 10], and $\hat{L} = \chi \Delta\Omega L$, where $\Delta\Omega$ is the effective solid angle subtended by the companion star to its primary, L the compact star's luminosity (in erg s^{-1}) and χ an attenuation factor which takes into account possible absorption/screening and scattering/reprocessing effects in the disk or corona surrounding the compact star (typically $\chi \lesssim 0.1$ for X-rays). Results of detailed hydrodynamical calculations of mass outflows in typical LMXBs show that both the irradiating fluxes and 'quality' of the spectra make possible the self-sustaining of the mass transfer [10]. In the simplest model of self-sustained evolution described here, the radiation-driven mass loss (corresponding to a luminosity $L^* \propto \beta |\dot{m}|_{rad}$, with β the fraction of the mass loss rate which is accreted by the primary star) can be either of order of $10^{18} - 10^{19}$ g s^{-1} or zero. The transition occurs at a value of the companion's mass m_c (hereafter, *the critical mass*) which corresponds to a sudden decrease of the accretion rate or even a permanent accretion turn-off if a msec pulsar has been produced [5, 7].

2.1. LIFETIME OF RD-LMXBs

Two features of the self-sustained evolutionary phase make the lifetime of a RD-LMXB substantially shorter than in standard models: (1) the mass loss is driven at a relatively large rate, and one which is weakly dependent on orbital evolution; and (2) the companion may contract inside its Roche lobe and still transfer mass. This last possibility depends crucially upon the quantity β, the amount of mass lost from the binary during the RD-evolutionary phase and upon the specific angular momentum of the mass loss.

The RD-LMXB lifetime is comprised of two phases, i.e., where the companion fills and underfills its Roche lobe. We have $\tau = \Delta t_{RL} + \Delta t_{NRL}$, where the time the Roche lobe is filled is $\Delta t_{RL} = (m_1/\langle|\dot{m}|_{rad}\rangle_{RL})(1 - m^*/m_1)$; and the time the Roche lobe is underfilled is $\Delta t_{NRL} = (m^*/\langle|\dot{m}|_{rad}\rangle_{NRL})\,(1 - m_2/m^*)$, with m_1 and m_2 respectively the initial and final companion's masses for $m_2 < m^* \lesssim m_1$, and $\langle|\dot{m}|_{rad}\rangle$ the corresponding average mass loss rate [7, 9]. Although the value of the RD-mass loss rate may depend upon the orbital evolution, here we will adopt the following numerical estimates: $\langle|\dot{m}|_{rad}\rangle_{RL} = \langle|\dot{m}|_{rad}\rangle_{NRL} \simeq 10^{-8}$ M$_\odot$ yr^{-1} $\equiv \dot{m}_8$. For LMXB's with degenerate companions, m_1 can be larger (for an initial $\beta \sim 1/3$) or even equal to m^* (for $\beta \simeq 1$). Where mass transfer is self-sustained and the companion always fills its Roche lobe, $m_2 = m_{c,RL}$, the critical mass being given by $m_{c,RL} \simeq (0.1M/\Sigma)/(1 - 0.1/\Sigma)$, with $\Sigma = (\xi_{-1} f_{-1} \beta \chi)^{3/2}$, where $\xi_{-1} = \xi/0.1$ (ξ is the conversion of accreted matter into energy), and $f_{-1} = f/(0.1\,\Upsilon)$ with Υ is a dimensionless quantity depending on the 'quality' of the irradiating spectrum (typically $\Upsilon \sim 1/2$ for soft gamma-rays and $\Upsilon \sim 10 - 100$ for X-rays) [7, 9, 10]. Alternately, after a Roche lobe underfilling evolutionary phase, $m_2 = m_{c,NRL} \simeq 0.6\,m^*[\xi_{-1} f\chi(m^*/0.04\ M_\odot)^{2/3}]^{-3/8}[1 + m^*/M]^{3/8}$, where we used $f \sim \Upsilon$ and $m^* \sim 0.04\ M_\odot$ appropriate to the evolution of the binary progenitor of PSR 1957+20 [7]. Applying the relevant formulae given above to LMXB's with degenerate companions, we find the lifetime of a RD-LMXB containing a companion with, say, $m_1 \sim 0.1\ M_\odot$ is $\tau_d \lesssim 10^7$ years for $\Upsilon\chi \sim 1$. In the case of LMXB's with degenerate companions underfilling their Roche lobe (as in the case of the progenitor of PSR 1957+20), $\Delta t_{NRL} \sim 10^6$ years;

(e.g., line cooling due to heavy elements) is in progress [London R. *et al.*, in preparation] as well as the calculation of RD-LMXB wind hydrodynamics [Tavani M. and Brookshaw L., in preparation].

for example, Cyg X-3 may be in this intermediate phase [12]. For the numerical estimate of the birthrate of very rapid LMXB's containing white dwarfs (see Sect. 4) we will use the lifetime $\tau_{vr,7} = \dot{m}_8^{-1}\tau_{vr}/(10^7 \text{ yrs})$.

Analogously, the inferred lifetime of an LMXB containing a lower main sequence companion whose outer layers are represented by a polytrope of index 3/2 is $\sim 10^7$ yrs [9]. For $\beta = 1$ we obtain $m_c/m^* \simeq 0.33(\xi_{-1}f\chi)^{-3/8} \cdot [(1 + m^*/M)/(m^*/0.4\,M_\odot)]^{1/4}$, and for $\beta = 2/3$ we have $m_c/m^* \simeq 0.11(\xi_{-1}f\chi)^{-3/4} \cdot [(1 + m^*/M)(m^*/0.4\,M_\odot)]^{1/2}$. In Sect. 4, for the lifetime of rapid LMXB's we will use $\tau_r' = \dot{m}_8^{-1}\tau_r/(3 \cdot 10^7 \text{ yrs})$.

2.2 ORBITAL EVOLUTION

The orbital evolution of an LMXB is determined by angular momentum loss (driven by gravitational radiation or a wind), as well as by mass loss from the companion star. For masses and orbital parameters of typical observed LMXBs, mass transfer mechanisms neglecting irradiation effects yield effective lifetimes and orbital timescales of order $10^8 - 10^9$ yrs [1]. For Roche lobe filling companions, the equation of the orbital evolution is particularly simple

$$\frac{\dot{P}_{orb}}{P_{orb}} = \frac{1}{2}(1 - 3n)\frac{|\dot{m}|_s}{m} \tag{1}$$

with $\dot{m}_s$ the mass loss rate of the standard evolutionary model driven by GR and MB [1], and n the effective stellar index of the companion which gives the stellar response to mass loss defined by $d\ln R = n\, d\ln m$, with R the companion radius. In the standard model $\dot{P}_{orb}/P_{orb} < 0$ for LMXBs with main sequence companions in thermal equilibrium with $n \simeq 1$, whereas $\dot{P}_{orb}/P_{orb} > 0$ for LMXBs with degenerate or subgiant companions with $n \simeq -1/3$. Evolution driven by the combined effects of GR and MB accounts successfully for the luminosity distribution of CVs.

Alternately, a self-sustained RD-mass loss from the companion (or from the outer edge of the disk) leads to an orbital evolution quite different from the standard model. The irradiated companion can transfer mass even if it does not fill its Roche lobe exactly . We obtain [9, 13]

$$\frac{\dot{P}_{orb}}{P_{orb}} = -3\left[1 - [(1 - \beta_1)\tilde{\alpha}_1 + (1 - \beta_2)\tilde{\alpha}_2](1 + q) - \beta q - \frac{1}{3}(1 - \beta)\frac{q}{1+q}\right]\frac{\dot{m}_{rad}}{m} + 3\left(\frac{\dot{J}}{J}\right) \tag{2}$$

where m is the mass of the companion star, $\dot{m}_{rad}$ the RD-mass transfer rate, $\dot{J}/J = (\dot{J}/J)_{GR} + (\dot{J}/J)_{MB}$ the fractional change of the orbital angular momentum J from gravitational radiation (GR) and magnetic braking (MB), $q = m/M$ the ratio of the companion mass to primary mass M, and $\tilde{\alpha}$ the specific angular momentum parameter defined by $\delta J = \tilde{\alpha}\delta m(1 - \beta)a^2 2\pi/P_{orb}$, where P_{orb} is the orbital period. The lower-indices '1' and '2' indicate two different possible contributions to the mass loss from the binary (e.g., from the inner edge of the accretion disk with $\tilde{\alpha}_1 \lesssim 0.1$ and from the orbital radius with $\tilde{\alpha}_2 \sim 1$). The dimensionless parameter $\tilde{\alpha}$ gives the effective angular momentum loss caused mass loss from the binary. For example, if mass is lost from the binary at the orbital distance a and if the the outflowing gas does not acquire any additional specific angular momentum, then $\tilde{\alpha} = 1$. For a self-sustained value of the mass loss rate the first term of Eq. (2) proportional to $\dot{m}_{rad}/m$ is the dominant one. Note that the quantity $\dot{P}_{orb}/P_{orb}$ is expected to be $\sim +10^{-6}$ yr^{-1} for RD-LMXBs with very low mass degenerate companions and $\beta \sim 1$

(possibly the case of Cyg X-3 [12]). In general, the time derivative of the orbital period of a RD-LMXB (Eq. (2)) can be either positive or negative depending primarily on the value of β and $\tilde{\alpha}$. The associated lifetime turns out to be $\sim 10^7$ years or less. The RD-orbital evolution can be very different from the evolution of the standard scenario as given by Eq. (1).

2.3 ORBITAL PERIOD 'GAP' OF LMXB's

If the radiation-enhanced mass loss and RD-response of the companion are neglected, Eq. (1) applies and binaries with main sequence companions and, say, $P_{orb} \sim 5$ hrs, have an irresistible tendency to evolve toward the 'period gap' with $\dot{P}_{orb} < 0$. If the mechanism producing the CV period gap (probably due to Roche lobe underfilling of the companion star evolved out of thermal equilibrium because of magnetic braking [1]) were applicable to the LMXB case then we would expect the CV and LMXB distributions to be similar. Fig. 1 shows that this is not the case, with a lack of LMXBs in the range 1 hr $\lesssim P_{orb} \lesssim 3$ hrs.

The orbital period gap can be 'crossed' relatively rapidly ($\tau \sim 10^6$ yrs) by RD-LMXBs containing degenerate companions [7], or never crossed and 'repelled' by RD-LMXBs with main sequence companions [9]. Preliminary calculations of the RD-orbital evolution with companion stars represented by polytropes of index 3/2 confirms that the main sequence companion can be repelled in correspondence with a broad range of P_{orb} [9]. Fig. 1 shows the schematic orbital behavior expected in the RD-evolution.

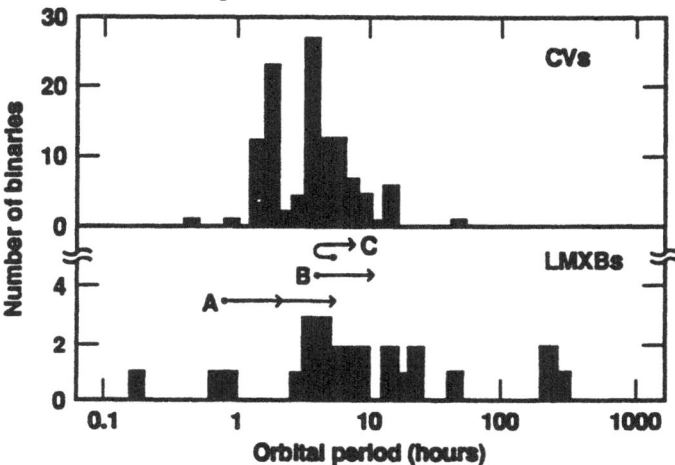

FIGURE 1   Distribution of orbital periods of CVs and LMXBs. Solid curves give the schematic behavior of the orbital RD-evolution [9]. *Line A* represents the rapid expansion of an LMXB with a degenerate companion for β in the range $1/3 \lesssim \beta \lesssim 1$; the timescale may be $\sim 10^6$ yrs. *Line B* and *line C* give the orbital evolution according to Eq. (1) of LMXBs with main sequence companions.

2.4 ORBITAL PERIODS OF BINARIES WITH RECYCLED PULSARS

If the mechanism of mass transfer is *not* driven by radiation during the evolution of LMXB's above the period gap (for $P_{orb} \gtrsim 3$ hours), the sudden quenching of mass trasfer occurring at $P_{orb} \sim 3$ hours may yield a millisecond pulsar whose radiation pressure could be large enough to stop any further accretion. In this case we expect all of the recycled pulsars with main sequence companions to be concentrated at $P_{orb} \sim 3$ hrs. Alternately, the expected orbital period distribution of RD-LMXBs after accretion turn-off is broad and expected to be in the range 3 hrs $\stackrel{<}{\sim} P_{orb} \stackrel{<}{\sim} 10$ hrs for LMXBs with main sequence companions. Future observations in the Galaxy of recycled pulsars with main sequence companions can in principle test the two evolutionary models. Doppler-induced spread caused by orbital motion in binaries with P_{orb} of order of a few hours and the possible presence of vaporized material (see sect. 3.2) make difficult the detection of this important class of binary millisecond pulsars. Pulsar searches to be performed with acceleration-algorithms are therefore strongly urged.

We note that the existence of one binary MSP with a very low mass remnant star in an orbit with $P_{orb} \simeq 9$ hrs (PSR 1957+20 [14]) already excludes the application of the standard evolutionary model to such a system[3].

3. Star-Vaporizing Pulsars

Even in the post-accretion phase, radiation produced by the relativistic wind emanating from millisecond pulsars can continue to vaporize and drive mass loss from their low-mass companions [5, 7, 9] leading to partial or even total evaporation of the companion star. Two binaries containing star-vaporizing pulsars (SVPs) are currently known.

Table 1

PSR	P_{spin} (msec)	P_{orb} (hours)	m_c ($M_\odot$)	L_P (erg s^{-1})	$\Delta L/L$	eclipse characteristics	Ref.
1957+20	1.60	9.2	~ 0.02	$1.5 \cdot 10^{35}$	$\sim 1/10$	stable	14
1744-24A	11.56	1.8	~ 0.1	?	1/3 - 1/2 - 1	highly variable	15

Table 1 gives the pulsar names, spin periods, orbital periods, companion masses, the inferred pulsar spindown luminosity L_P by assuming a neutron star moment of inertia 10^{45} g cm^2, the ratio of the average eclipse size ΔL over the total orbital circumference L, the eclipse characteristics and references. The ratio of the evaporation time scale $\tau_{evap.} = m_c/|\dot{m}|_{rad}$ and the spin-down time scale defined as $\tau_s = P/\dot{P}$ with P the pulsar spin period and $\dot{P}$ its time derivative can be written as

$$\frac{\tau_{evap}}{\tau_s} \simeq 0.8 \frac{P_{-3}^2}{(f/0.1)I_{45}} \left(\frac{m_c}{0.2 \ M_\odot} \right)^{1/3} \frac{1}{\Psi_c^2} \tag{3}$$

where P_{-3} is the spin period in milliseconds, I_{45} the neutron star's moment of inertia in units of 10^{45} g cm^2, Ψ_c the value of $\Psi = R/R_L$ at the accretion turn-off with R_L the Roche lobe radius. In Eq. (3) the total mass has been assumed to be $\sim 1.8 \ M_\odot$. Energy deposition in the irradiated companion's atmosphere by secondary radiation produced by the pulsar relativistic wind[4] gives $f \simeq 0.01\Upsilon$ [7]. The quality factor Υ depends on the details of the spectrum irradiating the companion star. The physics of radiative processes near the

---

[3]The orbital parameters of the other MSP in the process of vaporizing its companion PSR 1744-24A [15] have been probably influenced by star encounters in the core of the globular cluster Ter 5 where the binary system was formed.

[4]In the case of PSR 1957+20, whether the ratio $\tau_{evap.}/\tau_s$ is less than unity depends upon currently

contact discontinuity where the pulsar wind and electromagnetic radiation interact with the mass outflow is particularly interesting. The shape of the irradiating spectrum ultimately depends on the pulsar wind composition ($e^{\pm}$'s, ions) as well as on plasma effects due to the relativistic shock, synchrotron radiation of $e^{\pm}$-pairs and inverse Compton scattering against the soft background photons emitted by the irradiated face of the companion.

Even though complete evaporation of low-mass companions is a possible outcome of RD-evolution, only a subclass of MSPs are capable of producing a sufficiently large evaporative wind [9]. The RD-evolution of LMXB's makes plausible the existence of a relatively large number of MSP's in detached binaries containing lower main sequence stars heated and probably slightly evaporated by the pulsar wind. For $f \sim 0.1$, $m_c \sim 0.3$ $M_\odot$ and $P_{-3} \sim 10$ as expected by standard binary evolution at the upper edge of the period gap, Eq. (3) shows that these main sequence companions cannot be *completely* vaporized[5]. However, complete vaporization of the main sequence companion is possible if $P_{-3}/\Psi_c \sim 1$.

3.1 WIND HYDRODYNAMICS OF PSR 1957+20

The properties of the mass outflow of the eclipsing millisecond pulsar PSR 1957+20 have been studied with the help of a Smoothed Particle Hydrodynamics (SPH) code [18].

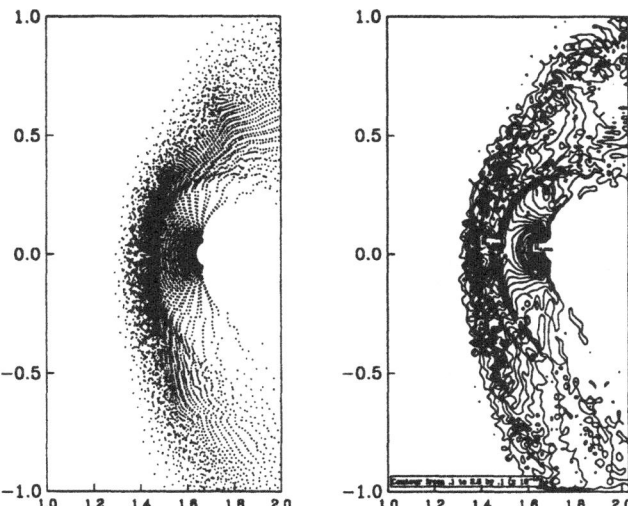

FIGURE 2 Example of a hydrodynamical calculation of the mass outflow in PSR 1957+20 in the corotating frame (from [18]). The mass loss rate from the half irradiated surface of the companion is $\dot{m} = 5 \cdot 10^{14}$ g s^{-1}, the gas temperature at the injection surface $T = 5 \cdot 10^6$ $^\circ$K, and the Mach number $\mathcal{M} = 1.5$

unknown details of the irradiating spectrum; we obtain $\tau_{evap}/\tau_s \simeq 0.8 f^{-1} I_{45}^{-1}$. For a synchtrotron spectrum at fixed electron energy, $\Upsilon \simeq 10$, and for a time integrated synchrotron spectrum of $e^{\pm}$pairs which lose most of their energy, $\Upsilon \simeq 100$ [7]. It is therefore plausible that the companion of 1957+20 would be completely evaporated even if $\Psi_c \simeq 1/3$. However, for the eclipsing pulsar in Ter 5 PSR 1744-24A we have $\tau_{evap}/\tau_s \gg 1$ [16]. We note that the discovery probability of binary pulsars not able to *completely* vaporize their companion is much larger than the probability of discovering systems with $\tau_{evap}/\tau_s \ll 1$.

[5]For a different view see [17]

In order to obtain physically meaningful results, pressure forces need to be considered in addition to gravitational and radiative forces. Fig. 2 shows an example of these calculations for a given set of initial conditions appropriate to a mass outflow which is thick to pulsar radiation except for the external leading particles near the bow shock region.

Note the initial unperturbed expansion of the gas followed by the formation of a strong shock at 2 - 3 companion star radii. The density profile as a function of orbital phase obtained for a given set of initial conditions can be compared with the density inferred from the eclipse properties at different frequencies [19]. An SPH calculation of the mass outflow [18] allows a sensible comparison among different input models, a selection of a range of plausible input quantities such as $\dot{m}_{rad}$, temperature, Mach number, and a detailed estimation of the effects yielding the remarkable timescale for orbital decay measured to be ~ 30 Myrs [19].

3.2 'HIDDEN' MILLISECOND PULSARS

The two currently known millisecond pulsars vaporizing their companion stars confirm the relevance of binary evolution driven by radiation from the primary relativistic star. The eclipse characteristics of the two systems are sufficiently different to make possible a classification of SVPs in terms of the properties of their radio-emission. PSR 1957+20 is the prototype of SVPs with stable eclipse properties (type I) whereas PSR 1744-24A shows a more erratic and sometimes intermittent behavior of radio-eclipses characteristics of a second class of SVPs (type II). In addition to these two SVP-classes, a third class of SVPs is expected on theoretical grounds (type III) [20]. The SVPs of type III are completely engulfed in the material provided by a Roche lobe underfilling companion star irradiated by the pulsar wind, and they are therefore 'hidden' pulsars because their radio-emission may be completely blocked by the evaporated material. In type III SVPs a steady state is reached and pulsar radiation pressure and the centrifugal barrier prevent accretion on the rapidly spinning neutron star. Binaries containing powerful MSPs and main sequence or white dwarf companions underfilling their Roche lobes are the most likely candidates for this class of type III SVPs. Particles in the pulsar wind irradiate most of their energy near the inner edge of the 'bubble' surrounding a hidden MSP. Depending on the characteristics of the pulsar wind and magnetic fields reflected at the inner bubble boundary, the radiation spectrum may range from soft X-rays to MeV-GeV γ-rays [20]. Some of the unidentified COS-B sources, soft X-ray transients and 'blue stragglers' may contain a SVP [20] and radio observations combined with ROSAT and GRO observations might provide evidence for the existence of SVPs of different types. Hidden MSPs may contribute to the resolution of the apparent discrepancy between the orbital period distribution of binary millisecond pulsars vs. their LMXB progenitors [20]. No MSP is currently known to be in a binary system containing a main sequence companion of mass 0.3 $M_\odot \lesssim m \lesssim 1$ $M_\odot$, even though most of the LMXBs have companion stars in this range.

3.3 SINGLE PULSARS IN GLOBULAR CLUSTERS

Several single pulsars are known in globular clusters in addition to PSR 1937+21 and PSR 1257+12 [e.g., 21]. Given the variety of stellar encounters in the cluster cores, it is not surprising that a relatively large fraction of 'recycled' pulsars are found without companions. We summarize here the main mechanisms for producing single recycled pulsars.

3.3.1 Disruptive Coalescence after tidal capture. Catastrophic collisions of neutron stars with passing cluster stars are possible in dense cluster cores. In one-third of all collisions the normal star is disrupted and a massive disk ($m_d \sim 0.1\ M_\odot$) is formed around the neutron star. However, not all of the disk material can be accreted because the accretion rate resulting from a massive disk is probably much larger than the Eddington limit [22]. The spin period of a spun-up neutron star depends of the total accreted mass m_a as follows $P_{-3} \simeq I_{45}^{3/4}(m_a/0.17\ M_\odot)^{-3/4}(M/(1.4\ M_\odot))^{-1/2}$, with M the mass of the neutron star. Since $m_a \ll m_d$ [22], the spin period of the resulting pulsar is likely to be in the range 0.1 - 1 sec if only $\sim 10^{-3}m_d$ is accreted. Remarkably, at least three relatively slow pulsars in dense clusters have periods in this range[6] without any evidence of a pulsar population of possible progenitors between 10 and 100 msec [21]. The pulsars that might have been formed by disruptive star coalescence are PSR 1754-20 (P = 288.6 msec) in NGC 6440, PSR 1820-30B (P = 378.6 msec) in NGC 6624, and PSR 1744-24B (P = 442 msec) possibly in Ter 5, which are all single and belong to clusters which have either dense or collapsed cores.

3.3.2 Binary Ionization. Stellar encounters resulting from 2-body and 3-body interactions are likely to produce single pulsars in dense cluster cores. The resulting timescale for binary ionization in the densest cores of globular clusters is $\sim 10^9$ years and the single pulsars found in M 15, 47 Tuc might have been formed by ionization of wide binaries with orbital periods $P_{orb} \gtrsim 100$ days [23]. However, the four single millisecond pulsars found in clusters with low-density cores cannot be produced by 3-body interactions.

3.3.3 Companion Vaporization. The existence of a single 3 msec pulsar in the cluster M 28 (PSR 1821-24) motivated both the study of star-vaporization driven by pulsar winds [5] and the evaluation of the binary ionization mechanism in globular clusters [23]. Binary ionization in the low-density cluster M 28 is unlikely, especially in view of the relatively small pulsar spin-down rate of PSR 1821-24 ($\tau \sim 3 \cdot 10^7$ years) [23]. Vaporization of the companion star by the pulsar wind is therefore a viable possibility for PSR 1821-24 as well as for PSR 1908+00 (P = 3.6 msec) in NGC 6760, PSR 1516+02A (P = 5.44 msec) in M 5, and PSR 1640+36 (P = 10.3 msec) in M 13, which are all MSPs in low-density clusters.

4. Statistics of Millisecond Pulsars and LMXB's

A statistical analysis of galactic MSP's gives a birthrate $BR(MSPs) \simeq 2 \cdot 10^{-5}yr^{-1}$ after taking into account the selection effects associated with radiopulsar surveys [2]. This birthrate is about 1000 times smaller than the galactic supernova formation rate and is too large by a factor $\sim 10 - 100$ to be in agreement with standard evolutionary models of LMXB's. This discrepancy is even more pronounced for the subgroup of 'rapid' low-mass binary pulsar systems (LMBP's), where the orbital period is $P_{orb} \lesssim 10$ days, and the birthrate $BR(rapid\ LMBPs) \simeq 1.5 \cdot 10^{-5}\ yr^{-1}$. It is appropriate to consider separately binaries defined as 'very rapid' ($P_i \lesssim 0.5$ days), 'rapid' (0.5 days $\lesssim P_i \lesssim 3$ days), and 'slow'

---

[6]The pulsar PSR 2127+11A in M 15 (P = 110.6 msec) might have been formed by accretion following star coalescence. Alternately, it might have been first spun-up in a binary subsequently ionized in the high-density core of M 15 and later spun-down by loss of electromagnetic radiation. This last possibility is suggested by the existence in M 15 of pulsars of intermediate spin periods such as PSR 2127+11B (P = 56.1 msec) and PSR 2127+11C (P = 30.5 msec).

($P_i \gtrsim 3$ days), with P_i the initial orbital period. From the estimates of section 2.1 we can calculate the birthrates of rapid LMXB's evolving due to the radiation-driven mechanism of mass transfer. We obtain [24] $BR(very\ rapid\ LMXB's) \simeq (7 \cdot 10^{-6}\ yr^{-1})f_w\tau_{vr,7}^{-1}$ and $BR(rapid\ LMXB's) \simeq (6 \cdot 10^{-7}\ yr^{-1})f_r\tau_r'^{-1}$, with $f_w \simeq 1/2$ the fraction of very rapid LMXB's with white dwarf and main sequence companions, and f_r the fraction of LMXB's with $P_{orb} \gtrsim 0.5$ days containing sub-giant companions as defined in Ref. [2]. If we compare the birthrate of very rapid LMXB's driven by radiation and of rapid MSP's we find that they are approximately equal. Given the current statistical uncertainties of pulsar surveys and LMXB number estimations the statistical MSP/LMXB discrepancy may therefore be resolved by very rapid RD-LMXB's [24]. We note that since the 'prompt' formation of MSP's after the AIC of white dwarf primaries possibly affects only a fraction of slow binaries, the AIC mechanism is not applicable in this context.

At present it is uncertain whether the MSP/LMXB discrepancy in globular clusters is the same as in the Galaxy [3] or smaller by a factor of ~ 10 [4]. The inferred MSP birthrate is $BR(cluster\ MSPs) \simeq (0.2 - 1.7) \cdot 10^{-6}$ yr^{-1}. From the measured value of $\dot{P}_{orb}/P_{orb}$ for 4U 1820-30 (see Table 2) and the large value of the mass loss rate of AC211 in the globular cluster M 15 [25], we obtain $BR(HML - LMXBs) \simeq 1 \cdot 10^{-6}$ yr^{-1}, if we consider only the two high-mass-loss (HML) binaries 4U 1820-30 and AC211. We note that despite their apparent low X-ray luminosities, LMXB's in globular clusters may have radiation-driven orbital timescales of order 10^7 yrs, similarly to the X-ray burster EXO 0748-676 (see Table 2). The RD-LMXB birthrate in globular clusters may well be $\gtrsim a\ few\ 10^{-6}$ yr^{-1}.

Table 2

LMXB	P_{orb} (hours)	$\dot{P}_{orb}/P_{orb}$ (yr^{-1})	timescale (years)	Ref.
Cyg X-3	4.82	$+(2.20 \pm 0.22) \cdot 10^{-6}$	$5 \cdot 10^5$	26
X 1822-371	5.57	$+(3.40 \pm 0.94) \cdot 10^{-7}$	$2.9 \cdot 10^6$	27
EXO 0748-676	3.82	$-(2.02 \pm 0.28) \cdot 10^{-7}$	$5 \cdot 10^6$	28
4U 1820-30	0.18	$-(1.08 \pm 0.19) \cdot 10^{-7}$	10^7	29

5. Observations vs. Theory of LMXB Orbital Evolution

Measuring $\dot{P}_{orb}$ for an LMXB requires a stable fiducial point in the X-ray light curve. Such a measurement is often difficult because the majority of LMXBs do not have any dips, eclipses or pulsations. Furthermore, LMXBs have been observed for about 20 years and the lower limit for the measurable $\dot{P}_{orb}/P_{orb}$ is of order 10^{-7} yr^{-1}. Therefore, no LMXB with measurable $\dot{P}_{orb}/P_{orb}$ is expected if the standard model applies to all LMXBs. However, non-zero measurements of $\dot{P}_{orb}$ for a relatively large number of LMXBs have recently become available. The quantity $\dot{P}_{orb}$ has now been measured for Cyg X-3 [26], X 1822-371 [27], EXO 0748-676 [28], and 4U 1820-30 [29], all of which have $P_{orb} \lesssim 5.6$ hours (see Table 2 and [30]). None of the measured values agrees with the predictions of the standard GR and MB models for the $\dot{P}_{orb}/P_{orb}$'s assumed to represent secular changes of orbital evolution. The inferred timescale for orbital evolution is about 100 times shorter than expected in all systems except 4U 1820-30. In the case of 4U 1820-30 with $P_{orb} \simeq 11$ minutes [31], the measured $\dot{P}_{orb}/P_{orb}$ has a sign *opposite* to what expected by binary evolution driven by GR with a degenerate and Roche lobe filling companion (cf. Eq.

(1) and [32]).

An interpretation of the measured $\dot{P}_{orb}/P_{orb}$ in terms of the RD-model of LMXB evolution has been proposed in [13]. The existence of a class of LMXBs with orbital timescales of order 10^7 years was indeed predicted on theoretical grounds when only the measurement of $\dot{P}_{orb}$ for Cyg X-3 was known [7]. Fig. 3 gives the main characteristics of the RD-model applied to the sources of Table 2.

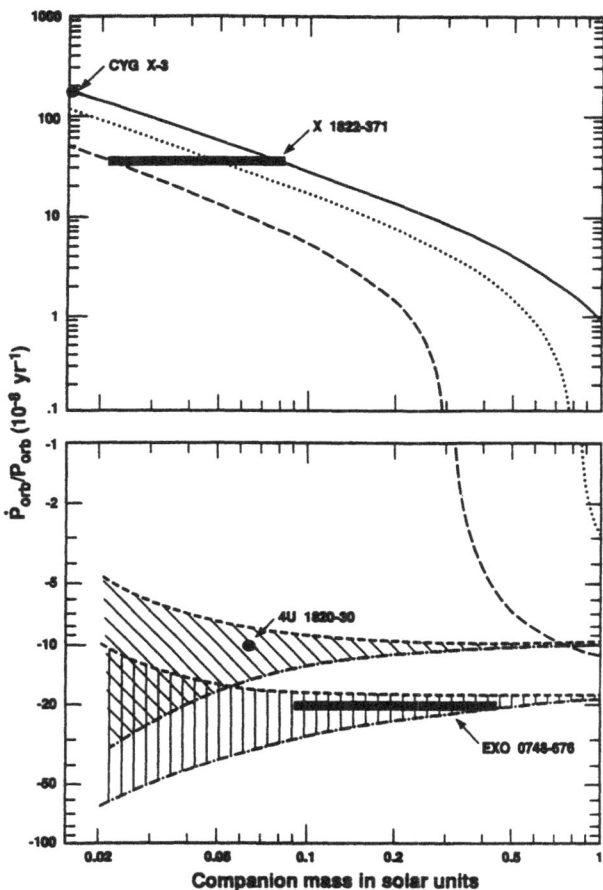

FIGURE 3 The quantity $\dot{P}_{orb}/P_{orb}$ computed for the RD-evolutionary model as a function of the mass of the companion for $\tilde{\alpha} = 1.1$. The upper part of the diagram has a positive $\dot{P}_{orb}/P_{orb}$. The three upper curves are computed for $\dot{m}_8 = 3$ and $\beta = 1$ (solid line), $\beta = 2/3$ (dotted line), and $\beta = 1/3$ (dashed line), respectively. The lower part of the diagram has a negative $\dot{P}_{orb}/P_{orb}$. The two shadowed areas give the allowed range for $\dot{P}_{orb}/P_{orb}$ with $\dot{m}_8 = 10$ and 20, respectively. Their boundaries are defined by the curves for $\beta = 1/10$ (short-dashed curves) and for $\beta = 1/20$ (dot-dashed curves). The horizontal thick marks in correspondence with the measured $\dot{P}_{orb}/P_{orb}$'s give the mass range for the companion stars.

Cyg X-3 fits the evolutionary scenario for a RD-LMXB containing a very low mass white dwarf companion [12, 9]. A strong wind is indeed observed in Cyg X-3 [33] and the observed orbital period change is in agreement with the inferred large wind mass loss rate. Even the source X 1822-371 has a positive and large rate of change of the orbital period [27]. This is in contradiction with the standard model if the LMXB contains a MS companion. Both the sign and the magnitude of $\dot{P}_{orb}/P_{orb}$ are different from the expected values (see Eq. (1)). However, as shown in Fig. 3, RD-LMXB's with lower main sequence companions may be characterized by a large positive value of $\dot{P}_{orb}/P_{orb}$ if $\dot{m}_8 \gtrsim 3$. Alternately, the X 1822-371 may contain a white dwarf companion [13], and Fig. 3 shows that in this case the likely mass is of order $m = (0.1\ M_{\odot})\dot{m}_8$. If both Cyg X-3 and X 1822-371 contain a very low-mass white dwarf, these systems can be naturally considered progenitors of binaries similar to the eclipsing pulsar PSR 1957+20. We note that both the sign and magnitude of $\dot{P}_{orb}/P_{orb}$ for Cyg X-3 and X 1822-371 suggest a stable RD-mass transfer with $1/3 \lesssim \beta \lesssim 1$ and $\Psi \lesssim 1$.

By contrast, EXO 0748-676 and 4U 1820-30 may belong to a class of RD-LMXB's in which a large mass loss from the binary plays a crucial role in determining the orbital evolution [13]. According to the RD-model, EXO 0748-676 and 4U 1820-30 are characterized by a relatively large mass loss from the binary with $\beta \sim 1/10$ and $\Psi \gtrsim 1$. For both EXO 0748-676 and 4U 1820-30 the RD-mass transfer is expected to be unstable ($\Psi \gtrsim 1$), in agreement with the large intensity fluctuations (by a factor of ~ 5) observed in EXO 0748-676 and 4U 1820-30. The RD-model allows two classes of RD-LMXBs characterized by different values of β [13]. Cyg X-3 and X 1822-371 belong to the high-β class, whereas EXO 0748-676 and 4U 1820-30 belong to the low-β class.

At present four out of fifteen LMXBs with $P_{orb} \lesssim 5.5$ hrs have measured $\dot{P}_{orb}/P_{orb}$'s. Of the total sample of ~ 100 LMXBs only ~ 36 binaries have well established orbital periods [e.g., 34]. A complete analysis of the selection effects affecting the measurements of $\dot{P}_{orb}/P_{orb}$ would be very useful in order to obtain an estimate of the true number of RD-LMXBs. As we have shown in Sect. 4 the existence of a class of short-lived LMXBs can be crucial for the resolution of the apparent statistical discrepancy between the MSP/LMXB birthrates. The discrepancy is resolved if more than a third of LMXBs are radiation-driven [24].

The detection of ionized mass outflow from LMXBs at the level of 10^{18} g s^{-1} and with temperature $T \gtrsim 10^{7\,\circ}$ K from the companion star and possibly the outer disk depends on complex details of both the geometry of the gas flow and the heating/cooling mechanism in the expanding corona. No clear indication of ionized mass loss has been obtained from known LMXBs thus far, with the possible and notable exceptions of Cyg X-3 [33] (a favorable case with $\beta \simeq 1$) and the source AC211 in the globular cluster M 15 [25]. AC211, a good candidate for a RD-LMXB, is probably characterized by a large mass outflow of order $10^{-7}\ M_{\odot}$ yr^{-1} [25] which simplifies the detection of emission/absorption lines Doppler-shifted by gas motion. We note here that line detection depends crucially on the geometry of the mass outflow as well as the total mass loss from the companion. In general for typical orbital parameters and wind velocities expected in RD-LMXBs, we can expect a relation of the type $\dot{m}_8 \simeq 10(L_{l,14})^{1/2}$, with $L_{l,14}$ the line luminosity in units of 10^{-14} erg cm^{-2} s^{-1} Å^{-1} (which corresponds to the current sensitivity limit for most LMXBs). Detection of Doppler-shifted lines from LMXBs with $\dot{m}_8$ between 1 and 10 may be particularly difficult, and this problem deserves a more detailed theoretical and observational analysis of known LMXBs.

References

[1] Rappaport, S., Verbunt, F., and Joss, P.C., 1983, *Ap.J*, **275**, 713.

[2] Kulkarni S.R., and Narayan, R., 1988, *Ap.J*, **335**, 755.

[3] Kulkarni, S., Narayan, R., and Romani, R.W., 1990, *Ap.J.*, **356**, 174.

[4] Fruchter, A., Goss, D., 1990, *Ap.J. (Letters)*, **365**, L63-L66.

[5] Ruderman, M., Shaham, J., and Tavani, M., 1989, *Ap. J.*, **336**, 507.

[6] Ruderman, M., Shaham, J., Tavani, M., and Eichler, D., 1989 *Ap. J.*, **343**, 292.

[7] Tavani, M., 1989, in *Proc. of the 23rd ESLAB Symposium*, ESA SP-296, Vol. 1, p. 241.

[8] London, R.A., McCray, R., and Auer, L.H., 1981, *Ap.J.*, **243**, 970.

[9] Tavani, M., 1990, submitted to *Ap.J*.

[10] Tavani, M., and London, R., 1990, to be submitted to *Ap. J.*.

[11] Shaham, J., 1991, these Proceedings.

[12] Tavani, M., Ruderman, M. and Shaham, 1989, *Ap. J.(Letters)*, **342**, L31.

[13] Tavani, M., 1991, *Nature*, **351**, 39-41.

[14] Fruchter, A.S., Stinebring, D.R., and Taylor, J.H., *Nature* , **333**, 237 (1988).

[15] Lyne, A.G., *et al.*, 1990, *Nature*, **347**, 650-652.

[16] Shaham, J., and Tavani, M., 1991, *Ap. J.*, in press.

[17] van den Heuvel, E.P.J. and van Paradijs, J., 1988, *Nature*, **334**, 227.

[18] Tavani, M., and Brookshaw, L., 1991, to be submitted to the *Ap. J.*.

[19] Ryba, M., and Taylor, J., 1991, submitted to the *Ap. J.*.

[20] Tavani, M., 1991, to be submitted to *Ap. J.*.

[21] Tavani, M., in *Relativistic Experiments in Space*, Rome 10-14 September 1990, eds. M. Demianski and R. Ruffini (Singapore: World Scientific).

[22] Verbunt, F., van den Heuvel, E.P.J., van Paradijs, J., and Rappaport, S.A., 1987, *Nature*, **329**, 312.

[23] Rappaport, S., Putney, Verbunt, F., 1989, *Ap.J.*, **345**, 210.

[24] Tavani, M., 1991, *Ap.J. (Letters)*, **366**, L27-L31.

[25] Naylor, T., Charles, P.A., Drew, J.E., and Hassall, J.M., 1988 *M.N.R.A.S.*, **233**, 285.

[26] Kitamoto, S., Miyamoto, S., and Matsui, W., 1987, *P.A.S.J.*, **39**, 259.

[27] Hellier, C., Mason, K.O., Smale, A.P., Kilkenny, D., 1990, *M.N.R.A.S.*, **244**, 39P.

[28] Parmar, A.N., Verbunt, F., Smale, A.P., and Corbet, R.H.D., 1991, *Ap.J.*, **366**, 253.

[29] Tan, J., et al., 1991, *Ap. J*, in press.

[30] Parmar, A.N., these Proceedings.

[31] Stella, L. White, N., and Priedhorsky, W., 1987, *Ap.J.(Letters)*, **315**, L49.

[32] Rappaport, S., Nelson, L.A., Ma, C.P., and Joss, P.C., 1987, *Ap.J.*, **322**, 842.

[33] Molnar, L.A., Kouba, J.A., and Raymond, J.C., 1990, *B.A.A.S.*, **22**, 1339.

[34] Ritter, H., 1990, *Astron. Astrophys. Suppl. Series*, **85**, 1179.

Editors' notes:

– For a detailed calculation of the shape of the mass outflow and the resulting mass and angular momentum loss from the PSR 1957+12 system: see M. Tavani and L. Brookshaw 1992, *Nature* **356**, 320–322.

– Recent IR observations of Cyg X-3 have shown its companion to be a helium-rich Wolf-Rayet like star, and therefore to be probably massive ($M \geq 4M_\odot$). The rapid increase in its orbital period is therefore most probably due to its very strong stellar wind mass loss, and the mechanism suggested in this paper will not be applicable to this source (cf. Van Kerkwijk *et al.* 1992, *Nature* **355**, 703–705).

A NEW MECHANISM FOR ANGULAR MOMENTUM LOSS IN VERY-LOW-MASS X-RAY BINARIES AND BINARY MILLISECOND PULSARS

M. BANIT and J. SHAHAM
Physics Department and Columbia Astrophysics Laboratory
Columbia University
538 West 120ʰ Street
New York, NY 10027
and
Institute for Theoretical Physics
University of California
Santa Barbara, CA 93106

Abstract. It is suggested that in some Low-Mass X-Ray Binaries and Millisecond Pulsar Binaries, companion winds that are excited by the radiation from the neutron star form only through the combined action of the radiation *heat* on the companion atmosphere and the radiation *force* on the slowly lifting wind. The winds so formed only leave from selective regions on the companion's illuminated surface as surface currents channel there relatively cool coronal matter from the whole illuminated area. Under suitable conditions, wind particles spend some time trailing the companion at close distances before taking off to escape from the system. This torques the binary into angular momentum loss that will be as efficient as the ones recently observed if the companion is bloated to Roche lobe dimensions.

I. Overview

We propose a new mechanism for the formation of companion winds by radiation from the neutron star (NS) in Low-Mass X-Ray Binaries (LMXBs) and in Binary Millisecond Pulsars (BMPs). The mechanism is at work when radiation intensity is of the order of the Eddington intensity appropriate for the particular binary and the particular matter-radiation interaction process, and when radiation heat alone is unable to lift matter off the companion's surface with the necessary escape velocity. A wind is then lifted off selective areas on the companion surface through the additional action of the radiation pressure on a slowly moving, relatively cool corona, fueled by the radiation heat. The unusually large orbital angular momentum changes, with timescales as short as $\precsim 10^7$ yrs, that have been observed in several LMXBs[1-4] and in the BMP 1957+20[5], can obtain when binary evolution in these systems is dominated by a sufficiently intense companion wind under the above circumstances. The companion may be sunk in its Roche lobe due to a bootstrapping[6-10]

401

E. P. J. van den Heuvel and S. A. Rappaport (eds.), X-Ray Binaries and Recycled Pulsars, 401–412.
© 1992 *Kluwer Academic Publishers.*

episode, during which it probably remains bloated and synchronised. When the initial heat-induced radial wind velocity is low, a high density region that trails the companion torques its orbital motion down while the wind passing through it picks up extra angular momentum. By contrast, the orbital period is made to increase by the wind with similar efficiency when the radial velocity is high. This new mechanism can also produce higher mass-flow rates.

Following the theoretical prediction of their existence[7] and their subsequent discovery in BMPs *1957+20*[11] and *1744-24A*[12], winds excited off companions in LMXBs and BMPs have been subject to intensive discussion. It is presently understood that these particular winds form because, quite generally, whenever heat is deposited in a stellar atmosphere there exists a pressure, P_{min}[13,14] below which no radiative or conductive[15] process in the atmosphere can cool it fast enough. Wind formation is therefore guaranteed below that pressure, i.e. above that point in the atmosphere, simply as a process of adiabatic expansion. To fully escape from an *isolated* spherical star, the wind speed at its sonic point must equal at least half the escape velocity from that point[9]; that puts a lower limit, $T_{s,min}$, on the sonic temperature. If the exciting radiation is such that it cannot be efficiently absorbed by the atmosphere at temperatures of order the value of $T_{s,min}$ for the stellar surface[13], an extended corona must be formed first, from the top of which particles can then escape at much lower escape velocity and temperature. Were the companion truly isolated, wind flow rates in such case would be very small[16] because only a pressure gradient can lift all the wind material up to the top of the corona, and therefore the residual pressure there, which is to supply the linear momentum of the wind, will be exponentially small compared to P_{min}.

An extra factor related to wind formation in *binaries* is the escape from the binary. If the wind does not have the required velocity it may end up being recaptured by the companion or forming an "excretion" disk[17,18], and chances for that increase when a corona is formed first.

Circumstances will, however, be quite different when the *pressure* of the neutron stellar radiation blowing on the forming corona is sufficiently large. In this case matter can, in principle, be lifted off the stellar surface at some velocity v_r that is below the escape velocity from the companion, and then secure the extra linear momentum it needs to escape out of the binary from the radiation field and from the companion's motion "under" it. For a steady state optically thick flow one may estimate the upper bound on the mass loss rate from the radiation pressure in such case by

$$\dot{m}_\omega \preceq 10^{18}(\frac{\psi}{10^{-2}})(\frac{L}{10^{38}erg/sec})(\frac{300km/sec}{v_{eb}}) \qquad \text{g/sec} \qquad (1)$$

where L is the luminosity of the neutron star, assumed to consist of relativistic particles (including photons), v_{eb} is the escape velocity from the binary and $4\pi\psi$ is the solid angle around the neutron star over which the radiation pushes the flow (possibly of the order of the eclipsing region in BMP1957+20) . Alternatively, one may estimate the mass loss rate due to gas pressure from conditions at the source, the companion's atmosphere[6]:

$$\dot{m}_\omega \preceq \frac{P_{min}A_s}{2c_s} \simeq 10^{18}\Lambda(\frac{\psi_s}{10^{-3}})(\frac{L}{10^{38}erg/sec})\left(\frac{300km/sec}{2c_s}\right) \qquad \text{g/sec} \qquad (2)$$

where

$$\Lambda = \left(\frac{P_{min} \mathcal{A}_s / L \psi_s}{6 \times 10^{-11} sec/cm} \right),$$

c_s is the sound speed at the sonic point, $\mathcal{A}_s$ the projected surface area of the star and $4\pi\psi_s$ is the solid angle that area subtends around the neutron star. When half the value of the escape velocity is substituted for the sound speed in Λ and for $\Lambda = 1$, Eq. (2) gives the canonical estimate[6] for radiation-excited mass flow off "isolated" companions; but here c_s need not be as high, the wind could remain cool and $\dot{m}_\omega$ can be larger. For light companions that almost fill their Roche lobe we may substitute[6] $\left(\frac{300 km/sec}{2c_s} \right) \sim 3$, so that the canonical estimate for $\dot{m}_\omega$ in this case will be that much higher.

Higher values of $\dot{m}_\omega$ can obtain when the neutron star radiation is beamed towards the companion. Such might also obtain when the companion spin is sufficiently high, thus increasing the initial speed of the wind [it seems, however, at present, that torques that spin-up/down the companion through the forming wind act only on the mass-loss timescale[19] and tidal synchronisation torques are far more effective[20]. We shall therefore only consider here synchronised rotation]. Still higher values of $\dot{m}_\omega$ can obtain when the illumination of the companion exceeds its own Eddington luminosity or when its resulting photospheric temperature exceeds the escape temperature from it, situations that are only likely to occur for planetary companions.

II. The Problem of the Short Orbital Timescales

It now appears that the orbital period-change time-scale in *1957+20* is $|P_{orb}/\dot{P}_{orb}| \sim 3 \cdot 10^7$ yrs[5] and the binary period P_{orb} *decreases*; hence (unless center-of-mass accelerations are involved), the orbit *decays*. A similar situation seems to exist in the LMXBs *4U1820-30*[1] and *EXO0748-676*[2], with binary period spin-up times of 10^7 and $5 \cdot 10^6$ years respectively, and in the LMXBs *CygX-3*[3] and *X1822-37*[4], which have binary period *spin-downs* with respective timescales of $5 \cdot 10^5$ and $3 \cdot 10^6$ years. We argue below that these LMXBs are likely to be VLMXBs (Very-Low-Mass-X-Ray-Binaries), i.e. systems with very low mass companions in which winds provide the predominant angular momentum loss mechanism.

To understand the meaning of such short orbital timescales we write, with $\frac{m}{M} \equiv q$ (m and M are the companion and the NS masses, respectively; for numerical purposes we use $M = 1.4 M_\odot$ in this paper) and upon assuming that the wind carries *all* the angular momentum loss,

$$\Psi \equiv -\frac{\dot{P}_{orb}/P_{orb}}{\dot{m}_\omega/m} = 3 \times \left[\frac{(1-\beta)(Q-1)}{(1+q)} - \frac{2(1-\beta)q}{3(1+q)} - \beta(1-q) \right], \qquad (3)^*$$

where β is the fraction of $\dot{m}_\omega$ that ends up being *accreted* on the neutron star ($\beta < 1$ for wind-dominated evolution) and where Q is the effective angular momentum per unit mass carried by the escaping particles, in units of the angular momentum per unit mass of the center-of-mass motion of the companion. Thus, without worrying yet about magnitudes, to only make $\Psi > 0$ it is required that

$$Q > \frac{1-\beta q^2}{1-\beta} + \frac{2}{3}q \geq 1 + \frac{2}{3}q. \qquad (4)$$

---

*Note that in (3) $\dot{m}_\omega$ is defined to always be a positive quantity.

We regard Eq. (4) to be a major clue to understanding the origin of the short timescales.

For a wind that leaves the companion through the part of its surface that faces the NS and which carries to infinity solely the angular momentum its particles had while *on the companion*, $Q < 1$ if the companion rotates synchronously, because these particles had the same orbital frequency as the companion but were moving at a *smaller* average distance from the center-of-mass. It is therefore clear that if Eq. (4) is to be satisfied, a physical process *must* exist in which the value of Q increases. [Note that ref. 21 uses the value of 1 for its parameter $\tilde{\alpha}$ to describe a wind that leaves with the specific angular momentum of the companion. In fact, using the notation of our present paper one finds that $\tilde{\alpha}\left(\equiv \frac{Q}{(1+q)^2}\right) < 1$ even when $Q = 1$, so that the conclusions of ref. 21 only hold for values $q \to 0$ and are thus not applicable, within the context of that same paper, to *either* of the two LMXBs with $\dot{P}_{orb} < 0$; see Fig. 1 of that paper]. A retrograde companion spin might push Q higher, if the spin were fast enough. More importantly, however, the particles in the wind could acquire a higher Q *after* they've left the companion, by the tidal interaction process that we propose here:

In general terms, this process can be thought of as stalling the wind's azimuthal motion in the corotating frame by balancing the Coriolis force against the companion's gravity, thus allowing the radiation to push matter out through the vicinity of the Lagrangian point L2. This raises the value of Q towards that of particles that corotate at L2. As we shall see in the following section, the initial lift-off speed v_r determines whether $Q > 1$ or $Q < 1$, so that $\dot{P}_{orb}$ values of either sign can readily obtain by the *same* mechanism. The orbital-period-change timescales turn out to be essentially of the same order as the mass-loss timescales, and could therefore be understood in terms of a very low companion mass and/or a low β value [β cannot be too low for LMXBs in a bootstrapping phase, however, because of the limited efficiency of the wind formation mechanism; see below].

III. The Simulations

A detailed hydrodynamical numerical code for analysing wind motion of so complex a geometry is not yet in the cards. To gain some physical insight, we nevertheless carried out a series of Monte Carlo simulations of free wind particles moving on ballistic trajectories. We briefly summarize here the results; the full report will be published elsewhere[19].

Particles were shot from the *inner* half-sphere of a synchronously rotating companion surface at constant flow density per unit projected area and constant radial velocity v_r. The radiation force was included everywhere (except in the companion's shadow) by giving the NS an effective mass of $M(1 - C_{rad})$[22], where $C_{rad} = 1$ represents the Eddington luminosity; for LMXBs, however, we have also carried out simulations for which we put $C_{rad} \equiv 0$ in the shadow region of the $(\alpha-)$ accretion disk. Trajectories were computed by a fifth-order self-adaptive Runge-Kutta scheme[23]. The initial orbital angular momentum of each particle was recorded and the increments to it due to the torques from the stars were added along its trajectory. For escaping particles, the terminal values were averaged to give Q. Other particles were either captured by the disk (if present) or recaptured by the companion. The latter group defined the companion surface area from which no wind emission is expected; upon returning, these particles also change the spin angular momentum of the companion and define the wind spin-up/down torque on it.

When trying to assess the simulations vis a vis the real situation two important points should be made, aside from the fact that particle collisions will dissipate part of the ballistic motion and could end up changing Q by changing the average trajectory: (i) In a real situation one expects that the "no-wind" regions will get somewhat hotter and establish lateral atmospheric (coronal) temperature and pressure gradients, along which rising matter in these regions will flow into the wind-emitting regions, not unlike the expected surface currents when L1 Roche Lobe overflow takes place. Therefore, we expect Eqs. (1) and (2) to remain essentially unaltered unless *no* surface region can emit a wind, in which case we do expect the wind flow to decrease substantailly. (ii) A real situation may involve an *optically thick* wind, for which the simple-minded radiation force that we used will be transmitted from the flow boundary through pressure gradient forces with a different spatial variation. While winds in question here are mostly optically thin to X-rays, they are optically thick in the radio for systems like 1957+20.

   Simulations for 1957+20. As we have seen earlier[10], for *1957+20* Eq. (4) translates to $Q \geq 1.01$, but the actual $Q - 1$ value required here should be much higher, not unlike the values needed for $\beta < 1$ LMXBs. We assume that pulsar radiation pressure drives *all* wind particles to infinity and $\beta = 0$. From the spin-down rate of the pulsar we estimate its luminosity to be $2 \cdot 10^{35} erg/sec$. For a companion radius of $10^{10} cm$ ($m = 0.02 M_\odot$) and binary separation of $1.7 \cdot 10^{11} cm$, the standard model gives [Eq.(2)] $\dot{m}_w \sim 2 \cdot 10^{15} \Lambda \ g/sec$. This seems to be a rather optimistic standard-model value for the wind-flow rate, unless the effective area involved, A_s, is larger than the projected surface area of the companion or else the companion is out of thermal equilibrium and happens to be close to filling its Roche lobe; then one could get an additional factor of ~ 12. The observed $\dot{P}_{orb}$ implies, by Eq. (3),

$$\left(\frac{\dot{m}_w}{2 \cdot 10^{15} g/sec}\right)(Q - 1.01) \simeq 8,$$

and unless $\dot{m}_w$ is $(2-6) \cdot 10^{16} \ g/sec$, Q needs to be very large indeed.

   Unless matter distribution in the system is very anisotropic and our line of sight is sufficiently above the orbital plane, the emergence of high frequency radio signals during eclipse time would permit $\dot{m}_w$ to actually be *no more* than $\sim 5 \cdot 10^{16} \ g/sec$ just by plasma frequency considerations; this takes us straight to the upper limit of Eq. (1) if ψ is of order 0.1. A combination of some pulsar radiation anisotropy, low c_s and a larger, non-equilibrium, companion radius could, in principle, come up to that factor in Eq. (2), but our simulations show that the plausible values for v_r and for Q [which could go as high as ~ 1.4, its L2 value for this system] still require a wind formation mechanism for which $\Lambda \succeq 5$ if the companion is detached and $\Lambda \succeq 1.5 - 3$ if it is in contact.

   Some of our results are presented in Figs. 1 and 2. To understand the effect of a small v_r compare the $C_{rad} = 4$ behaviour for $\frac{v_r}{v_e} = 1$ and for $\frac{v_r}{v_e} = 0.45$ (Fig. 2). In the former case, wind emission is symmetric for the front and back (with respect to the direction of motion) of the companion and $Q - 1 = -0.068$. In the latter, particles escape primarily from the back and the polar caps because all other particles are being pushed by the radiation right back into the companion; that wind pattern gives $Q - 1 = 0.26$ and the high Q of the wind can be seen by the *prograde* bending of the trajectories in panel (ii)b, caused by the companion gravity.

   The best simulations' results we found so far for *1957+20* are shown in the (iii) panels. Here $C_{rad} = 0.4$, $\frac{v_r}{v_e} = 0.86$ and the companion radius is $1.5 \cdot 10^{10} cm$, the upper limit found from photometry[24]. [If part of the wind were to form a geometrically thin equatorial disk around the companion, it is possible that only one half of the illuminated side of the

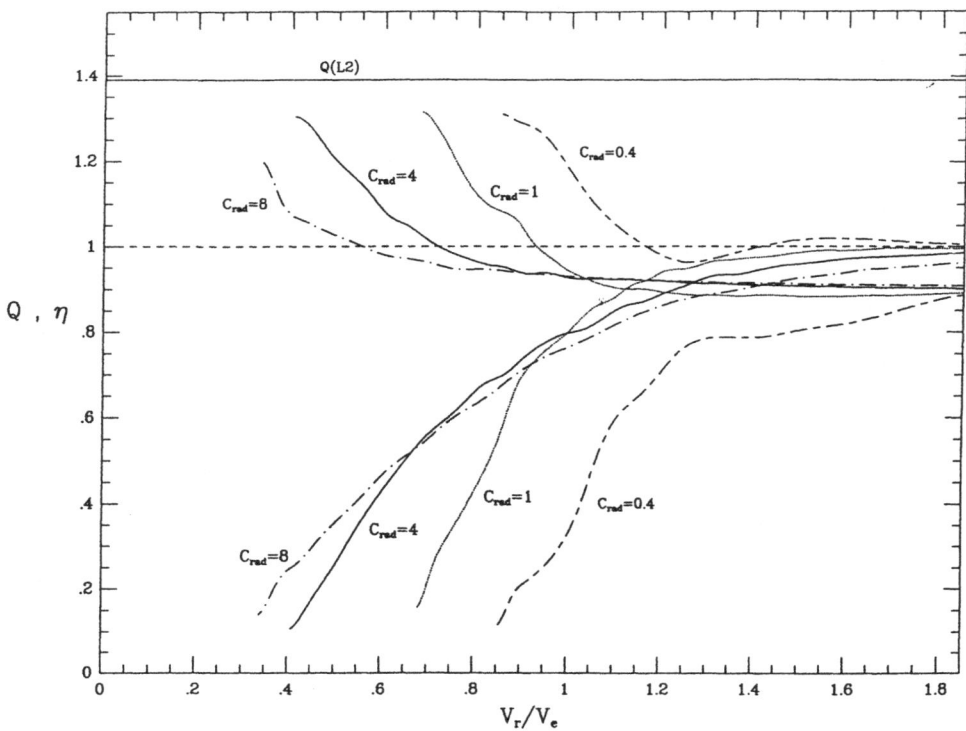

Figure 1. The values of Q [upper curves] and of the fraction, η, of the projected companion surface area from which the wind is emitted [lower curves], for some *1957+20* ballistic trajectories simulations. The binary parameters used here were $m = 0.02M_\odot$, $M = 1.4M_\odot$, stellar separation $a = 2.4R_\odot$, companion radius $0.16R_\odot$ (except in the $C_{rad} = 0.4$ curve, where the companion radius is $0.21R_\odot$) and synchronous rotation. The Roche lobe radius is $0.26R_\odot$. Notice that $Q - 1$ is negative for $\frac{v_r}{v_e}$ values around and above unity, but becomes positive for lower values. The curves will terminate on the left when v_r becomes too small for *any* particle to leave the companion (we terminated our own simulations already when η decreased to ~ 0.1 because of low particle statistics). $Q(L2)$ is the value of Q for particles that corotate at the Lagrangian point L2.

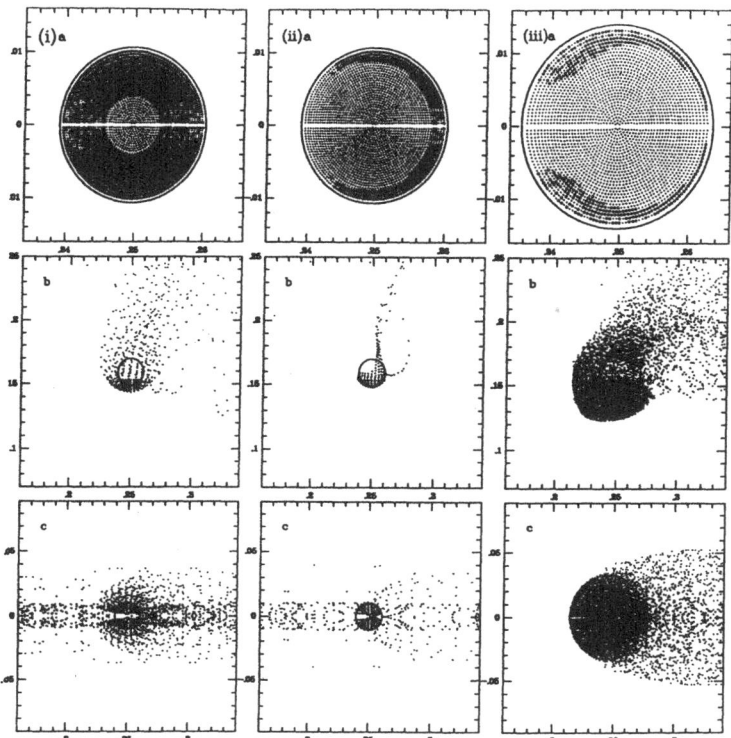

Figure 2. Wind geometries for simulations with binary parameters that are possible for *1957+20*. In (i) and (ii), all the stellar parameters are as in Fig. 1, with $C_{rad} = 4$ and with $\frac{v_r}{v_e}$ values of 1 [(i)a, (i)b and (i)c] and 0.45 [(ii)a, (ii)b and (ii)c]. In (iii), which we consider, at present, the best for *1957+20*, $C_{rad} = 0.4$, $\frac{v_r}{v_e} = 0.86$ and the companion radius is assumed to be larger, at $0.21R_\odot$, consistent with the upper bound of ref. 24. The "a" panels show, by dark dots, the wind emission region on the companion disk as seen from the NS [for (i) $\eta = 0.80$, for (ii) $\eta = 0.17$ and for (iii) $\eta = 0.14$]. The companion in these panels travels to the left. The "b" and "c" panels show snap-shots of particles viewed from directly above the binary plane ["b"; the NS is below and the companion moves to the left] and from the NS ["c"; the companion moves to the left]. All projected distances are normalized to $2\pi a(1 + q)^{-1}$. The voids seen in the companion equator are artifacts of the simulations and the black circles inside the coronas represent the boundaries of the companion. For the production of the snapshots in (iii), trajectories of escaping particles were truncated beyond ~ 2 orbital radii, to produce a realistic coronal contrast.

Notice the bending of trajectories *towards* the companion, best seen in the (ii) snap-shots, brought about by the dominance of the companion gravity over the Coriolis force when wind velocities are low, and causing Q in this particular simulation to be 1.26. For (i), Q=0.932 and for (iii), Q=1.28.

The *1957+20* eclipse can be best represented by a horizontal line that crosses (iii)c at a distance of ~ 0.015 units above the center of the companion. Notice that along this line there are at least 3 (5 if the line crosses the companion itself) jumps in column density, making the eclipse quite on-center for the lowest radio frequencies (with a somewhat fuzzy egress), but shifting its center towards the ingress, possibly in jumps, when the frequency increases.

companion is seen unobscured. This will raise the above upper limit by another factor of $\sqrt{2}$ to $2 \cdot 10^{10} cm$, the Roche lobe of the companion!]. Trajectories in this simulation are different from those in (ii) in that most escaping particles originate in the *front* of the companion. They still, however, spend more time *behind* it: they start out going around the non-illuminated side of the companion, with both its gravity and the Coriolis force pointing towards it, get delayed (or even form a loop) while in the region trailing it, and only then start their trip out. The $Q - 1$ value in these simulations turned out to be 0.28.

The observed symmetry of the eclipse in *1957+20* around the companion for the lower radio frequencies[11] indicates that our line of sight lies sufficiently close to the binary plane that it is intercepted at least by the corona, which, for low frequencies, is quite symmetric [Fig. 2, panel (iii)c]. The simulated column density decreases, possibly discontinuously, towards the eclipse egress, so that one may expect a shift of the eclipse center towards the ingress side as the radio frequency increases. The ingress phase itself seems to remain unchanged. Detailed hydrodynamics is expected to be of crucial importance to the complete understanding of the eclipse, however.

Simulations for LMXBs. It is not too difficult to understand *increasing* periods in LMXBs in view of the fact that matter is transferred from a low mass companion to a high mass NS. Examination of Eq. (4) shows that if q is small, realistic tidal Q values will require β to be small too. In Eq. (3) we substitute $\dot{m}_w = \dot{m}_{acc}\beta^{-1} = \dot{m}_E C_{rad}\beta^{-1}$ where $\dot{m}_{acc}$ is the accretion rate on the NS and where $\dot{m}_E$ is the relevant Eddington luminosity. We also define Ψ_o to be the value of Ψ for $\dot{m}_w = \dot{m}_{acc}$ and for $m = M$. With these we solve Eq. (3) for Q to find

$$Q' \equiv (1 - \beta)Q = 1 + \frac{2}{3}q + q\left[\frac{(1+q)\Psi_o}{3} - \left(\frac{2}{3}+q\right)\right]\beta, \qquad (5)$$

a linear function in β that we can plot along with simulations' results to solve for the system parameters. By Eq. (2), "standard" wind-formation efficiency constrains

$$\beta \succeq \beta_{min} \equiv \left(\frac{10^{-3}}{\psi_s}\right)\left(\frac{v_r}{300km/sec}\right)\Lambda^{-1} \qquad (6)$$

whenever the NS luminosity is fueled itself by that wind [i.e. in bootstrapping]; we therefore look in our simulations for $\left(Q', \beta\right)$ pairs that satisfy this constraint and Eq. (5).

We note, that the *slope* of Eq. (5) as a function of β is positive when $\Psi_o > 2 + \frac{q}{1+q}$, but when P_{orb} decreases at a slower rate or when it is altogether *increasing*, the slope of Eq. (5) goes negative. Fig. 3 shows the two cases, from which it is clear that for a negative slope, a simultaneous solution of the simulations' Q' curve and of Eqs. (5) and (6) exists for any $q \succeq 6 \cdot 10^{-5}\left(\frac{v_o}{300km/sec}\right)^3 f^{-\frac{15}{2}}\Lambda^{-3}$, where f is the Roche lobe filling factor, by radius, of the companion. Since solutions will involve small q's when Ψ_o is large, the LMXBs in question here seem self-consistently to be VLMXBs.

Fig. 4 investigates the two $\dot{P}_{orb} < 0$ LMXBs. Plotted as a function of q are the *largest* expected value of β, β_{max}, from Eq. (5) [the β value obtained from Eq. (5) when Q is the one for particles at L2]; and, from Eq. (6), the *smallest* expected value for β_{min} in the case of $v_r = v_e$ and $\Lambda = 1$, β_{min}' [taken at Roche lobe filling but without the extra factor of 3 in the estimate[6] of the wind mass-flow]. We see that for *both* LMXBs, a contradiction

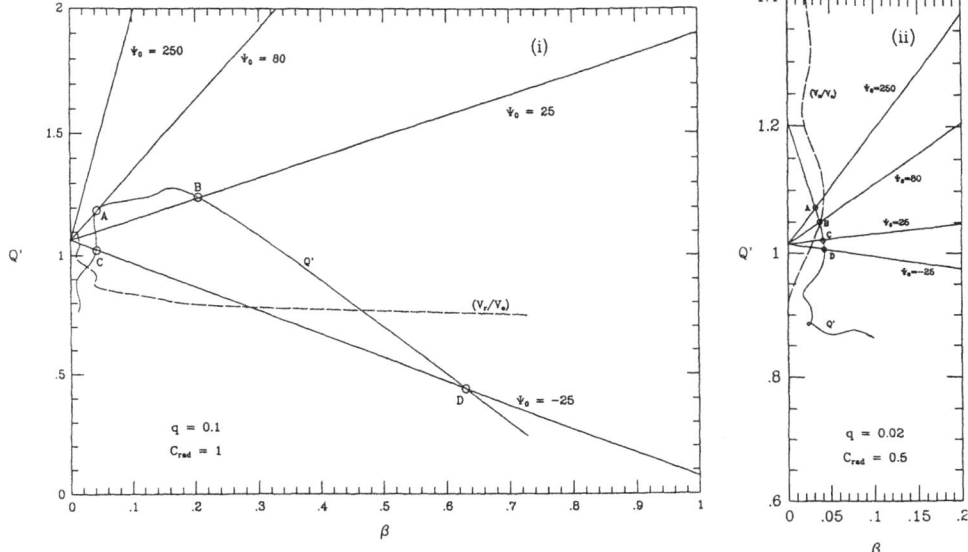

Figure 3. Simulations for *EXO0748-676* with a companion radius of $0.16R_\odot$, roughly the smallest radius that is still consistent with $8.3min$ eclipses during a $3.82hr$ orbital period. Results are shown for two companion masses: $0.14M_\odot$ [(i)], where the Roche lobe radius is $0.28R_\odot$, and $0.03M_\odot$ [(ii)], a companion which just fills its lobe. The dotted lines are the Q' results of the simulations: in (i) the disk is assumed to extend to 0.3 of the binary separation and $C_{rad} = 1$; in (ii) the fraction is 0.1 and $C_{rad} = 0.5$. The dashed lines depict the values of $\frac{v_r}{v_e}$ which generate each value of Q' and β in the runs. The solid curves plot Q' as a function of β from Eq. (5) for various values of Ψ_o: 250 [if the true luminosity is the average of the observed one], 25 [if the true luminosity is at the Eddington value but there is much obscuration in the system] and, for illustrative purposes, -25 [if the orbital period were to *increase* at the observed rate and at the Eddington luminosity]. Another Ψ_o value for which Eq. (5) is plotted is 80 [the lowest one that would still be consistent with bootstrapping in case (i), representing a $\dot{P}_{orb}$ value that is ~ 3 times larger than the average value observed or a luminosity of ~ 0.3 of the Eddington value or any combination]. For the A and C solutions in (i) we must have $\Lambda \succeq 3$; B and D of (i) are consistent with $\Lambda = 1$. All solutions in (ii) require $\Lambda > 3$.

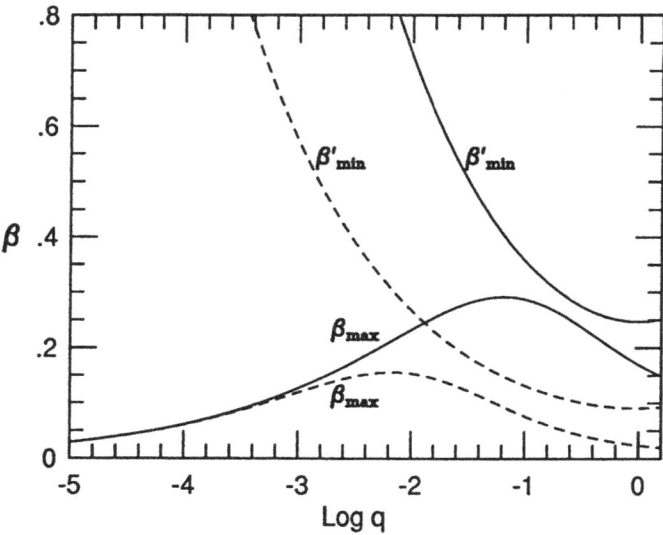

Figure 4. The values of β_{max} [obtained by solving Eq. (5) for the appropriate Q value of particles at the Lagrangian L2 point] and of β_{min}' [the β_{min} value from Eq. (6) with ψ_s calculated for a Roche-lobe-filling companion, $v_r = v_e$ and $\Lambda = 1$] as a function of q, for the two $\dot{P}_{orb} < 0$ LMXBs. The solid lines are for *EXO0748-676* ($\Psi_o = 250$), the dashed lines - for *4U1820-30* ($\Psi_o = 30$). Notice, that for each LMXB the lines do not even cross. For bootstrapping to be allowed, we must have $\Lambda \succeq 2$ in order to lower the β_{min}' curves.

$\left(\beta_{min}' > \beta_{max}\right)$ exists for all values of q. Unless $\Lambda \succeq 2$ or so, the inequality will not reverse for *any* q and bootstrapping will be ruled out. If Λ is large enough and solutions exist, they will typically involve, again, small values of q and β. A Roche-lobe-filling bloated companion could, thus, be consistent with bootstrapping

When $\dot{P}_{orb} < 0$, a sufficiently large Λ is still not a guarantee for an overall solution of the simulations' Q' curve and Eqs. (5) and (6). The existence of a solution will depend further on the actual capture cross section of the NS and on the true value of Ψ_o. In *EXO0748-676*, for example, $\Psi_o \sim 250$ if the true system luminosity is, indeed, equal to the average observed value of $1 \cdot 10^{37} erg/sec$. But if the transient nature of this system and its large flarings signify that we look at a NS that is obscured by a disk which only occasionally lets the full X-ray luminosity emerge into our line-of-sight, then the true luminosity may be at the Eddington level and $\Psi_o \sim 25$. From Fig. 3 we see that in the latter case, finding a consistent bootstrapping solution is straightforward with $\Lambda = 1$, while in the former it will be more difficult yet still possible, depending on the actual size of the disk. In the Figure we consider companions with a radius close to the minimum possible given the duration of the eclipse. We note, that in view of the fact that the ingress and egress widths found in *EXO0748-676* could arise from the radiation-heated corona[25] of our model, such an assumption about the companion radius is especially attractive. We see from Fig. 3 that it is actually possible for Ψ_o to be ~ 80 in the high state. Therefore, the magnitude of $\dot{P}_{orb}$ during the high state could be as high as ~ 3 times the average observed $\dot{P}_{orb}$, which might compensate for the smaller values expected during the low states, when the companion may see a reduced luminosity as well(and a reduced C_{rad} value).

IV. Conclusion

We propose a complex companion-wind configuration, the result of a low initial wind speed and a high radiation pressure, to account for the large orbital period changes in LMXBs and BMPs. While hydrodynamical calculations are unavailable at present, ballistic simulations are quite suggestive that under such conditions flows could torque the binary motion rather effectively. In particular, such wind patterns can naturally produce $\dot{P}_{orb} < 0$ values; these would be as large as the observed ones if the wind production efficiency is sufficiently high. Having the companion bloated out of thermal equilibrium and close to its Roche lobe[26] is a way of getting an effective $\Lambda \sim 3$ for the light companions, even though at present it seems that Λ needs to be somewhat (but not much) higher than that. To evolve while maintaining contact, the exponent of the companion's mass-radius relation should adjust to the instantenous Q' value. The most important questions now are how is that high efficiency obtained, how is v_r actually determined by the system parameters and whether or not the need for it to have these specific values is compatible with LMXB or BMP statistics. We hope to take these questions up in a future publication.

V. Acknowledgments

It is a pleasure to thank participants in the Workshop on Neutron Stars in Binary System, held at the ITP, for many stimulating discussions. The First-Class hospitality of the workshop Directors, Professors E.P.J. van den Heuvel and S.A. Rappaport, and of the Director and staff of the ITP is gratefully acknowledged. This work has been supported, in part, by NASA (Grants NAGW-1618 and NCC 5-37) and the NSF (Grant PHY89-04035).

VI. References

1. Tan, J. *et al.* *Astrophys. J.* (in the press, 1991).
2. Parmar, A.N., Verbunt, F., Smale, A.P. & Corbet, R.H.D. *Astrophys. J.* **366**, 253-260(1991).
3. Kitamoto, S., Miyamoto, S. & Matsui, W. *Publs. astr. Soc. Japan* **39**, 259-285(1987).
4. Hellier, C., Mason, K.O., Smale, A.P. & Kilkenny, D. *Mon. Not. r. astr. Soc.* **244**, 39P-43P(1990).
5. Ryba, M.F. & Taylor, J.H. *in Proc. NATO Santa Barbara ARW on X-Ray Binaries and the Formation of Binary and Millisecond Pulsars (eds. E.P.J. van den Heuvel & S.A. Rappaport)* (Kluwer Academic Publishers, 1991).
6. Ruderman, M., Shaham, J., Tavani, M. & Eichler, D. *Astrophys. J.* **343**, 292-312(1989).
7. Ruderman, M., Shaham, J. & Tavani, M. *Astrophys. J.* **336**, 507-518(1989).
8. Kluzniak, W., Ruderman, M., Shaham, J. & Tavani, M. *Nature* **334**, 225-227(1988).
9. Shaham, J. *in Proc. NATO Crete ASI on Neutron Stars - an Interdisciplinary Field (eds. J. Ventura & D. Pines)* (Kluwer Academic Publishes, 1991).
10. Shaham, J., this volume.

11. Fruchter, A.S., Stinebring, D.R. & Taylor, J.H. *Nature* **333**, 237-239(1988).

12. Lyne, A.G. *et al. Nature,* **347,** 650-652(1990).

13. London, R.A., McCray, R. & Auer, L.H. *Astropys. J.* **243,** 970-982(1981).

14. London, R.A. & Flannery, B.P. *Astrophys. J.* **258,** 260-267(1982).

15. Eichler, D. & Levinson, A. *Astrophys. J. (Lett).* **335,** L67-L70(1988).

16. Basko, M.M., Hatchett, S., McCray, R. & Sunyaev, R.A. *Astrophys. J.* **215,** 276-284(1977).

17. Webbink, R.F. *Astrophys. J.* **209,** 829-845(1976).

18. Pringle, J.E. *Mon. Not. r. astr. Soc.* **248,** 754-759(1991).

19. Banit, M. and Shaham, J. *Astrophys. J. (Lett.)* **388,** L19-L22 (1992).

20. Tassoul, J.L. *in Angular Momentum Mass Loss of Hot Stars (eds. L.A. Wilson & R. Stalio)* 7-32 (Kluwer Academic Publishers, 1990).

21. Tavani, M. *Nature* **351,** 39-41(1991).

22. Rasio, F.A., Shapiro, S.L. & Teukolsky, S.A. *Astrophys. J.* **342,** 934-939(1989).

23. Press. W.H., Flannery, B.P., Teukolsky, S.A. & Vetterling, W.T. *Numerical Recipes* (Cambridge University Press, 1986).

24. Djorgovski, S. & Evans, C.R. *Astrophys. J. (Lett).* **335,** L61-L65(1988).

25. Parmar, A.N., White, N.E., Giommi P. & Gottwald, M. *Astrophys. J.* **308,** 199-212(1986).

26. Podsiadlowski, Ph. *Nature* **350,** 136-138(1991).

X-RAY DIPPERS, ECLIPSING PULSARS AND THREE-DIMENSIONAL ORBITS

WŁODZIMIERZ KLUŹNIAK
Physics Department, Columbia University, New York, NY 10027 USA
and *Institute for Theoretical Physics, University of California Santa Barbara*
and *Subdepartment of Astrophysics, Oxford University*

ABSTRACT. Partial eclipses ("X-ray dips") and rapid orbital evolution of some accreting neutron stars can both be understood within the single framework of mass outflow from the binary system. The shape of partial X-ray eclipses is correlated with the binary period. The gradual eclipses in "ADC sources" resemble the radio eclipse in PSR 1957+20 and may be caused by a comet-like outflow. Shorter period sources exhibit more structure in their partial eclipses and may be similar to PSR 1744-24A. A new model is proposed for the geometry of the radio eclipse in the latter pulsar system.

1. Introduction

It should now be clear that the distinction between radio pulsars with low-mass binary companions and X-ray bright neutron stars with low-mass companions is a matter of the degree to which mass lost by the companion enters the sphere of influence of the neutron star (if no mass penetrates the light cylinder the neutron star can be a radio pulsar), rather than a matter of some fundamental qualitative difference. That is—contrary to the popular dichotomy whereby in "low-mass binary pulsars (LMBPs)[1]" the companion loses no mass while in "low-mass X-ray binaries (LMXBs)[1]" all the mass lost by the companion is captured by the neutron star—actually, in all binary systems composed of a detectable neutron star and a less massive companion, some matter lost by the companion can enter the Roche lobe of the neutron star and some matter can be ejected from the system. It is the motion of such matter in close binary systems and its observable signatures that form the focus of this contribution.

I will argue that there is a considerable similarity between eclipsing radio pulsar systems and their accreting counterparts, the so called X-ray dippers—some of which also show "true" eclipses, i.e. ones caused by the companion star. But first a note on word usage. In the interest of uniformity in treatment, I will allow any decrease in the flux from the source to be called an eclipse, provided that it is (presumably) caused by intervening matter and not by intrinsic fluctuations of luminosity of the actual source. In practice, any flux decrease which is related to (occurring at fixed phases of) orbital motion will be deemed to have satisfied that criterion. A partial eclipse is an eclipse in which the flux from the source does not go to zero, this could be either because only a part of an extended source is eclipsed totally or because the complete source is obscured by material of optical depth not sufficient to block out (absorb, scatter, reflect, smear any pulses) the radiation totally, or both.

---

[1] Those who have a nostalgic attachement to parts of speech other than nouns may wish to call elembee peas and elemex bees radio pulsars having a companion of lower mass and neutron stars accreting matter from a companion of low mass, respectively.

E. P. J. van den Heuvel and S. A. Rappaport (eds.), X-Ray Binaries and Recycled Pulsars, 413–424.
© 1992 *Kluwer Academic Publishers*.

2. Eclipsing radio pulsars I. PSR 1957+20

Two eclipsing radio pulsars are known at present. The first one to be discovered, PSR 1957+20, exhibits a rather symmetric eclipse, centered on the phase of superior conjunction of the pulsar, and lasting for about one tenth of the binary period (ref. 3). The eclipse is now understood to be caused by a plume of matter leaving the companion and swept back by pulsar radiation pressure (Fig. 1). The inferred appearance of the plume is reminiscent of a cometary tail.

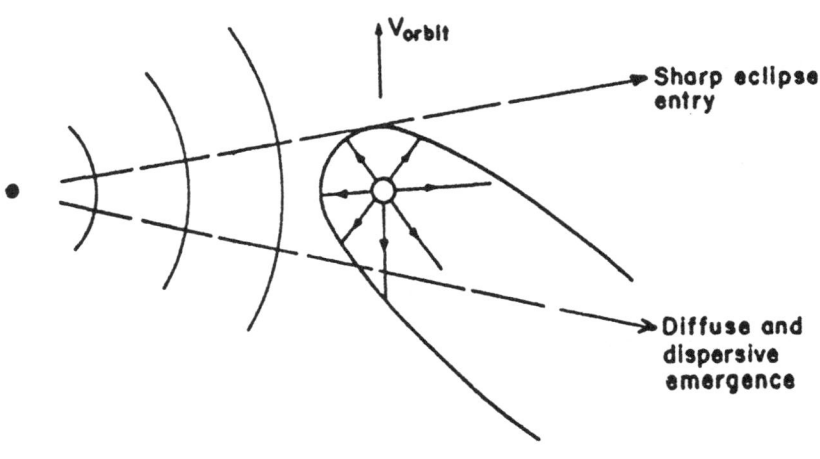

Figure 1. Bow shock and wake from the interaction of an evaporative plume with the kHz magnetic dipole radiation of a powerful millisecond pulsar. This figure is from ref. 8.

The plume is continuously fed by matter ablated (ref. 17) from that face of the companion which is irradiated by the pulsar. The prediction (ref. 7) of this model that the side of the companion facing the pulsar should be brighter has been borne out by optical observations (ref. 1).

There are several important lessons to be learned from the properties of this eclipsing pulsar. The first one is that even considerable mass loss from a binary companion need not totally obscure pulsed high-frequency radio emission. The pulsar is capable of pushing plasma away from its vicinity (as should have already been apparent from observations of the Crab pulsar). This is something of an embarassment for those theories of pulsar evolution (see ref. 20 for a review) which would have radio emission from young strongly magnetized binary pulsars hidden from view by wind from a main sequence companion.

The second lesson is that the classical Roche-lobe geometry is not necessarily relevant to systems with high radiation pressure. In such systems, the flow of matter is controlled not only by the attractive forces of gravity but also by the repulsive radiation force. In the optically thin regime, the repulsive central force caused by the pressure of radiation from the neutron star falls off (under some obvious assumptions) as the inverse square of the distance ($\sim 1/r^2$) and it is convenient to think of it as antigravity. A comparison (ref. 5) of "equipotential" contours with (Fig. 2b) and without (Fig. 2a) this antigravity term shows that symmetric, cometary-tail like (reminiscent of the eclipse geometry of PSR 1957+20), outflow from the low mass companion is a generic feature of binary systems with a strongly radiating primary (neutron star).

The third lesson is that details of the mass flow in the binary system may profoundly affect the evolution of the binary system. This statement is supported by the measured value, large in magnitude and negative in sign, of the rate of change of the orbital period of PSR 1957+20—the orbit decays on a timescale of $\sim 3 \times 10^7$y. A more extensive discussion of this point and of its possible implications can be found in the companion contribution (Kluźniak, Czerny, Ray 1991).

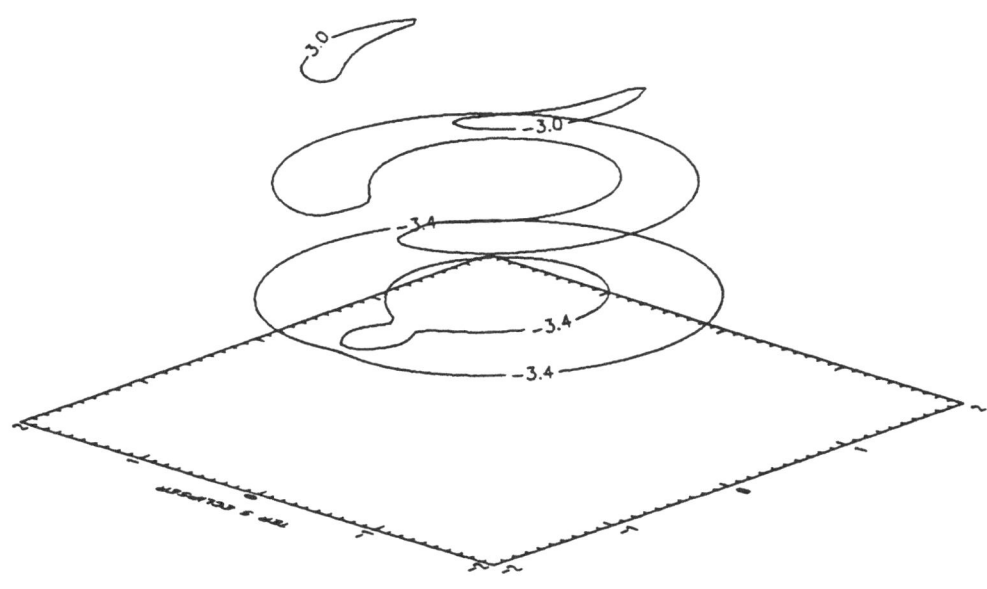

Figure 2a. Some contours of the Jacobi integral computed in the orbital plane for the restricted three-body problem with mass ratio of the two binary components equal to 0.063. The more massive component is located at the origin of the planar coordinate system shown, the companion is at $(-1, 0)$. The contours have been raised to bring out the analogy with equipotential contours. The Roche lobe structure is apparent in the lowest contour, the presence of Lagrange points L4 and L5 is apparent in the highest contour.

416

3. A conjecture.

I make the conjecture that partial eclipses ("dips" in the X-ray light curve) in low mass X-ray binaries are caused—as they are in the eclipsing radio pulsars—by that part of matter outflowing from the companion star which is not captured by the neutron star. The accretion disk and even accretion itself are incidental to the eclipse, dynamically important, perhaps, by providing radiation pressure and affecting the budget of mass and angular momentum transfer, but not being directly related to the obscuring material.

4. X-ray dippers I. Smooth symmetric eclipses.

Partially eclipsing accreting neutron stars were discovered long before their radio pulsar counterparts. Perhaps in part because of this, and in part by analogy with partial solar eclipse, symmetric partial eclipses have been interpreted as total obscuration of an extended X-ray source by the companion. This interpretation necessitated the introduction of a postulated scattering corona around the neutron star and its accretion disc (ref. 21 and references therein).

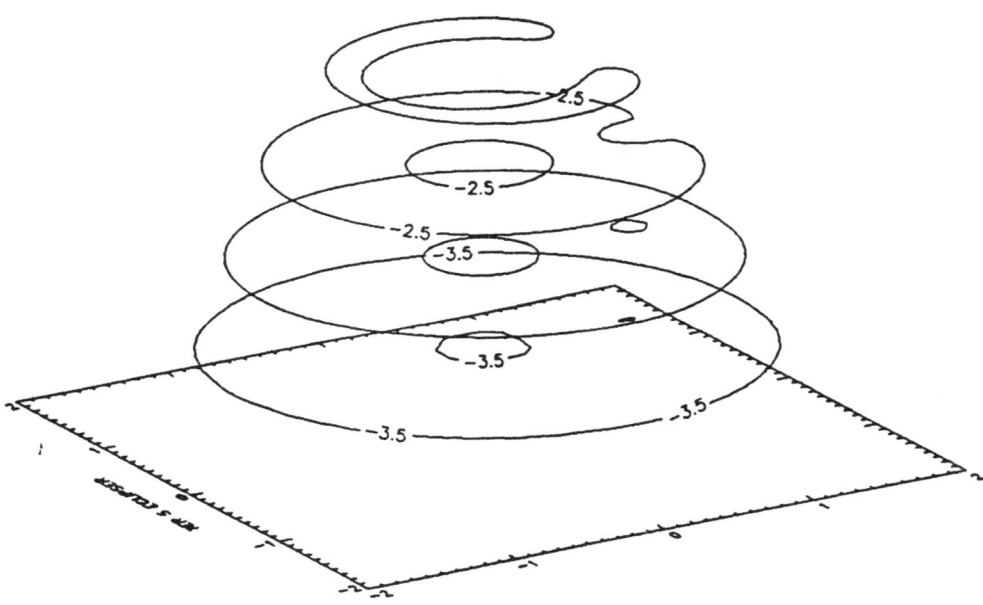

Figure 2b. Same as Fig. 2a, but with inclusion of a repulsive $1/r^2$ force from the more massive component. The repulsive force acts on the "third body" (test particle) but not on the companion, i.e. the binary orbital period and separation are unaffected by its presence. The strength of this "antigravity" term has been adjusted to correspond to exactly one half Eddington luminosity in the case when it originates in Thomson scattering by optically thin plasma of X-rays from the more massive component. Figure taken from ref. 5.

Interestingly enough, the possibility that the gradual and symmetric eclipse in the X-ray burst source XB 2129+47 is caused by an extended fluid structure around the companion star has been considered in the classic paper of McClintock *et al.* (1982). However, it was rejected out of hand on the grounds that such structure, of required size $\sim 2R_\odot$ in a system with a low-mass companion in a 5.3 h orbit, could not be in hydrostatic equilibrium. Today, when we are familiar with the dynamical structure of $\sim 1.4R_\odot$ diameter (the evaporative plume) in the eclipsing pulsar system of 9.2 h period, PSR 1957+20, the objection appears not to be insuperable.

It seems, that a hypothesis explaining symmetric gradual X-ray eclipses in terms of mass outflow from the companion (and from the system) must today be regarded as the conservative one, while the accepted explanation in terms of a postulated "accretion disk corona" (ADC) should be regarded as more speculative, being—as it is—*ad hoc* in nature. Among other difficulties, the ADC model must cope with the empirical fact that in one X-ray burster (XB 2129+47, $P_b = 5.3$ h; ref. 10) the eclipse is gradual, while in another (XB 0748-68, $P_b = 3.8$ h; ref. 12) the total eclipse is quite sharp ("U-shaped"), so presumably in one system the ADC exists while in the other, for reasons not known, it is absent. This observation is not equally fatal to the mass outflow model, because details of flow (with characteristic length scale comparable to the binary size) are more easily influenced by the orbital geometry of the system (e.g. sensitive to the orbital separation, hence to the orbital period) than the structure of an ADC would be.

A common thread of my argument is this: all models attempting to explain the shape of eclipses (dips) by phenomena occuring close to the neutron star require that the orbital period be imprinted—in a more or less artificial way—on a dynamical system in which the same period plays no role (orbital tidal forces are a negligible perturbation deep inside the gravitational well of the neutron star). In contrast, the flow pattern in the binary system of matter which is at distances from the neutron star comparable with the semi-major axis of the binary, and therefore is influenced by both bodies, depends in a natural way on the binary characteristics, such as the period.

5. X-ray dippers II. Asymmetric eclipses.

In addition to the symmetric eclipses, discussed above, the X-ray burst source XB 2129+47 shows (at other orbital phases) extended asymmetric partial eclipses (dips), similar to those observed in Cyg X-3. Some other sources, notably XB 0748-68, show rather sharper, numerous and more variable structures. All those dips are frequently interpreted as eclipses caused by a corrugated rim of the accretion disk (ref. 21 and references therein). Involved, multi-component models (including accretion disk corona) have been constructed to account for the variation of the light curve at different wavelengths.

While these empirical ADC models impressively fit the data, their physical basis is not secure. No physical mechanism has been proposed to account for the shape of the disk rim, for example. That is, it is easy to loft material (moving in Keplerian orbits around the neutron star, as those models assume) from the orbital plane, but it is not at all easy to bring it back into the plane with the required rapidity. For this reason, and others, Frank, King and Lasota (1987) proposed a different model. Their model has the advantage of imprinting the binary phase on the light curve and has the attraction of being unified in the sense that different classes of eclipses would be seen from the same source at different inclination angles, i.e. all the eclipsing LMXBs are supposed in that model to be intrinsically similar but observed at different angles.

A cursory examination of the period distribution (ref. 16) of the various types of eclipses in LMXBs reveals that the sharper erratic dips occur preferentially at shorter orbital periods, while the smoother partial eclipses ("ADC" objects, as they are denoted) occur in three sources located next to each other in a list ordered by the value of the orbital period. The three sources are XB 1822-37 with $P_b = 5.5\,\text{h}$, XB 2129+47 with $P_b = 5.3\,\text{h}$ and Cyg X-3 with $P_b = 4.8\,\text{h}$. This circumstance is fatal to the Frank, King and Lasota (1987) model, as the accidental and extrinsic inclination angle at which a binary system is observed is not expected to be correlated with the orbital period of the system (barring some unrecognized selection effect). As pointed out in the previous Section, the ADC model similarly fails to account for the orbital period distribution of these sources.

6. Orbital evolution. Effects of mass loss.

It is interesting to note that two of the three partially eclipsing sources have a measured time derivative of the orbital period and it has the same (positive) sign in both cases (the orbit expands). For 1822-37, $P_b/\dot{P_b} \sim 3 \times 10^6\text{y}$, while for Cyg X-3 $P_b/\dot{P_b} \sim 5 \times 10^5\text{y}$. Contrast this with the decaying orbit of the erratic dipper 0748-68, where $P_b/\dot{P_b} = -5 \times 10^6\text{y}$. The difference in sign of $\dot{P_b}$ can be made consistent with a difference in shape of the eclipse if our conjecture (Section 3) is accepted, but cannot be understood within the framework of any other of the proposed models of the partial eclipses.

In the limit of small companion mass, $m_2 << m_1$, and when loss of orbital angular momentum is caused chiefly by material losses,

$$\dot{J}/J = \beta(1 - f)\dot{m}_2/m_2, \tag{1}$$

where $0 \leq f \leq 1$ is that fraction of mass lost by the companion which is accreted by the neutron star, $\dot{m}_1 = -f\dot{m}_2$, and β is defined by eq. (1), the orbital period derivative is given by (compare Rappaport, Joss & Webbink 1982)

$$\dot{P_b}/P_b = -3[1 + f\beta - \beta]\dot{m}_2/m_2, \tag{2a}$$

$$\dot{P_b}/P_b = 3(fm_2/m_1)^{-1}[1 + f\beta - \beta]\dot{m}_1/m_1. \tag{2b}$$

In this limit, $\dot{P_b} > 0$ iff $(1 - f)^{-1} > \beta$, and

$$\dot{P_b} < 0 \text{ iff } (1 - f)^{-1} < \beta. \tag{3}$$

Thus, the sign of the orbital period derivative is quite sensitive[2] to the value of β (i.e. to the specific angular momentum of matter ejected from the system), and hence to details of mass flow.

Unlike $\dot{m}_2$, the accretion rate $\dot{m}_1 \geq 0$ can be inferred from the X-ray luminosity, hence eq. (2b) is more directly related to observable quantities than eq. (2a). It is obvious from eq. (2b) that the smaller the mass ratio of the companion to the primary, (m_2/m_1), the larger the orbital change, and that (except when $\beta = 1$ exactly) the rate of change of the orbital period increases with increasing mass loss from the system ($\dot{P_b} \propto f^{-1}$). At

---

[2] For $\beta = 1$, $\dot{P_b}/P_b = 3\dot{m}_1/m_2 > 0$ at any value of f and for any value of $m_2 << fm_1$.

the same time, if $\beta =$const> 1, the orbital period derivative will change its sign once (see relation [3]) at some value of $f = f_0$ as the fraction of mass accreted varies, $\dot{P}_b$ being necessarily positive for $f = 1$ (for any β) and negative for $f = 0$ (iff $\beta > 1$).

In the eclipsing pulsar 1957+20 there is no accretion, of course, $(f = 0)$ and by eq. (2a) the observed decay of the orbit $(P_b/\dot{P}_b = -3 \times 10^7 y)$ implies $\beta > 1$. If $\beta > 1$ were appropriate[3] also for Cyg X-3, the maximum possible mass of Cyg X-3 would be $m_2 = 3\dot{m}_1 P_b/\dot{P}_b$, i.e. $m_2 \leq 0.02 M_\odot$ (the upper limit being attained for $f = 1$) for a mass accretion rate of $10^{-8} M_\odot/y$, regardless of the actual value of β.

However, for $\beta < 1$ higher companion masses are allowed. In the limit of $f << 1$ (i.e. of large mass loss from the system), $m_2 = 3f^{-1}(1 - \beta)(P_b/\dot{P}_b)\dot{m}_1$. E.g. for $\beta = 0.7$, the value $m_2 = 0.05 M_\odot (f/0.1)^{-1}(\dot{m}_1/10^{-8} M_\odot\, y^{-1})$ is obtained for Cyg X-3, and $m_2 = 0.3 M_\odot (f/0.1)^{-1}(\dot{m}_1/10^{-8} M_\odot\, y^{-1})$ for XB 1822-37.

7. Eclipses and orbital evolution.

In the preceding section, I have outlined how the orbital evolution of several low-mass X-ray binaries can be understood in terms of mass loss from the systems. Future work will show whether the constraints, discussed above, on the fraction of mass lost from the system and on the angular momentum carried away by the mass lost are indeed satisfied for those systems.

Note that the mass outflow model (proposed by the Columbia group) also gives a satisfactory framework for understanding the X-ray light curve of Cyg X-3 as detailed modeling of the eclipse has shown (ref. 11). While similar work must be performed for the other partially eclipsing sources, at present there is every reason to trust that matter leaving the companion and orbiting the binary system will indeed modulate the light curve in the observed way also for the other two $\sim 5\,$h period sources (1822-37 and XB 2129+47).

8. Eclipsing radio pulsars II. The Terzan 5 eclipser.

In the previous sections the similarity between the eclipses in the pulsar 1957+20 and in some accreting neutron stars with periods $\sim 6\,$h were discussed. However, the partial eclipses of other sources discussed in Section 5 are quite different. In this Section, I point out that the eclipses observed in the radio pulsar 1744-24A are also quite different than in 1957+20. A simple model for the eclipse which is surprisingly successful in explaining major features of the eclipses in the Terzan 5 pulsar is presented in Section 9.

The binary system PSR 1744-24A contains an 11.6 ms pulsar in a cirular 1.8h orbit (ref. 9). The spin derivative, $\dot{P} \leq 10^{-20}$, is unusually low for this pulsar period and yields a spin-down power $I\omega\dot{\omega} \leq 10^{32}$erg/s, which is at least three orders of magnitude lower than the spin-down power of Fruchter's pulsar, PSR 1957+20. This implies that the radiation pressure on plasma in the PSR 1744-24A system is much lower than that in the other eclipsing pulsar.

There are three or four major distinguishing features of the radio eclipse in PSR 1744-

---

[3] This hypothethical assumption could be loosely justified by the similarity of the Cyg X-3 system to the XB 2129+47 system, which in turn shows eclipses similar to those of PSR 1957+20.

24A (ref. 9).

i) Typically the eclipses (at ~ 400 MHz) last for about one half the orbital period. Sometimes they are even longer, occasionally the pulsar is unobservable for several periods.

ii) The pulsar has been observed in mid-eclipse, i.e. at superior conjunction (phase = 0.25 in the convention of ref. 9), while it was eclipsed at neighboring phases.

iii) When the eclipse becomes longer, the ingress occurs earlier than usual while the phase of the egress remains unchanged.

iv) Before the pulsar disappears from view for a duration exceeding the orbital period, the eclipse length increases monotonically, the pulsar going into eclipses at progressively earlier phases [point iii) above] in each successive orbit.

Properties i) and ii) suggest that the system is observed at low inclination angle (i.e. the orbital plane is at a large angle to the line of sight). Indeed, it is easy to show that the most probable inclination angle, i, satisfies $\cot i \sim \pi t_E/P_b$, where t_E is the duration of the eclipse and P_b is the orbital period. Also, if the system were seen edge on ($i \sim \pi/2$) the pulsar would always be eclipsed at superior conjunction, contrary to observations.

Properties iii) and iv) strongly suggest motion of the eclipsing fluid in the direction of orbital rotation (in advance of the companion). Most likely, the fluid penetrates the Roche lobe of the pulsar. This is quite unlike the behavior of PSR 1957+20, where the egress is more ragged than the ingress, as expected for matter moving outwards under the influence of Coriolis forces (Fig. 1). Thus, property iii) seems to rule out, for this system, cometary-like models of the eclipse. In addition, property iv) suggests that the characteristic time scale of motion of the eclipsing material is comparable to the orbital period.

9. A three-dimensional model.

Here, I am concerned only with the geometry of the eclipsing medium and not the physical mechanism of the eclipse (pulse smearing or refraction or reflection, etc.).

In Fig. 3, I present a three-dimensional closed periodic trajectory discovered by Goudas (1963). The orbit is stationary in the binary frame, in which the coordinates of the two stars were chosen to be, for the mass ratio $m_1/m_2 = 9$, $(x, y, z) = (0, 0, -0.1)$ for the primary and $(0, 0, 0.9)$ for the secondary. The tick marks show time in units of $P_b/(2\pi)$. Note that this orbit has exactly the same period, P_b, as the binary system, but the projection of the trajectory onto the orbital plane of the binary (spanned by the x and y axes) is traversed twice in one period P_b.

The nearly triangular shape of the projection onto the xy plane is clearly related to the breaking of circular symmetry by the binary companion. This and higher order trajectories exhibit an exact reflection symmetry with respect to the binary orbital plane, and approximate n-fold rotational symmetry about the primary. The orbits (but not the direction of motion) are also symmetric about the plane (xz) perpendicular to the binary orbital plane and passing through both stars. Motion along the orbits has always the same sense as the orbital motion of the binary, it is positive (counterclockwise) in the xy plane, as can be seen in Fig. 3.

I propose that the eclipses in PSR 1744-24A are caused by a stream of matter centered (like beads on a wire) on the orbit exhibited in Fig. 3. The stream density is conjectured to be decreasing with increasing distance to the orbit and to be tracking the trajectory with lesser accuracy as the same distance increases. For reasons as yet unknown, the orbit is being fed continually preferentially on only one side of the binary orbital plane, thus leading to only one half (i.e. one loop) of the complete orbit being populated, the fluid being dissipated out of the orbit at the first self-intersection point (at dimensionless time π in Fig. 3). The line of sight to the observer is assumed to be inclined to the (positive) z axis at an angle of about $30°$.

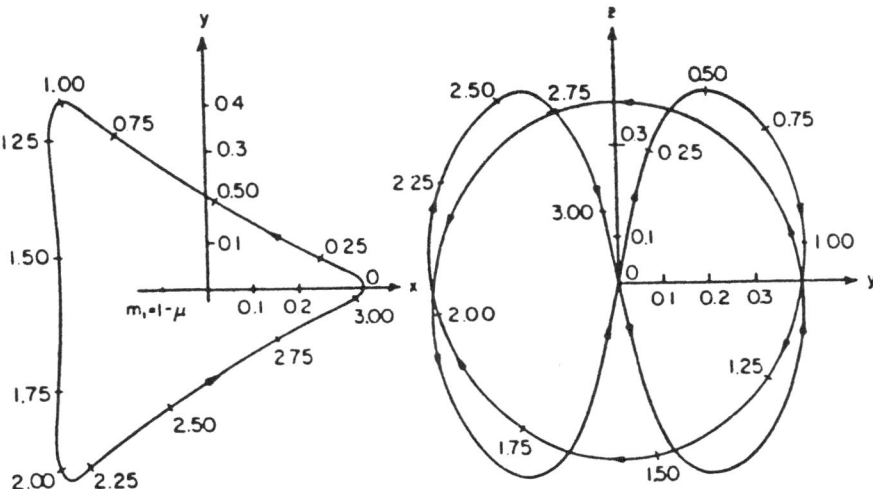

Figure 3. A three-dimesional doubly symmetric orbit found by Goudas (1963). Note that the trajectory consists of two loops, each of which can be obtained from the other by a reflection in the binary orbital plane (xy). Both stars are on the x axis, the pulsar would be at the position marked "$m_1 = 1 - \mu$." That self-intersection point of the orbit which is on the x axis lies between the two stars and is about 1.5 times more distant from the companion than from the neutron star. The figure is taken from ref. 18.

Occasionally, the other loop would also be filled with matter. As the fluid proceeds (and thickens) along the second half of the trajectory (the loop without tick marks in Fig 3.) it causes the ingress of the eclipse to occur earlier and earlier until the pulsar is eclipsed from all directions.

Thus, the typical length of the eclipse [property i) from Section 8] would be explained by organized motion of the eclipsing fluid close to a trajectory on which the gravitational forces of attraction, the centrifugal force and the Coriolis force all balance each other. This trajectory is quite distinct from other trajectories.[4] This accounts for the eclipses being reproducible even after marked fluctuations of their length. In the model presented

---

[4] In the restricted three body problem the subclass of closed orbits of interest here is discrete, unlike the continuous class of corresponding solutions of motion around a single gravitating body.

here, the flow of matter causing the basic (half orbital period) eclipse is only slightly affected by the changes in length of the eclipse, which are caused by matter flowing along a seperate part of the trajectory. Thus, the unusual constancy of the egress point as the eclipses become longer finds a natural explanation, as do the changes in the ingress phase[5] [properties iii) and iv)].

Strong support for the model presented here (or rather for its essential feature, motion of fluid along a well defined three-dimensional trajectory in the binary system) comes from property ii) of Section 8. It is immediately clear from Fig. 3 that the line of view to the pulsar of an observer stationed in the zz plane, but above (or below) the x axis would not be obstructed by the trajectory. Thus, the pulsar can reappear in the middle of eclipse.[6] This reappearance is a stumbling block for all other models considered to date, while it finds a natural explanation in our model.

9.1 SOME PREDICTIONS OF THE MODEL

1) As we have seen, according to this model of the eclipse, the second loop of the trajectory is usually empty of eclipsing fluid; at other times—when the eclipse is longer than usual— some of the matter in the system executes motion along this loop as well. It is reasonable to expect that in fact this loop is never completely devoid of matter, but usually the density of fluid following this part of the trajectory is rather low. Even in this low density region, the density of fluid along the trajectory is higher than elsewhere. Accordingly, even though the pulsar is not eclipsed during the corresponding half of the binary period, it is expected that when the line of sight intersects the trajectory, the increased density should manifest itself in increased dispersive delays or even perhaps a brief eclipse.

Hints of such behavior are in fact seen in the data. In Fig. 1 of Lyne *et al.* 1990 the pulsar is seen to have disappeared briefly at phase ~ 0.85 (mod 1) on 14Mar90. A disappearance or weakening at phase ~ 0.65 with possible delayed (in pulse phase) detection is apparent on 15Mar90 and 18Mar90.

2) The reader will have noted that the roles of the first and second loops of the trajectory in Fig. 3 would be reversed for an observer placed below (in the sense of negative values of the z coordinate) the binary orbital plane. That is, while in our postulated position above the orbital plane ($i < \pi/2$), our line of sight to the pulsar intersects the first loop of the trajectory above the binary orbital plane, thus leading to an eclipse centered on the phase of superior conjunction, an observer situated below the plane ($i > \pi/2$) would find her line of sight obscured by that part of the same loop which is on the side of the pulsar opposite to the companion. Therefore, the model predicts that some observers paradoxically see an eclipse centered on the phase of inferior conjunction of the pulsar, where the companion cannot possibly obscure the line of sight.

I predict that when a few more eclipsing pulsars of properties similar (in spin-down power and orbital period) to the one in Terzan 5 are discovered, one (or more) of them will exhibit eclipses centered on the phase of inferior conjunction of the pulsar.

---

[5] As noted in Section 8, this behavior is incompatible with the comet-like model of the eclipse proposed—following the limited success of similar models for the eclipse in PSR 1957+20, refs. 7, 8, 14—for PSR 1744-24A in ref. 15.

[6] The geometry of the eclipse is easier to visualize if a cloth ribbon is hung from (or moved along) a length of wire bent to the shape seen in Figure 3. A wire coat hanger will do nicely.

Interestingly enough, to a certain extent this is indeed the situation in three eclipsing nova-like cataclysmic variables, which in addition to a photometric eclipse (phase$\equiv$0, note the change in convention) show spectral absorption features at phases 0.4 to $\sim$ 0.62 (ref. 19). Perhaps not insignificantly, the three systems (V1315 Aquilae, SW Sextansis, DW Ursae Majoris) have orbital periods 201, 194 and 197 minutes which are fairly typical of the (neutron star) X-ray dippers and are in fact intermediate between the X-ray burster 0748-68 (Sections 5, 6) and the eclipsing pulsar PSR 1744-24A under discussion here.

10. Conclusions

Mass outflow from the secondary seems to provide an adequate framework for understanding eclipses and orbital evolution in both accreting and non-accreting neutron star binary systems. The dips in the X-ray light curve in some low mass X-ray binaries can be understood in the same way as the eclipses in radio pulsars, the variety of phenomena being similar in the two kinds of systems.

The eclipses in the Terzan 5 pulsar, PSR 1744-24A, are caused by fluid moving in a manner quite different from that of the eclipse-causing fluid in PSR 1957+20. Distinctive features of the Terzan 5 eclipsing pulsar can be understood in terms of motion along a well defined trajectory taking the fluid well out of the binary orbital plane.

Acknowledgments

It is a pleasure to acknowledge my debt to Mal Ruderman and Jacob Shaham who introduced me to the world of binary evolution with mass loss. Hospitality of Professor George Efstathiou at the Department of Physics, Oxford University is gratefully acknowledged. Research supported in part by NSF grants AST 86-02831 and PHY 89-04035.

References

1. Djorgovski, S. & Evans, C.R. 1988, ApJ 335, L61.

2. Frank, J., King, A.R. & Lasota, J.-P. 1987, A&A 178, 137.

3. Fruchter, A.S., Stinebring, D.R. & Taylor J.H. 1988, Nature 333, 237.

4. Goudas, . 1963, Icarus 2, 1.

5. Kluźniak, W., 1991. *Proceedings of the 15th Texas Symposium on Relativistic Astrophysics, in press.*

6. Kluźniak, W., Czerny, M. & Ray, A. 1991, this volume.

7. Kluźniak, W., Ruderman, M., Shaham, J. & Tavani, M. 1988, Nature 334, 225.

8. Kluźniak, W., Ruderman, M., Shaham, J. & Tavani, M. 1989, in *Timing Neutron Stars*, eds. H. Ögelman & E.P.J. van den Heuvel (Kluver: Dordrecht), pp. 641–645.

9. Lyne, A.G. *et al.* 1990, Nature 347, 650.

10. Mc Clintock, J. *et al.* 1982, ApJ 258, 245.

11. Molnar, L.A., Korda, J.A. & Raymond, J.C. 1990, Bulletin of the AAS 22, 1339.

12. Parmar, A.N. *et al.* 1986, ApJ 308, 199.

13. Rappaport, S., Joss, P.C. & Webbink, R.F. 1982, ApJ 254, 616.

14. Rasio, F.A., Shapiro, S.L. & Teukolsky, S.A. 1990, ApJ 342, 934.

15. Rasio, F.A., Shapiro, S.L. & Teukolsky, S.A. 1991, Nature , .

16. Ritter, H. 1990 A&A Suppl. 85, 1179.

17. Ruderman, M., Shaham, J. & Tavani, M. 1989, ApJ 336, 507.

18. Szebehely, . *Theory of orbits.*

19. Szkody, P. & Piché, F. 1990, ApJ 361, 235.

20. van den Heuvel, E.P.J. 1991, in *Neutron Stars: Theory and Observation*, eds. J. Ventura & D. Pines (Dordrecht: Kluwer) *in press.*

21. White, N. 1989, Astron. Astroph. Rev. 1, 89.

FROM MILLISECOND PULSARS TO LOW MASS X-RAY BINARIES

WLODZIMIERZ KLUŹNIAK‡, MICHAL CZERNY‡‡, ALAK RAY‡‡‡
‡*Physics Department, Columbia University, New York, NY 10027 USA*
and *Institute for Theoretical Physics, University of California Santa Barbara*
and *Subdepartment of Astrophysics, Oxford University*
‡‡*Space Research Center, Polish Academy of Sciences,*
ul. Bartycka 18, 00-716 Warszawa, Poland
‡‡‡*Tata Institute of Fundamental Research, Homi Bhabha Road,*
Bombay, 400 005 India

ABSTRACT. We have become convinced that many binary radio pulsars, and particularly the eclipsing pulsars, will become low-mass X-ray binaries (LMXBs) in the future. This conclusion is based on our analysis of the properties of the eclipsing pulsar 1744-24A in Terzan 5 and is supported by the reported decay of the orbit of the eclipsing pulsar 1957+20. We suggest that many LMXBs, and in particular most X-ray bursters in globular clusters, have descended from "millisecond" binary radio pulsars.

1. Introduction

Bisnovatyi-Kogan & Komberg (1974) predicted the discovery of *binary* radio pulsars arguing that some of the known binary X-ray sources will cease accretion in the future, leading to the turn-on of radio emission in the rapidly spinning neutron star. Subsequently, to explain the origin of the Hulse-Taylor binary pulsar, PSR 1913+16, Smarr & Blandford (1976) proposed a scenario incorporating this idea; they also pointed out that the spin period of a pulsar born in such a way should be related to its timing age (i.e. to the strength of the magnetic dipole).

Alpar *et al.* (1982), Fabian *et al.* (1983) and Radhakrishnan & Srinivasan (1984) were the first to suggest that some solitary (single) pulsars were also created in accreting systems. Alpar *et al.* applied a theoretical relation (now known as the "spin-up line"), expected from such a scenario, between the magnetic dipole of a pulsar and its spin period (Davidson & Ostriker 1973) to predict the time derivative of the known period of the then just discovered 1.6 millisecond pulsar 1937+21. The value of $\dot{P}$ subsequently observed differed from the predicted one only by a factor of two. This great success of the theory led to the nearly universal acceptance of the idea that binary pulsars and millisecond pulsars have in the course of their evolution passed through a phase of accretion. In particular, it is widely thought that millisecond pulsars (i.e. those with spin period $P \leq 20\,\mathrm{ms}$) are descendants of low mass X-ray binaries.

The orbital period of the Hulse-Taylor pulsar is known (e.g. Taylor & Weisberg 1989) to decay ($\dot{P}_b < 0$) on a timescale of $|P_b/\dot{P}_b| = 3.7 \times 10^8$y, in agreement with the predictions of Einstein's theory of general relativity (GR). The masses of the two stars are so close to each other ($1.442 \pm .003 M_\odot$ and $1.386 \pm .003 M_\odot$, *ibid*) that on evolutionary grounds (Smarr & Blandford 1976, Srinivasan & van den Heuvel 1982) it is believed that the companion to the pulsar is also a neutron star, a belief consistent with the lack of detection of an optical counterpart to the system. For this reason it is not expected that the system will

425

E. P. J. van den Heuvel and S. A. Rappaport (eds.), X-Ray Binaries and Recycled Pulsars, 425–436.
© 1992 *Kluwer Academic Publishers.*

evolve into an X-ray binary system, even though the stars are clearly going to coalesce in less than 10^9 years. We note, however, that from the observational point of view, a white dwarf companion is not yet excluded.

After the reported discovery of a radio pulsar in a binary system with orbital period of 32 minutes, it has been realised that a system similar to the Hulse-Taylor pulsar, but with a white dwarf companion, would be expected to evolve into an X-ray binary (Bisnovatyi-Kogan 1989; Rappaport, Putney & Verbunt 1989). The system in question, PSR 0021A-72A, may (Ables *et al.* 1989) or may not (Lyne 1991) exist, but Bisnovatyi-Kogan's (1989) main point, that some X-ray binaries may have been radio pulsars in the past, is well taken. It has been known for some time that a binary system composed of a neutron star and a white dwarf orbiting each other at a period $P_b < 8\,\mathrm{h}$ will be brought into Roche-lobe contact in less than a Hubble time by angular momentum losses predicted by GR.[1] The properties of the Hulse-Taylor pulsar leave no doubt that this prediction is correct, if the requisite system exists. Once the Roche lobe shrinks to the size of the companion star, transfer of matter through the inner Lagrangian point will be initiated. Below, in Section 4, we argue that the eclipsing pulsar PSR 1744-24A is such a system and show that complicating factors such as mass loss from that system do not affect the conclusion: PSR 1744-24A is going to be a low-mass X-ray binary.

Further, we argue that the eclipsing pulsar 1957+20 may become an LMXB in less than 10^8y. The evolution of this system is not driven by angular momentum losses to gravitational radiation in GR, but by mass loss from the system driven by energetic radiation from the pulsar (Kluźniak *et al.* 1988). To give credit where it is due, we note that such ablation of the companion was predicted by Ruderman, Shaham and Tavani (1989).

We also claim that the 41 minute X-ray binary 1627-67 may have in the past been a binary radio pulsar, although not a millisecond one, and may possibly yet again evolve into one. The neutron star in this system is at present a 7.7 s X-ray pulsar undergoing spin-up. We note that with the exception of the high latitude system Her X-1 all the Galactic X-ray pulsars have a combination of spin period and magnetic field placing them to the right (in the standard B vs P plane) of the line separating active radio pulsars from the radio-quiet ones.[2] If radio emission turns on and accretion turns off close to this "death line" the pulsar would be losing rotational kinetic energy at a rate similar to the eclipsing radio pulsar in Terzan 5. It would appear then, that 4U 1627-67 differs from PSR 1744-24A mainly in the strength of the neutron star magnetic dipole, the evolutionary history of the two binary systems having been possibly quite similar.

We show, in Section 5.1, that the "standard" theory of evolution of cataclysmic variables, when applied to LMXBs, predicts the existence of many binary millisecond pulsars with orbital periods of a few hours. Many of those will inevitably become low mass X-ray binaries again.

We point out, in Section 3, that the short orbital periods of some LMXBs (notably of X-ray bursters) are consistent with the evolutionary scenario discussed here (radio pulsars become LMXBs). We suggest that most, if not all, X-ray bursters in globular clusters

---

[1] In wider binaries, nuclear evolution of the companion may cause it to swell and overflow its Roche lobe. The binary pulsar 0820+02 could have acquired its present characteristics in this way in a low mass X-ray binary which itself evolved out of a binary radio pulsar system (e.g. Srinivasan & Bhattacharya 1987).

[2] We thank Dr. Kaiyou Chen for verifying this.

have descended from millisecond pulsars. This is certainly consistent with the fact that in globular clusters many millisecond pulsars have been discovered but only a few (ten) low-mass X-ray binaries are observed. The question of the origin of millisecond pulsars in globular clusters is a separate one and we do not address it in detail.

We would like to end this introduction on a socio-psychological note. The reigning doctrine, based on the dogma of field decay (Ostriker & Gunn 1969), would have strongly magnetized neutron stars transfigure themselves, in accreting binary systems, into weakly magnetized rapidly spinning radio pulsars (see e.g. van den Heuvel 1991 and references therein for a review). A striking illustration of this doctrine can be found in a recent paper by Romani (1990), who paints a figure of magnetic field decaying under the influence of accretion only after the neutron star crosses the "death line" separating active radio pulsars from "dead" ones (naturally, the pulsar is then "resurrected" and granted eternal life). In retrospect, it is clear that the doctrine implicitly requires reincarnation of pulsars, i.e. it implies that radio pulsars become X-ray binaries and then become radio pulsars again. In view of this, it is somewhat surprising that our explicit restatement (to be found in these pages) of an element[3] of this doctrine (that many low-mass X-ray binaries have descended from binary radio pulsars) has been greeted at this workshop with much skepticism. That reaction suggests to us that deep in their souls few workers, if any, are true believers in magnetic field decay.

2. Millisecond Pulsars In Globular Clusters.

We do not know the origin of millisecond pulsars in globular clusters. We take for granted that they exist and that their number is large. In the cluster 47 Tuc alone, at least ten millisecond pulsars are known, including ones with spin period 1.8 ms, 2.1 ms and 2.6 ms (Lyne 1991). It has been noted in this workshop (e.g. Kulkarni 1991) that the distribution of pulsars in globular clusters is not the one expected from a simple tidal capture model, i.e. the number of pulsars detected to date is not proportional to the square of the stellar density in the core of the clusters. This is in contrast to the distribution of LMXBs, which occur predominantly in clusters with high central core density and in many cases in core-collapsed clusters, in agreement with the tidal capture model (Verbunt & Hut 1987).

It is possible that the number of pulsars in globular clusters is much larger than the number expected from the theory that each of them has descended from a typical globular-cluster X-ray binary, and from the assumption that most of the angular momentum of the neutron star was gained by accreting matter in such a binary. If this were the case, it would suggest that if the LMXBs in globular clusters are genetically related to the millisecond pulsars at all, it is the radio pulsars that are progenitors of low-mass X-ray binaries and not the other way around.

In a previous paper, assuming that the analysis of birthrates performed by Kulkarni, Narayan & Romani (1990) was correct, we have suggested (Ray & Kluźniak 1990) that there were two classes of bright X-ray sources in globular clusters. One class would have contained sources with the lifetime τ_{XB}, inferred from the X-ray luminosity of LMXBs observed in globular clusters. A postulated second class of sources, which were assumed to be X-ray progenitors of millisecond pulsars in globular clusters, was supposed to have a

---

[3] To avoid possible misunderstandings we would like to state at the outset that our work does not assume in any way the decay of magnetic fields.

significantly shorter lifetime $\tau_{XX} < 10^{-2}\tau_{XB}$. In such a case, $> 90\%$ of the systems could have been of the short lived variety (giving rise to the radio pulsars), with the remaining $< 10\%$ of the X-ray sources still accounting for 90% of the actually observed X-ray binaries, simply because the probability of X-ray detection of the longer lived sources is enhanced over the probability of seeing the shorter lived variety by the factor τ_{XB}/τ_{XX} per source. This idea has later been used by Phinney and Kulkarni (1991), who made a specific choice for the two X-ray "channels."

Here we go further, and state that at present we see no reason to assume that millisecond pulsars in globular clusters had ever passed through a long lived X-ray binary phase. For example, collision of a compact star with an ordinary binary system may result in the disruption of one of the target stars and accretion of $\sim 0.1 M_\odot$, as suggested by Krolik, Meiksin and Joss (1984). If the compact star is a neutron star, nonstationary accretion of $\sim 0.1 M_\odot$ can occur on a hydrodynamical timescale, with the binding energy released mostly in a large neutrino flux (Zel'dovich, Ivanova, Nadezhin 1972).

However, we do not wish to argue that it is impossible for a millisecond pulsar to have been spun up in a low mass X-ray binary. Quite to the contrary, we will argue that it is not beyond imagination to have a binary neutron star system alternate a few times between the X-ray phase and the radio pulsar phase (the endpoint of such an evolution could be a single millisecond pulsar, but so could be the starting point). While this idea is not a necessary ingredient of our argument, if it were generally true, the relevant quantity of interest in statistics of pulsars and LMXBs would be the time spent in each phase rather than the birthrate of each type of system[4].

At any rate, we will present evidence that millisecond radio pulsars may give rise to low-mass X-ray binaries, particularly in the globular clusters, where it is possible for a single pulsar to become a binary through tidal capture.

3. X-ray Bursters In Globular Clusters And LMXBs.

Currently there are only two diagnostic features whose presence allows a definite identification of a given bright ($L \geq 10^{36}$erg/s) X-ray source as an accreting neutron star. They are X-ray bursts and coherent pulsations of the X-ray light curve. These two features are mutually exclusive, no X-ray burster is an X-ray pulsar and vice versa. There is little doubt[5] that the magnetic field of X-ray pulsars is strong (surface value $B \sim 10^{12}$G). The inference made is that in X-ray bursters the field is weaker ($B < 10^{10}$G?) and/or the rotation rate of the neutron star is higher than in X-ray pulsars. This seems to support the idea that neutron stars in LMXBs have similar properties to those in rapid radio pulsars, but so far there is no observational proof of that.

In the canonical view (Alpar et al. 1982), low-mass X-ray binaries contain rapidly spinning, weakly magnetized ($B \sim 10^8$–10^{10}G surface fields) neutron stars. We adopt this view here.

The data on binary periods strongly suggest that X-ray bursters typically have companions of extremely low mass, and that they may be very old systems. In the Galactic disk, the shortest binary periods have been measured for the low-mass X-ray binaries 4U 1627-67 (which is an X-ray pulsar with spin period 7.67 seconds) with $P_b = 41$ minutes and

---

[4] We thank Jim Applegate for a discussion of this point.

[5] For a review of X-ray binaries see e.g. Lewin & van den Heuvel 1983.

1915-05 (an X-ray burster) with $P_b = 50$ minutes (Ritter 1990). Of five systems with the next-lowest orbital periods, four are X-ray bursters (*ibid*). Among those four is the X-ray burster 0748-68 with orbital period $P_b = 3.8$ hours and its time derivative $\dot{P}_b = -9 \times 10^{-11}$ (Parmar 1991). The lowest orbital period ever measured belongs to the X-ray burster 1820-30 in the globular cluster NGC 6224: $P_b = 11$ minutes and $P_b/\dot{P}_b = -1. \times 10^7$y (Tan *et al.* 1991).

The periods and period derivatives of 0748-68 and of 1820-30 deserve attention. There is no question that an 11 minute period is possible only if the companion is more compact than a hydrogen degenerate dwarf and *a fortiori* more so than a main sequence star, see Fig. 1. The negative sign of $\dot{P}$ is then hard to understand for a Roche lobe filling companion (because $\dot{m} < 0$ and for cold degenerate dwarfs $dP_b/dm < 0$ in Roche lobe contact). This suggests strongly, that the system is undergoing "evaporative" evolution driven by mass loss from the system (Ruderman *et al.* 1989); that scenario allows the companion star to be detached from its Roche lobe (Kluźniak *et al.* 1988).

The fifty minute orbital period of 1915-05 is below the minimum value of 80 minutes found for a hydrogen burning companion by Paczyński and Sienkiewicz 1981. One possibility is that the system is similar to 1820-30, i.e. the companion is evolved. Another possibility is that the companion is a hydrogen star, but mass transfer in the past had been interrupted for a sufficiently long period of time for the star to have contracted to the equilibrium size appropriate for a cold star (see Fig. 1). Both these possibilities are in full accord with the scenario presented below, in Section 6, in which these X-ray bursters have passed through a phase in which they were millisecond binary pulsars,[6] similar to the eclipsing pulsars.

4. The Eclipsing Pulsars.

Two eclipsing pulsars are known at present. PSR 1957+20, a 1.6 ms pulsar in the Galactic plane with a $0.02 M_\odot/\sin i$ companion, has orbital period $P_b = 9.17$ h (Fruchter *et al.* 1988). The derivative of the spin period is $\dot{P} = 1.7 \times 10^{-20}$ and of the orbital period $\dot{P}_b = -(3.9 \pm 0.9) \times 10^{-11}$ (Ryba & Taylor 1991). PSR 1744-24A, an 11.6 ms pulsar in the globular cluster Terzan 5, is in a binary orbit of $P_b = 1.81$ h with a $0.1 M_\odot/\sin i$ companion (Lyne *et al.* 1990), $\dot{P} = (0.4 \pm 3.0) \times 10^{-20}$ and $\dot{P}_b = (-1.6 \pm 3.0) \times 10^{-11}$ (Taylor 1991). Below, we predict that $\dot{P}_b \leq -2 \times 10^{-13}$ for PSR 1744-24A.

Note that the measured value of $\dot{P}_b$ implies that the binary separation of the stars in PSR 1957+20 decreases on a timescale of $|a/\dot{a}| \approx 4 \times 10^7$y. Orbital decay on such rapid timescale can be understood if account is taken of angular momentum carried away from the system by matter ablated from the pulsar. GR losses can be neglected for this system on such short timescales and binary evolution is governed by the following relation valid in the limit of small companion mass (Kluźniak *et al.* 1988)

$$\dot{P}_b/P_b = \frac{3}{2}\dot{a}/a = 3(\beta - 1)\dot{m}/m, \tag{1}$$

see also Czerny and King 1988. The orbit can either expand or contract depending on the value of β. Here β is the specific angular momentum carried away by the wind (in units

---

[6] It would be hard to imagine a different scenario for the origin of 1915-05 if the mass losing secondary in that system were indeed a hydrogen star.

of that of the companion, i.e. $\dot{J}/J = \beta\dot{m}/m$, with J the orbital angular momentum of the system). Equation (1) assumes, as is appropriate for pulsar binaries, that no accretion is taking place. Clearly the companion is losing mass ($\dot{m} < 0$), hence, by eq. (1), $\dot{P_b} < 0$ only if $\beta > 1$. From eq. (1) and the data on the orbital decay timescale we conclude that the ablation model (Kluźniak et al. 1988; Ruderman, Shaham and Tavani 1989) can explain the observed evolution of PSR 1957+20 if $\dot{m} = m\dot{P_b}/[3P_b(\beta - 1)] \approx -6 \times 10^{-10} M_\odot/y$. The last (approximate) equality assumes that the system is observed at an inclination close to the orbital plane ("edge on") and is based on the (arbitrary) choice $\beta \approx 1.3$. The result is within the range of mass loss values estimated by Kluźniak et al. 1988, although we note that the spin-down luminosity of PSR 1957+20 is now known to be lower by an order of magnitude than the one assumed in that paper.

Armed with the above result, we can now place an upper bound on the expected mass loss rate from the second eclipsing radio pulsar system. The solid angle subtended by the Roche lobe of the $0.2M_\odot$ companion of PSR 1744-24A is approximately the same as that subtended by the white dwarf companion (Djorgovski & Evans 1988) of PSR 1957+20. The spin down luminosity, $(2\pi)^2 I\dot{P}/P^3$, of PSR 1744-24A is at least a factor of 500 lower than it is for PSR 1957+20, while the mass of the companion is larger by a factor of 5. If the relationship between the pulsar radiation flux at the companion and the mass loss were linear, we would have $\dot{m} \approx -1 \times 10^{-12} M_\odot/y$ and the shortest possible timescale for orbital change caused by ablation would be $|P_b/\dot{P_b}| \approx 10^{11}y$, unless β were hugely different from unity (eq. [1]). We note that the orbit can expand only if $\beta < 1$ and it is hard to imagine $\beta << 0$, therefore the computed timescale is approximately a lower limit to the ablation timescale for (a hypothetical) expansion of the orbit. This value, $10^{11}y$ vastly exceeds the timescale for orbital decay due to GR gravitational wave losses (Landau & Lifshitz 1962) which is about $4 \times 10^8 y$ for this system. We can safely conclude that PSR 1744-24 has a decaying orbit, like the Hulse-Taylor pulsar. If the companion is a helium degenerate dwarf, the eclipsing pulsar in Terzan 5 will evolve into a system similar to the eleven minute X-ray binary 4U 1820-30 in less than $10^9 y$. If the companion is a main sequence star, the system will evolve even sooner into an X-ray source similar to 1915-05.

As noted in Section 1, a similar suggestion (although not involving the eclipsing pulsars) has already been published by Bisnovatyi-Kogan (1989), who proposed that the X-ray source 4U1820-30 is a member of a postulated class of "second generation" of LMXBs whose precursor "was a recycled pulsar in a binary system of the type of PSR 0021-72A in the globular cluster 47 Tuc."

We conclude that both the eclipsing pulsars, albeit for different reasons, are evolving towards Roche lobe contact and hence both appear to be progenitors of low-mass X-ray binaries. The orbit of PSR 1744-24A should be decaying because of gravitational radiation losses. The orbit of PSR 1957+20 is decaying because the evaporative plume is presumably carrying away angular momentum in quantity per unit mass greater than the orbital angular momentum of the system.

5. Evolutionary Scenarios For The Eclipsing Pulsars

In the previous Section we predicted that in the present epoch the eclipsing pulsar in Terzan 5 has a decaying orbit ($\dot{P_b} < 0$). For completeness we now briefly discuss possible evolutionary schemes for the origin of this system.

5.1. STANDARD CV EVOLUTION

According to the standard theory of evolution (Spruit & Ritter 1983) of cataclysmic variables (CVs), i.e. of white dwarf binaries, the secondary detaches from its Roche lobe at a period of $P_b = 3\,$h because a) at $P_b > 3\,$h the secondary is out of thermal equilibrium due

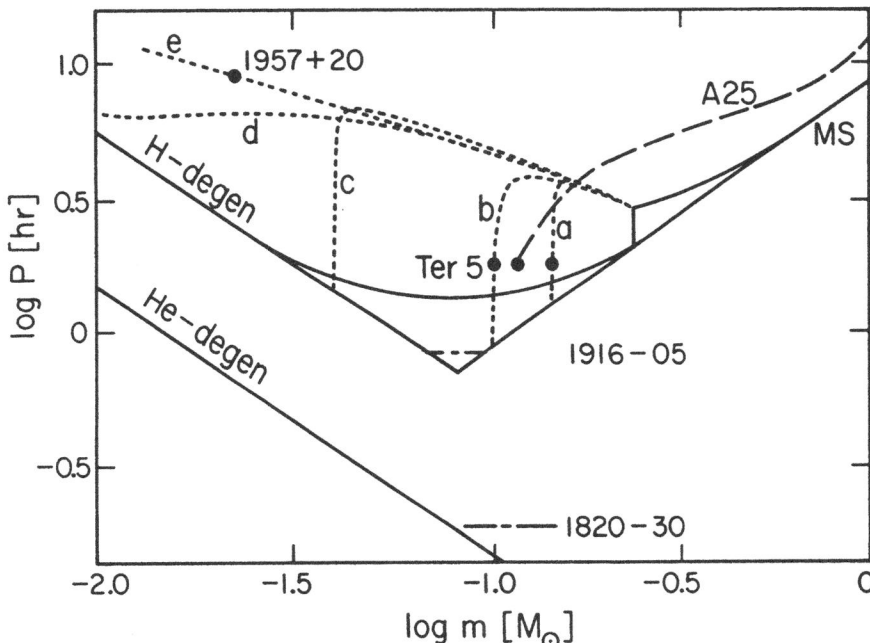

Figure 1. Possible evolutionary tracks of systems with "evaporative" mass loss. The decimal logarithm of the orbital period in hours is plotted versus the decimal logarithm of the mass of the companion in units of Solar mass. Likely location of the eclipsing pulsars (filled circles) as well as possible positions of the X-ray binaries 4U 1915-05 and 4U 1820-30 are also indicated (dash-dot-dash lines). The thick straight line segments correspond to systems with a main-sequence or a cold degenerate dwarf companion in Roche-lobe contact. According to the "standard" theory of their evolution, cataclysmic variables follow the thin curve (in the direction of decreasing companion mass, m). When this theory is applied to canonical LMXBs, the dotted tracks ensue, see Section 5 for details. The lines (a) through (e) differ only in the properties of the pulsar ablating its companion: in the strength of the magnetic dipole moment and in the initial value, P_0, of the rotational period of the neutron star. The values of P_0 and $\log(B/\text{Gauss})$, where $B \equiv \mu \times 10^{-18}\text{cm}^{-3}$, are respectively (a) 5.0 ms, 9.5; (b) 3.4 ms, 8.9; (c) 2.0 ms, 9.0; (d) 2.0 ms, 8.6; (e) 1.25 ms, 8.1. We assumed that 10% of the energy flux impinging on the companion is converted into kinetic energy of the evaporative plume, and we took $\beta = 0.86$ (see eq. [1]).

to rapid angular momentum losses caused by magnetic braking; and b) at $P_b = 3\,$h magnetic braking is turned off (or greatly reduced) and the star begins to contract to its equilibrium radius. The evolutionary track of this standard theory is given by the thin line in Fig. 1.

LMXBs differ from CVs only in the nature of the compact object, a neutron star in the former case and a white dwarf in the latter. Van den Heuvel & van Paradijs (1988) suggested that neutron star binaries depart from the CV evolutionary track at the edge of the "period gap," i.e. at $P_b = 3\,$h, because when mass transfer ceases, the neutron star becomes a pulsar and starts ablating the secondary (Ruderman, Shaham & Tavani 1989), giving rise to a track (such as the dotted line (e) in Fig. 1) passing through the edge of the gap and the present position of PSR 1957+20.

We point out that pulsars less energetic than 1957+20, i.e. those with a different value of magnetic field or a lower spin period (at the beginning of track (e) in Fig. 1) will not be able to reach the current position of 1957+20. Instead they will follow tracks (such as dotted lines (a), (b), (c) in Fig. 1) leading back to the line of Roche lobe contact. This is because after a certain time such pulsars ablate the companions at rates lower than the rate for orbital decay caused by GR gravitational radiation losses.

We stress that if the standard scenario for CV evolution is correct, X-ray binaries containing a neutron star with magnetic dipole $\mu \geq 10^{24}\,$G cm^3 and a main sequence star must evolve along tracks qualitatively similar to the ones described in the previous two paragraphs (i.e. similar to tracks (a) through (e) in Fig. 1). In short, the Spruit-Ritter theory, when applied to canonical LMXBs predicts the formation of binary radio pulsars with orbital periods $\sim$ few hours. Well tested predictions of general relativity then require that many of these pulsars evolve again into low mass X-ray binaries.

5.2. EVOLVED COMPANIONS

While pulsars with orbital parameters similar to those of PSR 1744-24A must be formed if the standard evolutionary scenario for CVs is correct, this actual system may have come about in a different way.

Pylyser & Savonije (1988, 1989) construct evolutionary tracks for binary systems with evolved companions. We note that some of those tracks lead to systems quite like PSR 1744-24A. One such track, Pylyser and Savonije's sequence A25, is shown in Fig. 1. The heavy dots at $P = 1.8\,$h correspond to the eclipsing pulsar 1744-24A for three choices of the inclination angle i in the interval [40°, 90°]. We note that if the Terzan 5 eclipsing pulsar evolved along this track it will[7] end up as a system similar to 4U 1820-30.

5.3 THE COLUMBIA SCENARIO

Kluźniak et al. (1988) pointed out that the "bootstrapped" or "self-excited wind" evolution envisaged by Ruderman et al. (1989) does not require that the companion is in Roche

---

[7] At the period $P_b = 2\,$h Pylyser and Savonije (1989) find a mass transfer rate of about $10^{-9} M_\odot/$y for sequence A25. While this is larger than the value we estimated in Section 4, there is no contradiction with our conclusion that the orbit is decaying. In fact, such a large mass loss rate is only possible if the system is undergoing a rapid decrease of binary period, e.g. because of magnetic braking. It is the possible importance of magnetic braking that led us to give an upper bound on $\dot{P}_b$ (lower bound on $|\dot{P}_b|$) in Section 4, rather than the single value predicted by GR.

lobe contact and suggested that the eclipsing pulsar 1957+20 is an end product of such evolution. That is, if one imagines that a part of the evaporative plume were captured and accreted by the neutron star, as it would have been prior to pulsar turn-on if such a plume had existed also in that phase, the system could have evolved to its current binary period and companion mass in an LMXB phase.[8]

Kluźniak *et al.* (1988) assumed that prior to this "bootstrapped" evolutionary phase the system evolved along the usual Roche-lobe contact line (e.g. the thick main sequence line in Fig. 1 down to $P_b \sim 2\,\mathrm{h}$ and then the thin Paczyński & Sienkiewicz [1981] curve to a period of $P_b \sim 80\,\mathrm{minutes}$). The subsequent track connecting the line of Roche lobe contact and PSR 1957+20 would pass very close to the current position of the Terzan 5 eclipser.

It is possible that evolution proceeds initially according to the "standard theory" discussed in Section 5.1. (dotted lines in Fig. 1) and then according to the Columbia scenario. Thus, a low mass X-ray binary would become a pulsar like 1744-24A, would then evolve to the bottom of track (b) in Fig. 1 to become an LMXB like 1915-05 and would then move up to the position of 1957+20, where the pulsar would turn on again. In the future, as discussed in Section 3., PSR 1957+20 may yet again become an LMXB and give rise to a single millisecond pulsar like the 1.56 ms PSR 1937+21. In this way, 1937+21 could have gone through three incarnations as an accreting X-ray source and three incarnations as a radio pulsar.

5.4 COLLISIONAL SCENARIOS

It is quite likely that PSR 1744-24A formed in a tidal capture of a massive white dwarf by a $> 0.8 M_\odot$ star which has evolved off the main sequence or in a head-on collision of a white dwarf and a giant (with subsequent accretion induced collapse) or in an exchange collision of a neutron star with a main sequence secondary. These scenarios will be considered in detail in a forthcoming paper. In some of these scenarios the system could have passed through an accreting phase (similar to the ones discussed above) at some time between the collision and the current epoch. A violent past of the PSR 1744-24A system is consistent with its presence in the globular cluster Terzan 5 and with its present location ten core radii away from the center of the cluster.

6. Overview

In Sections 2–5 we have outlined various theories and observations supporting the view that some binaries containing millisecond pulsars will evolve into X-ray binaries. If the companion star in such binaries is an evolved one, the resultant LMXB will have a short orbital period (example 4U 1820-30 with $P_b = 11\,\mathrm{minutes}$). If the companion is a main sequence star the period could be as low as $\sim 40\,\mathrm{minutes}$ (example, 4U 1915-05?). In

---

[8] As an aside we note that this Columbia scenario allows PSR 1957+20 to have come about by ablation of an evolved companion. Such ablation would have required too much energy to be possible in the radio-pulsar phase (see e.g. Czerny & King 1988), but if the ablation is occuring in the LMXB phase, as suggested by Ruderman *et al.* 1989 and Kluźniak *et al.* 1988, the released binding energy of accreted matter is more than adequate to evaporate the envelope of the evolved companion. In such a case the orbital period would have been decreasing also in the LMXB phase.

all cases the companion in the final LMXB stage would be very light and the system would be very old. Most X-ray bursters and especially the globular cluster ones seem to have those characteristics. The single source with a known binary period among globular cluster bursters, 4U 1820-30, has a negative value of $\dot{P}_b$, a result consistent with the ideas proposed here. If this measured value is the secular one, the system is definitely evolving towards the line of Roche lobe contact of a cold degenerate dwarf and there exists a theory (Ruderman *et al.* 1989, Kluźniak *et al.* 1988) allowing mass transfer without Roche lobe contact.

We noted that the evolutionary track outlined in Kluźniak *et al.* 1988 allows the Ter 5 pulsar to be the progenitor of a 1957+20 type system. More generally, a system could be reincarnated many times as a LMXB and a radio-pulsar binary, with the two phases alternating on a secular timescale.

7. Conclusions

Our analysis of the present state of the eclipsing pulsar in the globular cluster Terzan 5 and of its possible evolutionary histories indicates that this binary "millisecond" pulsar will in the future become a low mass X-ray binary. We propose that X-ray burst sources in globular clusters had as their progenitors binary millisecond radio pulsars similar to the eclipsing pulsars 1744-24A and 1957+20. This evolutionary sequence provides a natural resolution to the statistical difficulty posed by the large number of millisecond pulsars and small number of low mass X-ray binaries present in globular clusters. We propose that the few have descended from the many.

Acknowledgements

Conclusions overlapping our discussion in Section 4 of PSR 1744-24A have independently been arrived at and also presented at this workshop by Jacob Shaham and Marco Tavani. After completing this work we have been shown a preprint by Ergma & Fedorova (1991) who also claim that PSR 1744-24A is going to become a low mass X-ray binary.

We thank colleagues at this workshop for discussion and Professors E.P.J. van den Heuvel and G. Chanmugam for comments on the manuscript. This work is a continuation of our work on evolutionary scenarios for PSR 1744-24A presented as a poster in the NATO Workshop on Neutron Stars held in Agia Pelagia, Crete, September 2–14, 1990; W. K. thanks the organizers for financial support. M.Cz. thanks Prof. Martin Rees for hospitality at the Institute of Astronomy, Cambridge University in the fall of 1990; W.K. thanks Prof. George Efstathiou for hospitality at the Department of Physics, Oxford University in the same period. Research supported in part by NSF grants AST 86-02831, PHY 89-04035 and INT 87-15411-A01-TIFR.

References

Ables, J.G. *et al.* 1989, Nature, 342, 158.

Alpar, M.A., Cheng, A.F., Ruderman, M.A. & Shaham, J. 1982, Nature, 300, 728.

Bisnovatyi-Kogan, G.S. 1989, Astrofizika, 31, 567.

Bisnovatyi-Kogan, G.S., Komberg, B.V. 1974, Astron. Zh. 51, 373 (translated in Sov. Astron. AJ 1975, 18, 217).

Czerny, M. & King, A. 1988, MNRaS, 235, 33P.

Davidson, K. & Ostriker, J.P. 1973, ApJ 179, 585.

Djorgovski, S. & Evans, C.R. 1988, ApJ 335, L61.

Ergma, E.V. & Fedorova, A.V. 1991, Pis'ma Astron. Zh. *in press.*

Fabian, A.C., Pringle, J.E., Verbunt, F. & Wade, R.A. 1983, Nature 301, 222.

Fruchter, A.S. *et al.* 1990, ApJ, 351, 642.

Fruchter, A.S., Stinebring, D.R. & Taylor J.H. 1988, Nature, 333, 237.

Kluźniak, W., Ruderman, M., Shaham, J. & Tavani, M. 1988, Nature, 334, 225.

Krolik, J.H., Meiksin, A. & Joss, P.C. 1984, ApJ 282, 466.

Kulkarni, S. 1991, these proceedings.

Kulkarni, S., Narayan, R. & Romani, R. 1990, ApJ, 356, 174.

Landau, L.D. & Lifshitz, E.M. 1962, *The classical theory of fields*, (Oxford: Pergamon).

Lewin, W.H.G. & van den Heuvel, E.P.J., eds. 1983, *Accretion-driven X-ray sources*, (Cambridge University Press).

Lyne, A. 1991, these proceedings.

Lyne, A.G. *et al.* 1990, Nature, 347, 650.

Ostriker, J.P. & Gunn, J.E. 1969, ApJ 157, 1395.

Paczyński, B. & Sienkiewicz, R. 1981, Ap.J. 248, L27.

Parmar, A. 1991, these proceedings.

Phinney, E.S. & Kulkarni, S. 1991, Nature *in press.*

Pylyser, E. & Savonije, G.J. 1988, A&A 191, 157.

Pylyser, E. & Savonije, G.J. 1989, A&A 208, 52.

Radhakrishnan, V. & Srinivasan, G. 1984, in *Proc. 2nd Asian-Pacific Regional Meeting of the IAU, Bandung, August 1981*, eds. B. Hidayat & M. Feast (Jakarta: TIRA Pustaka), pp. 423–431.

Rappaport, S., Putney, A. & Verbunt, F. 1989, ApJ 345, 210.

Ray, A. & Kluźniak, W. 1990, Nature 344, 415.

Ritter, H. 1990 A&A Suppl. 85, 1179.

Romani, R. 1990, Nature 347, 741.

Ruderman, M., Shaham, J. & Tavani, M. 1989, ApJ 336, 507.

Ruderman, M., Shaham, J., Tavani, M. & Eichler, D. 1989, ApJ 343, 292.

Ryba, M. & Taylor, J. 1991, preprint.

Smarr, L.L. &Blandford, R.D. 1976, ApJ 207, 574.

Srinivasan, G. & Bhattacharya, D. 1987, in *The origin and evolution of neutron stars*, eds. D.J. Helfand & J.-H. Huang (Dordrecht: Reidel), pp. 109–119.

Srinivasan, G. & van den Heuvel, E.P.J. 1982, A&A 108, 143.

Spruit, H.C. & Ritter H. 1983, A&A. 124, 267.

Tan *et al.* 1991, submitted to ApJ.

Taylor, J.H. 1991, these proceedings.

Taylor, J.H. & Weisberg, J.M. 1989, ApJ 345, 434.

van den Heuvel, E.P.J. 1991, in *Neutron Stars: Theory and Observation*, eds. J. Ventura & D. Pines (Dordrecht: Kluwer) *in press*.

van den Heuvel, E.P.J. & van Paradijs, J. 1988, Nature, 334, 227.

Verbunt, F. & Hut, P. 1987, in *The origin and evolution of neutron stars*, D.J. Helfand & J.-H. Huang eds. (Dordrecht: Reidel), pp. 187–196.

Zel'dovich, Ya. B., Ivanova, L.N. & Nadezhin D.K. 1972, Astron. Zh. 49, 253 (translated in Sov. Astron. AJ 16, 209).

A DETERMINATION OF THE RADIO-LUMINOSITY FUNCTION AND RELATIVE NUMBER OF GLOBULAR CLUSTER PULSARS

R. S. FOSTER
National Research Council/NRL Cooperative Research Associate
Center for Advanced Space Sensing
Naval Research Laboratory
Washington, DC 20375 USA

M. TAVANI
University of California
Institute of Geophysics and Planetary Physics, LLNL, CA 94550 USA
Astronomy Department, UC Berkeley, CA 94720 USA

ABSTRACT. The currently known sample of radio pulsars discovered in twelve globular clusters allows a determination of the globular cluster pulsar radio-luminosity function and a comparison of different pulsar formation models. Even though these results are necessarily preliminary due to observational and statistical uncertainties, it is significant that the inferred slope of the radio-luminosity function of cluster pulsars is consistent with the slope determined for the field population. We consider four cluster pulsar formation models including two versions of the tidal capture model and two models independent of cluster core star densities. We find that the tidal capture model *does not* appear more favored than models that are independent of cluster star densities. For at least four globular clusters (M 28, M 53, Ter 5, and NGC 6440), the tidal capture model does not fully account for the number of detected pulsars in these clusters.

1. Introduction

Recent radio pulsar surveys have discovered a new population of radio pulsars in globular clusters. The existence of still active pulsars in globular clusters is interpreted as either a consequence of recent 'prompt' formation (possibly by accretion induced collapse of white dwarfs, see, e.g., Chanmugam and Brecher (1987); Grindlay and Bailyn (1989)) or a product of old neutron star 'recycling' (e.g., Alpar et al., 1982; van den Heuvel, 1988). Several studies of binary formation and interactions have been carried out mostly focused on the problem of determining the total number of pulsars in the globular cluster population (for a review see, e.g., Verbunt (1989)).

The main aim of this present work is twofold. We obtain: (1) a determination of the cluster pulsar radio-luminosity function (assumed to be a universal function for all cluster pulsars), and (2) a comparison of different formation models of cluster pulsars. In order to obtain a self-consistent solution for both the pulsar radio-luminosity function and the number of cluster pulsars we need to provide 'input' cluster weights obtained from models which assume different pulsar formation mechanisms. Whereas the radio-luminosity function turns out to be independent of the choice of pulsar formation models, the calculated cluster pulsar numbers are strongly dependent on the chosen model. We can therefore compare the result of our computed fractional numbers with the pulsar numbers expected from a direct application of the formation models. We considered two simple versions of the tidal capture model as well as two additional models which do not depend on star densities in the cluster cores. A longer presentation of this work is in Foster and Tavani (1991).

E. P. J. van den Heuvel and S. A. Rapaport (eds.), X-Ray Binaries and Recycled Pulsars, 437–442.

Our analysis is appropriate for the determination of a set of model-dependent *relative* number of pulsars for each cluster with positive detections. The method used in this analysis only considers clusters containing detected pulsars and *does not* address the problem of determining the absolute number of pulsars in each globular cluster. We do not account for the low radio-luminosity end of the radio-luminosity function (< 20 mJy kpc^2). The radio-luminosity range for field pulsars (which might be different from the range in globular clusters, e.g., Kulkarni, Narayan, and Romani 1990 [KNR]) is believed to extend to ~ 1 mJy kpc^2, and may break around a limiting luminosity of 3 mJy kpc^2 (Dewey, 1984). The current data set has several sources of uncertainty. The dominant sources are 1) inaccurate flux densities, 2) distance biases, and 3) incomplete knowledge of the pulsar distribution in clusters with no detected pulsars.

2. Radio-Luminosity Function and Relative Pulsar Numbers

In order to obtain unbiased distributions and take into account selection effects of current globular cluster pulsar surveys we use the computational method of Lyne, Manchester, and Taylor (1985) properly adapted to deal with pulsars in globular clusters. If the total number of pulsars in globular clusters were known, then the quantity $n(I) \cdot N_P$, where N_P is the total number of cluster pulsars (see KNR and Fruchter and Goss 1990 [FG]) would give an unbiased estimate of the pulsar number for the I-th globular cluster. Unlike the field population of pulsars, globular cluster pulsars are essentially at a known distance. Therefore, instead of pulsar 'densities' defined in LMT we need to consider the fractional numbers $n(I)$ which are different from zero only for those clusters with detected pulsars.

We use the observed distribution of pulsars as input for our numerical solutions: the number of pulsars in each globular cluster I is $N_I(I)$, and the number of pulsars in each radio-luminosity logarithmic interval L is $N_L(L)$. The two distributions $\Phi(L)$ and $n(I)$ are obtained by numerically solving the coupled equations

$$n(I) = \frac{N_I(I)}{W_J(I) \, \chi[L_{min}(I)]}, \tag{1}$$

$$\Phi(L) = \frac{N_L(L)}{\sum'_I W_J(I) n(I)}, \tag{2}$$

where we defined $\chi[L_{min}(I)] = \sum_{L_{min}}^{L_2} \Phi(L)$, and normalized the radio-luminosity function for each iteration so that $\sum_{L_1}^{L_2} \Phi(L) = 1$, with L_1 and L_2 the minimum and maximum pulsar luminosities of the actual sample of cluster pulsars. In our sample $L_1 \simeq 20$ mJy kpc^2 and $L_2 \simeq 2000$ mJy kpc^2. The primed sum in Eq. (2) is extended only to the clusters which satisfy the relation $L_{min}(I) \leq L$, where L_{min} is the minimum radio-luminosity (in mJy kpc^2) below which current radio surveys are not sensitive. The weights $W_J(I)$ are computed for each cluster according to the pulsar formation models specified below. For a simple interpretation of the results it is useful to redefine and normalize the dimensionless weights and pulsar fractional numbers so that $\sum_I W'_J(I) = 1$ and $\sum_I n'(I) = 1$, where the summation is limited to clusters with detected pulsars. (Note that since we deal only with fractional pulsar numbers the resulting $\Phi(L)$ and $n'(I)$ are independent of the normalization of the weights $W_J(I)$). The L-dependent effective weight $W_{eff}(L)$ (which takes into account both the formation model and the pulsar sensitivity limits of the surveys) has a 'step function' behavior when $L = L_{min}$, i.e., $W_{eff}(L) = 0$ when $L < L_{min}$, and $W_{eff}(L) = W'_J(I) \chi[L_{min}]$

when $L > L_{min}$. The only dependence on pulsar survey sensitivities and therefore cluster distances is in the quantity χ.

The cluster weights $W_J(I)$ have been determined for four different pulsar formation models labelled by the index J:

- Model (1) gives equal weight to all clusters and assumes a uniform distribution of pulsars independent of the number and density of cluster stars, $W_1(I) = $ constant.

- Model (2) assumes that the absolute visual magnitude of a globular cluster is proportional to the total number of cluster pulsars, $W_2(I) \propto 10^{0.4(4.85-MV)}$, where MV is the absolute visual magnitude of the cluster and 4.85 is the absolute solar visual magnitude.

- Model (3) assumes that the number of pulsars for each cluster is proportional to the total number of collisions among stars in the core of the corresponding cluster, $W_3(I) \propto (\rho_c^2 R_c^3/v)$, with ρ_c, R_c, and v, the core mass density, radius, and dispersion velocity, respectively. Models 3 and 4 are based on the mechanism of tidal capture *in the cluster core* (Fabian, Pringle and Rees, 1975) which has been considered to be instrumental to the formation of binaries containing millisecond pulsars (*e.g.*, Verbunt, Lewin, and van Paradijs, 1989).

- Model (4) applies a "retention" factor to the tidal capture model, $W_4(I) \propto (\rho_c^2 R_c^3/v) \cdot (1 - e^{-(v/v^*)^2})$, with v^* the escape velocity from the cluster.

All the globular cluster data, including core densities, escape velocities, and distance were obtained from the published data of Webbink (1985) except for the absolute visual magnitudes which were obtained from Djorgovski (1990). The weights $W_2(I)$ and $W_3(I)$ computed according to Webbink's or Djorgovski's data differ typically by a factor ~ 3. We take this factor as a measure of current uncertainty in determining cluster parameters.

3. Discussion

Starting with the set of detected radio pulsars in globular clusters we compute for each pulsar formation model the pulsar radio-luminosity function $\Phi(L)$ and the fractional pulsar numbers in a given globular cluster $n'(I)$. We let the numerical procedure continue until the bin uncertainty converged on a set of values with errors less than 1%. We then compare the $n'(I)$'s obtained from a numerical solution of Eqs. (1) and (2) with the fractional number of pulsars expected from pulsar formation models. As a measure of the statistical agreement/disagreement between the $n'(I)$'s and the expected fractional numbers $W'_J(I)$ we consider the ratio $\mathcal{R} = \log[n'(I)/W'_J(I)]$. The weights $W'_J(I)$ differ from the $W_J(I)$ only for a different normalization and they are normalized taking into account only clusters with positive pulsar detections. On the one hand, the radio-luminosity function is independent of the assumed weights $W'_J(I)$'s and ultimately depends only on the set of minimum luminosities $L_{min}(I)$'s. On the other hand, the ratios $\mathcal{R}$ are strongly dependent on the assumed weights.

We started with an initial distribution of 24 pulsar luminosities binned in 0.3 log units. Pulsar flux densities are generally uncertain by factors of 2 due to interstellar scintillation and globular cluster distances are uncertain by large amounts as well. This leads us to estimate that the assumed radio-luminosity of a single pulsar is uncertain by about 1 bin. We computed the errors in the determination of $\Phi(L)$ by repeatedly changing both the initial pulsar distribution with shifts of pulsar luminosities to the right of the corresponding L_{min}, and by changing the positions of L_{min}. Since we make no assumptions about the absolute number of radio pulsars in these clusters we were able to determine only the slope of the radio-luminosity function. The plot of the radio-luminosity function (Figure 1) is on a logarithmic scale between 20 mJy kpc² and 2000 mJy kpc² and it is normalized so that the maximum bin has an amplitude of 1 (0 in the logarithmic scale). The input radio-luminosity distribution is given as a histogram whereas the calculated radio-luminosity function is shown with its estimated errors, both in amplitude and the half width of the data bins. We determine a radio-luminosity function $\Phi(L) \propto L^{-1.2 \pm 0.3}$ binned per logarithmic interval.

The calculated fractional pulsar numbers are sensitive to particular models used to estimate the proper weights. Figure 2 shows the results of our calculations for the four different models having assumed the minimum radio-luminosity observable toward each cluster. Typical errors in the determination of the $\mathcal{R}$'s (equation (2) are of order of 20%, except for the core collapse clusters M 15 and NGC 6624 (Djorgovski and King, 1986) and the two clusters M 5 and NGC 6760 for which the uncertainties in the $n'(I)$'s amount to two orders of magnitude. Under the assumption of a large population of radio pulsars in globular clusters and an accurate model to determine the weights in Eqs. (1) and (2) we expect $\mathcal{R} \simeq 0$ on average. If $\mathcal{R}$ is much larger than zero then the probability of finding a radio pulsar in that globular cluster is small given the assumed model. If $\mathcal{R}$ is much less than zero then the model predicts that a population of radio sources has been missed and could be potentially detectable given the limiting radio-luminosity of the cluster. Given the uncertainties in both $N_I(I)$ and survey sampling we expect that a cluster is satisfactorily described by a given model if $-1 \lesssim \mathcal{R} \lesssim +1$.

The simple version of the tidal capture model considered in this paper accounts for the current pulsar numbers in eight out of a total of twelve clusters with pulsars, but there are four exceptions. The globular clusters Ter 5, M 53, M28, and NGC 6440 have either $\mathcal{R} < -1.3$ or $\mathcal{R} > +1.3$, respectively. The results of our analysis suggest the interesting task of performing additional pulsar searches especially in the aforementioned 'exceptional' clusters. For example, the tidal capture model 3 predicts the existence of yet-undiscovered pulsars in the clusters 47 Tuc, Ter 5, NGC 6440, and M28, while models 1 and 2 lead to the opposite prediction in Ter 5. A particularly important search would focus on the cluster M 53 (and all globular clusters with small tidal capture rates and no identified radio pulsars). The existence of one 33 msec binary pulsar in the cluster M 53 raises serious doubts about the applicability of the tidal capture model to this particular cluster. The possible discovery of additional pulsar(s) in M 53 (and other low weighted clusters) would question the application of the tidal capture model to low-density clusters. According to the tidal capture model, the formation rate of neutron stars in M 53 is 9.6×10^{-4} of all neutron stars formed in globular clusters. The cumulative weight (proportional to the detection probability) for all the clusters with tidal capture rates equal to or lower than M 53 is 1.3×10^{-2}. This weight is in sharp contrast with the absolute magnitude model (Model 2) where the cumulative weight for all clusters with production probabilities less than M 53 is closer to $\sim 50\%$.

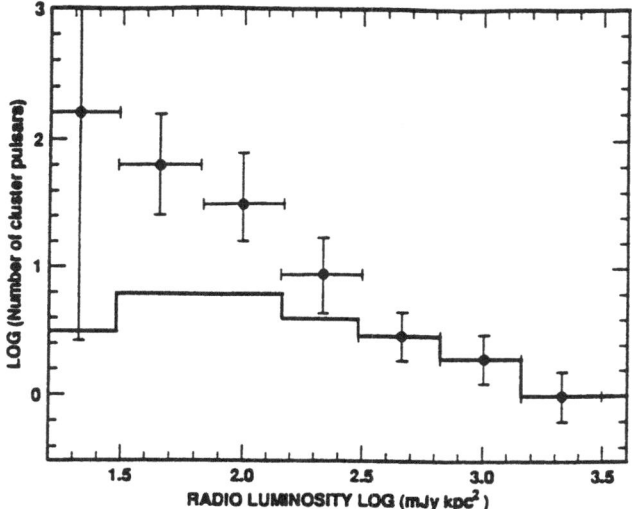

Figure 1 The globular cluster radio-luminosity function determined over the radio-luminosity intervals 20 mJy kpc² to 2000 mJy kpc² is plotted. The radio-luminosity function is normalised to 1 (0 in the log) in the bin containing the pulsar with largest radio-luminosity. The underlying histogram gives the actual distribution of detected cluster pulsars.

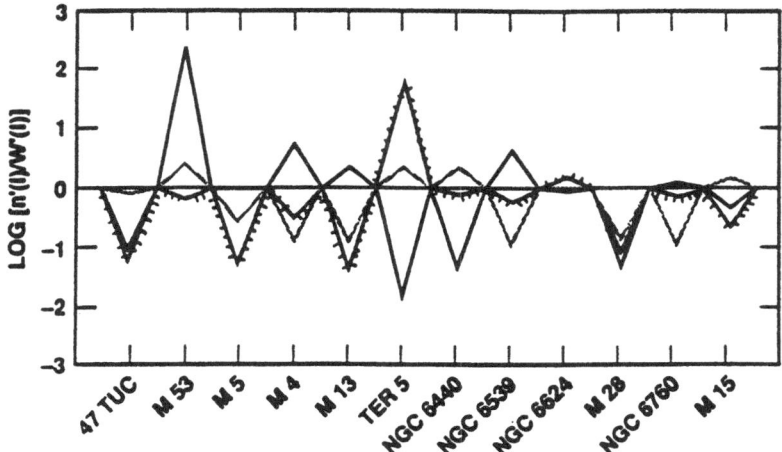

Figure 2 The quantity $\mathcal{R}$, i.e., the logarithm of the ratio between the fractional pulsar number $n'(I)$ (based on observations) and the theoretical values $W'(I)$ is shown for each cluster with detected pulsars and for different input models. Lines are drawn from each point corresponding to the position of $\mathcal{R}$ on the vertical axis to the mid-bin zero value in order to more effectively "guide the eye." Different lines refer to Model 1 (*solid-barred line*), Model 2 (*light-dotted line*), and Model 3 (*solid line*). The line corresponding to Model 4 is virtually the same as that one of Model 3 and has been omitted from the figure for clarity. We estimate that the $\mathcal{R}$ values have errors of ± 20%, except for M 5 and NGC 6760 where the errors in determining the corresponding $n'(I)$'s are close to two orders of magnitude (2 in the log). The source of this large uncertainty is related to the lack of accuracy in the determination of the radio-luminosity function in the lowest radio-luminosity bin. The value of the $\mathcal{R}$'s computed in model 3 for the two core collapse clusters M 15 and NGC 6624 are uncertain by a factor of order unity.

4. Conclusions

We find that the simple version of the tidal capture model considered in this paper does not satisfactorily describe the observed pulsar distribution in globular clusters. Alternative pulsar formation models (models 1 and 2) which are independent of *core* star densities are in better agreement with the observed pulsar distribution. in particular, model 1 seems the model which best describes the observed cluster pulsar distribution. The relevance of this pulsar formation model may be related to the role played by primordial binaries (e.g., Hesser et al., 1990; Mateo and Krzeminski, 1990) for the formation of cluster pulsars. More than one pulsar formation mechanism may operate for a given cluster. The tidal capture model *alone* is not enough to explain the observed pulsar distribution and more theoretical and observational work is necessary to clarify the many unsolved questions raised by the existence of pulsars in globular clusters.

Acknowledgements

We thank G. Djorgovski D. Backer, S. Kulkarni and A. Lyne A. Wolszczan and A. Lyne for useful sharing of data and information. We acknowledge the assistance of the NSF (grant AST-8719094), and of the U.S. DOE (contract W-7405-ENG-48). MT wishes to thank the International Center for Relativistic Astrophysics (I.C.R.A.) in Rome for hospitality.

References

Alpar, M.A., Cheng, A.F., Ruderman, M.A., and Shaham, J., 1982, *Nature*, **317**, 154.

Chanmugam, G. and Brecher, K. 1987, *Nature*, **329**, 696.

Dewey, R.J., 1984, Ph.D. thesis, Princeton University.

Djorgovski, S.G. 1990, *personal communication*.

Djorgovski, S.G., and King, I.R., 1986, *Ap.J. (Letters)*, **305**, L61.

Fabian, A.C., Pringle, J.E., and Rees, M.J., 1975, *M.N.R.A.S.*, **172**, 15p-18p.

Foster, R. S. and Tavani, M. 1991, *Ap. J. Lett.*, submitted.

Fruchter, A. S. and Goss, W. M. 1990, *Ap.J. (Letters)*, **365**, L63.

Grindlay, J. E. and Bailyn, C. D. 1989, *Nature*, **336**, 48.

Hesser, J.E., McClure, R.D., Fletcher, J.M., and Pryor, C., 1990, *Bull. Amer. Astron. Soc.*, **22**, Vol. 2, p. 1284.

Kulkarni, S. R, Narayan, R., and Romani, R. W. 1990, *Ap. J.*, **356**, 174.

Lyne, A. G., Manchester, R. N., and Taylor, J. H. 1985, *Mon. Not. R. astr. Soc.*, **213**, 613.

Mateo, M., and Krzeminski, W., 1990, *Bull. Amer. Astron. Soc.*, **22**, Vol. 2, p. 1284.

van den Heuvel, E. P. J. 1988, *Adv. Space. Res.*, **8**, 355.

Verbunt, F. 1989, in *Timing Neutron Stars*, ed. H. Ogelman and E. J. P. van den Heuvel (Dordrecht: Kluwer), p. 593.

Verbunt, F., Lewin, W. H. G., and van Paradijs, J. 1989, Mon. Notice R. astro. Soc., **241**, 51.

Webbink, R. F. 1985, *Dynamics of Star Clusters*, IAU **113**, 541.

CHAPTER 6

Pulsar Velocities

PULSAR VELOCITIES AND THEIR ORIGINS

V. Radhakrishnan
Raman Research Institute
Bangalore 560 080, India

ABSTRACT. The review covers the history of the measurements of pulsar velocities and the attempts to interpret them. Various correlations observed or claimed between the velocities and other pulsar properties are discussed and their relevance assessed. The picture one seems to be converging towards is that most pulsars come from binaries, roughly half of them when released have "orbital" velocities, and the others – the new born ones – have velocities that are a combination of orbital plus "kick" velocities. The physical origin of the kicks is obscure, but their existence seems well founded on observational grounds. An important offshoot of the attempts to understand pulsar velocities is the hypothesis, also made on other grounds, that the evolution of neutron star magnetic fields is intimately related to their rotational histories.

1. INTRODUCTION

The task assigned to me is to review the status of our knowledge and understanding of pulsar velocities and their origins. The evidence for the existence of substantial velocities comes from a variety of avenues, both observational and inferential, and I shall begin by summarising these. The earliest estimate obtained was from optical observations of the Crab pulsar. Trimble (1968) compared old and new plates and derived a value of approximately 110 km/sec for the proper motion of this object. In a remarkable paper, Gunn and Ostriker (1970) concluded from the 'z' distribution of 41 pulsars that they were high velocity objects (of order 100 km/sec) which were born in the galactic plane and had since moved away from it. Gott, Gunn and Ostriker (1970) proposed that the velocities could be of a binary origin as in the case of OB runaway stars discussed by Blaauw (1961).

2. RADIO OBSERVATIONS

Subsequent determinations of pulsar proper motions have almost all come from radio observations of different kinds. In an indirect method, based on interstellar scintillations, Galt and Lyne (1972) made the first such determination using observations obtained from two stations - one in Canada and one in Britain. Pulsar velocities can also be estimated from single station observations by using the rate of scintillation together with assumptions about the location, size and velocity of the interstellar inhomogeneities. The most extensive such observations have been carried out by Cordes (1986) for some 70 or so pulsars. They indicate velocities of the order of 100 km/sec with the tail of the distribution going up to a few hundreds of km/sec.

Timing of the pulses can also lead to estimates of the proper motions of pulsars. Manchester, Taylor and Van (1974) concluded from observations of PSR 1133+16 that it had a proper motion of the order of 380 km/sec. As the method depends on the regularity of pulsars and many were found to have irregularities it gradually fell into disfavour. But recently, as we heard at this meeting, schemes like TEMPO have been yielding interesting results on millisecond and other pulsars. The best measurements available today are, however, from radio interferometry carried out over many years at Jodrell Bank and also some similar observations from Australia. Lyne, Anderson and Salter (1982) observed 26 pulsars and obtained velocities with transverse components in the range

445

E. P. J. van den Heuvel and S. A. Rappaport (eds.), X-Ray Binaries and Recycled Pulsars, 445–452.
© 1992 *Kluwer Academic Publishers.*

of 10-370 km/sec. Bailes et al (1989, 1990) have made observations using the Parkes-Tidbinbilla interferometer, and among other things have measured the parallax and proper motion of PSR 1451-68 which together yield a transverse velocity of 88 ± 2 km/sec.

The most extensive interferometric measurements available today are from Harrison, Lyne and Anderson (1991) who have observed about 70 pulsars leading to an rms speed of 200 km/sec. The important results from this set of observations are that pulsar velocities can go all the way up to 1000 km/sec, the velocities measured interferometrically do not necessarily correspond with those obtained from scintillation measurements (as the latter involve the 'z' distance of the pulsar), and very interestingly that many pulsars with large characteristic ages are approaching the galactic plane from high latitudes rather than all going away from it as suggested by the earlier Jodrell study (Lyne, Anderson and Salter, 1982).

3. ACCELERATION MECHANISMS

Attempts to explain the origin of these now well established velocities go back almost to the earliest indication that pulsars may be high velocity objects. We may classify the various suggestions and hypotheses advanced into three categories - (i) prenatal (ii) natal and (iii) post-natal, with reference to the birth of the neutron stars. The first category assumes a binary origin for the pulsar and attributes the velocity to the orbital velocity of the progenitor which produced the neutron star. This is the same mechanism as that advanced for runaway OB stars by Zwicky as far back as 1957 and discussed at length by Blaauw in 1961 long before pulsars were discovered. As already mentioned, this was suggested as the operative mechanism by Gott, Gunn and Ostriker (1970).

The second category of mechanism supposes that a kick is delivered to the neutron star at birth, due for example, to an asymmetry in the supernova explosion. If the ejection of the envelope is asymmetric, then the conservation of linear momentum implies a kick to the neutron star. The origin of such an asymmetry is unclear, but the amount needed is very small because the mass of the ejecta is very much greater than the mass of the neutron star, and the velocity of the ejecta is typically thousands of km/sec.

Amongst the post-natal mechanisms may be mentioned the rocket effect proposed by Harrison and Tademaru (1975), and the asymmetric neutrino emission hypothesis advanced by Dorofeev, Radionov and Ternov (1984). In the rocket mechanism, the oblique pulsar magnetic dipole is displaced from the centre of the neutron star and is skew to the rotation axis. The consequent asymmetry in the dipole and quadrupole radiation propels the pulsar along its rotation axis soon after it is born. In the other mechanism it is the non-spherically symmetric emission of neutrinoes and conservation of linear momentum which leads to the pulsar accelerating from its position of birth.

4. CORRELATIONS

In choosing between the various alternative mechanisms listed in the last section, or searching for yet other explanations for the origin of pulsar velocities, one has relied on correlations between observable quantities which have sometimes illuminated and sometimes confused the issues. As an example may be mentioned the comparison of spin down and kinematic ages (τ_s and τ_k). For pulsars which have a proper motion measurement, a kinematic age can be estimated by assuming an origin for the pulsar in the galactic plane and neglecting the possible effect of an unmeasured radial component of velocity. (In odd cases, a pulsar can seem to be coming towards the galactic plane when in fact it is moving away from it and vice-versa.) In any case, what is important to note here is that there were many cases where the kinematic age so determined was much smaller than the spin down age, and this was one of the main arguments for supposing that pulsar magnetic fields decay spontaneously. As we shall see later, the question of field decay gets really intertwined with an understanding of the origin of velocities.

Another correlation or the lack of it is that between the position angle of the intrinsic polarisation in the middle of the pulse and the position angle of the proper motion direction in the sky. It is assumed here, following magnetic pole models for an explanation of the polarisation sweep observed in pulsars, that the mid-pulse intrinsic polarisation angle represents the projection of the rotation axis of the pulsar on the sky. A 100 per cent correlation will therefore be expected if the observed velocities were due to mechanisms such as the rocket propulsion one which operate along the rotation axis. Based on observations of 16 pulsars from the earlier Jodrell Bank study, Anderson and Lyne (1983) ruled out any such correlation.

On the other hand, it has been claimed by Pskovsky and Dorofeev (1989) that there is a correlation between the *magnitude* of the proper-motion and the angle between the rotation axis and the line of sight, in the sense that velocities are greatest when the rotation axis is in the plane of the sky. This would be expected naturally in the rocket type mechanism when the acceleration is along the rotation axis. Note that in the earlier test it is only the direction of the proper motion which is used, while in this case it is only its magnitude. And most of the derivations of the angle between the rotation axis and the sight line assume, it appears, the same polar cap model as in the test involving directions which involves no such derivation. The conclusions of the two sets of authors are therefore in direct contradiction. It seems to me that the uncertainties in the quantities involved in this test involving magnitudes are very large as can be seen from two plots of the same quantities published by these authors (1989, 1990) using virtually the same sample of pulsars. I find it hard therefore to take seriously this claimed correlation.

4.1. B-V Correlation

A quite unexpected and a very intriguing correlation found by Anderson and Lyne (1983) was that between the magnetic moments and transverse velocities of pulsars. When the velocities are plotted against the field it is seen that larger velocities are associated with larger values of the field. If a possible decay of the magnetic field is taken into account, the clear correlation of the velocities with the present values of the field is seen to weaken when plotted against the calculated initial values! It is hard enough to conceive of any physically plausible mechanism which could connect the velocities and initial dipole magnetic moments of neutron stars, but an enhanced "correlation" with the present day fields, lower in value than those of typical young pulsars, was intriguing indeed. Let me digress a little into some attempts to understand this correlation.

I noticed that if the velocities were plotted against spin-down age the distribution was suggestive of two groups of pulsars, one with high B and high V and the other with low B and low V. The separation was around a value of 10^7 yr for the spin-down age, which corresponds closely to the equilibrium spin up line at the Eddington accretion rate, as can be seen from a B-P plot. This led me to examine whether the "correlation" was really a consequence of the post-natal histories in binaries. The relatively long time spent in a low mass binary allowed the field to decay, and accretion caused the spin up of the low field pulsar from beyond the death line into the range of observable pulsars (Radhakrishnan, 1984).

A serious and basic problem with this line of thought was however, that in the break-up of a binary system consisting of an old spun up pulsar with a weak field and a newly born one with a high field, the $B - V$ correlation should have been exactly the opposite. To persist with this interpretation, the binary history would require the companion of the old low-field spun-up pulsar to disappear quietly, in a non-explosive manner. A possible scenario could be the spiralling-in of the neutron star during a common-envelope phase (Radhakrishnan, 1984; Radhakrishnan and Shukre, 1987).

As far as the high field pulsars are concerned it was shown (Radhakrishnan and Shukre, 1985) that the observed velocities could be satisfactorily reproduced on the assumption that they came from massive close binaries with relatively smaller time spans between the first and the second explosions so as not to cause a substantial decrease in the field of the first born neutron star. Even so, if pulsar fields did decay, there should be a slight correlation between B and V, again in the inverse sense from what was found by Anderson and Lyne. But perhaps it would be too small to be discerned given the limited sample and that too with only two components of the velocity being measured. It would be fair to say that at this stage we really had no understanding of the observed B-V correlation.

It should be pointed out that in the binary origin explanation put forward for the velocities of high field pulsars (Radhakrishnan and Shukre 1985) it was assumed that there were no other contributions like a kick at the birth of the neutron star. In fact it was claimed that it was unwarranted and unnecessary to invoke the existence of kicks as the observations could be well matched to what would be expected without them (Radhakrishnan, 1986). That was my picture of the velocities story about five years ago. Since then, however, this stand has become untenable for a number of reasons some ancient and others more recent.

5. ARGUMENTS FOR KICKS

The first suggestion that the velocities of pulsars could be due to a kick at birth from an asymmetric explosion was by Shklovskii (1970).

Kornilov and Lipunov (1984) and Tutukov et al (1984) concluded that a kick of 30-100 km/s must be given to the neutron star at the time of birth to explain the paucity of pulsars in wide binary systems.

A very significant argument that the parameters of the binary system containing PSR 1913+16 simply could not be reconciled with a perfectly symmetric second supernova explosion was advanced by Burrows and Woosley (1986). It was the precise determination of the component masses, made possible by measurements of the general relativistic effects, which gave strength to the argument claiming that there must be an asymmetry at the 3% level.

Blaauw (1985) advanced a strong argument that most pulsars must come from B2-B3 stars (6-10 $M_\odot$) to satisfy the statistics of their birthrate. As the orbital velocities of these stars in binary systems are simply not high enough, a kick must be invoked to explain the pulsar velocities.

Further support although indirect came from a Monte-Carlo simulation by Dewey and Cordes (1987) which seemed to fit the observations of pulsar velocities better when a kick velocity of ≈ 90 km/s was included.

The velocity of ≈ 970 km/s measured by Harrison et al (1991) for PSR 2224+65 cannot possibly be explained in terms of an orbital velocity in a binary system as it implies a period of the order of 1^m for the orbit!

All of the above considerations and more were incorporated into a very satisfactory and self-consistent explanation of both pulsar velocities and their correlation with magnetic moments put forward by Bailes (1989). This remarkable synthesis was achieved by making two unorthodox assumptions, (a) that most pulsars came from relatively wide binary systems, and (b) that the decay of the magnetic fields occurred only during an accretion phase. These assumptions together with a kick of 100-200 km/s at birth lead to a picture which agrees very well with observational numbers for the velocities, the fields, the birth rate statistics of both single and binary pulsars, their initial periods, and the existence of two classes of pulsars as a consequence of their binary history. In this picture, both the high B - high V and the low B - low V pulsars come from the same binary. The old pulsar has forgotten the kick it received at birth, lost its field during the accretion phase and has a low velocity corresponding to that in its orbit before release by the second supernova explosion. The new pulsar has a high field and also a high velocity due to the kick it received when it was formed and the binary disrupted.

The idea that field decay in neutron stars is associated with accretion on to them is not new and was in fact put forward by Bisnovatyi-Kogan and Komberg (1986) even before the first binary pulsar was discovered, but it has not had much currency. Extraordinarily, however, the recent work of Srinivasan et al (1990) provides a quantum-mechanical scenario for associating the decay of the magnetic field with accretion, and for non- decay at other times. In this new picture, the decay of the magnetic field is not because of the actual accretion of matter onto the neutron stars, but due to the concomitant slowing down of the rotation which expels the vortices in the superfluid interior. The vortices in turn drag out with them the magnetic flux from the superconducting core to the outer crust where ohmic decay can occur. The conclusions that follow as a result of this process are precisely those hypothesized by Bailes (1989) from purely observational considerations.

All in all, it is my view that as a result of the work of Bailes (1989) supported by that of Srinivasan et al (1990), we are very close to a satisfactory understanding of the origins and magnitudes of pulsar velocities. The most important new realisation, as I perceive it, is that of the fundamental connection between the rotational history of a neutron star and the evolution of its magnetic field not only in magnitude but even in form. And as we heard at this meeting from Ruderman (1991), the latter aspect can have profound implications for our understanding of the structure of the field in spun-up neutron stars, like the millisecond pulsars whose pulse widths, interpulses and other characteristics pose intriguing questions that we have barely begun to answer.

In the progress of our understanding of pulsars, it is my expectation that the intimate linkage between the rotational and magnetic evolution of neutron stars will play as important a part in the nineties as the realisation of a binary origin for most pulsars did in the eighties.

6. RECENT EVIDENCE FROM 'CURRENT' DISTRIBUTIONS

In the last part of my talk I would like to present some unpublished work by my colleagues Deshpande and Srinivasan which confirms impressively and extends the conclusions above regarding the origin of pulsar velocities. The study is based on a new analysis of pulsar "currents" similar to those of Phinney and Blandford (1981) and Vivekanand and Narayan (1981) but based on a total of eight surveys for which the survey parameters are known sufficiently well to make it possible to account for most of the selection effects. The parameters describing these surveys were taken from Narayan (1987) and Kulkarni and Narayan (1988) and the recipe for computing the selection effects was adopted from that given in Narayan (1987). Their analysis of the current distributions will be published in detail elsewhere but the main conclusions of Deshpande and Srinivasan are given below.

1. A lower limit for the time scale of spontaneous decay of pulsar magnetic fields is $\approx$ 18 M yr.

2. Many pulsars are born well away from the galactic plane with both long and short periods.

3. Most pulsars are born in binary systems, some tight and others wide.

4. Roughly half of all pulsars will have "orbital" velocities while the remaining half will have "orbital" plus kick velocities.

This study appears to provide evidence for pulars born over a range of galactic latitudes, showing that there is no mystery about pulsars moving towards the plane as found by Harrison et al (1991). They also find evidence for two classes of pulsars injected at characteristic ages of roughly 10^6 years and $10^{7.5}$ years. Since the latter also happen to be the low velocity, low field pulsars discussed by Bailes (1989), they are presumably the first-born pulsars from wide binaries. An analysis of the "current" of these pulsars as a function of the z-distance from the galactic plane suggests that they are injected predominantly at z $\approx$ 400−600 pc consistent with their binary origin. Another important new result of their analysis is that the pulsars "injected" at $\tau \approx 10^6$ yr must also be interpreted as the first-born from binaries. Again, this conclusion follows from the current distribution of these pulsars as a function of z-distance. The fact that there has been no field decay during the binary evolution suggests an origin from binaries in which the neutron star was not spun down significantly before the onset of mass transfer (Deshpande and Srinivasan tentatively identify them as tight binaries). Their attempt at accounting for the whole population of observed pulsars is represented in the following table which lists the different origins and characteristics.

Table

Fraction	Birth place	Velocity	B	P	Origin
$f_{S\star}$	Low z	High	High	Short	Single stars
f_{SNI}	Low z	High	High	Short	Disruption in first explosion
f_{SNI}	High z	High	High	Short	From massive run away stars
$\frac{1}{2}f_{tb}$	High z	High	High	Short	Second born in tight binaries
$\frac{1}{2}f_{tb}$	High z	Low	High	$P \approx 0.6$ s $\tau_{ch} \approx 10^6$ yr	First born in tight binaries
$\frac{1}{2}f_{wb}$	High z	High	High	Short	Second born in wide binaries
$\frac{1}{2}f_{wb}$	High z	Low	Low	$P \approx 0.3$ s $\tau_{ch} \approx 10^{7.5}$ yr	First born in wide binaries

A detailed analysis of the current distribution of pulsars indicates that the fraction of injected Pulsars $\frac{1}{2}(f_{tb} + f_{wb})$ is comparable to the fraction of short period pulsars. This implies that most of the pulsars have a binary origin.

It can be seen from the table that as far as the fields and velocities are concerned, the expected associations are not always high-high and low-low. Further, with reference to (4) above, that kicks add vectorially to orbital velocities and hence can work both ways. While the net velocity may in general be increased from the orbital one, as suggested by Bailes (1989), it could also be lowered as in the case of PSR 1913+16 to the point where the system remained bound after the second supernova explosion. Both the above considerations will cause a smearing of the B-V correlation. This led to the expectation, even before the data of Harrison et al (1991) became available, that the correlation should become less pronounced as the statistics improves. The data from all interferometric measurements to date is still suggestive of a weak correlation, as would be expected according to the Table above, but is considerably less pronounced than found by Anderson and Lyne from a sample of 26 pulsars. I think it is fair to say that we are getting close to an understanding of pulsar velocities and their correlations with other properties of the neutron stars.

ACKNOWLEDGEMENTS I thank my colleagues, C.S. Shukre for much help in the preparation of the talk, A.A. Deshpande and G. Srinivsan for kindly supplying me with information in advance of publication and R. Nityananda for suggestions to improve the manuscript.

REFERENCES

Anderson, B. and Lyne, A.G. (1983), Nature **303**, 597.

Bailes, M. (1989), Astrophys. J., **342**, 917.

Bailes, M., Manchester, R.N., Kesteven, M.J., Norris, R.P. and Reynolds, J.E. (1989), Astrophys. J., **343**, L53.

Bisnovatyi-Kogan, G.S. and Komberg, B.V. (1976), Soviet Astron. Lett., **2**, 130.

Blaauw, A. (1961), Bull. Astr. Soc. Netherlands, **15**, 265.

Blaauw, A. (1985), in W. Boland and H. van Woerden (eds), Birth and Evolution of Massive Stars and Stellar Groups, D. Reidel, Doredrecth, Holland, p.211.

Burrows, A. and Woosley, S.E. (1986), Astrophys. J., **308**, 680.

Cordes, J.M. (1986), Astrophy. J., **311**, 183.

Deshpande, A.A. and Srinivasan, G., (1991), to be published.

Dewey, R.J. and Cordes, J.M. (1987), Astrophys. J., **321**, 780.

Dorofeev, O.F., Radionov, V.N. and Ternov, I.M. (1984), JETP Let., **40**, 917.

Galt, J.A. and Lyne, A.G. (1972), Mon. Not. Roy. Astron. Soc., **158**, 281.

Gott, J.R., Gunn, J.E. and Ostriker, J.P. (1970), Astrophy. J., **160**, L91.

Gunn, J.E. and Ostriker, J.P. (1970), Astrophys. J., **160**, 979.

Harrison, E.R. and Tademaru, E.P. (1975), Astrophys. J., **201**, 447.

Harrison, E.R., Lyne, A.G. and Anderson B. (1991), these proceedings.

Kornilov, V.G. and Lipunov, V.M. (1984), Astron. Zh., **61**, 686.

Kulkarni, S.R. and Narayan, R. (1988), Astrophys. J., **335**, 755.

Lyne, A.G., Anderson, B. and Salter, M.J. (1982), Mon. Not. R. Astr. Soc., **201**, 503.

Manchester, R.N. Taylor, J.H. and Van, Y.Y. (1974), Astrophys. J. Lett., **189**, L119.

Narayan, R. (1987), Astrophys. J., **319**, 162.

Phinney, E.S. and Blandford, R.D. (1981), Mon. Not. Roy. Astron. Soc., **194**, 137.

Pskovsky, Yu.P. and Dorofeev, O.F. (1989), Nature **340**, 701.

Pskovsky, Yu.P. and Dorofeyev, O.F. (1990), J. Astrophy. Astron., **11**, p.507.

Radhakrishnan, V. (1984), in S.P. Reynolds, and D.R. Stinebring (eds), Proceedings of Millisecond Pulsars (NRAO, Green Bank, W.Va., USA).

Radhakrishnan, V. (1986), in J.P. Swings (ed.), Highlights of Astronomy 7, D. Reidel, Dordrecht, p.3.

Radhakrishnan, V. and Shukre, C.S. (1985), in G. Srinivasan and V. Radhakrishnan (eds), Proceedings of Workshop Supernovae, their Progenitors and Remnants (Bangalore), Ind. Acad. Sci., Bangalore, India, p.155.

Radhakrishnan, V. and Shukre, C.S. (1987), in F. Pacini (ed), Proceedings of the NATO Advanced Study Institute on High Energy Phenomena around Collapsed Stars (Cargese, Corsica, France), D. Reidel, Dordrecth, Holland, p.271.

Ruderman, M.A. (1991), these Proceedings.

Shklovskii, I.S. (1970), Astron. Zh., 46, 715.

Srinivasan, G., Bhattacharya, D., Muslimov, A.G. and A.I. Tsygan (1990), Curr. Sci., 59, 31.

Trimble, V. (1968), Astron. J., 73, 535.

Tutukov, A.V., Chugai, N.N. and Yungel'son, L.R. (1984), Sov. Astron. Lett., 10, 244.

Vivekanand, M. and Narayan, R. (1981), J. Astrophys. Astron., 2, 315.

Zwicky, F. (1957), in Morphological Astronomy, Springer-Verlag, Berlin, p.258.

STATISTICAL STUDIES OF THE LOCAL PULSAR POPULATION

Rachel J. Dewey
Pasadena, CA
USA

ABSTRACT. This paper briefly reviews the statistics of the observed pulsar population and the standard models of pulsar evolution that have been inferred from the observed population. It also addresses a number of as yet unresolved questions, in particular, whether magnetic fields decay, whether some pulsars are born with long periods, and what effects evolution in binary systems may have had on the population of single pulsars.

KEYWORDS: pulsars - stars: evolution - stars: neutron

1. INTRODUCTION:

Gunn and Ostriker (1970) (hereafter GO70) published the first major statistical analysis of the pulsar population more than twenty years ago; since then the number of known pulsars has increased more than 10-fold and statistical studies have continued, looking for increasingly subtle effects. Early studies were remarkably successful in identifying the basic features of the distribution and evolution of pulsars, though a major modification to GO70's scenario came with the discovery of millisecond pulsars and the development of binary spin-up theories. However, despite agreement on many elements of pulsar evolution, a number of important questions remain unresolved.

This paper briefly reviews the standard models of pulsar evolution and considers some of the questions addressed in recent work. It does not attempt a comprehensive discussion of pulsar statistics; more complete reviews can be found in Taylor and Stinebring (1986), Dewey (1989) and Srinivasan (1989). In addition it focuses only on "disk" pulsars (sometimes referred to as "field" pulsars), those not associated with globular clusters. Globular cluster pulsars are discussed extensively elsewhere in these proceedings.

2. BACKGROUND:

2.1 Modeling the pulsar population:

The approximately 500 known pulsars are only a small fraction of the total galactic pop-

E. P. J. van den Heuvel and S. A. Rappaport (eds.), X-Ray Binaries and Recycled Pulsars, 453–463.

ulation. Most statistical studies aim to model the overall galactic population and infer how pulsars evolve in time. This, however, involves a number of steps of extrapolation: from a pulsar's observed properties to its intrinsic characteristics, from the small fraction of the population detectable in pulsar searches to the entire population, and finally, from the present properties of this population to their evolution over time.

Pulsar searches reach only a small fraction of the galaxy, and a search's ability to detect a pulsar depends not only on the pulsar's luminosity and distance, but on its period, dispersion measure and location in the sky, resulting in complicated selection biases. In both obvious and subtle ways, properties typical of   the observed population may not be typical of   the population as whole. These issues are discussed in more detail by Lyne, Manchester and Taylor (1985) [LMT85] and Narayan (1987) [N87]. Despite these difficulties, much useful information can be obtained from the properties of the observed pulsar population.

2.2 Observed pulsar characteristics:

There are about 500 known disk pulsars. They are concentrated at low galactic longitudes (l) and latitudes (b), forming a wide disk population concentrated primarily in the inner portion of the galaxy. Periods (P) range from 1.55 ms to 4.3 s with most between 0.2 and 2 s. Period derivatives ($\dot{P}$) range from 1.2×10^{-20} s s^{-1} to 1.5×10^{-12} s s^{-1}, with 10^{-15} s s^{-1} typical. At 400 MHz, flux densities (S_{400}) range from ~ 1 mJy (the limit of recent surveys) to 5000 mJy. Dispersion measures (DM) range from 3 cm^{-3} pc to 1140 cm^{-3} pc. At (my) last count there were 10 binaries among the disk pulsars, amounting to 2% of the population.

Other properties can be inferred from the directly observable properties listed above. Distances (d) of 80 pc to 32 kpc are derived from dispersion measures using standard electron density models (e.g. LMT85). These, together with flux measurements, yield luminosity ($L = Sd^2$) estimates of 1 to 15000 mJy kpc^2 (roughly 10^{11} - 10^{15} erg s^{-1} Hz^{-1}) at 400 MHz. A number of important quantities are derived from P and $\dot{P}$: characteristic age, $\tau = P/2\dot{P}$, energy loss rate, $\dot{E} \sim \dot{P}/P^3$, and (assuming spindown due to magnetic dipole radiation), the dipole component of the magnetic field, $B \sim (P\dot{P})^{1/2}$. For the observed disk population: 1300 yrs $\lesssim \tau \lesssim 5 \times 10^9$ yrs, 10^{30} erg s$^{-1} \lesssim \dot{E} \lesssim 10^{38}$ erg s^{-1}, 1.4×10^8 G $\lesssim B \lesssim 2 \times 10^{13}$ G.

It is generally believed that pulsar luminosities should correlate with P and $\dot{P}$. GO70 proposed $L \sim B^2 \sim P\dot{P}$, and this relation has been used in many models of pulsar evolution (e.g. LMT85). Millisecond pulsar luminosities are incompatible with this law, and recent studies (e.g. Proszynski and Przybycien 1984) suggest that $L \sim \dot{P}^{1/3}/P \sim \dot{E}^{1/3}$ is a better description of the observed data. This latter relation has been used in many recent pulsar models; however, the scatter around even the best fit relation is very large (e.g. Fig 4. in Dewey [1989]).

Pulsar velocities provide information about pulsar evolution but are somewhat difficult to measure. There are currently about 35 proper motion measurements (Lyne, Anderson and Salter 1982, Bailes et al. 1990, Harrison, these proceedings), a handful derived from pulse timing (e.g. Rawley, Taylor and Davis 1988) and about 70 derived from observations of interstellar scintillation (Cordes 1986). Typical velocities are $100 - 200$ km s^{-1}, with values as low as 10 km s^{-1} and as high as 1000 km s^{-1} indicated by the data. A curious feature the velocity data is an observed correlation between v and $P\dot{P}$. This has been found in a number of samples (e.g. Anderson and Lyne 1983, Cordes 1986), but no convincing explanation of its origin has been proposed.

Finally the beaming fraction, and, in particular, its dependence on P and $\dot{P}$, is important in modeling the pulsar population, and difficult to measure. It can be studied via pulse shape and polarization data (*e.g.* Narayan and Vivekanand 1983 [NV83], Lyne and Manchester 1988 [LM88]); most models suggest that about 20% of all long period pulsars beam towards earth, but that the fraction may increase to nearly 100% in millisecond pulsars, possibly because the beams of these pulsars are fan-shaped rather than circular.

2.3 Modeling methods :

The aim of most statistical studies is to relate the number of observed pulsars with a given range of parameters to the total number of such pulsars in the Galaxy and to infer from this how pulsars evolve in time. The most common methods are, a) inverting the population by estimating the fraction of all galactic pulsars in a given parameter range detectable by one or more pulsar searches, and b) Monte Carlo methods where a model of the underlying galactic population is constructed and "searched", after which the model "observed" distribution is compared to the pulsars actually detected the by relevant searches. An interesting variant of the first approach is the "pulsar current" method (*e.g.* Vivekanand and Narayan 1981) which calculates the rate at which pulsars cross from periods shorter than P to periods longer than P. This method weights pulsars with large values of $\dot{P}$ more heavily than those with small $\dot{P}$; Narayan (1987) argues (persuasively), that it is more appropriate for studies of the birth rate and birth properties of pulsars because it gives lower weight to older, slowly evolving.

All these methods require search selection effects to be carefully accounted for, particularly when looking for subtle effects. Some studies have attempted to avoid this by constructing a flux-limited sample, but this is difficult to do, and, of necessity, excludes a significant fraction of the known pulsar population.

2.4 Standard model:

According to the standard picture of pulsar evolution, pulsars are remnants of massive ($\gtrsim 8\,M_\odot$) stars, formed in Type II supernovae. Angular momentum and magnetic flux are (approximately) conserved in the collapse so pulsars are born with periods of a few milliseconds, and high (10^{12} - 10^{13} G) magnetic fields; spindown evolution is dominated by magnetic dipole radiation. Magnetic fields decay on a $5 - 10\,$Myr timescale, as does luminosity, so pulsars are observable for $\sim 10-15\,$Myr and die when a combination of weak field and long period ceases to support radio radiation. Since massive stars lie close to the galactic plane, pulsars are believed to be born near the plane and receive a velocity kick at birth. Virtually all binary systems are disrupted by when the pulsar is formed.

Millisecond pulsars, with short periods and weak magnetic fields, and often members of binary systems, clearly do not fit into this scenario. Spin-up theories (*e.g.* Alpar *et al.* 1982, van den Heuvel 1984) were developed to account for the existence of millisecond pulsars and for the atypically small values of P and $\dot{P}$ found in binary pulsars. A standard scenario of "spun-up" (or "recycled") pulsars now exists, applying to binary systems not disrupted in the primary's supernova. This primary forms a neutron star which evolves as outlined above (though the presence of a companion may make it unobservable as a radio pulsar) until the secondary leaves the main sequence, expands, and overflows its Roche lobe. At this point the neutron star's period evolution becomes dominated by mass transfer from the secondary, and the pulsar is spun-up to an equilibrium period which depends strongly on the its magnetic field ($P_{eq} \sim B^{6/7}$). Once mass transfer ends the pulsar evolution continues as before, except with modified spin parameters.

2.5 $P\dot{P}$ diagram:

A useful means of picturing the course of pulsar evolution is the $P\dot{P}$ diagram. Figure 1a plots the 408 disk pulsars for which $\dot{P}$ has been measured. as well as a "spin-up line", corresponding to $P_{eq} \approx 2B_9^{6/7}$ where P_{eq} is measured in millisecond and B in units of 10^9 G, and a "death-line" corresponding to $\dot{P}/P^3 \approx 2 \times 10^{-17}\,\mathrm{s}^{-3}$. The exact location of the spin-up line, determined by the details of the mass transfer process and the structure of the neutron star's magnetic field, is somewhat uncertain (see Ghosh, these proceedings). Though there is some theoretical basis for the location of the death-line (*e.g.* Ruderman and Sutherland 1975), below which no radio pulsars are found, it is a mostly empirical concept. Figure 1b shows the same diagram, with three possible evolutionary tracks plotted: (a) evolution with no field decay, (b) evolution with exponential field decay on a 10 Myr time scale, and (c) like (b) but with spin-up after 20 Myr.

3. UNRESOLVED QUESTIONS:

Despite the success of the models outlined in §2.4, loose ends remain in our understanding of even "common-or- garden" (*i.e.* not millisecond) pulsars, in particular: 1) How (and why) do magnetic fields evolve? 2) What is the distribution of initial periods? 3) What effects does evolution in binary system have?

Recent statistical work addressing one or more of these questions includes Stollman (1987) [S87], Narayan (1987) [N87], Dewey and Cordes (1987) [DC87], Lyne and Manchester (1988) [LM88], Bailes (1989) [B89], Emmering and Chevalier (1989), Narayan and Ostriker (1990) [NO90], Sang and Chanmugam (1990) [SC90].

3.1 Magnetic field decay:

Field decay has been an accepted part of most pulsar models since GO70, despite a lack agreement on the underlying mechanism. Recently, however, the idea that the fields of all neutron stars decay exponentially has been increasingly questioned.

Most evidence for field decay comes from the distribution in P and $\dot{P}$ of the observed pulsar population. Most statistical studies (*e.g.* GO70, LMT85, NO90, though see Bhattacharya, these proceedings) show strong evidence for decreases in $P\dot{P}$ over a pulsar lifetime, compatible with exponential field decay on a 5-10 Myr timescale. Evidence for field decay also comes from the fact that many pulsars have characteristic ages significantly larger than their kinetic ages (z/v_z); in the presence of field decay, characteristic age overestimates true age (*e.g.* Lyne, Anderson and Salter 1982).

It has been suggested (*e.g.* Candy and Blair 1986, Blair and Candy 1989) that $P\dot{P}$ decreases because of the alignment of the magnetic and spin axes over time, because of an actual

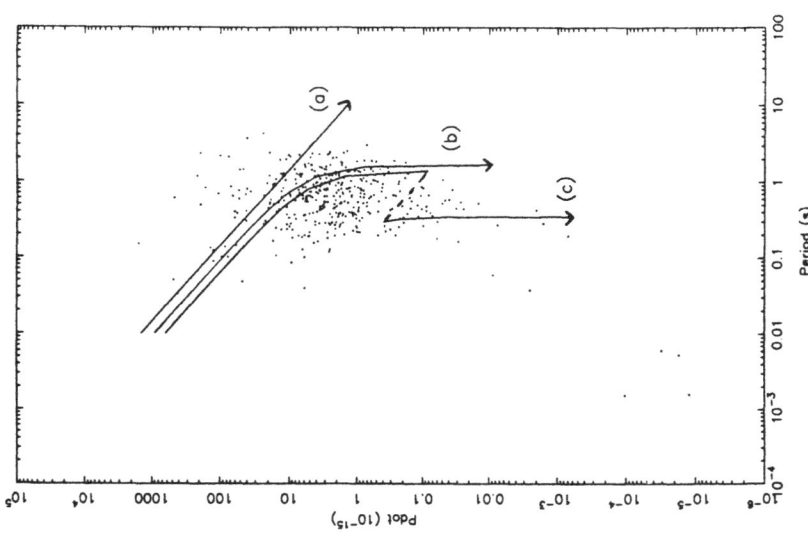

Figure 1b: Similar to Fig. 1a, but with three possible evolutionary tracks plotted: (a) no field decay, (b) exponential field decay with 10 Myr decay time, and (c) like (b) but with spin-up after 20 Myr. Binaries are not circled.

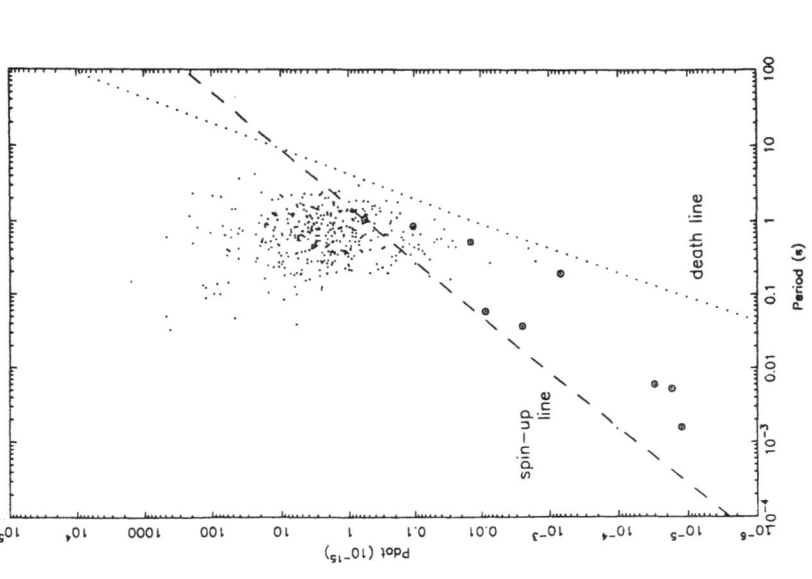

Figure 1a: P and $\dot{P}$ for the 408 disk pulsars with measured $\dot{P}$. Binary pulsars are circled. The "spin-up" line corresponds to $P_{eq} = 2B_9^{6/7}$ (P_{eq} in ms, B in 10^9 G). The "death-line" corresponds to $\dot{P}/P^3 = 2 \times 10^{-17}$ s^{-3}.

decrease in field strength. However, LM88, in a detailed study of pulse shape and polarization, concluded that alignment does not play a large role in period evolution. Perhaps most importantly, the existence of millisecond pulsars suggests that true field decay occurs in at least some pulsars: a neutron star must have a truly weak (not just aligned) dipole field in order to be spun up to millisecond periods.

However, there is considerable evidence that pulsar fields do not decay indefinitely. In particular, the existence of observable pulsars with very old binary companions (*e.g.* Kulkarni 1986, Wright and Loh 1986), and the very long lifetimes of millisecond pulsars derived from statistical studies (*e.g.* Srinivasan and Bhattacharya 1987), are understandable only if field decay eventually stops. Various explanations of this effect have been proposed: that field decay only occurs during mass transfer (*e.g.* Blondin and Freese 1986, B89), that field decay is not exponential (*e.g.* SC90), or simply that there is some residual field below which fields do not decay. Only the first of these, accretion-induced decay naturally explains why "residual" fields vary significantly from pulsar to pulsar (5^{10} for PSR 0655+64, 3×10^8 for PSR 1855+09).

SC90 investigated a potential mechanism (ohmic decay in the crust) for field decay and predict power-law, rather than exponential decay. They find, as do NO90 that such a law is compatible the statistics of the radio pulsar population. This model explains the apparent leveling off of field decay, but it is not yet clear if it can naturally explain the wide disparities in residual field.

Because the strongest evidence for substantial field decay is found in recycled pulsars, the idea of accretion induced field decay has been gaining increasing prominence. B89 explored this possibility, re-doing the kinetic vs. characteristic age analysis, taking into account that, for recycled pulsars, characteristic age cannot be compared directly to kinetic age. For pulsars below the spin-up line he compares the "release time", ($\tau_R = \tau - \tau_s$, where τ_s is characteristic age on the spin-up line) to the kinetic age, and finds only weak evidence for field decay in the absence of accretion. However, a more detailed analysis would be useful.

The field decay picture is further complicated by observations of X-ray pulsars suggesting that some old neutron stars have high (10^{12} G) magnetic fields (*e.g.* Makishima *et al.* 1988), including some that have undergone significant accretion (*e.g.* Verbunt, Wijers and Burm 1990). It remains unclear what mechanism could explain the existence of low magnetic fields in a significant number of radio pulsars and at the same time explain high fields in some old X-ray binaries.

3.2 Injection:

Injection refers to the hypothesis, first proposed by Vivekanand and Narayan (1981), that a significant number of pulsars are born (or first become observable) with periods of several hundred milliseconds (*i.e.* are "injected" into the pulsar population at long periods). As discussed below, there is considerable (though not overwhelming) statistical evidence supporting injection, but no direct evidence (such a long period pulsar in a young supernova remnant), and no convincing theoretical explanation of the effect.

3.2.1 Evidence pro an con:

The strongest argument for injection has been made by Narayan (1987) [N87]. Like

Vivekanand and Narayan (1981), N87 used the pulsar current method, but with a more complete analysis of the selection biases against short period, and concluded that approximately half of all pulsars are born with periods ~ 0.5 s. N87 also found that the effect was strongest for high magnetic field pulsars and hypothesied the period at birth might be directly related to the pulsars magnetic field. However, not using the pulsar current method, and making more conservative (possibly overly conservative) assumptions about the period dependence of luminosity and beaming fraction, LMT85 and S87 questioned the need for injection. Clifton and Lyne (1986) in their analysis of their search for short period pulsars also questioned the need for injection. Chevalier and Emering (1986) and EC89 found significant evidence for injection, though not at as pronounced a level as N87. Stokes *et al.* (1985, 1986), in analyzing the results of their searches for short period pulsar concluded that there was evidence for injection with luminosity laws such as $L \sim \dot{P}^{1/3}/P$ but not for $L \sim P\dot{P}$.

In an independent argument for injection, Bhattacharya and Srinivasan (1987) note that the birthrate of plerions (Crab-like supernova remnants) in the Galaxy is significantly lower than the pulsar birth rate. They argue that any pulsar born spinning rapidly will produce a plerion, so that if some pulsars do not produce a plerion at birth, is because they are spinning slowly.

The most ambitious recent analysis of the pulsar population was made by NO90, who compared the "observed" distributions predicted by 19 different models of the underlying population with the actual observed population. NO90 explicitly tested the hypothesis that there are two distinct pulsar populations, one born with short (~ 10 ms) periods, the other with longer periods. They found strong evidence for two populations (but see §3.2.2) and concluded that more than nearly half of all pulsars born with periods as long as 0.5-1 s.

In the various calculations the strength of the evidence for injection depends on the method used (pulsar current methods find stronger evidence) and on assumptions about the period dependence of the beaming fraction and luminosity. Models which assume $L \sim P\dot{P}$ and a period independent beaming fraction show no need for injection; those that assume $L \sim \dot{P}^{1/3}/P$ and a marked increase in the beaming fraction at short periods find that injection is required.

It is now generally accepted that $L \sim P\dot{P}$ is not compatible with observation and that $L \sim \dot{P}^{1/3}/P$ is a better model. However, there is evidence (*e.g.* S87) that this law flattens out at short periods, which may significantly affect injection calculations. Careful analyses of the short period pulsars discovered by Clifton and Lyne (1986) and Stokes *et al.* (1985) should help clarify this question. However, it will be hard to determine whether any observed deficiency of short period pulsars is due to their low luminosity or to their non-existence.

Similarly the question of the beaming fraction remains controversial. The high incidence of interpulses in millisecond pulsars, and studies of pulse shape and polarization (*e.g.* NV83 and LM88) suggest that the beaming fraction increases at short periods, and is nearly 100% at millisecond periods. However, LM88 find the increase less steep than do NV83, and question NV83's "fan-bean" picture. However, Narayan and Schaudt (1988) argue, from the statistics of pulsar in plerions, that very few short period pulsars are not beaming toward earth.

3.2.2 Two populations revisited:

As mentioned above, NO90 concluded that there are two distinct classes of pulsars, the 'F' population born with periods ~ 10 ms, and the 'S' population with periods $\sim 0.5 - 1$ s.

Surprisingly they found the strongest distinction between the two classes in their kinematic properties, the S population having high ($\sim 150\,\mathrm{km\,s^{-1}}$) velocities, and a large ($\sim 360\,\mathrm{pc}$) scale height at birth, and the F population having lower ($\sim 50\,\mathrm{km\,s^{-1}}$) velocities and a smaller ($\sim 120\,\mathrm{pc}$) scale height at birth. The populations also differed (though less markedly) in their magnetic fields: S pulsars averaging $8 \times 10^{11}\,\mathrm{G}$, and F pulsars, $1.3 \times 10^{12}\,\mathrm{G}$. Two curious features of these results can be noted. First, N87 found injection most pronounced for high field pulsars, yet NO90 found the fields of the injected S population to be lower than those of the F population. Similarly, though most of NO90's models included an explicit dependence of magnetic field on velocity (in order to account for the observed correlation between v and $P\dot{P}$, see §2.2), the high velocity S pulsars have, on average, weaker fields.

3.2.3 Summarizing:

Given current theories of the period dependence of beaming fraction and luminosity, there is substantial evidence for injection. N87 summarized three possible mechanisms for injection: a) that some stars lose a considerable amount of angular momentum before collapse, b) injection is induced by binary evolution – injected pulsars are those which have been shrouded by a binary companion during their early life, and c) late turn-on on magnetic fields as suggested by Blandford, Applegate and Hernquist (1983).

It is possible that more detailed studies of binary evolution will determine if the second (recycling) mechanism is viable. Confirmation of the first (pre-supernova angular momentum loss) requires either the observation of a long period pulsar in an obviously young supernova remnant[1] or a detailed theoretical explanation of how some (but not all) stars lose angular momentum before collapse. It is possible that evidence for the third mechanism (late B-field turn-on) might be found in pulse timing observations of young pulsars (Fruchter et al. 1988), but no such evidence is yet seen.

3.3 Binary effects:

The vast majority of pulsars are isolated objects, not members of binary systems and virtually statistical studies of pulsar evolution have dealt only with single pulsars, making no attempt to account for any effects of a companion. This approach is clearly an oversimplification; the majority of pulsar *progenitors* have at least one companion, and atypically small values of P, $\dot{P}$ among binary pulsars argue that these pulsars have been markedly affected by evolution with a companion.

It is therefore likely that some fraction of now-single pulsars have been affected by the presence of a now lost companion. Binary evolution has been invoked to explain a number of features of the general pulsar population: the large velocities, the observed correlation between v and $P\dot{P}$ (e.g. Radhakrishnan and Shukre 1985 [RS85], B89), injection (N87), and field decay (e.g. B89). However, modeling the effects of binary evolution (particularly accretion and spin-up) is complicated, and very few quantitative studies have been attempted.

---

[1] Note that Wolszczan, Cordes and Dewey (1991) conclude that their discovery of a 267 ms pulsar in the W44 remnant does not provide this evidence.

3.3.1 Kinematic effects:

One of the stronger conclusions about the effects of binary evolution was drawn by Dewey and Cordes (1987) [DC87]; they found that the high space velocities of pulsars could not be attributed solely to the break up of binaries, and that simultaneously explaining the high velocities and the low incidence of binaries, required some form of asymmetric kick in (or shortly following) the supernova explosion. When a symmetric supernova disrupts a circular binary, the two stars are released with speeds on the same order as their pre-explosion orbital speeds (RS85). For the pulsar velocity distribution to be explained solely by binary break-up, virtually all pulsars must originate in binary systems tight enough to undergo mass transfer – systems with $P_{orb} \lesssim 1000$ days before the start of mass transfer. Observations (*e.g.* Abt 1983) suggest that though about 85% of pulsar progenitors have at least one companion, only about 50% are in systems close enough to undergo mass transfer. The remaining pulsar progenitors are either single stars or members of wide binary systems; these will produce low velocity ($\lesssim 30\,\mathrm{km\,s^{-1}}$) objects unless asymmetries (or some other acceleration mechanism such as that of Harrison and Tademaru 1975) are invoked. In addition, though a scenario where all pulsar progenitors are members of close binary systems is compatible with the velocity distribution, DC87 DC87 found it to be incompatible with the low incidence of binary pulsars: such a model predicts that 25 - 50% of all pulsars in binary systems, in clear contradiction with the observed 2%.

At the time DC87 was published, the strongest evidence for asymmetric supernovae came from the work of Burrows and Woosley (1986) who concluded that the masses and orbital parameters of the PSR 1913+16 could best be explained by a small asymmetry in the second supernova. However, the recent discovery of PSR 1534+12 (Wolszczan 1991), a system very similar to 1913+16 except for its eccentricity of 0.27, provides stronger evidence. Unless very low mass He stars ($\sim 2.1\,M_\odot$) can explode as supernovae, the formation of a binary with the observed neutron star masses and small eccentricity, requires an asymmetry in the second supernova.

DC87 also examined the possibility that the observed correlation between velocity and magnetic moment might be induced by the variety of evolutionary paths in which pulsars are formed – in particular whether many low velocity, single pulsars may have been spun up by accretion. They found that, with standard models of binary evolution the number of pulsars formed by such paths was too small to induce an observable correlation between v and $P\dot{P}$. However B89, using a less orthodox model of binary evolution, presented a mechanism for producing a sizable number of low velocity, recycled single pulsars. B89 considers systems similar to Be x-ray binaries: wide ($P_{orb} \sim$ weeks to months) binaries containing a neutron star and a massive star. If the massive star explodes asymmetrically while the system is still wide, the old neutron star (with a somewhat decayed magnetic field) will be ejected with a (low) velocity comparable to its orbital speed, while the young, high field pulsar is ejected with a higher velocity due the supernova asymmetry. This model also helps explain the low incidence of binary pulsars: wide systems are more easily disrupted by asymmetric supernovae than narrow systems. However the assumption that mass transfer in massive x-ray binaries does not lead to a drastic decrease in orbital period is somewhat controversial, and a better understanding of the orbital evolution of such systems is needed before quantitative predictions can be made with the B89 model.

3.3.2 Non-kinematic effects:

Apart from the kinematic effects of binary disruption, the major effects of binary evolution are related to accretion spin-up. A thorough understanding of these effects requires both an understanding of magnetic field evolution and a better understanding of the mass transfer process. However, it is useful to consider what fraction of the pulsar population may have been recycled. The models of DC87 can be used to calculate the fraction of pulsars likely to be formed in paths where the system survives the explosion of the primary, undergoes mass transfer during the evolution of the secondary, and is finally disrupted by the explosion of the secondary. These models suggest that 2 - 15% of neutron stars may be formed in such paths, and that, due to the longer-than-average lifetimes of recycled pulsars, they may constitute a somewhat larger percentage of the observed pulsar population. To determine what fraction of of pulsars below the spin-up line were actually recycled S87 constructed a model of the pulsar population using only pulsars above the spin-up line (a sample presumably uncontaminated by recycled pulsars) and constructed from it a model of the "normal" pulsars below the spin-up line. Comparing this model to the observed population of pulsars below the spin-up line, he concluded that 35 - 50% of pulsars below the spin-up line (10 - 15% of the total population) are recycled. As discussed earlier, NO90 explicitly modeled the total pulsar population as two populations, and found that the the S population which can plausibly (though by no means unambiguously) be identified with recycled pulsars accounted for 20 - 50% of the observed pulsar population.

These calculations suggest a conundrum for further statistical studies, particularly those investigating fairly subtle effects such as injection: It is likely that too large a fraction of single pulsars are recycled for the effect to be ignored completely (or dealt with simply be excluding binary pulsars), but the effects of spin-up are too subtle for recycled pulsars to be unambiguously identified and excluded. Attempts to account for the effects of binary evolution will complicate statistical models of the pulsar population, but will probably be required in order for significant progress to be made in this area.

REFERENCES:

Abt, H.A. 1983, Ann. Rev. Astron. Astrophys. 21, 343.

Alpar,M.A., Cheng,A.F., Ruderman,M.A. and Shaham,J. 1982, Nature 300, 728.

Anderson,B. and Lyne,A.G. 1983, Nature 303, 597.

Bailes,M. 1989, Ap. J. 342, 917. [B89]

Bailes,M., Manchester,R.M., Kesteven,M.J. Norris,R.P. and Reynolds, J.E. 1990, M. N. R. A. S. 247, 322.

Bhattacharya, D. and Srinivasan, G. 1987, in High Energy Phenomena Around Collapsed Stars, ed. F. Pacini, (Dordrecht: Reidel), p.235.

Blair,D.C. and Candy,B.N. 1989 in Timing Neutron Stars, ed. H. Ogelman and E.P.J. van den Heuvel, (Dordrecht: Kluwer Academic Publishers), p.609.

Blandford,R.D., Applegate,J.H. and Hernquist,L. 1983, M. N. R. A. S. 204, 1025.

Blondin,K.M. and Freese,K. 1986, Nature 323, 786.

Burrows,A. and Woosley,S.E. 1986, Ap. J. 305, 680.

Candy,B.C. and Blair,D.G. 1986, Ap. J. 307, 535.

Chevalier,R.A. and Emmering,R.T. 1986, *Ap. J.* **304**, 140.

Clifton,T.R. and Lyne,A.G. 1986, *Nature* **320**, 43.

Cordes,J.M. 1986, *Ap. J.* **311**, 183.

Dewey,R.J. and Cordes,J.M. 1987, *Ap. J.* **321**, 780. [DC87]

Dewey,R.J. 1989, in *Timing Neutron Stars*, ed. H. Ogelman and E.P.J. van den Heuvel, (Dordrecht: Kluwer Academic Publishers), p.573.

Emmering,R.T. and Chevalier,R.A. 1989, *Ap. J.* **345**, 931.

Fruchter, A. S. *et al.* 1988a, *Nature* **331**, 53.

Gunn,J.E. and Ostriker,J.P. 1970, *Ap. J.* **160**, 979.

Harrison,E.R. and Tademaru,E. 1975, *Ap. J.* **201**, 447.

Kulkarni,S.R. 1986, *Ap. J. Letters* **306**, L85.

Kulkarni,S.R. and Narayan,R. 1988, *Ap. J.* **355**, 755.[NV83]

Lyne,A.G., Anderson,B. and Salter,M.J. 1982, *M. N. R. A. S.* **201**, 503.

Lyne,A.G., Manchester,R.N. and Taylor,J.H. 1985, *M. N. R. A. S.* **213**, 613.[LMT85]

Lyne,A.G. and Manchester,R.N. 1988, *M. N. R. A. S.* **234**, 477.[LM88]

Makishima,K. *et al.* 1988, *Nature* **333**, 746.

Narayan,R. and Vivekanand,M. 1983, *Astr. Ap.* **122**, 45.

Narayan,R. 1987, *Ap. J.* **319**, 162. [N87]

Narayan,R. and Schaudt,K.J. 1988, *Ap. J. Letters* **325**, L43.

Narayan,R. and Ostriker,J.P. 1990, *Ap. J.* **352**, 222.[NO90]

Proszynski,M. and Przybycien,D. 1984,in *Millisecond Pulsars* ed. S.P.Reynolds and D.R.Stinebring, (Green Bank: National Radio Astronomy Observatory), p.151.

Radhakrishnan,V. 1984,in *Millisecond Pulsars* ed. S.P.Reynolds and D.R.Stinebring, (Green Bank: National Radio Astronomy Observatory), p.130.

Radhakrishnan,V. and Shukre,C.S. 1985, ed. G.Srinivasan and V.Radhakrishnan, in *Supernovae, their Progenitors and Remnants*, (Bangalore: Indian Academy of Science), p.155. [RS85]

Rawley,L.A., Taylor,J.H. and Davis,M.M. 1988, *Ap. J.* **326**, 947.

Ruderman,M.A. and Sutherland,P.G. 1975, *Ap. J.* **196**, 51.

Sang,Y. and Chanmugam,G. 1990, *Ap. J.* **363**, 597.[SC90]

Srinivasan,G. and Bhattacharya,D. 1987, in *The Origin and Evolution of Neutron Stars* ed. D.J. Helfand and J.-H. Huang, (Dordrecht: Reidel), p.109.

Srinivasan,G. 1989, *Astron. and Astrophys. Rev.* **1**, 209.

Stokes,G.H., Taylor,J.H., Weisberg,J.M. and Dewey,R.J 1985, *Nature* **317**, 787.

Stokes,G.H., Segelstein,D.J., Taylor,J.H. and Dewey,R.J. 1986, *Ap. J.* **311**, 694.

Stollman,G.M. 1987, *Astr. Ap.* **178**, 143.[S87]

Taylor,J.H. and Stinebring,D.R. 1986, *Ann. Rev. Astron. Astrophys.* **24**, 285.

Verbunt,F., Wijers,R.A.M.J. and Burm,H.M.G. 1990, *Astr. Ap.* **234**, 195.

Vivekanand,M and Narayan,R. 1981, *J. Astrophys. Astron.* **2**, 315.

Wolszczan,A., 1991, submitted to *Nature*

Wolszczan,A., Cordes,J.M., and Dewey,R.J. 1991, submitted to *Astrophys. Jour. Lett.*

Wright,G.A., Loh,E.D. 1986, *Nature* **324**, 127.

CHAPTER 7

Neutron Star Interior Structure & Evolution

Neutron Star Structure and Superfluidity: A 1991 Perspective

David Pines
Department of Physics
University of Illinois at Urbana-Champaign
1110 West Green Street
Urbana, IL 61801

ABSTRACT. Our present understanding of neutron star structure and superfluidity is briefly summarized.

1. Introduction

The past few years have seen a resurgence of interest in the structure and evolution of neutron stars, and what can be learned about these from observations of pulsars, compact x-ray sources, and, possibly, γ-ray bursting sources. An incomplete list of the reasons for this renewed interest includes:

- The report (later withdrawn) of the observation of a 0.5 ms periodicity in 1987A. This led some theorists to posit the existence of quite new classes of neutron stars or to argue that the equation of state of existing neutron stars was so imperfectly known that such rapid rotation might be consonant with the neutron stars we know and love. Others, more confident of the current knowledge of the neutron matter equation of state and high densities, took the view that theory combined with observation had already rendered moot the existence of neutron stars which rotate that rapidly, and so one had only to wait for the observational results to be withdrawn or forgotten. That the second view prevailed can be taken as indicative of the maturity of neutron star physics, and or as a justification of the point of view that neutron stars can be used as cosmic hadron physics laboratories, in that meaningful constraints on hadron physical phenomena at temperatures and densities unachievable in the laboratory can be obtained from pulsar observations (Pines, 1980, 1991).

- Evolutionary questions posed by the discovery of millisecond pulsars, and by GINGA observations of magnetic field strengths in compact x-ray sources, which prompted a more careful look at the relation between pulsar periods and magnetic field strengths. For example, an extensive discussion among participants at a recent NATO Advanced Study Institute (Neutron Stars: Theory and Observation, held in Heraklion) led to a consensus that there is at present no observational evidence for the decay of pulsar magnetic fields in times $\sim 10^6$y, and that there exist a number of theoretical reasons for concluding that either magnetic fields in pulsars do not decay, or change only in consequence of pulsar spindown (see, for example, Ventura and Pines, 1991).

- The successful launch of ROSAT, and the forthcoming launch of GRO, which have been accompanied by a renewed interest in pulsar timing (cf Ögelman and Van den Heuvel, 1989), the

467

E. P. J. van den Heuvel and S. A. Rappaport (eds.), X-Ray Binaries and Recycled Pulsars, 467–471.
© 1992 *Kluwer Academic Publishers.*

468

discovery of many new glitching pulsars (Lyne and McKenna, 1990), and a renewed interest in pulsar cooling (and heating) (Tsuruta, 1992).

• Two glitching pulsars which have been "caught in the act" by alert observers (for the 1988 Vela glitch, see Flanagan, 1990; McCullough *et al.*, 1990); for the 1989 Crab glitch, see Lyne, Pritchard, and Smith 1991).

As might be expected, theorists have responded to this observational challenge by organizing workshops and advanced study institutes. Thus the 1988 Çesme ASI (Ögelman and van den Heuvel, 1989) gave rise to not one, but two meetings: Heraklion in 1990 (Ventura and Pines, 1991) and the present one, while a joint US/Japan workshop held in Kyoto in November, 1990, explored the structure and evolution of neutron stars (SENS '90), (see Pines, Tamagaki, and Tsuruta, 1992). Because the interested reader can find rather complete accounts of the reports at the Heraklion and Kyoto conferences in the above-cited proceedings volumes, I shall not attempt an overview here, but rather confine myself to an extremely brief summary of our present understanding of neutron star structure and superfluidity, and the extent to which observations of pulsar glitches and postglitch behavior provide information on the crustal neutron superfluid.

2. Neutron star structure and superfluidity

In Fig. 1 I show a cross section of a 1.4 M$\odot$ neutron star calculated using a moderately stiff equation of state for neutron matter at densities greater than the density of neutron matter, ρ_0. On the basis of mass determinations for millisecond pulsars orbiting neutron stars in binary systems $(1.32 \pm 0.03 \leq M/M_\odot \leq 1.442 \pm 0.003)$ we now have good reason to believe that most, if not all, pulsars possess masses close to 1.4M$\odot$, while the postglitch behavior of glitching pulsars provides a useful constraint on the stiffness of the high density neutron matter equation of state: it must be stiff enough to support a crust with an inertial moment which is $\geq 3\%$ of that of the star, a criterion which is met by the Bethe-Johnson equation of state used to construct Fig. 1.

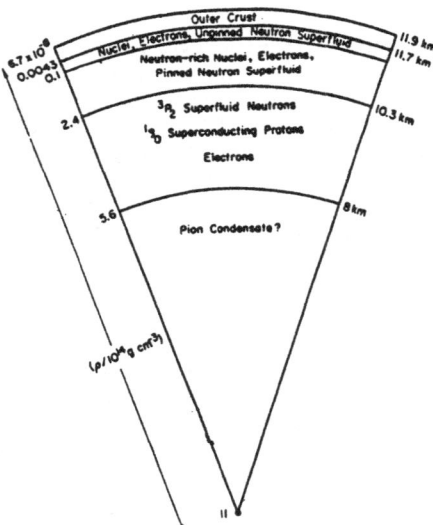

Figure 1. The cross-section of a 1.4M$\odot$ neutron star calculated using a moderately stiff (Bethe-Johnson) equation of state for neutron matter at densities $\geq \rho_0$.

The presence of neutron superfluid at densities, $.0015\ \rho_0 \leq \rho < \rho_0/2$ is firmly established theoretically, while both theory and postglitch observations strongly suggest that for densities, .

$0.04\ \rho_0 < \rho < \rho_0/2$, the vortices in that neutron superfluid are pinned to the crustal nuclei with which they coexist. Improved calculations of the energy gap for this $^1S_0$ superfluid have enabled theorists to place the limits shown in Fig. 2 for this important quantity, and I remind you that this crustal neutron superfluid possesses not only the highest transition temperature $(T_C \sim 0.7\ \text{MeV})$ of any superfluid thus far observed in the universe, but is also the most abundant.

As Ali Alpar and I have noted (Pines and Alpar, 1992), the composition of the quantum liquid core, and the precise density at which the transition from a solid crust to a quantum liquid interior takes place, continue to be somewhat uncertain, as does the question of whether there exists a narrow region of density near ρ_0 over which neutron matter might be normal, before one encounters the $^3P_2$ neutron superfluid produced by tensor forces in high density neutron matter.

While a pion condensate at densities $\geq 2\rho_0$ has seemed possible for over a decade, despite many theoretical efforts there remain enough uncertainties in the microscopic calculations that one cannot be sure it is realized in practice. The presence of proton superfluid in the quantum liquid core is also open to question, so that Fig. 1 should be regarded as a plausible scenario, to be confirmed by subsequent calculation or by observation.

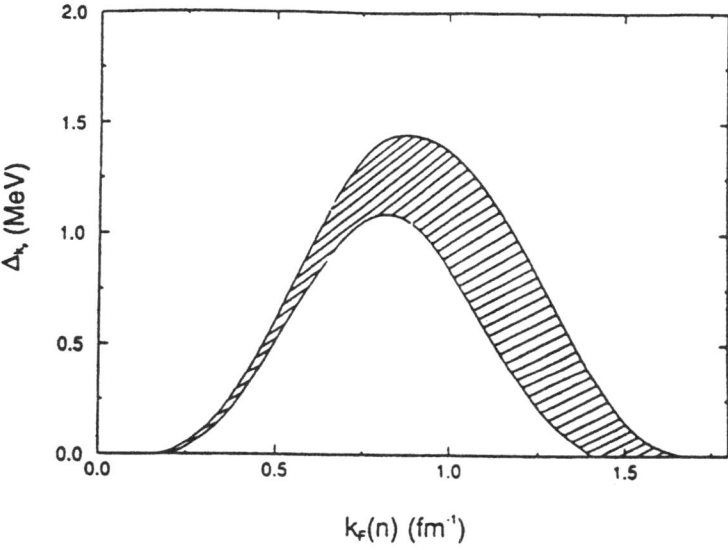

Figure 2. The calculated dependence of the $^1S_0$ gap, $\Delta(\text{MeV})$, on the density, expressed as function of the neutron Fermi wavevector, k_f. $k_f = 1.36\ \text{fm}^{-1}$ corresponds to a neutron density equal to that of nuclear matter, $\rho_0 = 0.17\ \text{fm}^{-3} \equiv 2.8 \times 10^{14}\ \text{gcm}^{-3}$.

3. The Vela Pulsar Glitches

Synoptic studies of the Vela pulsar following its "superglitch" in March, 1969 $[(\Delta\Omega_c/\Omega_c) \sim 2 \times 10^{-6}]$ have led to the observation of seven subsequent superglitches, of magnitude, $1.1 \times 10^{-6} \leq (\Delta\Omega_c/\Omega_c) \leq 3.1 \times 10^{-6}$, with sufficiently detailed coverage of its postglitch behavior that Alpar *et al.* (1992) have been able to identify four distinct post-glitch behaviors in the data. The analysis of the Christmas eve glitch, which was caught in the act by observers in Tasmania and South Africa, made possible the identification of a very rapidly relaxing region ($\tau \sim 0.4$d), and placed an upper limit on the time required for the crust and core to couple ($\tau_{cc} \leq 2$ min).

Alpar *et al.* (for a first report, see Pines and Alpar, 1992) have carried out a comprehensive re-evaluation of the postglitch relaxation following each of the eight superglitches observed between 1969 and 1988. They find that all the glitch data sets can be described in terms of three distinct components of short and intermediate time scale exponential relaxation, with relaxation times of 0.4d , 3.2d, and 32d, followed by a long-term recovery of the glitch-induced change in the spin-down rate that is linear in t. By making use of the vortex creep theory developed by Alpar *et al.* (1984a; 1984b), to treat the rotational dynamics of a superfluid with vortices pinned to crustal nuclei, they extract from observation a quite detailed map of the nature and extent of the pinned crustal neutron superfluid. Thus they:

- identify a plausible physical location within the star for the glitch,

- associate the presence of the three distinct relaxation times with the linear response of three distinct regions, corresponding to inertial moments, $(I_1/I) = 5.7 \times 10^{-3}$, $(I_2/I) = 1.5 \times 10^{-3}$, $0.51 \leq (I_3/I) \leq 1.2 \times 10^{-3}$, of the pinned neutron superfluid through which no sudden vortex motion occurred at the time of the glitch,

- conclude that the jump in response times from 3.2d to 32d reflects a transition from superweak pinning to weak pinning, with the 32d response reflecting the linear response of a weak pinning regime,

- associate the long term recovery with the non-linear response of regions of weak pinning.

This interpretation places constraints on the superfluid energy gap as a function of density, and the pinning energy as a function of the energy gap, constraints which are consistent with estimates of weak pinning parameters based on the best current microscopic energy gap calculations. Their work thus not only provides a detailed physical explanation of postglitch behavior, but also provides a direct link between phenomenological macroscopic theories of vortex motion and microscopic gap calculations. It places the presence of distinct pinned neutron superfluids in the crust on a firm observational footing, and yields a lower limit for both the physical extent of the pinned crustal superfluid region $[(I_p/I) \geq 3 \times 10^{-2}]$, and the crust of the neutron star $[(I_c/I) \geq 4 \times 10^{-2}]$. To the extent that the Vela pulsar is a representative neutron star, we may thus be in the somewhat surprising position of being able to map out much of the internal structure and dynamics of neutron stars some kiloparsecs away at least as well as we presently do for the earth and the rest of the solar system.

4. Acknowledgements

Much of the research carried out at Urbana and Aspen which is described here has been supported by NSF Grant PHY86-0377.

5. References

Alpar, M. A., Anderson, P. W., Pines, D. and Shaham, J.(1984a) *Ap. J.* **276**, 325.
Alpar, M. A., Anderson, P. W., Pines, D. and Shaham J. (1984b) *Ap. J.* **278**, 791.
Alpar, M. A., Chau, H. F., Cheng, K. S., and Pines, D. (1992), in preparation.
Flanagan, C. (1990) *Nature* **345**, 416.
Lyne, A., Pritchard, R. S., and Smith, F. G. (1991), in preparation.
McCullough, P. M., Hamiltonian, P. A., McConnell, D., and King, E. A.(1990) *Nature* **346**, 822.
McKenna, J. and Lyne, A. (1990) *Nature* **343**, 349.
Ögelman, H. and Van den Heuvel, E. (1989) Timing Neutron Stars, Kluwer Academic Pub.
Pines, D. (1980) *J. de Phys.* **41** C2-111.
Pines, D. (1991) in Neutron Stars: Theory and Observation, eds. J. Ventura and D. Pines, Kluwer Academic Pub.
Pines, D., Tamagaki, R., and Tsuruta, S. (1992) Structure and Evolution of Neutron Stars, eds. D. Pines, R. Tamagaki, and S. Tsuruta, Addison-Wesley Pub.
Pines, D. and Alpar, M. A. (1992) in Structure and Evolution of Neutron Stars, Addison-Wesley Pub.
Tsuruta, S. (1992) in Structure and Evolution of Neutron Stars, eds. D. Pines, R. Tamagaki, Addison-Wesley Pub.
Ventura,J. and Pines, D. (1991) Neutron Stars: Theory and Observation, Kluwer Academic Pub.

MAGNETIC FIELD EVOLUTION FROM NEUTRON STAR CRUST BREAKING

M. RUDERMAN
Physics Department and Columbia Astrophysics Laboratory †
Columbia University
538 West 120th Street
New York, NY 10027 and
Center for Astrophysics and Space Sciences
University of California
San Diego, La Jolla, CA 92093

ABSTRACT. Spinning-down (or up) neutron star crusts may be stressed beyond their yield strengths by crust neutron superfluid vortex line pinning. Such stresses may then move crustal plates and the magnetic field imbedded in them. Consequences can include continued magnetic moment decrease in dead spinning-down pulsars. Subsequent spin-up (e.g. by accretion from a companion) can lead to a variety of final spin periods and further reduction in magnetic dipole moment depending upon the initial pulsar magnetic field configuration. If the crust stress is relaxed by large scale cracking events these could cause pulsar timing glitches with magnitude and recurrence rates near those observed. In old or dead radio pulsars the sudden releases of stored elastic energy could give bursts of X-ray and gamma-rays whose number, energy, and rise time suggest those of Gamma-Ray Burst sources. The surface magnetic field of a spinning-down crust cracking neutron star may break up into large surface patches which move apart from each other but retain the original surface magnetic field. Pulsar spin-down torque observations would then reflect the decrease in average surface dipole field while the field strength inferred from a cyclotron resonance specrtral feature above a platelet would remain high and independent of stellar age.

1. INTRODUCTION

During the spin-up or spin-down of a rapidly rotating neutron star the neutron star's crust can become strongly stressed. One cause of possible large crust stress is pinning of vortices of the neutron superfluid which pervades the lower crust by nuclei of the crustal lattice embedded in it[1, 2, 3, 4]. This can cause the angular velocity change of the crust superfluid to lag that of a spinning-down or spinning-up crust and core. This velocity difference will grow and give an increasing force on the pinning crust nuclei until limited by any of three mechanisms:

1) Unpinning (and repinning) of pinned vortex lines. This is the basic assumption in the extensive analyses of Alpar *et al.*[2] on the "vortex creep" response of a neutron star crust to sudden spin-up events (glitches).
2) Breaking of the crust when such pinning forces cause the crust to become stressed beyond its yield strength[4, 5].
3) Movement into the pinned vortex region of parts of vortex lines with oppositely directed vorticity (e.g. vortex lines with S-like bends)[6]. These might relieve lattice stress without any unpinning or crust breaking.

---

† Permanent Address

473

E. P. J. van den Heuvel and S. A. Rappaport (eds.), X-Ray Binaries and Recycled Pulsars, 473–484.
© 1992 *Kluwer Academic Publishers.*

Because of the very high electrical conductivity of the lower crust, crust motion there may determine that of the pulsar's external magnetic field. However, certain properties of the lower crust ($\rho > 10^{13} gcm^{-3}$) have not yet been explored well enough to allow definitive predictions to be made. Of importance may be the microscopic distribution of atomic number of lower crust nuclei. The binding energy of magic number $Z = 40$ and $Z = 50$ nuclei are sufficiently close that the initial cooling phase of a young neutron star might freeze in almost random mixtures of these two nuclei especially near crust densities of $1 \times 10^{13} gcm^{-3}$ and $4 \times 10^{13} gcm^{-3}$ where sudden jumps are predicted for cold crusts in their lowest energy states. This could promote electron nucleus scattering ("impurity" scattering) to exceed electron phonon scattering in limiting the lower crust's electrical conductivity, and thus allow a much faster decay of magnetic field changes caused by a moving crust. Also not yet known well enough is the effect of changing nuclear species with depth (stratification) on the necessary backflow of crust matter above, through, or below any moving deep crust "plate". In the discussion below it is simply *assumed* that on time scales of interest magnetic field is frozen in the deep crust and the surface field linked to it moves with the deeper field. This implies that if much crustal backflow occurs *above* the moving lower crust the eddy diffusion time through this backflowing layer is small (both because crust electrical conductivity drops with density and because such possible backflowing layers would be thinner than the moving crust below whose density extends from 10^{13} to $10^{14} gcm^{-3}$). Even more important may be the neglect here of additional stresses needed to maintain backflow. Crust yield motion is simply assumed to begin just when it would in a thin hollow shell with the same dimensions, strength, and tangential stresses as those of the deep crust where vortex pinning stress is strongest. This evolution of the surface magnetic field of the star is just that at the core surface in the model of Srinivasan *et al.*[7] (unless stress falls below the crust's yield strength). In this paper we shall summarize causes and some possible consequences of neutron star crustal plate motion which follow when crusts are strained beyond their yield strength and the above assumptions are valid.

2. PINNING STRESS IN A NEUTRON STAR CRUST

The neutron superfluid which fills the space between lattice nuclei mimics uniform rotation through a quasiparallel array of quantized vortex lines. In most of the crust, where ρ is between 10^{13} and $10^{14} g\ cm^{-3}$, these vortex lines are expected to be pinned to crust nuclei[2, 1, 4]. When the crust spin rate and crustal neutron superfluid rotation rate are equal there is no shear stress on the lattice to which the vortex lines are pinned. Otherwise there is a force density on the lattice within the pinning region of the crust[5],

$$\mathcal{F} = 2\omega \times (\mathbf{\Omega}_n \times \mathbf{r})\rho_n, \tag{1}$$

for a lattice superfluid rotation velocity difference

$$\omega \equiv \mathbf{\Omega}_n - \mathbf{\Omega}. \tag{1'}$$

Here ρ_n is the internuclear superfluid neutron density, $\mathbf{\Omega}_n$ is its average angular velocity, $\mathbf{\Omega}$ is the angular velocity of the lattice and of the neutron superfluid vortices pinned to them, and $\mathbf{r}$ is the distance from the spin axis. Only part of this lattice body force can be balanced by pressure from the deformed lattice and its degenerate electrons together with the gravitational force. The rest must be balanced by lattice shear strength. The lattice elastic yield strength will be exceeded when (Ref. 5 Eq. 18b) ω exceeds the crust breaking limit

$$\omega_B \sim 10^{-1} \left(\frac{P}{10^{-3}s}\right) \left(\frac{\theta_{max}}{10^{-2}}\right) s^{-1}. \tag{2}$$

Here θ_{max} is the maximum change in length per unit length which a neutron star crust can support (under tension or compression) before exceeding its elastic yield strength. If ω reaches ω_B the lattice yields to plastic flow, crumbling, or cracking, and will not support any significant further increase in stress (unless all backflow is suppressed). Possible vortex unpinning from the crustal lattice nuclei could, in principle, keep ω from reaching ω_B.

From various estimates of the magnitudes of pinning forces we conclude the following[6].

1) Even if $\theta_{\max}$ for the large scale structure of the crust is 10^{-2} and the crustal lattice is always randomly aligned at a vortex core, lattice yield strengths are exceeded before unpinning when $P \lesssim 10$ ms.

2) Arguments for microcrystal alignment at vortex cores suggest extending the lattice breaking before unpinning regime to periods $P \lesssim 10^2$ ms.

3) If the lattice has large scale "fault planes" it can be much weaker than the microcrystals of which it is formed and the above range of pulsar spin periods may be greatly underestimated. An effective $\theta_{\max} \sim 10^{-4}$ gives a range of spin periods $P < 10$ s for lattice breaking spin-down.

3. STRENGTH OF A NEUTRON STAR'S CRUST

In the deep crust region where neutron vortex line pinning is strongest ($\rho \sim 4 \cdot 10^{13} \mathrm{g} \ \mathrm{cm}^{-3}$) the melting temperature of the "Coulomb" lattice $T_m \sim 4 \cdot 10^9 {}^\circ \mathrm{K}$, and the lattice shear modulus $\mu \sim 2 \cdot 10^{29} \mathrm{dyne} \ \mathrm{cm}^{-2}$. In this region the neutron superfluid transition temperature, $T_c \sim 10^{10} {}^\circ \mathrm{K} \sim T_m$, much greater than the deep crust temperature of neutron stars more than a few decades old.

In the lower crust lattice quantum effects are unimportant and the dimensionless strain θ in the lattice can depend upon stress (σ) only in the form

$$\sigma = \mu\theta f(\theta, \ T/T_m, \ \delta), \tag{3}$$

where δ is a dimensionless measure of the filling factor and form of a crystal's complex of dislocations. Stress experiments on near Coulomb metallic microcrystals such as Li or Mg at ordinary densities may be extrapolated by Eq. (3) to the superdense crystalline matter of a neutron star crust to yield the following features for neutron star crust microcrystals:

a) $\sigma = 3\theta\mu$ until an elastic strain limit $\theta_{\max}$ is reached;

b) $\theta_{\max} \sim 5 \cdot 10^{-3}$ at $T \sim 10^{-1} T_m$;

c) $\theta_{\max} \sim 10^{-2}$ for $T \ll T_m$;

d) $\theta_{\max} \ll 10^{-2}$ for $10^{-1} T_m \ll T < T_m$;

e) transition from continuous strain to a brittle response to stress over a small temperature range (of order $10^{-2} T_m$) near $T \sim 10^{-1} T_m$.

When $\theta > \theta_{\max}$, σ remains near $3\theta_{\max}\mu$: the crystal lattice grains yield and continue to deform without much further increase in stress. When $T \lesssim 10^{-1} T_m$ a crystal grain yields discontinuously: it continually but eratically "breaks" on a scale $\Delta\theta < \theta_{\max}$.

A crucial and unresolved problem is not so much the extrapolation through more than 13 orders of magnitude from the density of small laboratory crystals to that within the deep crust of a neutron star, but rather extrapolating from the behavior of very small single crystals or grains to that of an entire crust which extends over 10^{17} lattice constants. How does a possibly brittle neutron star crust break when stress grows slowly to that at which crust matter must yield? Does a crust develop faults along which it repeatedly slips at much lower stresses than $3\theta_{\max}/\mu$? What happens at the discontinuities in atomic number and lattice spacing which exist in the layered crust? Does the crust yield only by very many microscopic breaks so that its large scale respone still resembles the plastic flow of a hot crust or does it yield by rarer large scale cracking? For the crust as a whole an effective $\theta_{\max}$ as small as 10^{-4} may be plausible.

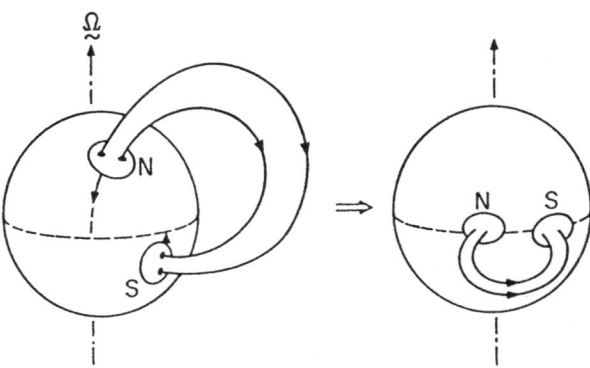

FIGURE 1. Evolution of the surface magnetic field of a short period spinning-down neutron star when equatorial zone magnetic field line reconnection is ignored[5].

4. CRUST MOTION AND MAGNETIC FIELD EVOLUTION

If crust yield strength is exceeded the strongly conducting metallic lower crust of a spinning down neutron star may continually move away from the spin axis toward the equator. Just how the crust achieves this motion is not clear. Forced crust flow toward the equator, where shear strain is a maximum, would have to be balanced by crust flow away from the equator. This might possibly be achieved by neutron star crust analogues of terrestrial plate subduction or by analogous superduction. Alternatively the forced equatorial crust "liquification" and equatorial zone vortex destruction may be a source of backflow of crust "fluid" through cracks which it makes and fills in the moving plates. The crust motion would cease when spin-down induced stresses no longer exceed the crust's yield strength. In Ref. 6 Eq. (18) this is shown to happen when the spin period (P) exceeds $\hat{P}$ given by

$$\hat{P} \sim 10 \left(\frac{10^{-4}}{\theta_{max}} \right)^{1/2} \text{ s} \tag{4}$$

if stresses needed to maintain backflow are ignored.

If imbedded magnetic field moves with the lower crust, magnetic pole strength would accumulate in the equatorial zone as shown in Fig. 1. (We assume that the crust aove this most strongly stressed part moves with the deep crust, but this may need detailed investigation.) If this was the end state of their evolution then radiopulsars older than one or two spin-down times would be expected to have much larger angles between their spin axes and dipole moments than younger ones. Such an evolution toward orthogonal rotators is not observed in canonical radiopulsars. If surface magnetic poles do not accumulate in the equatorial zone they must find partners there and recombine with them. Because of the high conductivity of the lower crust the recombination is not achieved through ohmic decay. Rather, azimuthal crust breaking from the connecting strong magnetic field itself must contribute to brining opposite poles together. Such recombination will be effective mainly for strongly magnetized $(B \gtrsim 10^{11}\text{G?})$ neutron stars

A predicted spin-down driven evolution of average B in canonical solitary radiopulsars with $\theta_{max} = 10^{-4}$ is shown in Fig. 2, where $\hat{P} = P_3 = 10$ s. Neutron stars born in binary systems may spin-down and then be spun-up in the same binary system. In a globular cluster, solitary pulsars evolving initially as indicated in Fig. 2 may subsequently be captured to form a binary. Further evolution of the binary or of the companion star can lead to accretion powered spin-up of the neutron star.

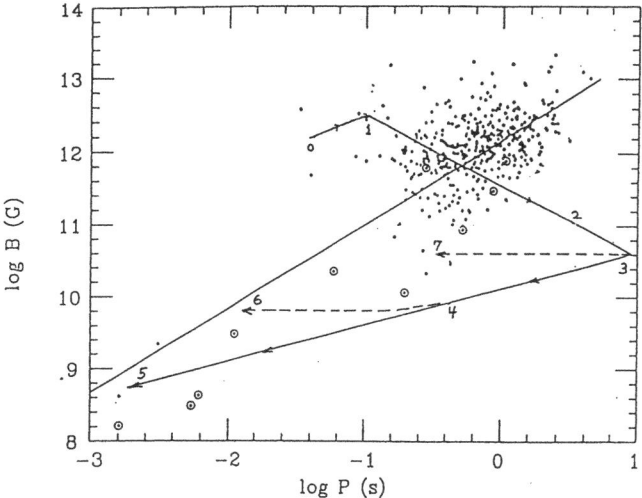

FIGURE 2. Evolution of dipole magnetic fields for a weak crust with $\theta_{max} = 10^{-4}$.

The evolution of the neutron star's surface magnetic field during such a spin-up phase depends upon the initial field distribution when the spin-up begins. Crustal plate motion should now be away from the spin equator toward the spin poles. Three possible initial configurations and the evolution which results are shown in Figs. 3 – 5 and Fig. 2. In Fig. 3, with N and S magnetic polar caps in opposite hemispheres, the dipole moment becomes aligned by spin-up. It also increases very modestly. In Fig. 4, the "sunspot configuration" with both magnetic polar caps in the same hemisphere, these two polar caps are pressed together at the same spin axis pole by spin-up crust motion. If large scale crust cracking does not occur the configuration formed after a long spin-down from period P_0 to P_3 (cf., Fig. 2) may consist of a near uniform N-pole distribution over the hemisphere with S-poles at the equator which have not yet recombined because of the absence of N-pole partners there. In a subsequent spin-up, field evolution from crust movement still leads to the same final state as that indicated in Fig. 4. As this sort of sunspot configuration star is spun-up from P_3 to P plate motion causes the stellar magnetic dipole to be diminished by a factor $(P/P_3)^{1/2}$. The evolved field would be very inhomogeneously distributed over the star's surface, very strong at the small polar cap (if $P/P_3 \ll 1$) and almost zero elsewhere. When the magnitude of the neutron star *average* dipole field $\langle B \rangle$ is measured by observing the pulsar spin-down rate after accretion has ceased, this average would be reduced from the initial average $\langle B_3 \rangle$ when $P = P_3$ according to

$$\langle B \rangle \sim \langle B_3 \rangle \left(\frac{P}{P_3} \right)^{1/2} . \tag{5}$$

If $P/P_3 \ll 1$ all poles and moments of B would be pressed into a small (radius $\sim R\langle B \rangle / \langle B_3 \rangle$) and thus nearly plane dipolar cap at the spin axis. In the absence of a specially symmetric initial configuration (e.g., axial symmetry) the star's evolving magnetic dipole should become orthogonal to its spin as $\langle B \rangle$ diminishes according to Eq. (5). In Fig. 5 we have an intermediate initial configuration with some of the flux from the North polar cap (N) reentering the stellar surface in the same (upper) hemisphere from which it was emitted (S_1), while the rest reenters in the opposite (lower) hemisphere (S_2). Here spin-up would lead to a dipole whose orientation and magnitude depend upon the relative fluxes into $S_1(\Phi_{S_1})$ and $S_2(\Phi_{S_2})$. If

$$\Phi_{S_2} \ll \left(\frac{P}{P_3} \right)^{1/2} \Phi_{S_1} \tag{6}$$

478

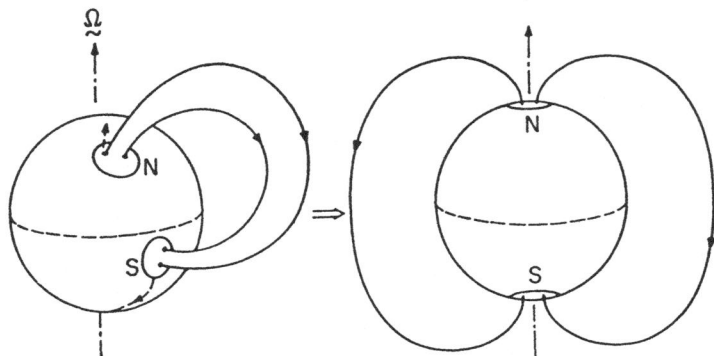

FIGURE 3. Evolution of the surface magnetic field of a short period spinning-up neutron star when flux lines initially connect the two spin hemispheres.

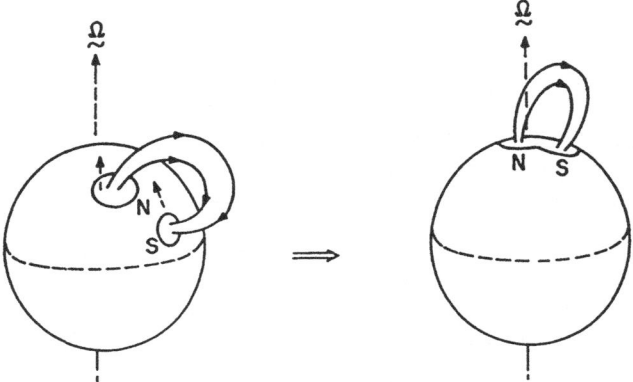

FIGURE 4. Evolution of the magnetic field of a spinning-up short period neutron star when all flux leaving a hemisphere reenters the same hemisphere ("sunspot" configuration).

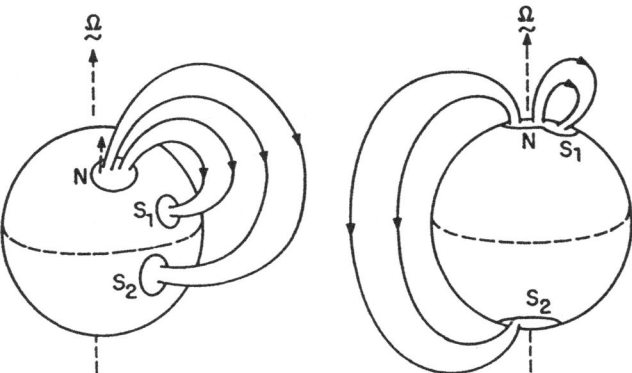

FIGURE 5. Evolution of the magnetic field of a spinning-up short period neutron star with an initial flux configuration which is intermediate between that of Figs. 5 and 6.

the spun-up star's dipole becomes essentially that of Fig. 5, an orthogonal dipole much smaller than its original size. If

$$\Phi_{S_2} \gg \left(\frac{P}{P_3}\right)^{1/2} \Phi_{S_1} \tag{7}$$

the spun-up star's field evolves as in Fig. 4 into a somewhat larger aligned dipole. If

$$\Phi_{S_2} \sim \left(\frac{P}{P_3}\right)^{1/2} \Phi_{S_1} \tag{8}$$

the spun-up star's dipole is reduced in magnitude with no special final orientation relative to the stellar spin. Figure 2 shows possible evolutionary paths with a weak ($\theta_{max} = 10^{-4}$) crust. The spin-down segment $\overline{0123}$ is discussed in ref. 6.

The possible spin-up evolution of the average (dipole) field shown in Fig. 2 for various field configurations for spin-up of a weak crust star begins at the period $P_3 = 10$ ms. A pure sunspot-like initial configuration leads to the $\overline{345}$ path and the dipole becomes increasingly more orthogonal as $\langle B \rangle$ decreases. The path $\overline{37}$ is followed for the geometry of Fig. 3 and the dipole moment becomes more aligned as spin-up continues. The intermediate path $\overline{346}$ is followed when the initial period P_3 is such that Eq. (11) is satisfied. However, at $P \sim P_4$ Eq. (8) becomes appropriate. Further large spin-down of $\langle B \rangle$ no longer occurs as accretion driven spin-up brings P toward the accretion spin-up equilibrium line. In this case the magnetic dipole moment is mainly orthogonal as point 4 is approached along the path $\overline{34}$, but becomes aligned during further spin-up along $\overline{46}$. If spin-up stops along $\overline{46}$, or if the pulsar spins-down to lie along that segment after reaching the accretion spin-up equilibrium line, the dipole may be oriented anywhere between alignment and orthogonality, with alignment favored nearest the spin-up line. When the point 4 occurs at $P < 10$ms even the lowest $\langle B \rangle$ shortest P pulsar could have non-orthogonal dipole moments. Comparisons of these neutron star period and magnetic field evolutions with observations are given in ref. 6.

5. CRUST CRACKING AND PERIOD GLITCHES

We turn now to some of the special phenomena and structures which result when the crust is assumed to yield to growing global stress with large scale "cracking"; parts of the crust are assumed to slip suddenly to relieve some significant fraction of an otherwise growing crustal stress[10, 4]. For warm crusts such large cracking would probably not occur because the yielding is more likely to be by plastic flow. It was noted above that a transition from plastic flow to a more brittle response might be expected near the same ratio of transition temperature (T_t) to crystal melting temperature (T_m) as that measured in the laboratory for (near) Coulomb crystals such as Li or Mg; then $T_t/T_m \sim 10^{-1}$. The transition regime is over a very small fraction of T_m. For neutron star crusts relevant T_m are several times $10^{9}°$K and the transition may be expected to occur at around the Crab pulsar inner crust temperature. It almost certainly has already occured in the older Vela pulsar crust where the temperature at the base has been estimated[11] to be $T = 1.3 \times 10^{8}°$K so that $T/T_m \sim 3 \cdot 10^{-3}$. However, it should be emphasized that the crust might well everywhere limit stress entirely by microscopic crumbling even when each grain responds rather brittlely. The assumption of relatively large scale crust cracking is a hypothesis which must still be supported mainly by comparisons of its consequences with neutron star observations. A significant fraction of the growing strain from forced crust motion is assumed below to be relieved in such sudden cracking which relieves the maximally strained crust ($\theta \sim \theta_{max}$) by $\Delta\theta$. Because the spin-down induced crustal stresses are directed away from the spin axis poles, the sudden $\Delta\theta$ motions of stressed crust generally also move crust outward (except in the equatorial zone). A sudden outward motion of part of the crustal lattice also carries outward those crustal neutron superfluid vortices which are pinned to the moving lattice's nuclei. In this way crustal neutron superfluid which was prevented from spinning down with the crust because its vortices could not move outward suddenly reduces the angular frequency lag ω between its rotation speed Ω_n and the rotation speed of the crustal lattice Ω. The decrease in Ω_n from an outward displacement of vortices by $R\Delta\theta/2$ is $\Delta\Omega_n = -\Delta\theta\Omega_n$. The crust must, of course, recoil by spinning-up, a motion which is quickly communicated to and shared with the rest of the star. Then the final shared crust spin-up is

$$\Delta\Omega = \frac{\Delta\theta\Omega_n I_n}{I - I_n}, \tag{9}$$

with I the moment of inertia of the whole neutron star and I_n that of the suddenly spun-down crustal neutron superfluid. If we identify $\Delta\Omega$ with the observed spin-up glitches of Vela-like radiopulsars, put $\Omega \sim \Omega_n$, and use a canonical estimate $I_n \sim 10^{-2}I$, then

$$\frac{\Delta\Omega}{\Omega} \sim 10^{-2}\Delta\theta. \tag{10}$$

For Vela glitches $\Delta\Omega/\Omega \sim 10^{-6}$. This gives $\Delta\theta \sim 10^{-4}$ for Vela's crust. The interval between glitches (τ_g) should be the length of time needed for further spin-down to rebuild the strain $\Delta\theta$ relaxed by the crust cracking:

$$\tau_g \sim \frac{\Delta\theta}{\dot{\Omega}/\Omega}. \tag{11}$$

The combination $(\Delta\Omega/\Omega)(1/\tau_g)$, called the "glitch activity" rate by McKenna and Lyne[12], is independent of $\Delta\theta$ and θ_{max}. For pulsars rotating with $P > 10$ ms[10]

$$\frac{\Delta\Omega}{\Omega} \cdot \frac{1}{\tau_g} = \frac{\dot{\Omega}}{\Omega} \frac{I_n}{I} \sim \frac{10^{-5}\text{yr}}{(\text{Age}/10^3\text{yrs})} \tag{12}$$

The model result of Eq. (12) is compared to observed glitch activity rates of young pulsars in Table 1. Agreement is quite reasonable for the Vela-like family of 10^4yr old pulsars. We would attribute the very low glitch activity in the younger Crab family solely to the fact that their lower crusts are warmer and respond to growing stress mainly by plastic flow[4,12]. A recent estimate[11] for the internal Crab pulsar temperature gives $T \sim 4 \cdot 10^{8\circ}\text{K} \sim 10^{-1}T_m$. It would appear from Table 1 that the transition from crust plastic flow to a more brittle response takes place after about 1,700 yrs, the estimated age of PSR 0540. From laboratory results shown in ref. 10 that transition would be expected during a relatively small interior temperature drop $\Delta T \sim 10^{-1}T \sim \text{several} \cdot 10^{7\circ}\text{K}$. The radiopulsars much older than Vela should continue to have glitches, but with much longer interglitch intervals τ_g so that few have been observed so far. The observed glitch repetition rates give an effective $\Delta\theta \sim 2 \cdot 10^{-4}$ for Vela and $4 \cdot 10^{-5}$ for PSR 1737.

Table 1. Glitch Activity in Young Radiopulsars. *The estimated glitch activity of PSR 0540 is based upon the reported $\Delta\Omega/\Omega \sim 10^{-5}$ for the only glitch observed in the decade since its discovery[13]. All other data entries are taken from McKenna and Lyne[14].

Pulsar	Age (10^3 yrs)	Glitch Activity (10^{-7}yr^{-1})	
		Observed	Eq. (11)
0531	1.2	0.1	80
1509	1.5	~ 0	70
0540	1.7	10^*	60
0833	11	8	9
1800	16	?	6
1737	20	4	5
1823	21	5	5

6. EVOLVING MAGNETIC FIELD STRUCTURE IN SPINNING DOWN NEUTRON STARS

If a post-Crab pulsar crust relaxes with large scale cracking the crust breaks up into "platelets" [5, 10]. "New" matter flows into cracks between platelets. This new matter differs in important ways from the old crust matter of a platelet. It does not have a large magnetic field while the original crust magnetic flux remains frozen in the platelets. Continued spin-down induced cracking would be expected to take place at platelet boundaries and beyond in the interplatelet crustal matter. The platelets themselves should then retain their integrity and their original magnetic fields during spin-down. Such evolution of the surface magnetic field of a spinning-down pulsar crust is indicated in Fig. 6. When the crust is so warm that it stretches plastically without large cracking (i.e., Crab-like or younger) the original crust is stretched (and thus thinned) by the motion toward the equator. As a young pulsar spins-down from its initial spin period and cools, the initial polar cap fields initially spread with the highly conducting crust in which they are imbedded to fill much of each hemisphere. At $P = P_c$ cracking begins; thereafter, platelets move apart until the spin-down period reaches the $\hat{P}$ of Eq. (8). Each platelet has frozen into it the field $B(P_c)$ which it had when cracking first began. This field should be quite uniform over a platelet surface since it came from the initial stretching of the field of a relatively small polar cap. Then in the spin-down regime $P_c < P < \hat{P}$ while the *average* surface dipole field $\langle B(P) \rangle$ decreases inversely with P[7, 6],

$$\langle B(P) \rangle = \langle B(P_c) \rangle P_c / P, \tag{13}$$

the rather uniform platelet fields (B_p) remain frozen at

$$B_p = \langle B(P_c) \rangle \sim B(\text{Vela}) \sim 3 \cdot 10^{12} \text{G}. \tag{14}$$

Therefore measurements of B in older spun-down neutron stars should give conflicting results. Those based upon the magnitude of the magnetic dipole moment (spin-down in radiopulsars, accretion torques in X-ray pulsars) should give smaller values of B than those inferred from surface cyclotron resonance features in the X-ray spectra of X-ray pulsars and Gamma-Ray Burst sources). The predicted extreme uniformity of a platelet field could also make a cyclotron resonance feature much narrower than would be expected otherwise.

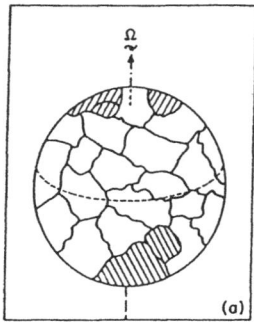 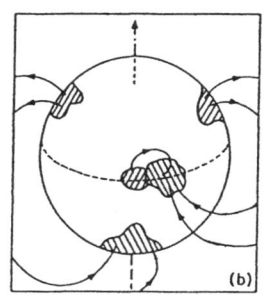

FIGURE 6. Evolution of surface platelets and magnetic field after crust cracking begins. a) An initial configuration of cracks; the magnetic field is continuously distributed through the surface. b) Separation of original platelets (hatched) after further spin-down and equatorial zone subduction. At the original platelets B maintains the value it had in a). $B \sim 0$ at new interplatelet crust.

7. GAMMA-RAY BURSTS

If radiopulsar glitches are a manifestation of crust cracking, they are associated with sudden releases of stored elastic strain energy in the deep crust. The total rate of such events from the neutron star population of the Galaxy, the amount of energy released per glitch, and the initial timescale for that release are all near what seem to be required for the sources of observed gamma-ray bursts (GRBs).

Because efficient generation of gamma-rays from this elastic energy release may be high only in the older, longer period, neutron stars which have relatively charge-starved magnetospheres[14], observed burst sources may be limited to dead radiopulsars which no longer have steady copious sources of $e^{\pm}$ pairs in their magnetospheres. This is an old neutron star population which should, therefore, be distributed broadly, a kiloparsec or more above and below the Galactic disk since the younger active radiopulsars typically have very high velocities (~ 150 km s^{-1}). Because the so far observed GRB sources seem to be isotropically distributed about us only those neutron stars nearer than about a kiloparsec should be in the candidate source population. These could number 10^7, about 10^{-2} of all Galactic neutron stars. Since GRBs are observed at a rate of about 10^2yr^{-1} each candidate source must give $10^2 \text{yr}^{-1} \times 10^{10} \text{yrs} \times 10^{-7} = 10^5$ GRBs.

The total number of glitches expected during the spin-down of a pulsar from P_c (the spin period at which crust cracking begins) to $\hat{P}$ (the spin period of Eq. (4) at which crust cracking ceases) is

$$N_g = \frac{1}{\Delta\theta} \ln\left(\frac{\hat{P}}{P_c}\right) \sim 10^5. \tag{15}$$

The numerical value assumes $P_c = 5 \cdot 10^{-2}$s (midway between Crab and Vela periods), $\hat{P} = 10$s (Eq. 4 with $\theta_{max} = 10^{-4}$), and $\Delta\theta = 5 \cdot 10^{-5}$ (for PSR 1737). The glitches will continue after radiopulsar turnoff. Each sudden release of elastic strain energy from the (lower) crust of thickness ℓ in a glitch is

$$\mathcal{E}_g \sim \ell R^2 \mu \theta_{max} \Delta\theta \sim 10^{39} \left(\frac{\theta_{max}}{10^{-3}}\right) \left(\frac{\Delta\theta}{10^{-4}}\right) \ell_5 \text{ ergs}, \sim 10^{38} \text{ergs}. \tag{16}$$

An equivalent energy release comes from the sudden slowdown of crust neutron superfluid rotation.

Models for the response of a pulsar magnetosphere to a sudden onset of large amplitude kHz vibrations of surface magnetic field suggest that particle depeltion is a necessary condition for high efficiency extreme relativistic particle acceleration. Large GRBs may accompany glitches only in magnetospheres starved of charge because outer magnetosphere pair production and γ-ray emission has turned off or, probably more significantly, because there is no longer the steady $e^{\pm}$ production of canonical radiopulsars which have not yet reached the canonical "death line." This would restrict strong GRB sources to slowly rotating extinct radiopulsars, as already suggested by Blaes et al.[14] If the needed large GRB source population consists mainly of dead radiopulsars, P_c in Eq. (15) should be replaced by the radiopulsar extinction period of about 2 s. With $\theta_m \sim 10^{-4}$, the smallest value compatible with the limit $\theta_m > \Delta\theta$, Eq. (15) gives $N_g = 5 \cdot 10^4$, near (but perhaps disturbingly less than) the above estimate of 10^5 GRB's per neutron star. The elastic energy released in each glitch-GRB is given by Eq. (16). If 10^{38} ergs is ultimately converted into energetic radiation from the neutron star's suddenly disturbed magnetosphere, the observed GRB fluence would be $\mathcal{E}_g/4\pi/(1 \text{ kpc})^2 \sim 10^{-6}\text{erg cm}^{-2}$, typical of that in a GRB. The timescale for initiating this energy release (Δt) is that for a crack (which moves with the deep crust shear wave velocity $v_s \sim 2 \cdot 10^8 \text{cm s}^{-1}$) to propagate through the lower crust thickness (ℓ). The rise time $\Delta t \sim \ell/v_s \sim 5 \cdot 10^{-4}$s, is consistent with GRB observations. The time structure of the whole GRB depends upon unknown details of how a crack network develops within the crust. After the crack, the released elastic energy is initially in the form of crust shear vibrations with frequencies ($\sim v_s/\ell$) in the kHz regime. Blaes et al.[14] have shown how such high frequency elastic vibrations reach the stellar surface as Alfven waves with amplitudes greatly amplified from those of the initial deep crust oscillations. In the GRB model of Blaes et al., elastic energy release by a mid-crust mechanical instability is replenished by stress from continued cold mass accretion onto the star surface rather than spin-down as suggested here. Spin-down powered crust cracking is deeper and thus can involve greater elastic energy storage and release. Many details of the Blaes et al. model can, however, be carried over.

No GRBs were observed during the Vela radiopulsar's glitches, implying not much more than 10^{38}erg s^{-1} into any glitch associated GRB from this pulsar. The model's $\mathcal{E}_g \sim 10^{38}$erg might have escaped detection. Moreover, even if the plasma density in Vela's magnetosphere were only the minimum Goldreich-Julian charge density needed to keep the near magnetosphere electric field in the star's corotating reference frame near zero, potential GRB emission from this pulsar could be suppressed just by the inductive potential drop generated by moving ambient magnetosphere charge[10].

It is a pleasure to thank P. Jones for very many enlightening discussions about neutron star crusts and superfluids. I am also grateful for conversations with A. Alpar, B. Bhattacharya, R. Blandford, E. van den Heuvel, G. Taylor, R. Epstein, P. Goldreich, F. Lamb, D. Pines, J. Rankin, V. Radharishnan and J. Sauls and for the hospitality of S. Tsuruta, R. Tamagaki, N. Shibazaki, R. Dalitz and the Department of Theoretical Physics of the University of Oxford. This research has been supported in part by NSF-89-01681 and is contribution number 461 of the Columbia Astrophysics Laboratory.

REFERENCES

[1] P. Anderson and N. Itoh, Nature, 256, 25 (1975).

[2] A. Alpar, P. Anderson, D. Pines and J. Shaham, Ap.J., 282, 791 (1984).

[3] R. Parker, Phys. Rev. Letters, 28, 1080 (1972).

[4] M. Ruderman, Ap.J., 203, 213 (1976).

[5] M. Ruderman, Ap.J., in press (1991).

[6] M. Ruderman, "Neutron Star Plate Tectonics II," submitted to Ap.J. (1991).

[7] G. Srinivasan, D. Bhattacharya, A. Muslimov, and A. Tsygan, Current Science, **59**, 31 (1990).

[8] A. Alpar, S. Langer and J. Sauls, Ap.J., **282**, 533 (1985).

[9] J. Sauls, *Timing Neutron Stars*, H. Ögelman and E.P.J. van den Heubel (eds.) (J. Kluwer Academic Publishers, Dordrecht, 1989).

[10] M. Ruderman, "Neutron Star Plate Tectonics III," submitted to Ap.J. (1991).

[11] A. Alpar, K.-S., Cheng, and D. Pines, University of Illinois preprint (1990).

[12] J. McKenna and A. Lyne, Nature, **343**, 349 (1990).

[13] H. Ögelman, G. Hasinger, and J. Trümper, IAU Circ. 5162 (1991).

[14] O. Blaes, R. Blandford, P. Goldrecih, and P. Madav, Ap.J., **343**, 839 (1989).

CHAPTER 8

Accretion Disks & Disk-Magnetosphere Interactions

DIAGNOSTICS OF DISK-MAGNETOSPHERE INTERACTION IN NEUTRON STAR BINARIES

PRANAB GHOSH AND FREDERICK K. LAMB
Departments of Physics and Astronomy
University of Illinois at Urbana-Champaign
1110 W. Green St., Urbana, IL 61801
and
Institute for Theoretical Physics
University of California
Santa Barbara, CA 93106

ABSTRACT. The interaction between the magnetospheres of accreting neutron stars and accretion disks plays a key role in determining the properties of many accretion-powered neutron star X-ray sources and the recycled binary and millisecond rotation-powered pulsars. Here we show that the behavior of the horizontal branch quasi-periodic intensity oscillations in low mass X-ray binaries and the correlation between the magnetic fields and periods of binary and millisecond pulsars are sensitive probes of the state of the inner disk.

1. Introduction

Accreting magnetic neutron stars are now recognized as the central power sources in a large number of astrophysical systems that produce high-energy radiation (see Ghosh and Lamb 1991a). Such systems are thought to include not only accretion-powered X-ray pulsars (see Joss and Rappaport 1984; Lamb 1989a; Nagase 1989) but also most low-mass X-ray binaries (LMXBs) (see White 1989; van der Klis 1989; Lamb 1989b). The γ-ray burst sources may also be accreting magnetic neutron stars (see Lamb 1988). Finally, accretion is believed to be the mechanism by which some rotation-powered pulsars have been "recycled" by being spun up again after spinning down during their first lives as rotation-powered pulsars. Most rotation-powered pulsars in binary systems as well as solitary rotation-powered pulsars with weak magnetic fields and millisecond spin periods are currently thought to have been recycled (see van den Heuvel 1991).

In binary systems containing neutron stars, plasma may flow from the companion star onto the neutron star as the result of several processes. When the companion star has a high mass ($M \gtrsim 15M_\odot$), plasma can be transferred to the neutron star via a radiation-pressure-driven wind, Roche-lobe overflow of the atmosphere of the

E. P. J. van den Heuvel and S. A. Rappaport (eds.), X-Ray Binaries and Recycled Pulsars, 487–510.

companion, or a combination of these two processes. When the companion star has a low mass ($M \lesssim 2M_\odot$), plasma can be transferred via Roche-lobe overflow of the envelope of the companion or via a wind produced by heating of the outer layers of the companion by X-rays from the neutron star.

Mass transferred via Roche-lobe overflow has a large specific angular momentum with respect to the neutron star and will therefore produce an extensive accretion disk around it. Mass transferred via a stellar wind may have sufficient specific angular momentum to form a small accretion disk around the neutron star, although the extent and time development of such disks is highly uncertain at present (Shapiro and Lightman 1976; Davies and Pringle 1980; Soker and Livio 1984; Livio et al. 1986a,b; Ho 1988). Recent numerical simulations (Taam and Fryxell 1988; Sawada et al. 1989; Blondin et al. 1990; Matsuda et al. 1991) suggest that the sign of the angular momentum of the accreting matter may reverse quasi-periodically, causing the direction of the plasma circulation around the neutron star to reverse.

When an accretion disk forms around a neutron star with a significant magnetic field, the interaction between the disk and the magnetic field changes the structure of the inner disk from what it would be for accretion onto a nonmagnetic neutron star or black hole. If the stellar magnetic field is sufficiently strong ($B \gtrsim 10^7$ G for accretion rates $\lesssim 10^{18}$ g s^{-1}), magnetic stresses terminate the Keplerian disk flow before the accreting plasma reaches the stellar surface, forming a magnetosphere around the neutron star. The accreting plasma may then flow along field lines within the magnetosphere toward the magnetic poles of the star, fall as blobs or fingers "between" field lines over a larger portion of the stellar surface, or both. The physics of the transition from Keplerian disk to magnetospheric flow plays an essential role in determining such key properties of disk-accreting neutron stars as the location of the inner edge of the disk, the accretion torque on the star, and the electromotive force available to accelerate charged particles.

Here we consider several models of the inner disk and show that the so-called horizontal branch intensity oscillations (HBOs) observed in many of the luminous LMXBs and the magnetic fields and periods of recycled pulsars are both sensitive diagnostics of conditions in the inner disk. In particular, we show that within the beat-frequency model of the HBOs and a model of the disk-magnetosphere interaction (Ghosh and Lamb 1978, 1979a,b, hereafter GL), the intensity dependence of the HBO frequency (see van de Klis 1989, 1991; Lamb 1989b, 1991a) can be used to probe the physical state of the inner disk. We show further that the distribution of recycled pulsars in the magnetic field vs. period plot (see van den Heuvel 1991) can be used to probe the structure of the inner disk and the critical stellar rotation rate at which the accretion torque vanishes, as suggested by White and Stella (1987, 1988). There are preliminary indications that at least one popular model of the inner disk is inconsistent with the data currently available.

In §2 we summarize the essential physics of the interaction between the accretion disk and the magnetosphere. In §3 we show how the predicted frequency of the HBOs as a function of X-ray intensity depends on the state of the inner disk and compare the predictions of various models of the inner disk with the observed frequency-intensity relation. We summarize the basic properties of the accretion torque on a neutron star in §4. In §5 we compare the spin-up lines predicted by various models of the inner disk with the observed distribution of pulsars that are thought to have been recycled. Our conclusions are summarized in §6.

2. Physics of the Disk-Magnetosphere Interaction

The interaction between the stellar magnetic field and the highly conducting plasma in the accretion disk is a complex process that surely has both steady and episodic components. The steady differential motion between the disk plasma and the star twists and pinches the magnetic field lines connecting the star with the disk. GL emphasized that the twist of the field lines caused by this differential rotation cannot increase indefinitely. Instead, the growth of the magnetic free energy and the electric current density will eventually be halted by one or more of several possible processes. GL specifically stressed the importance of magnetic flux reconnection in the disk and magnetosphere and turbulent diffusion in the disk (see also Kuijpers 1990). Other possible processes that may be important include microinstabilities and the formation of double layers (Lamb and Ghosh 1991). Such processes can greatly enhance the dissipation of electrical currents in small regions of the magnetosphere, thereby reducing the magnetic free energy (see Spicer 1982).

2.1 STEADY FLOW MODEL

In GL, we argued that understanding the *time-averaged* properties of the disk-magnetosphere system in terms of a steady flow model would be an important first step in clarifying the nature of the disk-magnetosphere interaction. In developing such a model, we found it conceptually useful to divide the transition zone between the disk and the inner magnetosphere into two parts, based on their different characteristics.

We showed that there is a broad *outer transition region* in which the stress of the stellar magnetic field—like the effective viscous stress in the disk—transports angular momentum, but is not strong enough to cause the disk plasma to depart significantly from Keplerian circular motion. As a result, the angular-velocity profile remains nearly Keplerian in this region. In the inner part of the outer transition region, the magnetic stress greatly exceeds the effective viscous stress but is still too weak to disrupt the Keplerian motion. The stellar magnetic field is gradually screened by electrical currents induced in the disk.

We showed further that there is a narrow *inner transition region*, which has the character of a *boundary layer*. In this region the magnetic stress is so strong that it not only greatly exceeds the effective viscous stress but also disrupts the Keplerian flow, creating an inward magnetospheric flow. In our original model, strong screening currents circulating in the disk-magnetosphere boundary layer (and in those portions of the accretion bundle near the boundary layer) screen the stellar magnetic field by a facor ~5 in a radial distance much less than r.

In order to describe within an MHD model the time- and spatially-averaged effect of the time-dependent and possibly small-scale dissipative processes that limit the electrical current and magnetic field strength (see above), we introduced an effective electrical conductivity σ_{eff}. Assuming as a first approximation that σ_{eff} is isotropic, we showed that the assumption of a steady state determines its value. We argued that since the relevant dissipative processes are expected to be self-adjusting, such a description of the time-averaged behavior of the accretion flow is plausible.

From this starting point, we developed a quantitative model of the electric, magnetic, and velocity fields and the thermal structure of the inner and outer transition regions. We showed that the boundary layer forms at the cyclindrical radius ϖ_0 determined by conservation of angular momentum, that is,

$$\frac{B_p B_\phi}{4\pi} 4\pi \varpi^2 \Delta\varpi \approx \dot{M} \varpi v_K . \tag{2.1}$$

Here B_p and B_ϕ are the poloidal and azimuthal components of the magnetic field that is coupled to the disk flow, ϖ is the cylindrical radius, $\Delta\varpi \ll \varpi_0$ is the radial width of the boundary layer, $\dot{M}$ is the mass accretion rate, and v_K is the Kelerian velocity. We showed further that the radial width $\Delta\varpi$ of the boundary layer is given by

$$\Delta\varpi \approx \frac{c^2}{4\pi\sigma_{\text{eff}} \, v_{\varpi 0}} , \tag{2.2}$$

where $v_{\varpi 0}$ is the radial velocity of plasma in the boundary layer, and that the physical variables describing the accretion flow change near ϖ_0 on this length scale.

In the following sections, we shall make use of the radius ϖ_0 as a probe of conditions in the inner part of the disk. The value of ϖ_0 as a diagnostic arises from the fact that it depends on the thermal state of the inner disk, as well as the dynamics and electrodynamics of the boundary layer. The thermal state of the inner disk determines the dominant cooling mechanism(s) and opacity sources in the boundary layer, whether the boundary layer is optically thin or thick, and, finally, whether it is gas-pressure-dominated (GPD) or radiation-pressure-dominated (RPD). The dynamics of the boundary layer enter via the requirements that radial and azimuthal momentum be conserved while the electrodynamics enter through the structure of the electrical current system that determines the extent to which the magnetic field is twisted and screened within the boundary layer. Here we focus on the effects of the thermal state of the inner disk on ϖ_0.

2.2 MODELS OF THE INNER DISK

In GL, we were concerned primarily with the interaction of the magnetospheres of accretion-powered pulsars with disk flows. These neutron stars are thought to have dipole magnetic fields $\sim 10^{11}$–10^{12} G. As a result, even for the largest mass accretion rates thought to be relevant ($\sim 10^{17}$–10^{18} g s^{-1}), the magnetic field terminates the Keplerian flow at $\sim 10^3$–10^4 km. We therefore considered neutron star magnetospheres immersed in the "middle" region of a Shakura-Sunyaev accretion disk, which is optically-thick and GPD, with electron scattering dominating free-free absorption (Shakura and Sunyaev 1973, hereafter SS; see also Shakura 1972).

However, in recent years it has become clear that some accreting neutron stars may have much weaker but still dynamically important magnetic fields. For example, during the past decade attention has focused on the possibility that the rapidly rotating rotation-powered pulsars with relatively weak ($B \sim 10^9$ G) dipole magnetic fields seen in binary systems and sometimes as solitary pulsars were "recycled" in binary systems by accreting mass at near-Eddington rates during a previous, X-ray–emitting phase (Alpar *et al.* 1982; Radhakrishnan and Srinivasan 1981, 1984).

More recently, observational evidence and theoretical arguments have led to the tentative conclusion that the most luminous LMXBs are neutron stars with relatively weak ($B \sim 10^8$–$10^9 G$) dipole magnetic fields accreting at near-Eddington rates (see van der Klis 1989, 1991; Lamb 1989b, 1991a). Indeed, these LMXBs are thought to be the progenitors of the recycled binary and millisecond pulsars (see Bhattacharya and van den Heuvel 1991).

The inner radius of the Keplerian disk flow around such weakly magnetic neutron stars accreting at such high rates is expected to be $\sim$15–50 km. Thus, if the disk flow were described by the SS model, the neutron star magnetosphere would be immersed in the "inner" RPD region of the SS model (White and Stella 1987). However, the SS disk model is known to be thermally and viscously unstable in this region (Pringle, Rees and Pacholczyk 1973; Lightman and Eardley 1974; Shakura and Sunyaev 1976; Pringle 1976; Piran 1978). Various classes of alternative models have therefore been suggested.

In one class of alternative models, the SS assumption that the shear stress is proportional to the total pressure (the so-called α-model) is replaced by the asumption that the shear stress is proportional to some combination of the gas and total pressures (Cunningham 1973; Taam and Lin 1984; Szuszkiewicz 1990 and references therein). These RPD models are generally more stable than the RPD region of the SS model. In particular, the so-called β-model (Cunningham 1973), in which the viscous stress is proportional to the gas pressure alone, is stable to both thermal and viscous modes, although how such a scaling may arise physically remains somewhat uncertain (but see Stella and Rosner 1984).

In another class of models, the disk is assumed to be optically thin to absorption in the vertical direction. In such disks the ions are typically much hotter than the electrons, both species are generally much hotter than in corresponding optically-thick disks, and the disk is GPD. Examples of such "two-temperature" disks include models in which the electrons are cooled by Comptonizing a copious flux of soft photons coming from an external source (see Shapiro, Lightman, and Eardley 1976, hereafter SLE) and models in which the dominant cooling mechanism is bremsstrahlung which, at high accretion rates, is Comptonized by the electrons (White and Lightman 1989, 1990). Being GPD, these models are stable to the viscous mode. However, the thermal stability properties of these models are still somewhat uncertain (Pringle 1976; White and Lightman 1989, 1990; Lightman 1990). Although originally developed as alternatives to the "inner" region of the SS model, many GPD models could be extended to larger radii, where they would provide alternatives to the "middle" and, perhaps, the "outer" regions of the SS model. We shall come back to this point in §5.

2.3 INNER RADIUS OF THE DISK

The boundary-layer techniques described in GL can be adapted to the inner, RPD region of the SS accretion disk model, as well as to alternative inner-disk models, to obtain expressions for the inner radius ϖ_0 of the disk (Ghosh and Lamb 1991b; but see §6). Following SS, geometrically-thin accretion disks are traditionally divided into qualitatively different regions and asymptotic solutions are then obtained for the physical variables well within each region. In these asymptotic solutions, the physical variables are power-law functions of the central mass, the mass flux, and the radial coordinate. As a result, when ϖ_0 lies well within an asymptotic region and these solutions are used, ϖ_0 is a simple power-law function of the mass accretion rate $\dot{M}$, the stellar magnetic moment μ, and the stellar mass M, that is,

$$\varpi_0 \propto \dot{M}^a \mu^b M^c , \qquad (2.3)$$

where the values of the exponents a, b, and c depend on the particular model of the inner disk being considered.

Table 1 lists the values of the exponents a, b, and c obtained for the various disk models described above, including the "middle" region of the SS disk (denoted 1G). For the 1G model, Table 1 lists the exact values of the exponents rather than the slightly different approximate values given in GL, to emphasize the fact that the physical arguments leading to the scaling given in GL are fundamentally different from the those leading to the expression for the Alfvén radius in spherical accretion (Lamb, Pethick and Pines 1973), which happens to have almost the same scaling. The similarity of the scalings produced by these two very different flows does *not* indicate that the magnetosphere-accretion flow interactions are similar (Lamb 1989a; Ghosh and Lamb 1991a), contrary to what has often been assumed in the literature (see, *e. g.*, van den Heuvel 1977; Henricks 1983). The results for the 1R disk model shown in Table 1 differ from those given by White and Stella (1987, 1988) because these authors calculated ϖ_0 incorrectly, using the vertical thickness of the unperturbed disk rather than the vertical thickness of the boundary layer.

The results listed in Table 1 show that different models of the inner disk lead to different values of a, b, and c, so that these exponents provide "signatures" of the different models. All one-temperature, optically-thick, RPD models have the *same* signature regardless of the prescription used for the effective viscous stress, because the viscous stress is negligible compared to the magnetic stress within the boundary layer, by definition. Thus, for these models the location of the inner edge of the disk is, to a first approximation, independent of the viscous stress prescription used.

3. HBO Frequency vs. Intensity as a Diagnostic

The discovery and study of QPOs in the most luminous LMXBs has contributed in important ways to our understanding of the nature of these LMXBs (see Lewin, van Paradijs, and van der Klis 1988; van der Klis 1989, 1991; Lamb 1989b, 1991a). The QPOs in these sources are named after the spectral state in which they are

Table 1. Scaling of the Innermost Radius of the Disk

| | $\varpi_0 \propto \dot{M}^a \mu^b M^c$ | | |
Disk Model	Value of a	Value of b	Value of c
1T Opt thick GPD (1G)	-0.25	0.58	-0.21
1T Opt thick RPD (1R)	-0.15	0.51	-0.13
2T Opt thin GPD Compt brems (2B)	-0.48	0.57	0.05
2T Opt thin GPD Compt soft photon (2S)	-1.70	0.80	0.73

strongest (see Hasinger 1987, 1988a,b; Lamb 1988a,b, 1989a,b). The spectral states are themselves identified with the three branches of the Z-shaped track that these LMXBs trace out in X-ray color-color diagrams (Hasinger 1987; 1988a,b; Hasinger, Priedhorsky and Middleditch 1987; Schulz, Hasinger and Trümper 1989). Here we focus on the behavior of the so-called horizontal-branch oscillations (HBOs) as a diagnostic of the structure of the inner disk, within the context of the magnetospheric beat-frequency model of the HBOs (Alpar and Shaham 1985; Lamb et al. 1985; Shibazaki and Lamb 1987) and the GL model of the disk-magnetosphere interaction.

3.1 BEAT-FREQUENCY MODEL

The observed centroid frequencies of the HBOs fall in the range $\sim$20–55 Hz and are strongly positively correlated with the 2–10 keV X-ray intensity of the source (van der Klis et al. 1985; Hasinger et al. 1986; Stella et al. 1987). In the beat-frequency model, the HBOs are explained as follows. The neutron star rotates with a spin frequency ν_s in the same sense as the Keplerian disk flow, which circulates with frequency ν_{K0} at the inner edge ϖ_0 of the disk. Instabilities in the inner disk and the interaction of the disk plasma with the neutron star magnetosphere create plasma density and magnetic field fluctuations in the boundary layer at ϖ_0. These fluctuations circulate with the local Keplerian frequency ν_{K0} (see Lamb et al. 1985; Shibazaki and Lamb 1987). A given plasma or magnetic field fluctuation therefore reappears at the same magnetospheric azimuth with frequency

$$\nu_B \equiv \nu_{K0} - \nu_s , \qquad (3.1)$$

which is the beat frequency between ν_{K0} and ν_s. Hence the mass flux from the boundary layer to the star, which is modulated by the interaction between the rotating magnetospheric field and the fluctuations in the plasma and the magnetic field in the boundary layer, varies quasi-periodically with frequency ν_B, causing the accretion luminosity and the X-ray intensity to vary with this frequency. This quasi-periodic intensity variation is the observed HBO.

Equations (2.3) and (3.1) may be combined to yield

$$\nu_{K0} = (\nu_{HBO} + \nu_s) \propto \dot{M}^\alpha \mu^\beta M^\gamma \,, \tag{3.2}$$

where the exponents α, β, and γ in equation (3.2) are related to the exponents a, b, and c of equation (2.3) by

$$\alpha = -\tfrac{3}{2}\,a\,, \quad \beta = -\tfrac{3}{2}\,b\,, \quad \gamma = \tfrac{1}{2} - \tfrac{3}{2}\,c\,. \tag{3.3}$$

The 2–15 keV X-ray intensity I may not be proportional to the mass accretion rate $\dot{M}$ for a variety of reasons (Lamb *et al.* 1985; Lamb 1988a). These include changes in the geometry of the emission region, the spectrum of the emitted X-rays, and the amount of obscuration with $\dot{M}$, and possible ejection of matter after it has passed through the boundary layer. If the relation between the X-ray intensity I and the accretion rate $\dot{M}$ can be represented by $I \propto \dot{M}^\lambda$ for the range of I under consideration, it follows from equation (3.2) that

$$(\nu_{HBO} + \nu_s) \propto I^{\alpha/\lambda} \mu^\beta M^\gamma \,. \tag{3.4}$$

The deviation of the exponent λ from unity is then a measure of the possible deviation of the X-ray intensity I from strict proportionality to $\dot{M}$.

3.2 FREQUENCY VS. INTENSITY

Equation (3.4) shows that within the framework of the beat-frequency model of HBOs and the GL model of disk-magnetosphere interaction, measurements of the HBO frequency of a given source as a function of its intensity can be used as a diagnostic of the physical state of the disk. In particular, a plot of $\log(\nu_{HBO} + \nu_s)$ vs. $\log I$ should produce a straight line of slope α/λ, if the magnetosphere is immersed in an asymptotic region of the disk. The slope of this line depends on the disk model whereas its intercept with the vertical axis depends on the mass, magnetic moment, distance, and spectrum of the source (the latter enters via the X-ray bolometric correction). Thus, if the stellar rotation frequency ν_s of the source were known, one could use such a plot to obtain a variety of other information about the source. Unfortunately, no periodic oscillations have so far been detected in any of the Z sources, and hence their spin frequencies are unknown.

Even so, one can still obtain some information about an individual source from HBO frequency-intensity data. To see what constraints are possible, consider what happens as one varies the *assumed* stellar spin frequency ν_s' in a plot of $\log(\nu_{HBO} + \nu_s')$ vs. $\log I$. According to equation (3.4), the shape of the curve depends on ν_s', but always has a slope less than that of the $\log(\nu_{HBO})$ vs. $\log I$ curve. When ν_s' is equal to the true spin frequency ν_s, the curve becomes a straight line, provided that the magnetosphere is immersed in an asymptotic region of the disk. The slope of this straight line depends on the physical state of the inner disk. Changing the assumed mass, magnetic moment, distance, or bolometric correction simply shifts the line up and down. In practice, the $\log(\nu_{HBO} + \nu_s')$ vs. $\log I$ curve displays little curvature even if ν_s' is not equal to ν_s, unless the range of intensities considered is relatively large. Thus, the principal effect of changing the assumed spin frequency is to change the slope of the $\log(\nu_{HBO} + \nu_s')$ vs. $\log I$ curve.

The data for each source are a set of ν_{HBO} vs. I measurements. Our discussion shows that if the model discussed here is correct, one should be able to choose a spin frequency ν_s' for each source so that a plot of $\log(\nu_{HBO} + \nu_s')$ vs. $\log I$ produces an approximately straight data track. The slope of this track will depend on the spin frequency assumed, but cannot exceed the slope of the $\log(\nu_{HBO})$ vs. $\log I$ track (Stella 1988). In principle, only the correct spin frequency will produce a straight line, and the slope can then be compared with the slope predicted by a given disk model. In practice, the track appears approximately straight for a fairly wide range of assumed spin frequencies, because data are available only for a relatively small range of intensities. One can still check whether the data from a given source are consistent with a particular disk model by attempting to adjust the assumed spin frequency so that the slope of the $\log(\nu_{HBO} + \nu_s')$ vs. $\log I$ track is equal to the slope predicted by the disk model. If this is possible, the model is allowed by the data; if not, the model can be rejected as a description of the disk in that source.

Although one can compare data from a single source with the predictions of a disk model in this way (see below), the range of observed intensities for most sources is so limited that the results are not very significant. We therefore adopt a different approach, which makes use the entire body of currently available frequency-intensity data. In order to pursue this approach, we must make the additional assumption that the inner disks in all the Z sources being considered are in the *same* physical state. This hypothesis is consistent with our current understanding of the Z sources.

Given this "universal structure" hypothesis, one can check whether the data from all sources in a given collection are consistent with a particular model of the inner disk. One does this by attempting to adjust the assumed spin frequencies of the sources so that the slopes of the $\log(\nu_{HBO} + \nu_s')$ vs. $\log I$ tracks are the same and equal to the slope predicted by the disk model being considered. If it is possible to find a set of spin frequencies such that the slopes of these tracks *are* the same for all the sources, the structure of the inner disk may be the similar in all of them. If in addition the slope is consistent with the slope predicted by the disk model being considered, the the model remains viable; if not, the model must be rejected.

The results of this analysis can be displayed compactly if the data from the different sources are combined by translating the various $\log(\nu_{HBO} + \nu_s')$ vs. $\log I$ tracks vertically or horizontally until they coincide. Operationally, this can be done by dividing the observed intensity I by a scale factor I_0. Within the model, this amounts to compensating for differences in the stellar mass and magnetic moment, the distance, and the X-ray spectrum. One then has a single plot that displays both the consistency of the data from different sources and its consistency with the model being considered. As more, and more accurate, data become available, one can use this plot to test the validity of the universal structure hypothesis as well

Table 2. QPO Diagnostics

| Disk Model | Slope α | Best-fit Value of ν_s (Hz) | | | |
		GX 5−1	Cyg X-2	GX 17+2	GX 340+0
1T Opt thick GPD (1G)	0.37	120	120	200	120
1T Opt thick RPD (1R)	0.23	200	200	350	200
2T Opt thin GPD Comp brems (2B)	0.73	48	48	100	48
2T Opt thin GPD Comp soft photon (2S)	2.6	—[a]	—[a]	—[a]	—[a]

[a]No universal track with a slope as large as 2.6 could be constructed (see text).

as the viability of various disk models within the framework of the beat-frequency model of HBOs and the GL model of the disk-magnetosphere interaction.

3.3 RESULTS AND DISCUSSION

We have applied the procedure just described to all currently available HBO frequency-intensity data, which are from the four sources GX 5−1, Cyg X-2, GX 17+2, and GX 340+0. The disk models that we have considered are the one-temperature, optically-thick GPD model (hereafter denoted 1G); one-temperature, optically-thick RPD models (denoted 1R); the two-temperature, optically-thin model with cooling by Comptonized bremsstrahlung (denoted 2B); and the two-temperature, optically-thin GPD model with a soft photon source (denoted 2S). Table 2 lists the slopes α predicted by each of these models. For the present analysis, we assume $\lambda = 1$.

It was possible to construct "universal tracks" with the slopes predicted by the 1R, 1G, and 2B disk models. These tracks are shown in Figure 1. The corresponding spin frequencies are listed in Table 2. These spin frequencies are all consistent with our present understanding of the spin evolution of the neutron stars in the luminous LMXBs and the properties of these sources.

It was not possible to construct a universal track with a slope as steep as the slope of 2.6 predicted by the 2S disk model. The universal track with the steepest slope that can be constructed from this data set has a slope ≈ 1.8. The values of ν_s' that produce this track are 0, 0, 30, and 0 Hz for the sources GX 5−1, Cyg X-2, GX 17+2, and GX 340+0, respectively. The vaues of ν_s' for GX 5−1, Cyg X-2, and GX 340+0 are much lower than the spin frequencies expected in these sources on evolutionary and other grounds. For example, current evolutionary scenarios suggest that the neutron stars in the luminous LMXBs have been accreting at a high rate for 10^7–10^8 yrs, during which they would have accreted sufficient angular momentum to spin them up to much higher rates (see §4 and Bhattacharya and

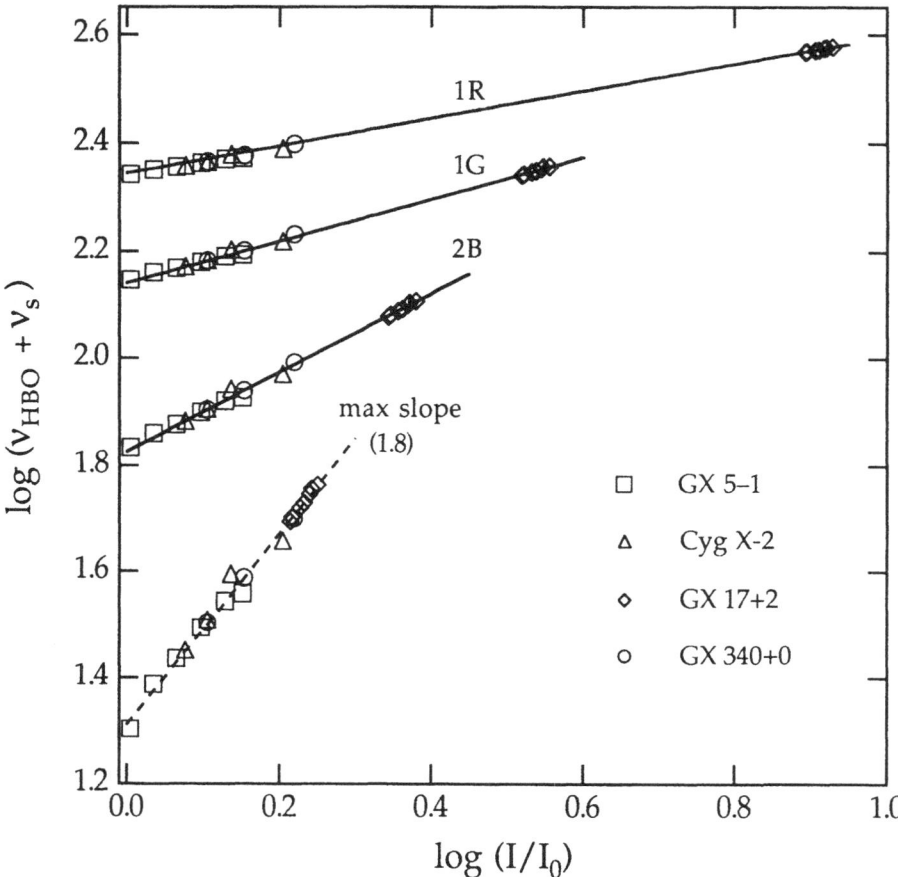

Fig. 1. Plot showing how HBO frequency-intensity data from a collection of sources can be used to test models of the inner disk. Shown are solid lines with the slopes 0.23, 0.37, and 0.73 predicted by the one-temperature, optically-thick, RPD model (labeled 1R); the one-temperature, optically-thick, GPD model (labeled 1G); and the two-temperature, optically-thin model with cooling by Comptonized bremsstrahlung (labeled 2B). Superposed on these lines are "universal tracks" with the same slopes, constructed as described in the text using HBO data from GX 5−1, Cyg X-2, GX 17+2, and GX 340+0. Also shown is the universal track with the largest slope ($\alpha \approx 1.8$) that could be constructed from the data; the corresponding dashed line labeled "max slope" is meant only to guide the eye. No universal track could be constructed with a slope as steep as 2.6, the slope predicted by the two-temperature, optically-thin model with cooling by external soft photons. The GX 5−1 data are from van der Klis *et al.* (1985); Cyg X-2, Hasinger *et al.* (1986); GX 17+2, Stella *et al.* (1987) and Penninx *et al.* (1990); GX 340+0, Penninx *et al.* (1991). For GX 17+2, we converted (only) the *Ginga* PC data given by Penninx *et al.* to a frequency-intensity relation, using Fig. 2 of that reference (for more details, see Ghosh and Lamb 1991b).

van den Heuvel 1991). Also, X-ray intensity oscillations produced by oscillating or rotating beams are suppressed by escape-time effects only if the frequency is sufficiently high (Brainerd and Lamb 1987; Kylafis and Klimis 1987; Lamb 1988a). Moreover, detection of periodic oscillations is more difficult at high frequencies (see van der Klis 1989, 1991). Thus, the low current upper limits on the amplitudes of any periodic intensity oscillations in these sources also suggest that their spin frequencies are relatively high. The universal track of steepest slope is therefore probably not relevant for the luminous LMXBs. If the spin frequencies of the neutron stars in these sources are $\gtrsim 10$ Hz, as expected, the steepest universal track has a slope less than 1.8, and hence is in greater disagreement with the 2S model.

To summarize, the 2S disk model seems to be inconsistent with the beat-frequency model of the HBOs and the GL model of disk-magnetosphere interaction, for $\lambda = 1$. We caution, however, that λ may not be unity and that other models of the disk-magnetosphere interaction would in all likelihood predict different slopes, even for $\lambda = 1$. Moreover, the disk models considered here are not valid if the mass accretion rate approaches the Eddington critical rate, which may occur when Z sources are at the right end of the horizontal branch (see van der Klis 1989, 1991; Lamb 1989b, 1991a).

Our analysis also shows that models in which the HBO frequency is the Keplerian frequency at the inner edge of the disk (see Lamb 1988a) are inconsistent with the universal structure hypothesis. To see this, note that the beat-frequency model becomes equivalent to the Keplerian frequency model if the neutron star spin frequency in the beat-frequency model is set equal to zero. However, it is then impossible to construct a universal track from the entire data set. An additional parameter that affects the slope of the track, such as the spin frequency in the beat-frequency model, is needed in order to construct a universal track.

4. Accretion Torques

Accretion of disk plasma by a magnetic neutron star produces an accretion torque on the star (Pringle and Rees 1972; Lamb, Pethick and Pines 1973; GL). In GL, we described how this torque can be evaluated by calculating the integral of the angular momentum flux density over a suitable surface enclosing the neutron star. We showed that the accretion torque N_s on the star can be expressed as

$$N_s \approx n(\omega_s) N_0 \,, \tag{4.1}$$

where the reference torque

$$N_0 \equiv \dot{M}(GM\varpi_0)^{1/2} \tag{4.2}$$

is the torque that would be produced by accretion of matter with the specific angular momentum corresponding to a circular Keplerian orbit at the inner edge

ϖ_0 of the disk and the dimensionless torque function n depends primarily on the fastness parameter

$$\omega_s \equiv \nu_s/\nu_{K0} = (\varpi_0/r_c)^{3/2}, \qquad (4.3)$$

which is a measure of the dynamical importance of the neutron star spin rate. Here r_c is the corotation radius, at which matter in circular Keplerian orbit corotates with the star $(\nu_K(r_c) = \nu_s)$.

4.1 DEFINITION OF THE CRITICAL FASTNESS

The stresses that contribute to the accretion torque are generally of three types: material, magnetic, and viscous. GL discussed the contributions of each type when evaluated on a surface near the inner disk and the resulting qualitative behavior of the dimensionless torque function $n(\omega_s)$. In brief, the torque contributed by the material stress always acts to spin up the star and is closely equal to N_0 whereas the torque contributed by the viscous stress is negligible. The torque contributed by the magnetic stress is caused by shearing of the magnetic field lines connecting the star and the disk and consists of two parts. The first part, which is due to stellar field lines that interact with the disk between ϖ_0 and r_c, acts to spin up the star, since these field lines are twisted in the same direction as the stellar spin. The second part, which is due to the stellar field lines that interact with the disk outside r_c, acts to spin down the star, since these field lines are twisted in the opposite direction to the stellar spin. GL argued that when the corotation radius r_c is sufficiently close to the inner edge of the disk at ϖ_0, the second part of the magnetic torque will dominate the sum of the first part and the torque contributed by the material stress, producing a net spin-down torque on the star. The sign of the total accretion torque thus depends on the position of the corotation radius relative to the inner edge of the disk.

Equation (4.3) shows that the ratio ϖ_0/r_c can be written in terms of the fastness parameter ω_s, and hence that the sign of the total accretion torque depends on ω_s. For slow rotators $(\omega_s \ll 1)$, the total torque is positive. For very fast rotators $(\omega_s \approx 1)$, the magnetic torque is so negative (due to the dominance of the contribution from the field lines outside r_c) that it more than offsets the material torque, and the total torque is negative. The accretion torque therefore vanishes at some intermediate fastness ω_c, called the *critical* fastness. The value of the critical fastness is much more sensitive to details of the disk-magnetosphere interaction than is the behavior of the radius of the inner edge of the disk (GL).

4.2 ESTIMATES OF THE CRITICAL FASTNESS

GL estimated the size of B_ϕ by balancing its amplification (by differential rotation) and its reduction (by reconnection). In their original approach, GL assumed that B_ϕ is amplified by azimuthal shear on the time scale $\gamma_a|\nu_K - \nu_s|$, where γ_a is a numerical factor of order unity, and reduced by reconnection on the time scale $2h/\xi v_A$, where $2h$ is the thickness of the disk, v_A is the Alfvén speed in B_ϕ, and

$\xi(\sim 0.1\text{--}1)$ is a dimensionless parameter that characterizes the rate of reconnection in the disk and the outer magnetosphere. The resulting steady-state value of B_ϕ is

$$|B_\phi| = \left(\frac{\gamma_a}{\xi}\right) 4\pi h (4\pi \rho)^{1/2} |\nu_K - \nu_s|, \tag{4.4}$$

where ρ is the plasma density in the reconnection region. This estimate for B_ϕ gave $\omega_c \approx 0.35$.

Several authors (Lamb 1978; Ghosh 1982; Lamb 1984; Wang 1987; Zylstra 1988) have subsequently pointed out that the azimuthal component is generated by shearing of the poloidal component B_z of the magnetic field by the differential rotation and hence that a better estimate of the rate of amplification would be $\gamma_a |\nu_K - \nu_s| B_z$; this is also the estimate one obtains from the induction equation. Using this prescription, the steady-state value of B_ϕ is

$$|B_\phi| = \left(\frac{\gamma_a}{\xi}\right)^{1/2} \left[4\pi h (4\pi \rho)^{1/2}\right]^{1/2} |\nu_K - \nu_s|^{1/2} |B_z|^{1/2}. \tag{4.5}$$

This estimate of B_ϕ behaves similarly to the estimate (4.4) in many ways, but generally gives smaller values of the magnetic pitch in the outer transition zone and therefore a smaller (in magnitude) spin-down contribution to the torque (Lamb 1989a; Ghosh and Lamb 1991a). Its asymptotic behavior at large radii is also significantly different (Ghosh and Lamb 1991b). As a result, it gives a critical fastness larger than the original estimate of GL.

Wang (1987) used an expression similar to equation (4.5) to calculate the critical fastness and obtained a value very close to unity. However, Wang's result is inaccurate in two respects. First, he neglected screening of the poloidal field B_z by currents flowing in the transition zone and the outer magnetosphere; as a result, the calculation is not self-consistent. Second, his expression for the accretion torque diverges as the stellar spin rate goes to zero ($\omega_s \to 0$) and therefore cannot be correct, at least in this limit.

At present, we are recalculating the azimuthal and poloidal magnetic fields in a more self-consistent way. The results of these calculations will be reported elsewhere (Ghosh and Lamb 1991b).

Given the current uncertainty in the critical fastness given by the GL model and its sensitivity to the modeling of the disk-magnetosphere interaction, in §5 we shall consider a plausible range of values for ω_c. In particular, we shall assume that ω_c cannot exceed unity, since this would disrupt a steady flow, and that it is greater than 0.25. We believe the latter bound is conservative, in the sense that the actual value of ω_c is unlikely to be smaller.

5. The Spin-up Line as a Diagnostic

Some 500 rotation-powered pulsars are now known. The intrinsic periods P and period derivatives $\dot{P}$ of about 400 have been measured with reasonable accuracy. In discussing the evolution of pulsars, it is customary to plot $\log(P\dot{P})^{1/2}$ vs. $\log P$ and to interpret $(P\dot{P})^{1/2}$ as a measure of the strength of the dipole magnetic field of the pulsar (see Srinivasan 1989; Lamb 1991b; Bhattacharya and van den Heuvel 1991). Figure 2 shows such a B–P diagram. Not included in Figure 2 are those pulsars in globular clusters with $\dot{P}$'s that appear to be significantly affected by acceleration in the cluster gravitational potential (see Taylor 1991; Lyne 1991).

5.1 RECYCLING

GL discussed the importance of the critical fastness in explaining the observed spin-up and spin-down behavior of accretion-powered X-ray pulsars and the large number of relatively long-period X-ray pulsars (see also Elsner, Ghosh, and Lamb 1980). Lamb (1981) pointed out that the critical fastness determines the asymptotic spin rate achieved by a neutron star that has been accreting at the Eddington rate for a long time, and called attention to the likelihood that the moderately high spin rate of the relatively weak-field binary pulsar PSR 1913+16 reflects spin-up to the critical fastness during an earlier accretion phase (see also Bisnovatyi-Kogan and Komberg 1975; Smarr and Blandford 1976; Radhakrishnan and Srinivasan 1981).

Following the discovery of millisecond pulsars, it was suggested that these pulsars, like PSR 1913+16, have been "recycled" by being spun up during a previous, accretion-powered X-ray emission phase in which the neutron star accretes mass at a rate near the Eddington critical rate (Alpar *et al.* 1982; Radhakrishnan and Srinivasan 1984; for reviews, see Srinivasan 1989 and Bhattacharya and van den Heuvel 1991). Supporting this hypothesis is the fact that the binary and millisecond pulsars share several characteristics that set them apart from the bulk of pulsars. For example, about 40% of pulsars with periods less than 12 ms are in binary systems, whereas only about 3% of all known pulsars are in binaries (see van den Heuvel 1991). Also, the binary and millisecond pulsars generally have much weaker inferred dipolar magnetic fields than other pulsars. More specifically, most binary and all millisecond pulsars have inferred dipolar fields less than 4×10^{10} G, whereas 96% of all known pulsars have inferred dipolar fields greater than 3×10^{11} G (see van den Heuvel 1991). It is therefore thought that the binary and millisecond pulsars form a separate class of pulsars that have been spun up by accretion.

At present, 21 pulsars are known to be in binary systems while 34 pulsars are known to have periods less than 100 ms. Many pulsars in the first group are also in the second. Including pulsars in either group, 38 pulsars are presently candidates for having been recycled (Taylor 1991). We caution that hypotheses other than recycling have also been advanced to explain this class of pulsars (see Lamb 1991b).

5.2 SPIN-UP LINES

According to the accretion torque theory described in §4, the spin period of a neutron star that is accreting at a fixed rate will tend toward the equilibrium spin period

$$P_{\text{eq}} = \omega_c^{-1} \nu_{K0}^{-1}. \tag{5.1}$$

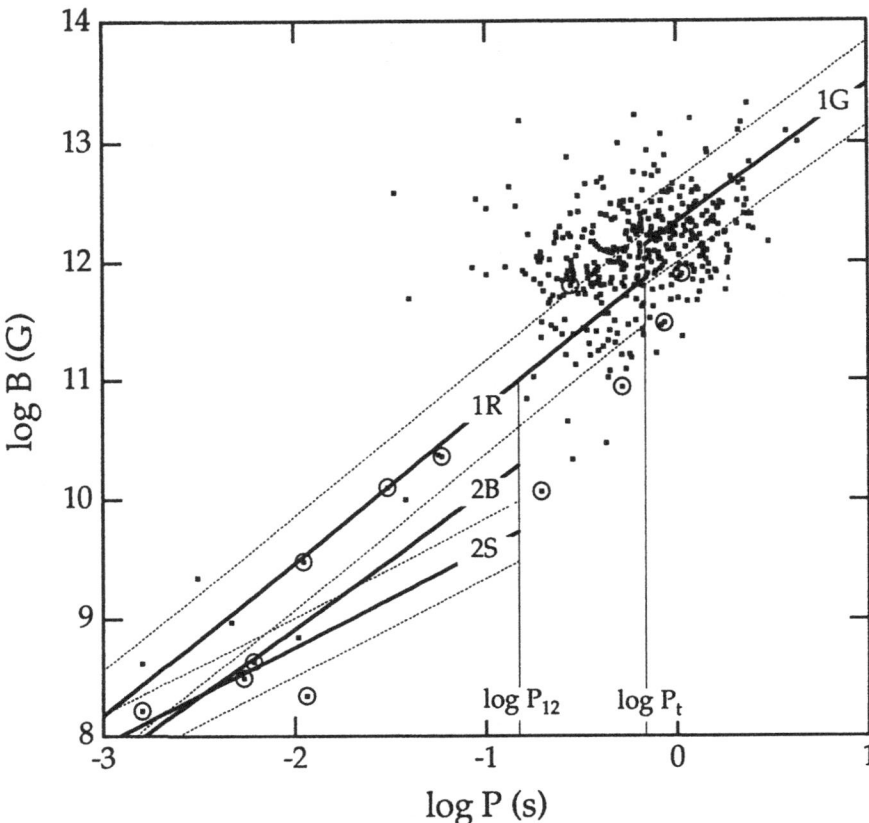

Fig. 2. Plot showing how the observed spin periods and inferred magnetic fields of rotation-powered pulsars can be used to test the consistency of various disk models with the recycling hypothesis. Shown are 409 pulsars; the 12 binary pulsars are encircled. The heavy solid lines labeled by model names are the spin-up lines predicted by the 1G, 1R, 2S, and 2B disk models, if the critical fastness ω_c is 0.5. Spin-up lines assuming $\omega_c = 1$ and $\omega_c = 0.25$ (thin dotted lines) are also shown for the 1G, 1R, and 2S models, to indicate the uncertainty in these spin-up lines caused by the uncertain value of ω_c. The spin-up line predicted by the 2B model is similarly uncertain; the corresponding lines assuming $\omega_c = 1$ and $\omega_c = 0.25$ have been omitted only to make the diagram more readable. The vertical line labeled P_t indicates the location of the transition from the 1G to the 1R model, while the line labeled P_{12} indicates the estimated position of the transition from the 1R to the 2S or 2B models (see text). The pulsar data are from Bhattacharya and van den Heuvel (1990), Kulkarni et al. (1991), Prince et al. (1991a), Prince et al. (1991b), Taylor (1991), and Wolszczan (1991).

Assuming that the mass accretion rate through the boundary layer is equal to the Eddington critical rate and that ν_{K0} scales with $\dot{M}$, μ, and M according to equation (3.2), the equilibrium spin period can be written

$$P_{\mathrm{eq,E}} = \omega_c^{-1} \mu_{30}^{-\beta} \left(M/M_\odot \right)^{-\gamma} R_6^{-\alpha} \, P_0 \,. \tag{5.2}$$

Here μ_{30} is the magnetic moment of the neutron star in units of 10^{30} G cm^3 and R_6 is the stellar radius in units of 10^6 cm. P_0 sets the scale of the equlibrium spin period and is determined by the constant of proportionality in equation (3.2).

Expression (5.2) shows that the equilibrium spin period of a neutron star accreting at the Eddington critical rate depends not only on the disk-magnetosphere interaction and the physical state of the inner disk, through ω_c and β, but also on the mass and magnetic moment of the star. Thus, such a star will approach a definite spin period only if the accretion torque causes the spin period to evolve toward the equlibrium period in a time much shorter than the time scales on which accretion changes the mass and magnetic moment of the star. However, as the spin period approaches the equilibrium spin period, the magnitude of the accretion torque rapidly decreases and hence the time required to change the spin period grows longer, whereas the time scale on which the mass of the star increases remains almost unchanged. Thus, the star never reaches its equilibrium spin period.

Even though the spin period of an accreting neutron star does not approach a fixed value, it will not become shorter than $P_{\mathrm{eq,E}}$, evaluated using the mass and magnetic moment of the star at the end of the accretion phase, provided that the mass flux through the boundary layer does not exceed the Eddington mass flux. Equation (5.2) shows that for a neutron star of given final mass (and hence radius), $P_{\mathrm{eq,E}}$ is a power-law function of its magnetic moment (assuming that ω_c is not a function of the magnetic moment). This function therefore defines a straight line in the B–P diagram. If the weak-field binary and millisecond pulsars are indeed recycled, they should reappear on or below this "spin-up line", and should now all lie below it, assuming that their magnetic moments have not increased. Like all pulsars, they are also expected to lie above the higher of the so-called "death line", below which the radio luminosity is so low that pulsars are very difficult to detect, and the so-called "Hubble line", below which the spin-down time exceeds the time since the Big Bang (see van den Heuvel 1991). As these latter lines do not concern us here, they have been omitted from Figure 2.

The slope $-\beta^{-1}$ of the spin-up line and its intercept P_0 both depend on the model of the disk-magnetosphere interaction and the state of the inner disk, and both are therefore useful diagnostics (White and Stella 1987, 1988; Ghosh and Lamb 1991a; Lamb and Ghosh 1991). Table 3 lists the values of $-\beta^{-1}$ and P_0 given by the GL model of disk-magnetosphere interaction for the four disk models discussed in §2.

In Figure 2, we show spin-up bands for three of the four disk models. These bands were generated by assuming that the critical fastness lies between 0.25 and 1.0 (see §4). The central line within each band corresponds to $\omega_c = 0.5$. For the 2B disk model, we show only the central line of the band, in order to reduce the complexity of the figure. The positions of the bands also depend on the mass and radius of the neutron star. Here we have adopted the values $M = 1.4 M_\odot$ and $R_6 = 1$ for all the bands. The band for each model is shown over the region where the model may be valid.

The regions of validity of the various disk models were determined as follows. The transition between the 1G and the 1R models takes place at the radius r_t where the gas pressure P_g equals the radiation pressure P_r in the disk (SS; Treves,

Table 3. Spin-Up Line Characteristics

Disk Model	Slope $(-\beta^{-1})$	P_0 (s)
1T Opt thick GPD (1G)	1.15	0.349
1T Opt thick RPD (1R)	1.30	0.585
2T Opt thin GPD Compt brems (2B)	1.17	2.92
2T Opt thin GPD Compt soft photon (2S)	0.83	51.4

Maraschi, and Abramowicz 1988). For accretion at the Eddington rate, r_t is given by (Ghosh and Lamb 1991a)

$$r_t = 7.45 \times 10^7 (M/M_\odot)^{1/3} R_6^{16/21} \text{ cm} . \qquad (5.3)$$

In the B-P diagram, the transition between the 1G and 1R models thus takes place at the period

$$P_t = \nu_K(r_t)^{-1} \omega_c^{-1} , \qquad (5.4)$$

where $\nu_K(r_t)$ is the Keplerian frequency at r_t and ω_c is the critical fastness discussed in §4.

The criterion that determines the radius r_{12} of the transition between one- and two-temperature regions is not fully understood. SLE argued that the viscous mode in the 1R disk becomes unstable "suddenly and violently" at a certain critical radius, and hence that the disk undergoes a "phase change" from a cool, one-temperature state to a hot, two-temperature state within a narrow transition region about the critical radius. If so, the critical radius is a good estimate of r_{12}. SLE showed that in constant-α disks, the critical radius is given implicitly by the criterion $P_r = \frac{3}{2} P_g$. However, in their numerical work SLE used the more conservative criterion $P_r = 3 P_g$, which we also use here. From the properties of SS disks, one can show that

$$r_{12} \approx 0.35 \, r_t . \qquad (5.5)$$

In the B-P diagram, the transition from the 1R to the 2S or 2B disks takes place at a period P_{12} given by a relation analogous to equation (5.4), with r_t replaced by r_{12}.

Figure 2 shows that the spin-up lines predicted by the different disk models do not intersect at the appropriate spin periods. The (rather small) mismatch between the 1G and 1R models at P_t is expected, given that both "models" are actually asymptotic solutions valid only well inside or outside r_t (see §2). As a result, if a physical variable such as the temperature, density, or disk thickness is calculated from both models at any radius near r_t, the two values of the physical variable will disagree, typically by a factor $\lesssim 2$. The mismatch in the 1R and 1G spin-up lines near P_t simply reflects this. A more exact calculation of the SS disk structure would give a continuous spin-up *curve* (rather than a spin-up *line*), smoothly joining the 1R line, which is valid at small P, to the 1G line, which is valid at large P.

The much larger mismatches between the 1R and the 2S or 2B spin-up lines near P_{12} are more significant. These mismatches stem from real, physical mismatches between the corresponding disk solutions, mismatches that do not seem to have been fully appreciated in the literature. At any given radius, the two-temperature disk models are qualitatively different from the one-temperature model, and generally give very different values for the physical variables. Thus, in order for the flow to go from the one-temperature solution to a two-temperature solution, there must be a "sudden and violent" transition, to use SLE's phrase. The widely different positions of the spin-up lines predicted by the one- and two-temperature disk models near the period P_{12} in the B-P diagram simply reflect the very large differences between these models at the conjectured transition radius r_{12}.

It does not seem to be widely appreciated that hot, two-temperature disks, like the 2S and 2B models, are valid alternatives to the one-temperature SS disk not only in the "inner region" but also at *larger* radii. This point was discussed briefly by SLE, for the 2S model. A similar argument can be made for the 2B model. Thus, the spin-up lines obtained by extending the 2S and 2B spin-up lines shown in Figure 2 to spin periods greater than P_{12} represent possible alternatives to the 1R spin-up line at these periods, unless convincing arguments can be given as to why real disks should choose the one-temperature solution at these radii. We are investigating this question, and will report our results elsewhere (Ghosh and Lamb 1991b). In the present work, we tentatively adopt the conventional view that the flow makes an abrupt transition from a one- to a two-temperature solution near r_{12}.

5.3 RESULTS AND DISCUSSION

Figure 2 shows that a composite spin-up line constructed from the spin-up lines predicted by the 1G and 1R disk models is consistent with the observed periods and period derivatives of the supposedly recycled pulsars, given the uncertainty in our knowledge of the critical fastness. In contrast, the spin-up line predicted by the 2S model appears to disagree sharply with these observations. Both the slope and the large value of P_0 (see Table 3) contribute to this disagreement. The spin-up line predicted by the 2B model also appears inconsistent with the positions of several of the fastest millisecond pulsars, even when allowance is made for the uncertainty in the critical fastness, primarily because of the moderately large value P_0 predicted by this model.

We caution that other models of the disk-magnetosphere interaction would in all likelihood predict somewhat different spin-up lines. Moreover, several approximations used in constructing all existing models of the inner disk fail if the accretion rate through the inner disk approaches closely or exceeds the Eddington critical rate. Whether such high mass fluxes are to be expected during the final spin-up phase of the recycling scenario is uncertain.

The spin-up line used in almost all previous work on pulsar recycling (see, *e. g.*, Srinivasan 1989; Wolszczan *et al.* 1989; Bhattacharya and van den Heuvel 1991; van den Heuvel 1991) is approximately the same as the spin-up line found here for the 1G disk model and $\omega_c = 0.5$. This is, however, fortuitous. The spin-up line used in most previous work was drawn using an expression analogous to equation (5.1), with $\omega_c = 1$ and the Kepler frequency evaluated at the magnetospheric radius for spherically-symmetric radial accretion at the assumed rate (van den Heuvel 1977; Henrichs 1983; Bhattacharya and van den Heuvel 1991). This radius probably is not relevant to the spin-up of neutron stars in LMXBs, which are thought to be accreting from disks (see Lamb 1989a; Ghosh and Lamb 1991a). The similarity of the results stems from the fact that the radius of the inner edge of the disk for the 1G model of disk flow happens to scale with the stellar magnetic moment and mass flux approximately (but not exactly) the same as the radius of the magnetosphere for spherically-symmetric radial flow and is somewhat smaller. As noted above, this is not due to any similarity in the way in which these two flows interact with the magnetosphere, but is merely a coincidence. As a result of this coincidence, the spin-up line used in most previous work happens to be similar, for spin periods greater than P_t ($\approx 800\,\mathrm{ms}$), to the spin-up line predicted by the 1G model of disk flow. Thus, even though the spin-up line derived by assuming spherically-symmetric radial inflow probably is not relevant, the numbers obtained from it are not dissimilar to those predicted by the 1G model of disk flow, *for spin periods periods $\gtrsim 800\,ms$.*

However, the spin-up line at such large spin periods is not relevant to the recycling hypothesis. Hence, the spin-up line derived assuming spherically-symmetric radial inflow is almost always extended to spin periods much smaller than P_t. At such periods this naive approach is not only conceptually incorrect but also gives very misleading numbers, because the 1G disk model is not valid for the small magnetospheric radii involved, while the disk models that may be valid predict scalings that are quite different from the scalings of the magnetospheric radius for spherically-symmetric radial accretion. More generally, use of a single spin-up line over the full B-P diagram tacitly assumes that the radius where the accretion flow couples to the magnetic field of the star has the same value and scales in the same way for all accretion rates and stellar magnetic moments of interest. This is not expected, on quite general grounds, for neutron stars accreting from a disk. Instead, the scalings for magnetospheres immersed in the inner region of the disk is expected to differ from the scalings for magnetospheres that extend to much larger radii, leading to different spin-up lines in different parts of the B-P diagram.

6. Conclusions

We have shown that the HBOs observed in many of the most luminous LMXBs and the periods and magnetic fields of recycled pulsars are sensitive probes of conditions near the inner edges of accretion disks around weakly magnetic neutron stars and of the disk-magnetosphere interaction. In particular, we have shown that the intensity-dependence of the HBO frequency can be used, within the beat-frequency model of HBOs and the GL model of disk-magnetosphere interaction, as a diagnostic of the physical state of the inner disk. The position of the spin-up line in relation to the positions of "recycled" pulsars in the magnetic field vs. period diagram is also useful as a diagnostic of conditions in the inner disk and the disk-magnetosphere interaction.

Our preliminary results suggest that the two-temperature model of the inner disk with cooling by Comptonization of soft photons may be inconsistent both with the observed behavior of the HBOs and with the recycling hypothesis. The two-temperature model of the inner disk with cooling by Comptonized bremsstrahlung may be inconsistent with the recycling hypothesis. However, we caution that the original form of the boundary-layer technique developed by GL, which we have used here, may not be accurate when used with hot, two-temperature disks. These and related problems are being investigated further. The results will be reported elsewhere.

It is a pleasure to thank J. Arons, R. Hamilton, G. Hasinger, L. Jin, M. van der Klis, J. Kuijpers, S. Kulkarni, J. McClintock, M. C. Miller, J. van Paradijs, T. Prince, M. Ryba, J. Taylor, and N. E. White for helpful discussions. We are grateful to the Institute of Theoretical Physics at Santa Barbara and to the directors E.P.J. van den Heuvel and S. Rappaport of the Workshop on Neutron Stars in Binary Systems for their kind hospitality. P. G. also thanks R. H. for help in drawing the figures. This research was supported in part by NASA grant NAGW 1583 and NSF grant PHY 86-00377 at Illinois and by NSF grant PHY 89-04035 at Santa Barbara.

7. References

Alpar, M. A., and Shaham, J. 1985, *Nature*, **316**, 239.

Alpar, M. A., Cheng, A. F., Ruderman, M. A., and Shaham, J. 1982, *Nature*, **300**, 728.

Bhattacharya, D., and van den Heuvel, E. P. J. 1990, personal communication.

Bhattacharya, D., and van den Heuvel, E. P. J. 1991, *Phys. Rept.*, in press.

Bisnovatyi-Kogan, G. S., and Komberg, B. V. 1975, *Soviet Astr.*, **18**, 217.

Blondin, J. M., Kallman, T. R., Fryxell, B. A., and Taam, R. E. 1990, *Astrophys. J.*, **356**, 591.

Brainerd, J., and Lamb, F. K. 1987, *Astrophys. J. (Letters)*, **317**, L33.

Cunningham, C. T. 1973, Ph.D. thesis, University of Washington.

Davies, R. E., and Pringle, J. E. 1980, *Mon. Not. R. astr. Soc.*, **191**, 599.

Elsner, R. F., Ghosh, P., and Lamb, F. K. 1980, *Astrophys. J. (Letters)*, **241**, L155.

Ghosh, P. 1982, Invited talk presented at the Annual Meeting of the Astronomical Society of India, Gorakhpur, India, November, 1982.

Ghosh, P., and Lamb, F. K. 1978, *Astrophys. J. (Letters)*, **223**, L83 (GL).

_____. 1979a, *Astrophys. J.*, **232**, 259 (GL).

_____. 1979b, *Astrophys. J.*, **234**, 296 (GL)

_____. 1991a, in *Neutron Stars: An Interdisciplinary Field*, ed. J. Ventura and D. Pines (Dordrecht: Kluwer Academic Publ.), in press.

508

_____. 1991b, in preparation.

Hasinger, G. 1987, *Astron. Ap.*, **186**, 153.

_____. 1988a, in *Physics of Compact Objects*, ed. N. E. White and L. G. Filipov (*Adv. Space Res.*, **8**), p. 377.

_____. 1988b, *Physics of Neutron Stars and Black Holes*, ed. Y. Tanaka (Tokyo: Universal Academy), p. 97.

Hasinger, G., Langmeier, A., Sztajno, M., Trümper, J., Lewin, W.H.G., and White, N. E. 1986, *Nature*, **319**, 469.

Hasinger, G., Priedhorsky, W. C., and Middleditch, J. 1989, *Astrophys. J.*, **337**, 843.

Henrichs, H.F. 1983, in *Accretion-Driven Stellar X-Ray Sources*, ed. W.H.G. Lewin and E.P.J. van den Heuvel (Cambridge University Press), p. 393.

Ho, C. 1988, *Mon. Not. R. astr. Soc.*, **232**, 91.

Joss, P. C., and Rappaport, S. A. 1984, *Ann. Rev. Astr. Ap.*, **22**, 537.

Kuijpers, J. 1990, in *Active Close Binaries*, ed. C. Ibanoğlu and I. Yavuz (Dordrecht: Kluwer Academic Publ.), in press.

Kulkarni, S. R., Anderson, S. B., Prince, T. A., and Wolszczan, A. 1991, *Nature*, **349**, 47.

Kylafis, N. D., and Klimis, G. 1987, *Astrophys. J.*, **323**, 678.

Lamb, D. Q. 1988, in *Nuclear Spectroscopy of Astrophysical Sources, AIP Conf. Proc. No. 170*, ed. N. Gehrels and G. H. Share (New York: American Institute of Physics), p. 265.

Lamb, F. K. 1978, unpublished.

_____. 1981, in *IAU Symp. 95, Pulsars*, ed. W. Sieber and R. Wielebinski (Dordrecht: Reidel), p. 357.

_____. 1984, Lecture series presented at the Workshop on Astrophysical Magnetospheres, Taos, New Mexico, August 1984.

_____. 1988a, in *Physics of Compact Objects: Theory versus Observations*, ed. L. Filipov and N. White (Oxford: Pergamon), in press.

_____. 1988b, Talk presented at the Los Alamos Workshop on QPOs, La Cienega, N. M., September 1988.

_____. 1989a, in *Timing Neutron Stars*, ed. H. Ögelman and E.P.J. van den Heuvel (Dordrecht: Kluwer Academic Publ.), p. 649.

_____. 1989b, in *Proc. 23rd ESLAB Symp. on X-ray Astronomy*, ed. J. Hunt and B. Battrick (ESA SP-296), p. 215.

_____. 1991a, in *Neutron Stars: An Interdisciplinary Field*, ed. J. Ventura and D. Pines (Dordrecht: Kluwer Academic Publ.),in press.

_____. 1991b, in *Frontiers in Stellar Evolution*, ed. C. Geiger and D. Lambert (Astron. Soc. Pacific), in press.

Lamb, F. K., and Ghosh, P. 1991, in *Particle Acceleration Near Accreting Compact Objects*, ed. J. van Paradijs and M. van der Klis (Amsterdam: Royal Netherlands Academy of Sciences), in press.

Lamb, F. K., Pethick, C. J., and Pines, D. 1973, *Astrophys. J.*, **184**, 271.

Lamb, F. K., Shibazaki, N., Alpar, M. A., and Shaham, J. 1985, *Nature*, **317**, 681.

Lewin, W.H.G., and Joss, P. 1983, in *Accretion-Driven Stellar X-Ray Sources*, ed. W.H.G. Lewin and E.P.J. van den Heuvel (Cambridge: Cambridge University Press), p. 41.

Lewin, W.H.G., van Paradijs, J., and van der Klis, M. 1988, *Space Sci. Rev.*, **46**, 273.

Lightman, A. P. 1990, personal communication.

Lightman, A. P., and Eardley, D. M. 1974, *Astrophys. J. (Letters)*, **187**, L1.

Livio, M., Soker, N., de Kool, M., and Savonije, G. J. 1986a, *Mon. Not. R. astr. Soc.*, **218**, 593.

_____. 1986b, *Mon. Not. R. astr. Soc.*, **222**, 235.

Lyne, A. G. 1991, this volume.

Matsuda, T., Sekino, N., Sawada, K., Shima, E., Livio, M., Anzer, U., and Börner, G. 1991, *Astron. Ap.*, submitted.

Nagase, F. 1989, *Pub. Astr. Soc. Japan*, **41**, 1.

Penninx, W. *et al.* 1990, *Mon. Not. R. astr. Soc.*, **243**, 114.

Penninx, W. *et al.* 1991, *Mon. Not. R. astr. Soc.*, **249**, 113.

Prince, T. A., Anderson, S. B., Kulkarni, S. R., and Wolszczan, A. 1991a, *Astrophys. J. (Letters)*, in press.

Prince, T. A. *et al.* 1991b, in preparation.

Piran, T. 1978, *Astrophys. J.*, **221**, 652.

Pringle, J. E. 1976, *Mon. Not. R. astr. Soc.*, **177**, 65.

Pringle, J. E., and Rees, M. J. 1972, *Astron. Astrophys.*, **21**, 1.

Pringle, J. E., Rees, M. J., and Pacholczyk, A. G. 1973, *Astron. Ap.*, **29**, 179.

Radhakrishnan, V. and Srinivasan, G. 1981, Paper presented at the 2nd Asian-Pacific Regional Meeting of the IAU, Bandung, Indonesia.

_____. 1984, *Proc. 2nd Asian-Pacific Regional Meeting of the IAU*, ed. B. Hidayat and M. W. Feast (Jakarta: Tira Pustaka), p. 423.

Sawada, K., Matsuda, T., Anzer, U., Börner, G., and Livio, M. 1989, *Astron. Ap.*, **231**, 263.

Schulz, N. S., Hasinger, G., and Trümper, J. 1989, *Astron. Ap.*, **225**, 48.

Shakura, N. I. 1972, *Astron. Zh.*, **49**, 921 [Engl. transl. *Sov. Astron.–AJ*, **16**, 756].

Shakura, N. I., and Sunyaev, R. A. 1973, *Astron. Ap.*, **24**, 337.

_____. 1976, *Mon. Not. R. astr. Soc.*, **175**, 613.

Shapiro, S. L., and Lightman, A. P. 1976, *Astrophys. J.*, **204**, 555.

Shapiro, S. L., Lightman, A. P., and Eardley. D., M. 1976, *Astrophys. J.*, **204**, 187.

Shibazaki, N., and Lamb, F. K. 1987, *Astrophys. J.*, **318**, 767.

Smarr, L. L. and Blandford, R. D. 1976, *Astrophys. J.*, **207**, 574.

Soker, N., and Livio, M. 1984, *Mon. Not. R. astr. Soc.*, **211**, 927.

Spicer, D. S. 1982, *Space Sci. Rev.*, **31**, 351.

Srinivasan, G. 1989, *Astr. Ap. Rev.*, **1**, 209.

Stella, L. 1988, *Memorie della Società Astr. Italiana*, **59**, 185.

Stella, L., and Rosner, R. 1984, *Astrophys. J.*, **277**, 312.

Stella, L., Parmar, A. N., and White, N. E. 1987, *Astrophys. J.*, **321**, 418.

Szuszkiewicz, E. 1990, *Mon. Not. R. astr. Soc.*, **244**, 377.

Taam, R. E., and Fryxell, B. A. 1988, *Astrophys. J.*, **327**, L73.

Taam, R. E., and Lin, D. N. C. 1984, *Astrophys. J.*, **287**, 761.

Taylor. J. H. 1991, this volume.

Treves, A., Maraschi, L., and Abramowicz, M. 1988, *Pub. Astron. Soc. Pacific*, **100**, 427.

van den Heuvel, E.P.J. 1977, in *Proc. 8th Texas Symp. Relativistic Ap.* (*Ann. NY Acad. Sci.*, **302**), p. 14.

————. 1989, in *Timing Neutron Stars*, ed. H. Ögelman and E.P.J. van den Heuvel (Dordrecht: Kluwer Academic Publ.), p. 523.

————. 1991, in *Neutron Stars: An Interdisciplinary Field*, ed. J. Ventura and D. Pines (Dordrecht: Kluwer Academic Publ.), in press.

van der Klis, M. 1989, *Ann. Rev. Astr. Ap.*, **27**, 517.

————. 1991, in *Neutron Stars: An Interdisciplinary Field*, ed. J. Ventura and D. Pines (Dordrecht: Kluwer Academic Publ.), in press.

van der Klis, M., Jansen, F., van Paradijs, J., Lewin, W.H.G., van den Heuvel, E.P.J., Trümper, J., and Sztajno, M. 1985, *Nature*, **316**, 225.

Wang, Y.-M. 1987, *Astron. Astrophys.*, **183**, 257.

White, N. E. 1989, *Astr. Ap. Rev.*, **1**, 85.

White, N. E., and Stella, L. 1987, *Mon. Not. R. astr. Soc.*, **231**, 325.

————. 1988, *Nature*, **332**, 416.

White, T. R., and Lightman, A. P. 1989, *Astrophys. J.*, **340**, 1024.

————. 1990, *Astophys. J.*, **352**, 495.

Wolszczan, A. 1991, *Nature*, **350**, 688.

Wolszczan, A., *et al.* 1989, *Nature*, **337**, 531.

Zylstra, G. 1988, Ph.D. thesis, University of Illinois at Urbana-Champaign.

QUASI PERIODIC OSCILLATIONS AND SOUND

M. Ali Alpar
Physics Department
Middle East Technical University
Ankara 06531
Turkey

Jacob Shaham
Physics Department Columbia University
and Columbia Astrophysics Laboratory
New York
NY 10027 USA

Quasi periodic oscillations (QPOs) [1,2] provide a unique diagnostic for low mass X-ray binaries. Different classes of QPOs characterize different source states as reflected in the x-ray spectral properties [3]. The basis of the systematic behaviour is likely to lie in the simplest global physical properties of these systems starting with the kinematics. This paper aims at delineating some basic determining properties, including some new suggestions, that may serve to build a general framework for a theoretical discussion of the QPO phenomenon. Its approach will be that of a workshop report on current research rather than a general review of the subject.

The motivation of the beat frequency model [4] for the first discovered QPO's was the notion of an evolutionary link between millisecond pulsars and low mass X-ray binaries. The searches for millisecond periods from LMXB's had not (and still have not) been successfull, but the discovery of QPO's in the 30-50 ms range immediately brought to mind the possibility that these objects were indeed neutron stars in the process of being spun up by accretion [5,6] towards an equilibrium between the rotation rate ν_* of the neutron star and the rotation rate ν_K of the accretion disk's inner boundary. Interpreting the QPO frequency as the beat frequency between these two frequencies provides an explanation for the observed correlation between the QPO frequency and the count rate. This model gives a magnetic field at the neutron star surface in the range 10^9 G, and a stellar rotation period of order 10 milliseconds. Both of these are in the right range for millisecond pulsars. QPOs were observed also from a high mass X-ray binary [7]. Interpreting these as beat frequency QPOs gives stellar rotation periods and magnetic fields in agreement with observations, lending strong support to the beat frequency model.

---

*Also: Institute for Theoretical Physics, University of California, Santa Barbara, CA 93106

E. P. J. van den Heuvel and S. A. Rappaport (eds.), X-Ray Binaries and Recycled Pulsars, 511–517.
© *1992 Kluwer Academic Publishers.*

A crucial requirement for the realization of this model is the breaking of azimuthal symmetry with respect to the rotation axis. If the accretion flow in the disk's inner boundary region interacting with the star's magnetosphere possessed azimuthal symmetry there would be no structure to beat against the magnetosphere [8]. The clumping in the accretion flow invoked to realize this symmetry breaking has been described as random shots. This naturally leads to low frequency noise (LFN) whose strength is correlated with that of QPOs [8]. Such correlation is not universal and LFN is not present in the normal branch and flaring branch spectral states; a more generalized approach to QPOs and to symmetry breaking in terms of wave packets may prove fruitful.

The normal branch QPOs are particularly interesting as they have a rather narrow 5-8 Hz frequency band in all sources. It is likely that this band reflects a universal physical state and its intrinsic frequencies (rather than beat frequencies between distinct components). The X-ray spectrum is comptonized and the spectral changes as a source moves up and down the normal branch can be described simply as changes in the degree of comptonization due to changes in the electron scattering optical depth τ_{es} [9,10,11]. The dependence of the amplitude and phase of the normal branch QPOs on photon energy is nicely explained if these QPOs are oscillations of τ_{es} [10,11].

What makes τ_{es} oscillate? The universality of the 5-8 Hz QPO frequency band suggests that the normal branch state is characterized by a universal phenomenon, generally believed to be the approach of the luminosity to the Eddington value and the consequent thick accretion disk configuration. It has been proposed that the oscillations in optical thickness arise from a feedback between radiation and infalling matter in a spherically accreting fraction of the flow with slow dynamical timescales due to radiation pressure [10,11]. Such oscillations arise when the luminosity $L = (1 - \epsilon)L_E$ is within about $\epsilon = 10\%$ of the Eddington value L_E and the radial fraction of the mass accretion rate is about 4ϵ. The radial oscillation model does not involve the rotation in the system.

Here we describe a picture [12] that assigns a fundamental role to rotation through its effect on sound modes. We start with the recognition that radiation pressure will lead to the formation of subsonic regions and allow for the presence of sound modes [3]. If there is sound in a rotating medium like a thick accretion disk, the density and consequent optical thickness oscillations will have a natural band of frequencies imposed by rotation. For sound waves traveling in a plane perpendicular to the rotation axis in a rigidly rotating medium the dispersion relation [13] is:

$$\omega^2 = (2\Omega)^2 + c_s^2 k^2 \tag{1}$$

where ω is the sound frequency, Ω the rotation rate, c_s the speed of sound and k the wavenumber. This dispersion relation reflects the dominant effect of the Coriolis force which, like any long range correlation, leads to a low frequency cut-off at long wavelengths. (Other instances of this phenomenon are the plasma frequency, the Higgs mass and the Jeans mass.) Here we see the effect of rotation in the simplest case. Equation (1) can be modified to include the complications of a finite medium, wave propagation at arbitrary angles from the rotation axis, modes with nonzero azimuthal wavenumber m and differential rotation. The dispersion relation can also be employed to study instabilities and growth rates of modes. Papaloizou and Pringle have investigated the general dispersion relation and its instabilities [14]. They find an analogy with the Klein-Gordon equation, with rotation playing the role of finite mass. In the case of differential rotation $(2\Omega)^2$ is replaced with $\kappa^2 = 2\Omega(2\Omega + r\partial\Omega/\partial r)$. For Keplerian rotation $\kappa = \Omega$. Sound waves propagating parallel to the rotation axis are purely acoustic with zero frequency in the long wavelength limit. For propagation at angle θ from the rotation axis, two bands of real frequency exist, the acoustic band with $0 \leq \omega \leq \kappa \, cos(\theta)$ and an upper band with $\omega \geq \kappa$. There is a gap between frequencies ω and $\kappa \, cos(\theta)$. We use the model dispersion relation

$$c_s^2 k^2 = \frac{\omega^2(\omega^2 - \kappa^2)}{\omega^2 - \kappa^2 cos^2(\theta)} \qquad (2)$$

which has the basic features of the full dispersion relation and is obtained from it for $m = 0$. The gap in this dispersion relation brings to mind the depletion in the QPO power spectra between the QPO peak and the low frequency noise (LFN), when present. The obvious suggestion is to construct power spectra by folding a density of states in the wavevector $\vec{k}$. The relation between the QPO peak and the suppressed LFN of the normal branch can then be modelled in terms of the two branch dispersion relation and the density of states including its dependence on the angle θ between $\vec{k}$ and the rotation axis. The QPO peak has a finite width rather than including all frequencies in the upper band from the cutoff to infinity. This is because short wavelengths (high frequencies) with random phases and azimuths will average out in the observed effect of the optical thickness oscillations while long wavelengths of the order of the system size dominate, leading to a frequency band around the long wavelength limit of the upper band, at $\omega \approx \kappa$.

A thick accretion disk configuration obstructing part of the radiation from the neutron star has been proposed by van der Klis et al [15]. A numerical model calculated by Eggum, Coroniti and Katz [16] for a supercritical thick disk around a black hole has many features that may be relevant for a thick disk model in the present context. If the normal branch QPOs are indeed caused by sonic perturbations, one must identify a sonic region in the thick accretion disk. We assume the thick disk configuration depicted in Figure 1. The radiation from the star counteracts gravity in the inner regions of the disk hump but changes direction as it emerges normal to the outer surface of the hump. There is accreting matter also in the funnel, with a continuous transition from disk material into this corona as part of the accretion flow in the disk loses some of its angular momentum to radiation. Both media are optically thick to electron scattering and the part of the beam that scatters through the disk is effected by the optical thickness oscillations of the disk which carry the rotation

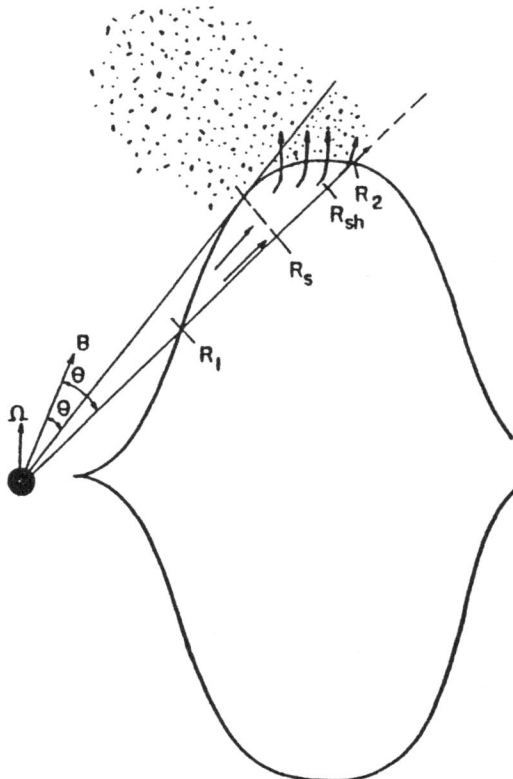

Figure 1: The thick disk. The sonic surface is at R_s. A part of the X-ray beam entering at R_1 is comptonized in the disk, the rest in the corona. The radiation transfer is roughly radial in the inner parts and bends towards the surface normals in the outer disk. R_{sh} denotes the shock.

as our model dispersion relation indicates. Material in the corona may also have the imprint of rotation both because of residual angular momentum and because it will be coupled to oscillations in the disk. The presence, on the normal branch, of 55 Hz horizontal branch QPOs together with the 6 Hz QPOs [16] indicates, according to the beat frequency model, that the Kepler frequency in the innermost edge of the disk in the plane is not affected by radiation pressure. Hence the radiation from the neutron star must be beamed to avoid the inner boundary region. The transition to low Kepler velocities associated with the effect of radiation pressure in the disk hump is depicted in Figure 2. Interpreting the observed 6 Hz frequency band as the long wavelength limit of the upper band of our dispersion relation we have in the keplerean case

$$\kappa = \Omega = (\epsilon GM/R^3)^{1/2} \approx 6Hz \times 2\pi \tag{3}$$

where M is the neutron star mass and R the radial distance from the neutron star. Taking $R = 3 \times 10^7 cm$ gives $\epsilon = 0.27$ for a one solar mass neutron star. A radial drift rate can be defined through

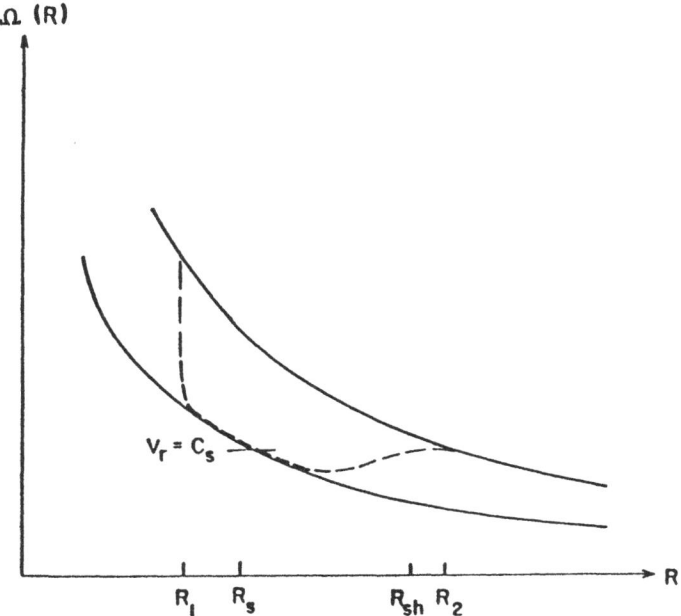

Figure 2: Rotation frequencies in the thick disk. The upper curve shows the "bare" Keplerean frequencies, and the lower curve the frequencies reduced by radiation force. The dashed line shows qualitatively the actual orbital frequencies in the disk.

$$v_r \equiv \alpha(r)\Omega(R)r \qquad (4)$$

where r is the cylindrical radius from the rotation axis and $\alpha \leq 1$ defined here is not the thin disk viscosity parameter. A subsonic region exists where $v_r \leq c_s$. For an isothermal disk the condition $v_r = c_s$ gives $\alpha = 0.1(kT(keV))^{1/2}/r_7$. The observed spectra have been fitted with $kT \approx 1keV$ [10,11]. We adopt this temperature here as the isothermal temperature of the disk.

We thus have a subsonic region, a musical instrument, so to speak, which is tuned by rotation and whose sound we perceive through its optical properties. How are the modes excited to macroscopic amplitudes? A possible source of the sound is a shock [17] at the outer boundary of the subsonic region where supersonic accretion arrives from the outer part of the accretion disk (see Figure 2). From the model we find [18] that the fastest growing modes have growth times of the order of the rotation period and the real and imaginary parts of the wavenumber k are of order κ/c_s for these modes. With $\kappa = \Omega \approx 12\pi \ rad/s$ and $c_s = 3 \times 10^7 cm/s$ we obtain wavelengths $\lambda \approx 5 \times 10^6 cm$ which are commensurate with the dimensions of the disk hump region.

At $L \approx L_{Edd}$ slow v_r and large density give prohibitively large values of τ_{es} if the entire flow is traversed. If the source of the QPOs is the radially accreting corona this is avoided as only a small part of the total mass accretion rate is in the radial flow. The present model avoids this problem because the initially beamed radiation scatters through only a section of the disk hump. Taking a section from $R_1 = 2.5 \times 10^7 cm$ to $R_2 = 3.5 \times 10^7 cm$ gives $\tau_{es} \approx 17$. All radiation is comptonized, most of it in the corona. The fraction comptonized in the disk hump carries the 6 Hz imprint of rotation.

In this picture we expect an interesting effect of source inclination. It is believed that for the bright LMXB's we look normal to the orbital plane. For sources where we happen to look into the plane in such a way that we see the radiation scattered through the disk hump but not the radiation scattered by the corona we would see $L \approx 10^{35} - 10^{36} ergs/s$ with a high 6 Hz QPO fraction. These would be transient sources as they would be observed when on the normal branch. The discovery of such sources would improve the count of LMXB's.

To summarize, we propose to model QPOs in terms of wave packets. The normal branch QPOs arise in a subsonic region formed by radiation pressure in a thick disk caused by and have a frequency spectrum characterized by rotation. The horizontal branch QPOs (not discussed here) require the study of MHD perturbations in the inner disk boundary.

Acknowledgments

This work was supported by a NATO Collaborative Research Grant, by Grants NASA NAGW 1618 and NSF INT 89-18665 with partial support at the ITP from the NSF under Grant No.PHY89-04035 supplemented by funds from NASA.

References

[1] Lewin,W.H.G.,van Paradijs,J.,and van der Klis,M. (1988) *Space Sci. Rev.* **46**, 273.

[2] van der Klis,M. (1989) *Ann. Rev. Astr. Ap.* **27**, 517.

[3] Hasinger,G. (1987) *Astr.Astrop.* **186**,153.

[4] Alpar,M.A. and Shaham,J. (1985) *Nature* **316**, 239.

[5] Alpar,M.A.,Cheng,A.F.,Ruderman,M.A. and Shaham,J.(1982) *Nature* **300**, 728.

[6] Radhakrishnan,V. and Srinivasan,G.(1982)*Curr.Sci.* **51**, 1096.

[7] Angelini,L.,Stella,L. and Parmar,A. (1989)*Ap.J.* **346**, 906.

[8] Lamb,F.K.,Shibazaki,N.,Alpar,M.A., and Shaham,J. (1985) *Nature* **317**, 681.

[9] Stollman, G., Hasinger, G., Lewin, W.H.G., van der Klis, M., and van Paradijs, J.(1987) *Mon. Not. R. Astr. Soc.* **227**, 7p.

[10] Fortner,B.,Lamb,F.K. and Miller,G. (1989) *Nature* **342**,775.

[11] Lamb,F.K. (1991) *Ap.J.*, in the press.

[12] Alpar,M.A.,Hasinger,G.,Shaham, J. and Yancopoulos, S. (1991) *Astr. Astrop.* submitted.

[13] Chandrasekhar,S. (1961) *Hydrodynamic and Hydromagnetic Stability*, Clarendon, Oxford, pp.589-91.

[14] Papaloizou,J.C.B. and Pringle,J.E. (1984) *Mon.not.R.astr. Soc.* **208**, 721.

[15] van der Klis,M.,Stella,L.,White,N.J.,Jansen,F.,and Parmar,A., (1987)*Ap.J.* **316**, 411.

[16] Eggum, G.E., Coroniti, F.V. and Katz, J.I. (1988) *Ap. J.* **330**, 142.

[17] Hasinger,G.,van der Klis,M., Ebisawa,K.,Dotani,T. and Mitsuda,K. (1990) *Astr.Astrop.* **235**, 131.

[18] Landau,L.D. and Lifshitz,E.M. (1975) *Fluid Mechanics*, Pergamon, Oxford, p.332.

[19] Yilmaz,A. and Alpar,M.A. (1991) in preparation.

"EXPERIMENTAL" DISCOVERY OF A TILT INSTABILITY IN STANDARD ACCRETION DISKS AROUND LUMINOUS X-RAY SOURCES

J.A. PETTERSON, J.P. GREENBERG and R.C. IPING*
San Diego Supercomputer Center
University of California at San Diego
La Jolla, CA 92093
        and
Institute for Theoretical Physics
University of California
Santa Barbara, CA 93106

ABSTRACT. Experiments performed at the San Diego (super) computer visualization lab, using a computergraphics model of the standard Shakura-Sunyaev type accretion disk in an X-ray binary system have demonstrated the tendency of these disks to be unstable to the development of warped deformations, when located around an X-ray source of very high luminosity. Experimental determination of the threshold luminosity above which the instability is activated gave 1–10% of Eddington Luminosity, depending on binary parameters. The instability may provide a suitable mechanism for causing accretion disks to tilt out of the binary plane, thereby causing them to start precessing. These results suggest that disk precession may be a very common phenomenon in high luminosity X-ray sources; much more common than suggested by the small number of presently known cases of the precessing disk phenomenon.

1. Introduction

1.1 THE PRECESSING DISKS IN Her X-1 AND SS 433

Nearly two decades after the discovery of the long term (35 day) period in the binary X-ray source Hercules X-1 (Tananbaum et al., 1972), it is now generally believed that a tilted, and probably also twisted, retrogradely precessing accretion disk is the cause of the group of effects referred to collectively as the "35 day cycle". It is, however, still controversial exactly why the accretion disk in this binary system precesses; i.e., what the precise physical mechanism is by which the precession operates. Nevertheless, some partial answers can be given already.

It is known, for instance, that disk precession may be expected to occur due to tidal forces from the companion star if the disk is tilted out of the binary plane (Katz, 1973), and that this mechanism can naturally reproduce a precession period of 35 days in the Her X-1 system if tidal forces over the entire disk, from inner to outer edge, as well as viscous coupling between all parts of the disk are included in the calculation (Merritt and

---

*presently at Dept. of Natural Sciences, University of Guam

E. P. J. van den Heuvel and S. A. Rappaport (eds.), X-Ray Binaries and Recycled Pulsars, 519–525.

Petterson, 1980). It is also known that such a model with tidally driven precession of the disk is expected to produce a somewhat erratic '35 day clock', due to the variability of the disk size related to variability in the mass transfer rate. It has been shown (Boynton *et al.*, 1980) that such clock properties fit the stochastic properties of the observed 35 day cycle well. Other mechanisms, in which the disk precession is driven by a precession in either the companion star (Roberts, 1974) or in the neutron star (Truemper *et al.*, 1986) have also been proposed, but they tend to have considerable difficulty circumventing the observational indication that the 35 day clock does not keep time as accurately as one would expect of a stellar precession. Moreover, these mechanisms find themselves contradicted by arguments from respectively tidal evolution (which should have aligned the rotation axis of the companion star) and the history of accretion in the system (which should have aligned the neutron star rotation axis). The simplest, and seemingly entirely adequate assumption to make for Her X-1 therefore seems to be that *the disk is the only thing in this system that precesses, driven by tidal forces from the companion star.* The sole remaining mystery then is how the disk got tilted out of the binary plane before it could start to precess. We provide a possible answer to this question in the present paper.

Previous suggestions for tilting mechanisms have been very few: Crosa and Boynton (1980) proposed a radiative feedback mechanism between disk tilt and mass outflow from the inner Lagrangian point, and also hinted at the possibility of tilting the outermost disk edge viscously by the rest of the disk, but neither of these proposals were worked out to any degree of detail. Katz' (1973) suggestion that the disk rim may be tilted due to an anomalous outflow event that was not symmetrical about the binary plane predicted that the tilt would decay away in about one decade, which we now know not to have happened. Thus the need for a tilting mechanism with detailed properties and predictive power has remained.

The peculiar object SS 433 can provide us with some further clues: If we may assume that the precession of its jets is guided by a precession of the innermost parts of its accretion disk, which is not proven at this time but seems most reasonable, then we seem to have a system here in which the *entire* accretion disk, from inner to outer edge, *is precessing as a single unit* (optical lightcurves show that the outer parts of the disk precess in unison with the jets). A similar conclusion can be drawn for Her X-1 from the fact that it has a "short on" stage in the middle of the "off" part of the 35 day cycle. The natural explanation of this stage is that it occurs when the observer on earth can peek at the neutron star underneath the disk rim when the disk is showing its "backside" to us. Since pulsations *can* be discerned during "short on", the neutron star should be in direct view at this time, which can only occur if the entire disk (including the inner edge) is tilted out of the binary plane. The complete disk must therefore be precessing, and it must be precessing at one and the same rate, because the "short on" occurs at a phase in the 35 day cycle which has not noticeably changed in at least a decade. Further evidence that the innermost parts of the Her X-1 disk are tilted and precessing in unison with the rest of the disk is provided by pulse profile variations during the 35 day cycle (Petterson, Rothchild, and Gruber, 1991). The tilting mechanism we need for the disks in these two binary systems is therefore preferably one that can tilt the disk over a large range of different radii. Some reliance on the viscous cohesion between all parts of the disk is thereby admissible, but it must be remembered that the purely viscous coupling between inner- and outermost disk edges is not strong. A tilting mechanism which can aid this coupling will therefore have an advantage in explaining observed precession properties.

1.2 DISK TWISTING

A disk which is tilted out of the binary plane over a substantial range of radii must be twisted (Petterson, 1975, 1977). This is an inescapable consequence of the strongly differential precession of disk "rings" ($\omega(r) \sim r^{3/2}$, for ring radius r), induced by the orbiting companion star. Since the disk in Her X-1 extends from radii of 10^8 cm to radii exceeding 10^{10} cm, precession frequencies of disk rings would differ by no less than 3 orders of magnitude, if rings were allowed to move independently. A twisted shape would therefore develop rapidly, causing viscous forces between rings to increase until they stop further "wind up", and thereby make the disk precess as a single unit. The equilibrium shape of this unit would of course still be twisted.

Because the induced precession of disk rings is in the retrograde direction at all radii, the disk as a whole must also *precess retrogradely*, in agreement with observational indications. But because the precession of the outermost ring is *fastest* (it is closest to the precession inducing companion star!) the disk's outermost rings would be expected to precess *ahead* of the other disk rings if other influences could be ignored. The fact that it is possible for Her X-1 to obtain observational evidence that the outermost ring is in fact lagging, not leading the disk in its precessional motion (*e.g.*, Petterson, 1977; Crosa and Boynton, 1980) is therefore an indication that other influences should *not* be ignored. One of these, radial drift, was already shown to have importance (see last two references, above). Another one, illuminative effects due to the X-ray source (Petterson, 1978), turns out to be even more important, as the remainder of this paper will show.

2. Experimental Investigation of X-Ray Binaries in the (Super) Computer Laboratory

Astronomy does not have the luxury of being able to manipulate the objects it studies, and experimental investigation of real X-ray binaries is therefore out of reach. However, it is possible to make up for this disadvantage to some degree with the use of computer simulations of models.

Computers have of course been used for the testing of theoretical models in astrophysics for many years. What makes computer testing with the latest generation of equipment suddenly so interesting and potentially useful, is the fact that explosive increases in computational power of supercomputers combined with the extremely rapid rate of evolution of computergraphics is beginning to make computertests possible which in realistic visual detail and dynamical accuracy are beginning to approach the quality of "real" experimental tests of objects in the laboratory.

Moreover, when the numbercrunching supercomputer with attached graphics workstation is also equipped with a "real-time" input devise which allows the manipulation of parameter values while the dynamical computation proceeds, then "experiments" can be performed that nature itself could never perform, and this makes the (super) computer lab even more useful.

2.1 MODELS AND APPROXIMATIONS USED

Details of the "experimental" set-up in the visualization lab used for the here described experiments on X-ray binaries were published elsewhere (*e.g.*, Iping and Petterson, 1990).

We employed a thin accretion disk model (Shakura and Sunyaev, 1973) with standard properties $(\dot{M}, M, \alpha,$ etc.), and a twist structure described by a standard twist equation (*e.g.*, Hatchett *et al.*, 1981). We then studied the dynamical evolution of the twist structure under various circumstances.

3. "Experimental" Results

We followed the twist evolution of thin accretion disks with different initial twist and tilt structures. This structure is described by two functions, $\beta(r)$ and $\gamma(r)$, giving respectively the inclination (tilt) and precession angles of individual rings at radius r, with r varying from innermost to outermost disk edge (see also Petterson, 1977).

As long as we choose $\gamma(r)$ to be a monotonically *increasing* function of r for the initial disk, β values tended to increase in time at all radii, until an "equilibrium" value for $\beta(r)$ was reached. This equilibrium value was independent of our initial conditions, as far as we could tell (we choose $\beta = 1°$, $\beta = 5°$, $10°$, and $20°$ (for all r) as initial configurations), but was strongly dependent on the assumed X-ray luminosity. When we chose $\gamma(r)$ to be a monotonically *decreasing* function of r for the initial disk, β values tended to decrease rapidly, leaving us with an accretion disk *in the binary plane* in (usually) less than one precession time.

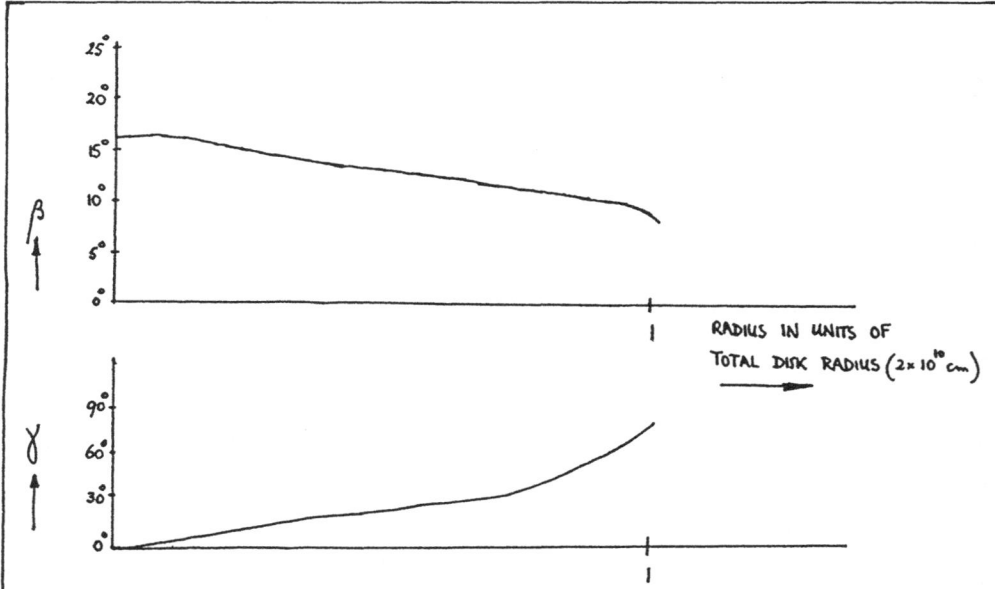

Figure 1. Equilibrium twist structure of Her X-1 disk, illuminated by an X-ray source of $10^{30}W$ (10^{37} erg/sec). Note that outermost disk edge is tilted slightly *less* than most other disk "rings" (probably due to boundary condition that new matter is transferred to disk *in* the binary plane. Precession angle γ is monotonically *increasing* with radius, making the outermost ring lagging in the precessional motion if it is retrogradely moving.

The timescale on which $\beta(r)$ increased or decreased, strongly dependent on the assumed X-ray luminosity, tended to become shorter as the value of L_x increased. This behavior strongly suggested that the disk was unstable to tilting perturbations due to the presence of the strong X-ray illumination. Subsequent experimenting with the value of L_x showed us that the instability was activated only at very high X-ray luminosities ($10^{30}W$ and higher for Her X-1 binary parameters). Figure 1 shows the equilibrium tilt and twist configuration for a disk scaled to Her X-1 parameters under illumination by a $10^{30}W$ ($= 10^{37}$ erg/sec) X-ray source, with a viscosity parameter α of value unity. (Different choices for α gave different equilibrium twist structures; typically $\beta(r)$ decreased as α increased.)

4. Interpretation and Explanation of Results

With the help of our "experimental" accretion disk in the form of a computer model, which we could manipulate until we understood how the instability worked, we have been able to produce the following simple explanation of the above described "behavior" of the accretion disk:

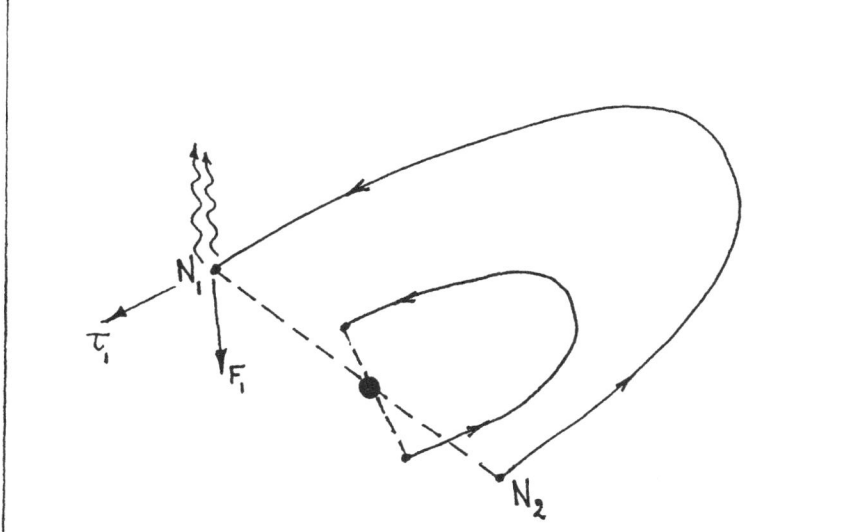

Figure 2. Shadowing mechanism of disk ring by other ring at smaller radius, which causes a nonaxisymmetric effect for twisted disk, leading to an increase of inclination angle of the outer ring (arrows on rings indicate rotation sense of matter in the disk, wavy lines reemitted radiation, F_1 the radiation reaction force, and τ_1 the corresponding torque on the ring matter at N_1).

Twisted disks in which $\gamma(r)$ increases monotonically with radius are structures in which the X-ray shadow of every disk ring onto its *outer* neighbors causes non-axisymmetrical illumination patterns which cause torques on the disk rings that increase their inclination. Figure 2 shows two rings in such a type of twisted disk. The presence of the inner ring causes the X-ray illumination of the point N_2 of the outer ring to be less than that of the point N_1 (due to shadowing). Assuming that the X-rays are absorbed by the disk, and reemitted perpendicularly to the local disk surface, a radiation reaction torque will occur at N_1 but not (or less) at N_2, causing the inclination angle β of the outer ring to increase. For monotonically decreasing $\gamma(r)$ with radius, the inner ring would shadow N_1 more than N_2, thus setting up a torque which decreases the inclination angle β of the outer ring. This mechanism could operate for a large range of values for L_x, but it can of course only operate *effectively* as a tilting mechanism if it is strong enough to overcome the everpresent viscous tendency to *reduce* the inclination of disk rings in a twisted disk (the viscous effect is always reducing $\beta(r)$, regardless of the form of $\gamma(r)$, unless $\gamma(r) = 0$). This seems to explain why there is a threshold value for the X-ray luminosity, below which the instability does not work. This threshold value lies approximately at the value for L_x where radiative forces can overcome the viscous ones (*i.e.*, one or two orders of magnitude below the Eddington limit).

Illuminative effects on the deformed accretion disk can of course also affect the rate at which disk rings are induced to precess. We refer the reader to a previous publication (Iping and Petterson, 1990) for a presentation of our findings on this subject.

5. Conclusions

Our computer experiments have led us to the discovery of an instability in standard thin accretion disks, which cause these disks to tilt out of the binary plane at all radii where the precession angle $\gamma(r)$ is a monotonically increasing function of r, which may include a large range of radii, possibly the entire disk. Since small perturbations of a nearly flat disk could grow due to this instability, it is reasonable to assume that many disks in bright X-ray binaries might be subject to this instability, therefore get tilted, and presumably then start precessing. The observational signature of precessing disks which were tilted by this mechanism is that they will have a twist shape described by a $\gamma(r)$ function which increases with radius. These precessing disks should therefore have *lagging* outer edges, as seems to be the case for the Her X-1 disk.

6. Acknowledgements

We are very grateful to the San Diego Supercomputer Center for making available to us equipment and computer time that made the here reported research possible. This research was supported in part by the National Science Foundation under Grant No. PHY89-04035.

7. References

Boynton, P.E., Crosa, L.M., and Deeter, J.E. (1980) *Ap. J.*, **237**, 169.
Crosa, L.M. and Boynton, P.E. (1980) *Ap. J.*, **235**, 999.
Hatchett, S., Begelman, M., and Sarasin, C. (1981) *Ap. J.*, **247**, 677.

Iping, R.C., and Petterson, J.A. (1990) *Astron. Astrophys.*, **39**, 221.

Katz, J.I. (1973) *Nature Phys. Sci.*, **246**, 87.

Merritt, D. and Petterson, J.A. (1980) *Ap. J.*, **236**, 255.

Petterson, J.A. (1975) *Ap. J. (Letters)*, **201**, L61.

— (1977) *Ap. J.*, **218**, 783.

— (1978) *Ap. J.*, **226**, 253.

Petterson, J.A., Rothchild, R. and Gruber, D. (1991) *Ap. J.*, in press.

Roberts, W.J. (1974) *Ap. J.*, **187**, 575.

Shakura, N.I. and Sunyaev, R.A. (1973) *Astron. Astrophys.*, **24**, 337.

Tananbaum, H., Gursky, H., Kellogg, E.M., Levinson, R., Schreier, E., and Giacconi, R. (1972), *Ap. J. (Letters)*, **174**, L143.

Truemper, J., Kahabka, P., Oegelman, H., Pietsch, W., and Voges, W. (1986) *Ap. J. (Letters)*, **300**, L63.

6 HZ QUASI-PERIODIC OSCILLATIONS FROM LOW-MASS X-RAY BINARIES: THE SOUND OF AN ACCRETION DISK?

M. Ali Alpar
Physics Department, Middle East Technical University
Ankara 06531, Turkey
and
Institute for Theoretical Physics, University of California
Santa Barbara, CA 93106, USA

Günther Hasinger
Max-Planck Institut für Extraterrestrische Physik
8046 Garching bei München, Germany

Jacob Shaham
Physics Department and Columbia Astrophysics Laboratory
Columbia University, New York, NY 10027, USA
and
Institute for Theoretical Physics, University of California
Santa Barbara, CA 93106, USA

and

Sophia Yancopoulos
Physics Department and Columbia Astrophysics Laboratory
Columbia University, New York, NY 10027, USA

Quasi-periodic oscillations (QPOs) at frequencies of 5-8 Hz are a common feature of low-mass X-ray binaries (LMXBs) while in their normal branch (NB) states[1]. We propose that these are due to long wavelength sonic perturbations in a rotating, thick, accretion disk present in these systems at the time. The frequency of such perturbations is related to the rotation frequency and the vorticity in a simple manner and is generally of the order of the rotation frequency.

The fact that NB QPO frequencies from different sources all fall in a narrow range (5-8 Hz) suggests that the NB spectral state corresponds to a universal source configuration. A natural suggestion is that such configuration corresponds to accretion at the Eddington rate.

E. P. J. van den Heuvel and S. A. Rappaport (eds.), X-Ray Binaries and Recycled Pulsars, 527–535.

The observed low frequencies arise naturally as the dynamics is slowed down by radiation pressure when the luminosity is near the Eddington value[2]. Oscillations in optical thickness of a Comptonizing medium[3-5] provide a natural means of modulating the emergent flux at frequencies of the order of the dynamical frequencies in that medium. Such optical thickness oscillations provide a simple and elegant explanation for the observed dependence of the phase and amplitude of the oscillations on photon energy[4,5]. In the model of Refs. [4] and [5], the oscillations arise from feedback between the luminosity and the infalling matter in a radial part of the accretion flow. The frequency is determined by the radial accretion timescale, i.e. the free fall time in the presence of near-Eddington flux. This radial feedback mechanism requires a tuning of the luminosity and the oscillation amplitude.

We propose that the 6 Hz oscillations in optical thickness find a natural alternative explanation in terms of the rotation of the accretion disk. We assume that as the accretion rate approaches the Eddington value, the disk thickens and obstructs part of the radiation from the neutron star[6] and that this transition in the disk causes the observed transition of the source from the horizontal branch (HB) to the NB spectral state. The radiation impinging on the thickened disk is Comptonized and isotropized there. As in the earlier models[4,5], the time dependence of the optical thickness oscillations in the Comptonizing medium is again imprinted on the emergent radiation. However in the present model, part of the Comptonization occurs in the thick disk itself and the QPO signature acquired by the fraction of the radiation that is scattered through the disk [Fig. 1] represents the intrinsic sound spectrum of the disk.

It is a fundamental consequence of rotation that the dispersion relation for sound waves acquires a new branch, characterized by a *finite* minimum frequency at the long wavelength limit that is determined by the rotation rate[7]. This low frequency cutoff for long wavelengths is typical of systems with long range correlations, as exemplified by the plasma frequency and by the Jeans and Higgs masses. For sound waves in a rotating medium it is the Coriolis force that leads to such correlations. Rotation thus provides a natural, universal and stable selection of similar frequencies for systems with similar structure and similar rotation tares.

If the medium were to rotate rigidly, the dispersion relation for waves propagating in the plane perpendicular to the rotation axis would be

$$\omega^2 = (2\Omega)^2 + c_s^2 k^2 \quad (1)$$

where ω is the sound frequency, Ω is the rotation frequency, c_s is the speed of sound and k the wavenumber. The dispersion relation is more complicated in the presence of differential rotation, for wave propagation at arbitrary angles to the rotation axis and for modes that break azimuthal symmetry (non-zero m). Papaloizou and Pringle[8] have investigated the general dispersion relation with particular emphasis on unstable modes. We shall here work with a simple generalization of Eq.(1), given in Eq.(3), which retains the basic features of the full dispersion relation on the following counts:

(i) As Papaloizou and Pringle[8] point out, the instabilities and general properties of the dispersion relation are analogous to the properties of the Klein-Gordon equation with rotation playing the role of the finite mass. This "mass term" $(2\Omega)^2$ is already present in the simplest case of the dispersion relation in a rotating medium, that given in Eq (1).

(ii) In the presence of differential rotation, the term $(2\Omega)^2$, which is the square of the

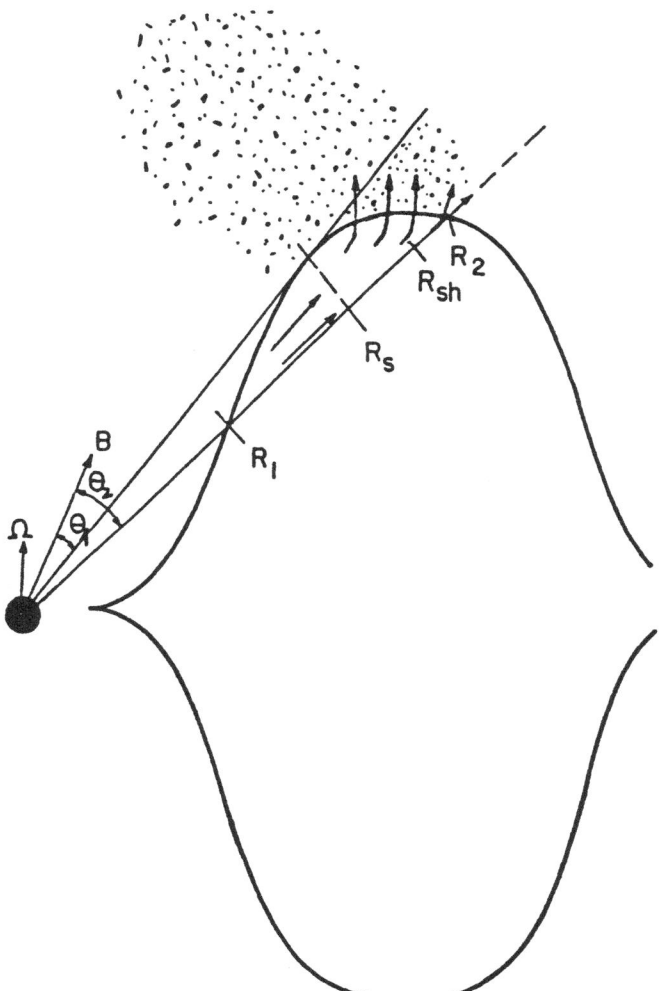

Fig. 1. The thick disk. The shaded region around the hump is the subsonic region. The sonic surface is represented by R_s. Part of a wide X-ray beam between θ_1 and θ_2 enters the disk at R_1. This radiation is scattered and Comptonized in the disk's hump. The rest of the beam may be Comptonized in the corona. The lines through the disk indicate the direction of radiation transfer, which is roughly radial in the inner parts and bends towards the surface normals in the outer disk. R_{sh} denotes the shock separating the outer region and the subsonic region.

vorticity for uniform rotation, should be replaced by

$$\kappa^2 = 2\Omega(2\Omega + r\frac{\partial\Omega}{\partial r}) \tag{2}$$

where r is the distance from the rotation axis. In the special case of Keplerian rotation $\kappa = \Omega$, while for rigid body rotation $\kappa = 2\Omega$. Eq.(2) can be obtained as a low frequency cutoff similar to that in Eq.(1) by taking the long wavelength limit of the dispersion relation discussed in Ref. [8].

(iii) Sound waves propagating parallel to the rotation axis feel no rotational effects and are purely acoustic, with zero frequency for the long wavelength limit. Only for propagation transverse to the rotation axis is there a low frequency cutoff as in Eq.(1). For a direction of propagation at an angle θ from the rotation axis there are two bands of real frequencies, one around zero frequency and another at frequencies above κ. The two bands are separated by a gap between frequencies $\kappa\cos\theta$ and κ. It is very appealing to associate the two bands with the QPO peak and the "very low frequency noise" in observed power spectra and to interpret the power minimum between these two features as the gap in the dispersion relation averaged over θ. This two-band structure is present in Eq.(3).

(iv) We concentrate here on the m=0 modes for simplicity. We postulate that higher m modes are not as easily excited. Dispersion relations for them are obtained by replacing κ with $\kappa + m\Omega$ in the absence of differential rotation and by more complicated transformations when differential rotation is present. Our simple model dispersion relation will be

$$c_s{}^2 k^2 = \omega^2(\omega^2 - \kappa^2)/(\omega^2 - \kappa^2\cos^2\theta) \tag{3}$$

for a particular θ. This has the basic features of the full dispersion relation and can be derived from it for $m = 0$.

We propose that the 5-8 Hz quasi-periodic oscillation frequencies reflect the low frequency cutoff imposed by rotation on the upper branch of a dispersion relation like that in Eq.(3) for sound in a subsonic region of the accretion disk. In the NB spectral state of a LMXB, mass accretion rates are close to the Eddington value and the disk has a thick configuration in response to the radiation pressure. There is evidence that the radiation from the star is beamed and does not impinge on the inner boundary of the disk. This is indicated by the observation[9] of 55Hz QPOs during the NB state of CygX-2.

According to the beat frequency model[10] the 55 Hz QPO frequency is the difference between the Keplerian frequency in the inner edge of the accretion disk and the rotation frequency of the neutron star, and the above observations therefore imply the "bare" (i.e. not reduced by radiation pressure) Keplerian frequency at the inner boundary of the disk out to some radius R_1 where the stellar beam intersects the disk [see Fig. 1]. Above R_1 the hump of the disk geometrically obstructs a portion of the radiation from the star. Matter in the hump is optically thick to electron scattering. Radiation transfer whithin the disk is radially outward for the inner side of the hump facing the neutron star, reducing the effective gravitational constant G to ϵG, while near the outer surface of the hump the radiative transfer bends towards surface normals and is no longer effective against the radially inward pull of gravity. Thus the radiation pressure is important in the hump of the thick disk, absent in the inner disk below R_1 and not effective against gravity in the outer disk. There is thus a region of minimum effective gravity and minimum Keplerian frequency associated with the location of the hump, where ω^2 goes from $\epsilon GM/r^3$ to GM/r^3 [M is the mass of the neutron star; see Figs. 1 and 2].

In an isothermal disk with uniform speed of sound $c_s = \left(\frac{5kT}{3\mu}\right)^{\frac{1}{2}}$, where T is the temperature, k the Boltzmann constant and μ the mass per pressure particle in the disk, this region of slowest dynamical timescales can become subsonic. We then have a filter where sound, that is density and consequent optical thickness perturbations, has a low frequency cutoff imprinted by the (slow) rotation frequencies. Perturbations of wavelengths small compared to the size of the system will not be important in the overall signature of the thick disk filter as they average out due to random times of occurrence and random azimuthal phases in the diak. It is the longest wavelength perturbations, with sizes comparable to the size of the entire subsonic region in the hump, that determine the overall characteristics of the region at any given time. This qualitative argument also suggests the dominance of $m = 0$ modes over the higher m modes. Thus, frequencies around the cutoff determined by the rotation rate are the effective ones for driving density and optical depth oscillations of the subsonic region. The dimension of this region, R, is comparable to the radial distance from the neutron star a: which the hump is located; based on standard disk models, we take it to be a few $10^7 cm$.

In positing a subsonic hump region in a thick disk as the origin of the 6 Hz QPO frequencies we have introduced a musical instrument, so to speak, whose frequency spectrum is tuned by rotation and whose sound we perceive through its optical properties. How does one play it, i.e. excite its modes to the observed level, which cannot be ascribed to *random* spontaneous perturbations? The QPO amplitude reflects, in addition to the actual sound amplitude in the disk, also the fraction of the beamed radiation which geometrically has to sample the 6 Hz optical thickness oscillations, so a rather large sound amplitude is indicated. The spontaneous appearance of such amplitude is likely to be due to a shock between the supersonic outer disk and the subsonic hump region [see Figs. 1 and 2]. Perturbations arriving from the supersonic side are amplified when they cross the shock boundary[11]. If the modes are to match finite amplitude boundary conditions at the shock boundary R_{sh} as well as at the inner boundary of the subsonic region at R_s, the imaginary part of the wave number is of order $1/R$. In the absence of solutions for a finite accretion disk with the appropriate boundary conditions, we turn to the model dispersion relation of Eq.(3), which refers to propagation in an *infinite* medium. Examining Eq.(3) for the imaginary part of the frequency we find[12] that for the fastest growing modes both the real and the imaginary parts of the wavelength are of order $2\pi c_s/\kappa$. Typical values for c_s for a temperature of 1 keV and $\kappa = \Omega = 12\pi$ rad/s give wavelengths of order 10^7 cm. These are indeed the longest possible in the hump because R is of that order too. For all directions of propagation the fastest growing modes have $Im(k)R < 3\pi R_7$, with R_7 measuring R in units of 10^7 cm. Also for the wavelength λ, $\lambda_7 > 0.67$ where λ_7 measures λ in units of 10^7 cm. The growth times for these modes are of order κ^{-1} and the real part of the frequency is of order κ.

The Keplerian frequency at reduced gravity ϵG and distance R from the star is given by

$$\Omega^2 = \epsilon GM/R^3. \qquad (4)$$

For Keplerian rotation, $\Omega = \kappa = \omega_{min}$. Taking $\omega_{min} = 2\pi \times (6Hz) = 12\pi$ rad/s and a one solar mass neutron star, we find

$$\epsilon = 0.29\left(\frac{R}{3 \times 10^7 cm}\right)^3 \qquad (5)$$

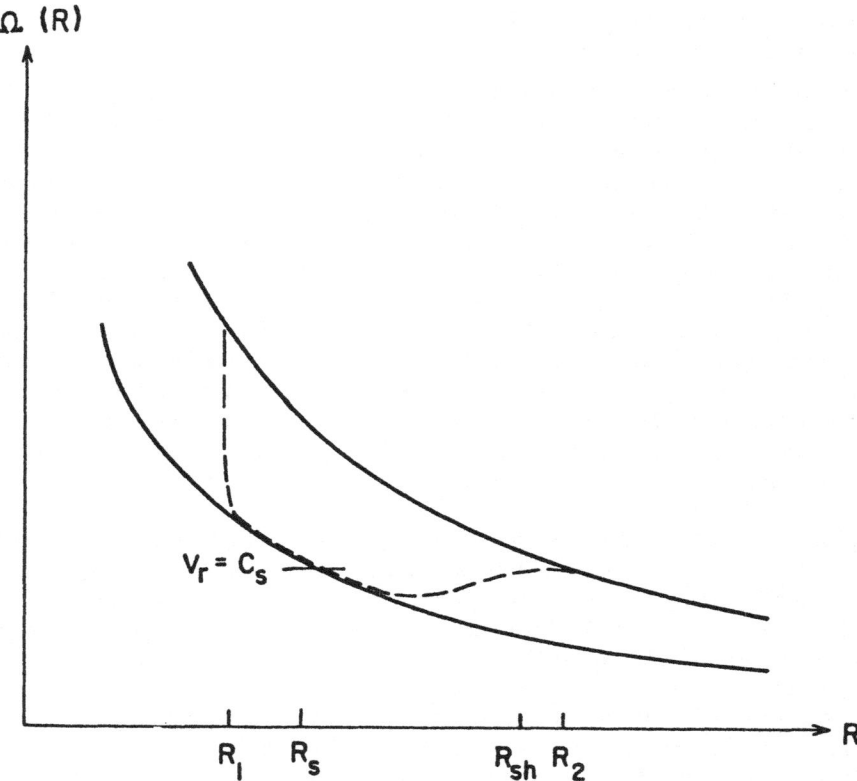

Fig. 2. The run of Keplerian frequencies through the thick disk. The upper curve shows the "bare" Keplerian frequencies and the lower curve the frequencies reduced by radiation force in the radial direction. The dashed line shows qualitatively the actual orbital frequencies in the disk. From R_1, where the beam enters the disk, and throughout the hump, the frequencies are reduced by the beam radiation force while in the outer disk the frequencies become bare Keplerian again, as the radiative flux bends out of the radial direction.

so that with reasonable sizes for the thick disk, fine tuning of the luminosity to the Eddington value is not required.

The location of the susonic region of the hump is determined by setting the radial inflow velocity v_r to the sound speed c_s. Let us scale v_r with the Keplerian velocity via $v_r = \alpha(r)\Omega r$, where α is defined by this relation (and is *not* the viscosity parameter of the thin "α disk" models). With r_s representing a mean value for r on the sonic surface and with $\Omega(r_s) = 12\pi$, we find

$$\alpha(r_s) = 0.11[kT(keV)]^{\frac{1}{2}}(\frac{r_s}{10^7 cm})^{-1}. \tag{6}$$

As is the case with any circumstellar or coronal matter with the appropriate optical depth τ_{es} of order 10, electron scattering isotropizes beamed radiation[13] and washes out pulsed signals at the neutron stellar rotation frequency[14]. Such matter also Comptonizes the radiation from the neutron star. The dependence of the amplitude and phase of the NB QPOs on photon energy has been already explained in terms of oscillations in the optical thickness of a Comptonizing medium[4,5]. In Ref.[4] the observed $70 msec$ hard lag in the 6 Hz QPOs is interpreted as a $180°$ phase difference plus a $10 msec$ hard lead, fixing τ_{es} to be about 20 over a distance of $3 \times 10^7 cm$. According to the same work, the observed variations of the amplitude and phase indicate an electron temperature of 1 keV. We invoke this same Comptonization model, with the same values for T and τ_{es}, the only difference being that the Comptonizing medium is part of the thick accretion disk instead of being a radial accretion flow and that the dynamical source of the 6 Hz oscillation lies in the rotation rather than in the free fall. Detailing our model requires as yet unavailable knowledge of thick disks and their Comptonization abilities, but it seems that even though in our geometry not all photons will scatter, most directions *will* "see" a sufficient number of multiply scattered photons to render the conclusions of Ref.[4] valid here as well.

At luminosities close to the Eddington value, inward flows slow down and optical depths through an isotropic accretion flow geometry or through an entire disk become prohibitively large. When Comptonization in a radially accreting corona is postulated[4,5] one can avoid the high optical thickness by putting only a small fraction of the accretion flow into the radial structure, dumping the rest into a separate accretion disk. In our thick disk model, large Comptonizing optical depths can be avoided if radiation from the star is beamed in such a way that its path traverses only part of the hump and not the entire length of the disk. As we noted above, the beat frequency interpretation of the 55 Hz oscillation observed on the NB lends support to such a notion. The optical thickness of the disk to electron scattering is given by

$$\tau_{es} = \int_1^2 d\ell\sigma_{es}(\ell)n(\ell)$$

where σ_{es} is the electron scattering cross section, ℓ is the distance along the path, 1 and 2 are the entry and exit points of the beam into and out of the hump and $n(\ell)$ is the particle density,

$$n(r) = \dot{M}/(2\pi r^2 f(r)v_r(r)m_p).$$

We use $f(r)$ to denote the height-to-radius ratio of the disk and $\dot{M}$ is the mass accretion rate. Taking a path from (spherical) radius $R_1 = 2.5 \times 10^7 cm$ to $R_2 = 3.5 \times 10^7 cm$ [see

Fig.1] and using the sonic surface values of ϵ (Eq.(5)) and α (Eq.(6)) we find

$$\tau_{es} = 2(1-\epsilon)cR_*(1+f^2)^{\frac{3}{2}}(R_1{}^{-\frac{1}{2}} - R_2{}^{-\frac{1}{2}})/(f\alpha(\epsilon GM)^{\frac{1}{2}}) \simeq 17,$$

where R_* is the neutron stellar radius and c is the speed of light. A similar optical depth could actually obtain over a longer path because α is likely to increase as we move from the hump to the inner disk. This is because the effective gravity has an unbalanced component along the inner disk surface (the funnel[15]) which causes v_r to increase.

Obscuration by a thick disk was already invoked by van der Klis et al.[6] to explain the 6 Hz oscillations; however, in the present model we do not necessarily involve a physical oscillation of the disk boundaries in and out of the line of sight. A spherically symmetric corona is likely to be present as well, in and above the funnel region, that is responsible for most of the Comptonized radiation but not for the QPO. We underline the basic effect of *rotation* as the reason for the "universality" of the 6 Hz band. To the extent that an Eddington luminosity determines a universal disk structure, the difference between various sources should reflect their inclinations, since these determine the fraction of the radiation scattered and Comptonized in the hump that we receive and its contribution to the total flux we see.

The present model also casts the relation between disk thickness, inclination and observability of orbital timescales in a new light: some years ago, the paucity of observed orbital variations in LMXBs was explained in terms of a selection effect[16], namely that thick accretion disks shield the X-ray sources at high inclinations hence any observed source must be viewed from a near normal direction and orbital variations would be strongly quenched. We now have, however, evidence for orbital binary periods from several LMXBs, including some 6 Hz sources[17]. The neutron stars do seem to be observable in spite of the thick disks, presumably via scattering and Comptonization in a corona around the star and above the disk. Here we propose scattering also by the hump, to produce a 6 Hz modulated component of the scattered radiation. High inclination sources would be detectable during their NB states even if a central corona were absent, provided their disks thicken up enough to scatter radiation out of the shadow of any outer rim structures. Such sources would have Comptonized spectra with luminosity values that are a few percent Eddington's, if the relative 6 Hz power in presently known 6 Hz sources is a measure of the hump's contribution. The number of such high inclination sources should increase with the angular width of the hump as seen from the neutron star. Their detection would lend support to the present model and help resolve the present statistical discrepancy[18] between the estimated numbers of LMXBs and their possible descendants[19,20], the millisecond pulsars, by "finding" the "missing" LMXBs.

Direct evidence for the 6 Hz rotation frequencies in the disk, in the form of Doppler shifted lines, would give crucial support for this model. Systematic correlations of NB branch behavior and 6 Hz signatures with inclination would also support and detail the model. Such evidence would constitute observational information on the existence and nature of thick accretion disks.

It is possible that 6 Hz density oscillations in a subsonic region manifest themselves also as quasi-periodic oscillations in the accretion rate onto the star and hence in the luminosity from the star, as the 6 Hz band is also imprinted on the density perturbations that cross the sonic surface to be advected to the star. It will be interesting to look for a

6 Hz qusiperiodicity during the HB state, perhaps contributing to the low frequency noise, as evidence for an emergent thick disk.

JS and SY gratefully acknowledge partial support from NASA, through grant NAGW-1618. MAA and JS gratefully acknowledge partial support from the NSF, through grant PHY89-04035.

REFERENCES

1. Lewin, W.H.G., van Paradijs, J. and van der Klis, M., Space Sci. Rev. **46**, 273-377 (1988).

2. Hasinger, G., Astr. Astrophys. **186**, 153-158 (1987).

3. Stollman, G., Hasinger, G., Lewin, W.H.G, van der Klis, M. and van Paradijs, J., Mon. Not. R. Astr. Soc. **227**, 7p-12p (1987).

4. Lamb, F.K., Astrophys. J., in the press (1991).

5. Fortner, B., Lamb, F.K. and Miller, G.S., Nature **342**, 775-777 (1989).

6. van der Klis, M., Stella, L., White, N.E., Jansen, F. and Parmar, A.N., Astrophys. J. **316**, 411-426 (1987).

7. Chandrasekhar, S., "Hydrodynamic and Hydromagnetic Stability", pp. 589-591 (Clarendon Press, Oxford 1961).

8. Papaloizou, J.C.B. and Pringle, J.E., Mon. Not. R. Ast. Soc. **208**, 721-750 (1984).

9. Hasinger, G., van der Klis, M., Ebisawa, K., Dotani, T. and Mitsuda, K., Astr. Astrophys. **235**, 131-146 (1990).

10. Alpar, M.A. and Shaham, J., Nature **316**, 239-241 (1985).

11. Landau, L.D. and Lifshitz, E.M., "Fluid Mechanics", p. 332 (Pergamon Press, 1975).

12. Yilmaz, A. and Alpar, M.A., in preparation (1991).

13. Kylafis, N.D. and Klimis, G.S., Astrophys. J. **323**, 678-684 (1987).

14. Lamb, F.K., Shibazaki, N., Alpar, M.A. and Shaham, J., Nature **317**, 681-687 (1985).

15. Eggum, G.E., Coroniti, F.V. and Katz, J.I., Astrophys. J. **330**, 142-167 (1988).

16. Milgrom, M., Astron. Astrophys. **172**, L1-L2 (1977).

17. Mason, K.O., in Proc. 23rd ESLAB Symposium on X-ray Astronomy, in the press (1991).

18. Kulkarni, S.R., Narayan, R. and Romani, R., Astrophys. J. **356**, 174-179 (1990).

19. Alpar, M.A., Cheng, A.F., Ruderman, M.A. and Shaham, J., Nature **300**, 728-730 (1982).

20. Radhakrishnan, V. and Shrinivasan, G., Curr. Sci. **51**, 1096-1099 (1982).

MAGNETIC FLARES NEAR BLACK HOLES [1]

MARTIN VOLWERK, JAN KUIJPERS[2] and ROELAND VAN OSS
Sterrekundig Instituut
Rijksuniversiteit Utrecht
P.O. Box 80 000
3508 TA Utrecht
The Netherlands

ABSTRACT. Magnetic flaring interactions are investigated between a black hole and an ionized accretion disk. For the electrodynamics in the presence of a black hole we use the $3 + 1$-split as described by Thorne, Price and Macdonald (1986). This allows us to treat the problem in a quasi-classical manner.

We assume that magnetic fields created in the disk and extending into a force-free corona, form a connection between the disk and the black hole's horizon as one of the footpoints is transported across the innermost stable orbit. Relative motions between plasma at both ends of the link set up an electrical circuit formed by the linking magnetic fields, the well-conducting disk and the horizon. We assume that the Alfvén crossing time of the circuit is less than the shearing time. Then the coronal part of the circuit evolves along a sequence of force-free states.

If the total resistance of the circuit (including the horizon part) is large enough a steady state may be possible with a slipping circuit and an azimuthal magnetic field component which is smaller than the poloidal component. In general however a steady state is not possible because of the large magnetic Reynolds number of such an astrophysical system. Then kinetic energy, either from the rotating hole or from the accreting matter, is stored periodically in the form of force-free magnetic fields and released in field relaxations by reconnection. These magnetic flares resemble those in a system which consists of a magnetic neutron star and an accretion disk (Aly and Kuijpers, 1990; Kuijpers, 1990). We consider both a non-rotating and a maximally rotating black hole and derive under what conditions magnetic flares can be expected. We find that the energy release in the form of a sequence of magnetic explosions can be a substantial fraction of the Eddington luminosity.

References

Aly, J. J. and Kuijpers, J. 1990, *Astron. Astrophys.* **227**, 473.

Kuijpers, J. 1990, in *Active Close Binaries*, ed. C. Ibanoğlu, NATO ASI, Kluwer Acad. Publ., p. 761.

Thorne, K.S., Price, R.H. and Macdonald 1986 *Black Holes: The Membrane Paradigm*, Yale Univ. Press, New Haven.

---

[1]Manuscript in preparation for *Astron. Astrophys.*

[2]Also at CHEAF (Centrum voor Hoge Energie Astrofysica), P.O. Box 41882, 1009 DB Amsterdam

E. P. J. van den Heuvel and S. A. Rappaport (eds.), X-Ray Binaries and Recycled Pulsars, 537.
© 1992 *Kluwer Academic Publishers.*

SUMMARY

CONCLUDING REMARKS

Virginia Trimble
Physics Department   and   Astronomy Program
University of California   University of Maryland
Irvine CA 92717   College Park MD 20742

1. INTRODUCTION

A bit more than 20 years ago, Jorge Sahade began the concluding remarks at the first meeting devoted entirely to interacting binaries (IAU Colloquium # 6) by saying that it had brought together "those who had worked hard on the subject, those who had thought deeply about it, and a few who had done both." Contemplating the number of preprints that have changed hands over the past week, I am inclined to add a third category, "those who have written extensively about it." At that 1969 meeting, binaries with compact components appeared only in the very last discussion section where Robert E. Wilson proposed that one might learn about novae by studying the class of X-ray sources of which the prototype was Cyg X-2 (firmly claimed by him as a binary, on the basis of evidence that would now seem very flimsy). He was clearly right, though the work is still "in progress." There were 21 official participants at that Elsinore gathering, of whom about half are still active in binary star astronomy, but none of whom attended the present meeting.

A considerably larger number, about 85, of us have gathered in Santa Barbara. For at least three participants, this was their first conference, and I hope we have succeeded in making them feel welcome, for the now wide-flung subject of binary star astronomy will remain healthy just as long as the flow of "new blood" into it brings the oxygen of new ideas and fresh viewpoints. For many or most of us, it was the first conference we attended under the fairly immediate shadow of war, and we are both grateful to those who made somewhat difficult journeys to participate and mindful of those who were unable to do so. I shall remember it as the first meeting from which I had to phone my stockbroker, as well as the first at which Roger Romani had to explain my own data to me.

Our scientific memories will vary, but all are likely to include a feeling of shock at first sight of the table of properties of pulsars in 47 Tuc shown by Andrew Lyne. The simplest way to make such a sharply peaked distribution is to truncate one that risely steeply toward short periods with observational effects (which do cut in at 3 msec or so). But the real answer is likely to be more complex.

E. P. J. van den Heuvel and S. A. Rappaport (eds.), X-Ray Binaries and Recycled Pulsars, 541–547.
© 1992 *Kluwer Academic Publishers.*

2. FORMATION OF NEUTRON STARS IN CLOSE BINARIES

Formation mechanisms for X-ray binaries and binary and millisecond pulsars seem a logical starting point; they were also the core of the workshop, because the most serious outstanding problem seems to be making
enough of them, expecially the single and wide binary pulsars in globular
clusters, without overproducing something else. All three of Shakes-
peare's ways of getting something appear in the scenarios discussed.
Some are born that way, as neutron stars with close companions, some achieve it through accretion-induced collapse, and some have it thrust upon
them through capture, collision, or exchange of components.

The born-in-situ channel, discussed by van den Heuvel and Webbink,
requires an initially rather wide system, so that the core of M_1 can grow
to the 2.2 M_o or thereabouts required for core collapse, but not so wide
that there is no interaction, and a relatively low initial mass ratio.
This channel naturally produces high mass X-ray binaries and, eventually,
double NS binary pulsars. It can be stretched downward in mass to take
in Her X-1 and (if the minimum helium core mass for core collapse is
small enough) slightly less massive secondaries; but the extremes of the
LMXRB range present problems.

Accretion-induced collapse, addressed by Nomoto, Woosley, and Webbink, requires even more restrictive initial conditions, in order for
the accreting white dwarf to burn and retain the accreted material, rather than igniting it explosively and expelling more than was accreted
or even disrupting the entire star. If helium is accreted, then white
dwarfs more massive than 0.8 M_o (and the right accretion rate) suffice;
while for hydrogen gas accretion, the white dwarf must come within a few
percent of the Chandrasekhar limit. Though Sirius B is close to 1.0 M_o,
single field white dwarfs apparently occupy a narrow mass range around
0.6 M_o, and the larger values found for some CVs are poorly determined.
Webbink's suggestion that the few true recurrent novae (nuclear explosions only decades apart) have white dwarfs $\gtrsim 0.97$ M_{ch} and are sufficient
in number to give rise to the field LMXRBs is intriguing. The inventory
of RNe has about 4+1 systems in it at any given time, though the membership varies (Webbink's list had only U Sco in common with one from 1978).

Finally, the dense conditions in the cores of globular clusters,
especially during and after core collapse, favor formation of binary and
spun-up neutron stars through (a) stellar collisions, (b) two-and-three
body capture processes, and (c) exchange of a neutron star for one component of a primordial binary, as variously addressed by Hut, Ruffert, and
Phinney. Some of these can produce single msec pulsars without a preceed-
ing LMXRB stage, thus possibly solving one of the statistical puzzles
presented by the cluster pulsars. Similar processes for other kinds of
stars ought to produce large numbers of CVs (a current search for which
was described by Grindlay) and anomalously bright white dwarfs. Precious
little is known about globular cluster white dwarfs, but Richer and Fahlman's candidates in M71 are actually fainter than field WDs of the same
color, indicating higher than average mass, which is at any rate consistent with their being merger products (but then where are the comparable
number expected of normally produced WDs in the cluster?).

I think that all of these pathways probably reach the desired goal some of the time, that no one of them accounts for all of the systems, and that the available volume in the phase space of initial conditions is awkwardly narrow for most of them.

Several of the scenarios would be easier to assess if we knew more about the initial and current binary populations in globular clusters. The field stellar population is normally described as including about 30% binaries (capable of interaction), with a fairly flat distribution in log a over the physically possible range, and a distribution of mass ratios consistent with random selection from a standard initial mass function (I don't necessarily believe all of this). Much less can be said about the cluster population. Phinney presented the evidence contradicting earlier suspicions that there were essentially no primordial binaries in the clusters, and suggested 10% as a reasonable fraction of stars in pairs that could both interact and remain bound over the cluster lifetime. Clearly this number is very uncertain, and little is known about the distribution of periods or mass ratios within the population, either for the clusters as formed or now. Kinds of investigations that might yield futher information include radial velocity measurements, searches for additional main sequence eclipsing binaries (W UMa's), examination of statistics of stars above the main sequence in an HR diagram (though blended images act exactly like real binaries), and searches for cataclysmic variables.

3. ARE THE INVENTORIES COMPLETE?

From time to time, one has the impression that there are several different possibilities for something, but none of them is entirely satisfactory. For instance, formation of binary neutron stars in situ, by capture, and by accretion-induced collapse all presented problems in accounting for the globular cluster pulsars. Production via star exchange with primordial binaries may resolve them. Several other cases from the recent past illustrate the virtues of expanding inventories of possibilities from three to four or thereabouts.

An historical example, discussed by Taam and Webbink, is the augmentation of mass transfer cases A, B, and C by common envelop processes to account for close systems that must originally have been much wider. More recently, the ways of driving angular momentum and mass transfer -- nuclear evolution, gravitational radiation (which goes back at least to a paper by Kraft et al. in 1964), and magnetic braking -- have expanded to include radiation driving. And there is a closely related additon to the possible range of fates of the companion. Besides unbinding, merging, or orbiting forever, M_2 can also boil away.

At this meeting, we heard several possible expansions of the repertoire for killing off magnetic fields of neutron stars. Ohmic decay and accretion-induced decay (presumably a result of heating) have been joined by field line transport associated with vortex transport, that is, spin-down-induced decay, as presented by Bhattacharya. Complementarily, some neutron stars may have begun their lives with fields below 10^{10} G. Tanaka suggested that N(B) in conjection with other properties of neutron stars was more suggestive of two populations than of decay. Or, as Shaham proposed, core fields may be only 10^8 G from the beginning, so that only a crustal field needs to decay.

Discussions of pulsar velocities sometimes leave out one of the (more or less) well-established ways of making other kinds of run-away stars. In addition to a boost from disrupting a binary system or a kick from a supernova explosion, stars can also be given high velocities by dynamical processes in the clusters in which they formed.

Finally our inventory of compact objects has remained fixed at three for a very long time. Black holes go back to the late 18th century, white dwarfs to the very beginning of this one, and even neutron stars to work by Baade and Zwicky in 1933-34. The flurry of activity on stars made of strange quark matter or pion condensate, prompted by the false alarm of a 0.5 msec pulsar in 1987A has died away, and it is perhaps time to ask, on a non-emergency basis, whether there may not be other configurations possible. S. Bahcall has made one proposal along these lines (bound hadronic matter). And Zwicky always insisted on the reality of "object hades," more compact than a neutron star, but well away from infinite density. Incidentally, if the title of this section led you to expect some words of wisdom on whether our inventories of objects in the various categories discussed at Santa Barbara are complete, the conventional answer is yes for the powerful high and low mass XRBs, no for the transients, and absolutely no for pulsars, searches for which are limited by distance, pulse period, dispersion, beaming, and probably other things. Personally I even wonder about the bright XRB completeness.

4. APPLICATIONS

Binaries with compact components yield several kinds of information useful to other branches of astronomy. First, some neutron star masses can be measured. Those for which improved data were presented include:

$$1913+16 = 1.4409 \qquad + \qquad 1.3875$$
$$1855+09 = 1.27 \ (+0.23, -0.15) \ + \qquad 0.233 \ (+0.26, -0.17)$$
$$2127+11C = 1.33 \qquad + \qquad 1.38$$
$$1534+12 = 1.36 \pm 0.03 \qquad + \qquad 1.32 \pm 0.03$$
$$2303+46 = 2.8 \ \pm 0.1 \ \text{(total)}$$

In each case, the pulsar is given first, and the secondary is presumably another neutron star, except for 1855+09. The narrowness of the mass range is remarkable, even in comparison with expected initial masses. Good masses for accretion-powered X-ray sources (whose masses could/should have grown by several tenths of a solar mass) would be of enormous interest, but await models (or tricks) adequate for extracting center-of-mass velocities from the smear of lines due to M_2, disk, stream, and hot spot.

Second, X-ray binaries provide the most convincing evidence for the existence of black holes in the real world. McClintock and Tanaka reviewed the four well-known candidates, Cyg X-1, LMC X-1 and 3, A0620-00, and a handful of others (with little overlap between their lists), some of them newly-discovered Ginga soft X-ray transients. The number of these according to Tanaka, implies a total galactic reservoir of 5000 black hole binaries.

Third, millisecond pulsars test aspects of general relativity not probed within the solar system. The orbit period evolution of 1916+13 is just what it should be if the quadrupole formula for gravitational radiation is correct; and geodetic precession of its pulse profile has perhaps been seen. According to Wolszczan, 1534+12, with an expected period near 680 yr, will eventually be a more persuasive case, because the angle between spin axis and orbit axis is known.

Fourth, many aspects of neutron stars in general and binaries in general are illuminated by the combination. These include the temporal evolution of magnetic field strength and geometry and temperature, mass transfer and common envelop processes, and pseudo-red-giants with neutron star cores.

Finally, statistics and properties of X-ray and pulsar binaries provide evidence on (a) asymmetries of supernova explosions (Radhakrishnan, Dewey, Wolszczan), (b) the radial dependence of mass-to-light ratio in M15 (indicating many white dwarfs and neutron stars in the core, but no single massive black hole; Phinney), and (c) disks, jets, coronae, winds, magnetospheres, and other structures that our systems share with active galaxies (Arons, Lamb, Ghosh, Pacini). One item that may be interestingly transferrable is the range of configurations presented by Lamb and Ghosh for disk-magnetosphere iteractions different from the familiar image of too-tight belt confining squishy midsection.

5. POTPURRI

5.1 Neat New Things

A large number of items that came up at the workshop were new, at least to me, and seem worth remembering. The names are those who mentioned, or "posted", them.

a. Some of the high velocity pulsars more than 100 pc from the galactic plane are moving toward the plane (Harrison).

b. The supernova remnant GS5.4 has probably been rejuvenated as a plerion by PSR 1758-24 plowing into the shell (Frail and Kulkarni).

c. Globular cluster millisecond pulsars have fewer interpulses than field msec objects (Johnston, Navarro, et al.).

d. The number of msec pulsars in a cluster scales as about the square root of central density (Foster, Tavani), which is probably an artifact of the summation of several formation processes (Phinney).

e. The slowing down index of 0540-69 is probably 2.7 rather than 2.0 (several speakers), and it has had a giant glitch (Truemper).

f. Vela never gets a chance to relax completely (Pines).

g. A 1400 GHz pulsar survey found 9 pulsars (of 100 new ones) with slowing-down ages less than 30,000, more than doubling the supply of these young objects (Lyne).

h. Five of the six LMXRBs in the SMC and LMC are very soft and may constitute a new and abundant class of msec progenitors, possibly in a common envelope phase (Truemper and others).

i. The 455 keV (positron annihilation?) line in the Crap Nebula has been confirmed by Figaro (Pacini).

j. Neutron stars may experience plate tectonic processes that move both crust and field lines around (Ruderman).

k. All globular cluster X-ray binaries are weird (Parmar).

l. The large number of black hole candidate Ginga transient sources (Tanaka) must mean that these systems live longer than neutron star systems, that there are also many more SN systems than currently catalogued, or that core collapse (at least in binaries) makes many more black holes and fewer neutron stars than conventionally believed; this would not be terribly difficult to arrange, given the difficulties of making shock ejection happen at all (Woosley).

5.2 Things to Look for

In a few cases, one gets the impression that a vexing issue would yield to a very small amount of observational data of just the right sort. Items that one might look for in this spirit include:

a. Radio pulsations from LMXRBs in low states (Rappaport)

b. A pulsar in SN 1987A

c. The redshifted Fe line in additional X-ray binaries and gamma ray bursters (McClintock)

d. Another SS 433 and another Geminga (testing whether this one is e.g., the closest msec pulsar)

e. A third member of the 1957+20 and 1744-24A class, on the grounds that one is a discovery, two is a confirmation, and three is a well known class of astrophysical objects

f. Confirmation of a 10^{12} G field from cyclotron lines in 1E2259+ 586 as a test of the rapidly-rotating white dwarf model (Kulkarni), thereby leaving the object as either a single neutron star rotating too slowly to produce the luminosity we see or as a binary with no current evidence for a companion and some fairly tight limits on a hypothetical one

g. ROSAT projects in progress: thermal emission from old pulsars and cores of historic SNe; additional very soft sources and optical identifications; diffuse interstellar emission from old pulsar nebulae; thermal X-ray halo of the Crab Nebula (Truemper)

5.3 Things We Would Like to Know

Undoubtedly all of us arrived at the workshop with questions in mind and left with some of them answered, but the supply more than replenished by new, and with luck more profound, questions. These are some of my favorites; many of them were asked more or less rhetorically by one or more of the speakers.

a. What do WDX2 and NSX2 mergers look like?

b. Why are there no XRBs with companion masses between about 1.5 and 8 solar masses?

c. Why are so many neutron stars in massive XRBs such slow rotators?

d. What is the rotation period of the neutron stars in typical low mass systems? The few we know directly are long (1-122 s), but all four are high B, unusual systems.

e. Is there any accurate measure of pulsar ages; and what does it measure -- time since B reached maximum strength, time since maximum rotation spee, time since core collapse, or what?

f. Why are two of the four good black hole candidates in the LMC?

g. Why are the EUV/ultrasoft X-ray sources found so far all in the Magellanic Clouds, and is this related to question e somehow?

h. If there are really 5000 black hole systems in the Milky Way (mostly low mass ones) are there really 10^4 of 10^5 neutron star ones?

i. In QPO sources, why is the neutron star magnetic field correlated with the evolutionary state of the secondary?

j. (Where) are there white dwarfs more massive than $1.1 M_o$?

k. What causes the rapid positive and negative changes in orbit period of LMXRBs? Is it useful to remember that some examples of many other classes of binaries behave similarly?

l. What is the real range of neutron star masses and does it overlap with the white dwarf one? I have a 15 year old bet with Ken Brecher on this topic.

m. Is there a real gap around 10^{10} G in the distribution of neutron star field strengths?

n. And, more generally, what is the real evolution of B(t), and which physical processes contribute to its rise and fall under which circumstances?

ACKNOWLEDGEMENTS

My involvement in binary stars really dates back to the 1969 Elsinore IAU Colloquium, and I remain grateful to the organizers and participants for allowing me to gate crash. Kip Thorne and Ed van den Heuvel asked the questions that led to my investigating binary mass ratios for various kinds of systems; and Ed is thanked also for the invitation to participate in this workshop, financial support being provided by NATO and by the IRS through a Schedule A deduction.

Authors Index

Object Index

Subject Index

The manufacturer's authorised representative in the EU is Springer
Nature Customer Service Centre GmbH, Europaplatz 3, 69115 Heidelberg,
Germany. If you have any concerns regarding our products, please
contact ProductSafety@springernature.com

Printed and bound by CPI Group (UK) Ltd, Croydon, CR0 4YY

24/04/2026

02096316-0011